中国农业科学院
兰州畜牧与兽药研究所科技论文集
（2016）

中国农业科学院兰州畜牧与兽药研究所　主编

中国农业科学技术出版社

图书在版编目（CIP）数据

中国农业科学院兰州畜牧与兽药研究所科技论文集. 2016 / 中国农业科学院兰州畜牧与兽药研究所主编. —北京：中国农业科学技术出版社，2020.6
 ISBN 978-7-5116-4715-3

Ⅰ. ①中… Ⅱ. ①中… Ⅲ. ①畜牧学—文集 ②兽医学—文集 Ⅳ. ①S8-53

中国版本图书馆CIP数据核字（2020）第068859号

责任编辑 闫庆健 阿米娜·加玛力
责任校对 贾海霞

出 版 者	中国农业科学技术出版社
	北京市中关村南大街12号　邮编：100081
电　　话	（010）82106632（编辑室）（010）82109702（发行部）
	（010）82109709（读者服务部）
传　　真	（010）82106625
网　　址	http：//www.castp.cn
经 销 者	各地新华书店
印 刷 者	北京富泰印刷有限责任公司
开　　本	880mm×1 230mm　1/16
印　　张	36.75　彩插16面
字　　数	1 192千字
版　　次	2020年6月第1版　2020年6月第1次印刷
定　　价	170.00元

━━━━━ 版权所有·翻印必究 ━━━━━

《中国农业科学院兰州畜牧与兽药研究所科技论文集（2016）》编委会

主 任 委 员：杨志强　张继瑜

副主任委员：刘永明　阎　萍　王学智

委　　　员：高雅琴　梁春年　梁剑平　李建喜
　　　　　　李剑勇　李锦华　刘丽娟　潘　虎
　　　　　　时永杰　田福平　杨博辉　严作廷
　　　　　　杨　晓　曾玉峰　周　磊

主　　　编：杨志强　张继瑜　王学智　杨　晓
　　　　　　周　磊

副　主　编：刘永明　阎　萍　曾玉峰　刘丽娟

主要撰稿人：高雅琴　梁春年　梁剑平　李建喜
　　　　　　李剑勇　李锦华　潘　虎　时永杰
　　　　　　田福平　杨博辉　严作廷

前　言

近年来，在中国农业科学院科技创新工程的引领下，研究所的科研水平快速提升。我所科研人员和管理人员不但有工作上的热情，更有对工作认识上的高度和对学科理解上的深度。他们在紧张繁忙的实践活动中，笔耕不辍，将自己的研究成果写成论文。这不单是科研人员和管理人员的工作总结、过程记录，更是他们智慧的结晶，最终成为研究所的一笔宝贵财富。

为了珍惜这笔财富，加强优秀论文的交流与传播，营造更加浓厚的学术氛围，促进科研水平和管理水平的提升，切实推进研究所的科技创新，科技管理处搜集了2016年研究所科研人员公开发表的论文编印成《中国农业科学院兰州畜牧与兽药研究所科技论文集》第五卷，共131篇。由于时间仓促，可能还有论文未能收录，敬希鉴谅！

编者

2020年6月

目 录

Simultaneous Determination of Diaveridine, Trimethoprim and Ormetoprim in Feed Using High Performance Liquid Chromatography Tandem Mass Spectrometry
................YANG Yajun, LIU Xiwang, LI Bing, LI Shihong,
　　　　　KONG Xiaojun, QIN Zhe, LI Jianyong （1）

The Coordinated Regulation of Na$^+$ and K$^+$ in Hordeum Brevisubulatum Responding toTime of Salt Stress
............WANG Chunmei, XIA Zengrun, WU Guoqiang, YUAN Huijun,
　　　　　WANG Xinrui, LI Jinhua, TIAN Fuping,
　　　　　ZHANG Qian, ZHU Xinqiang, HE Jiongjie,
　　　　　KUMAR Tanweer, WANG Xiaoli, ZHANG Jinlin （17）

Genome-wide Association Study Identifies Loci for the Polled Phenotype in Yak
................LIANG Chunnian, WANG Lizhong, WU Xiaoyun,
　　WANG Kun, DING Xuezhi, WANG Mingcheng, CHU Min,
　　　　　XIE Xiuyue, QIU Qiang, YAN Ping （33）

Integrated Analysis of the Roles of Long Noncoding RNA and Coding RNA Expression in Sheep (*Ovis aries*) Skin during Initiation of Secondary Hair Follicle
.... YUE Yaojing, GUO Tingting, YUAN Chao, LIU Jianbin,
　　GUO Jian, FENG Ruilin, NIU Chune, SUN Xiaoping, YANG Bohui （45）

Genetic Diversity and Phylogenetic Evolution of Tibetan Sheep Based on mtDNA D-Loop Sequences
... LIU Jianbin, DING Xuezhi, ZENG Yufeng, YUE Yaojing, GUO Xian,
　　GUO Tingting, CHU Min, WANG Fan, HAN Jilong, FENG Ruilin,
　　SUN Xiaoping, NIU Chune, YANG Bohui, GUO Jian, YUAN Chao （63）

Evaluation of Crossbreeding of Australian Superfine Merinos with Gansu Alpine Finewool Sheep to Improve Wool Characteristics
.... LI Wenhui, GUO Jian, LI Fanwen, NIU Chune （83）

Microwave-assisted Extraction of Three Bioactive Alkaloids from *Peganum harmala* L. and Their Acaricidal Activity against *Psoroptes cuniculi* in Vitro
.... SHANG Xiaofei, GUO Xiao, LI Bing, PAN Hu,
ZHANG Jiyu, ZHANG Yu, MIAO Xiaolou (101)

Synthesis and Pharmacological Evaluation of Novel Pleuromutilin Derivatives with Substituted Benzimidazole Moieties
.... AI Xin, PU Xiuying, YI Yunpeng, LIU Yu, XU Shuijin,
LIANG Jianping, SHANG Ruofeng (122)

Short Communication: N-Acetylcysteine-mediated Modulation of Antibiotic Susceptibility of Bovine Mastitis Pathogens
.... YANG F., LIU L. H., LI X. P., LUO J. Y., ZHANG Z.,
YAN Z. T., ZHANG S. D., and LI H. S. (136)

Influences of Season, Parity, Lactation, udder Area, Milk Yield, and Clinical Symptoms on Intrammary Infection in Dairy Cows
.... ZHANG Z., LI X. P., YANG F., LUO J. Y.,
WANG X. R., LIU L. H., LI H. S. (140)

Acaricidal Activity of Oregano Oil and its Major Component, Carvacrol, Thymol and p-cymene Against *Psoroptes cuniculi* in Vitro and in Vivo
.... SHANG Xiaofei, WANG Yu, ZHOU Xuzheng, GUO Xiao,
DONG Shuwei, WANG Dongsheng, ZHANG Jiyu,
PAN Hu, ZHANG Yu, MIAO Xiaolou (151)

Lowering Effects of Aspirin Eugenol Ester on Blood Lipids in Rats with High Fat Diet
.... KARAM Isam, MA Ning, LIU Xiwang, KONG Xiaojun,
ZHAO Xiaole, YANG Yajun, LI Jianyong (158)

Characterization of the Complete Mitochondrial Genome Sequence of Wild Yak (Bos mutus)
.... LIANG Chunnian, WU Xiaoyun, DING Xuezhi, WANG Hongbo,
GUO Xian, CHU Min, BAO Pengjia, YAN Ping (166)

Determination of Antibacterial agent Tilmicosin in Pig Plasma by LC/MS/MS and its Application to Pharmacokinetics
.... LI Bing, GONG Shiyue, ZHOU Xuzheng, YANG Yajun,

LI Jianyong, WEI Xiaojuan, CHENG Fusheng,
NIU Jianrong, LIU Xiwang, ZHANG Jiyu (169)

Evaluation on Antithrombotic Effect of Aspirin Eugenol Ester from the View of Platelet Aggregation, Hemorheology, TXB_2/6-keto-$PGF_1\alpha$ and Blood Biochemistry in Rat Model

·················· MA Ning, LIU Xiwang, YANG Yajun, SHEN Dongshuai,
ZHAO Xiaole, MOHAMED Isam,
KONG Xiaojun, LI Jianyong (183)

Multi-Residue Method for the Screening of Benzimidazole and Metabolite Residues in the Muscle and Liver of Sheep and Cattle Using HPLC/PDAD with DVB-NVP-SO_3Na for Sample Treatment

········ XIONG Lin, HUANG Lele, SHIMO Shimo-Peter, LI Weihong,
YANG Xiaolin, YAN Ping (199)

Treatment of the Retained Placenta in Dairy Cows: Comparison of a Systematic Antibiosis with an Oral Administered Herbal Powder Based on Traditional Chinese Veterinary Medicine

················ CUI Dongan, WANG Shengyi, WANG Lei, WANG Hui,
LI Jianxi, TUO Xin, HUANG Xueli, LIU Yongming (210)

The Complete Mitochondrial Genome of Hequ Tibetan Mastiff Canis Lupus Familiaris (Carnivora: Canidae)

·············· GUO Xian, PEI Jie, BAO Pengjia, YAN Ping, LU Dengxue (220)

Molecular Characterization and Phylogenetic Analysis of Porcine Epidemic Diarrhea Virus Samples Obtained from Farms in Gansu, China

···························· HUANG M.Z., WANG H., WANG S.Y.,
CUI D.A., TUO X., LIU Y.M. (223)

Association of Genetic Variations in the ACLY Gene with Growth Traits in Chinese Beef Cattle

····················LI M.N., GUO X., BAO P.J., WU X.Y., DING X.Z.,
CHU M., LIANG C.N., YAN P. (233)

PPARα Signal Pathway Gene Expression is Associated with Fatty Acid Content in Yak and Cattle Longissimus Dorsi Muscle
.... QIN W., LIANG C.N., GUO X., CHU M., PEI J., BAO P.J., WU X.Y., LI T.K., YAN P. (244)

Genetic Characterization of Antimicrobial Resistance in *Staphylococcus aureus* Isolated from Bovine Mastitis Cases in Northwest China
.... YANG Feng, WANG Qi, WANG Xurong, WANG Ling, LI Xinpu, LUO Jinyin, ZHANG Shidong, LI Hongsheng (254)

Quantitative Structure Activity Relationship (QSAR) Studies on Nitazoxanide-based Analogues Against Clostridium Difficile in Vitro
.... ZHANG Han, LIU Xiwang, YANG Yajun, LI Jianyong (262)

Evaluation of the Acute and Subchronic Toxicity of Ziwan Baibu Tang
.... XIN Ruihua, PENG Wenjing, LIU Xiaolei, LUO Yongjiang, WANG Guibo, LUO Chaoying, XIE Jiasheng, LI Jinyu, LIANG Ge, ZHENG Jifang (276)

Evaluation of the Acute and Subchronic Toxicity of *Aster tataricus* L. F.
.... PENG Wenjing, XIN Ruihua, LUO Yongjiang, LIANG Ge, REN Lihua, LIU Yan, WANG Guibo, ZHENG Jifang (287)

Comparative Proteomic Analysis of Yak Follicular Fluid During Estrus
.... GUO Xian, PEI Jie, DING Xuezhi, CHU Min, BAO Pengjia, WU Xiaoyun, LIANG Chunnian, YAN Ping (305)

Syntheses, Crystal Structures and Antibacterial Evaluation of Two New Pleuromutilin Derivatives
.... SHANG Ruofeng, XU Shuijin, YI Yunpeng, AI Xin, LIANG Jianping (316)

Flavonoids and Phenolics from the Flowers of *Limonium aureum*
.... LIU Yu, SHANG Ruofeng, CHENG Fusheng, WANG Xuehong, HAO Baocheng, LIANG Jianping (326)

The Role of Porcine Reproductive and Respiratory Syndrome Virus as a Risk Factor in the Outbreak of Porcine Epidemic Diarrhea in Immunized Swine Herds
.... HUANG Meizhou, WANG Hui, WANG Shengyi, CUI Dongan, TUO Xin, LIU Yongming (328)

Crystal Structure of 14-［（1-benzyloxycarbonyl-amino-2-methylpropan-2-yl）sulfanyl］Acetate Mutilin, $C_{34}H_{49}NO_6S$
.... SHANG Ruofeng, YI Yunpeng, LIANG Jianping (339)

The Complete Mitochondrial Genome of *Ovis ammon darwini*（Artiodactyla: Bovidae）
.... MAO Hongxia, LIU Hanli, MA Guilin, YANG Qin, GUO Xian, ZHONG Lamaocao (345)

荷斯坦奶牛乳腺组织冻存及乳腺上皮细胞原代培养技术改进
.... 林 杰，王旭荣，王 磊，张景艳，王学智，孟嘉仁，杨志强，李建喜（348）

6株牛源副乳房链球菌的分离和鉴定
.... 李新圃，罗金印，杨 峰，王旭荣，李宏胜（357）

苦马豆素的来源、药理作用及检测方法研究进展
.... 黄 鑫，梁剑平，高旭东，郝宝成（364）

牦牛胎儿皮肤毛囊的形态发生及E钙黏蛋白的表达和定位
.... 佘平昌，梁春年，裴 杰，褚 敏，郭 宪，阎 萍（374）

五氯柳胺口服混悬剂的制备及其含量测定
.... 张吉丽，李 冰，司鸿飞，郭 肖，朱 阵，尚晓飞，周绪正，张继瑜（383）

牦牛角性状候选基因的筛选
.... 佘平昌，吴晓云，梁春年，褚 敏，丁学智，阎 萍（396）

基于GC-MS技术的蹄叶炎奶牛血浆代谢谱分析
.... 李亚娟，王东升，张世栋，严作廷，杨志强，杜玉兰，董书伟，何宝祥（404）

牛源金黄色葡萄球菌耐药性与相关耐药基因和菌株毒力基因的相关性研究
.... 杨 峰，王旭荣，李新圃，罗金印，刘龙海，张 哲，王 玲，张世栋，李宏胜（417）

SAA与HP在奶牛活体和离体炎性子宫内膜上皮细胞中表达的研究
………… 张世栋，董书伟，王东升，闫宝琪，杨　峰，严作廷，杨志强（424）

金黄地鼠动脉粥样硬化模型的病理观察和生化指标分析
………… 马　宁，杨亚军，刘希望，秦　哲，孔晓军，李世宏，李剑勇（433）

青蒿素的来源及其抗鸡球虫作用机制研究进展
……………………… 黄　鑫，郭文柱，高旭东，郝宝成，梁剑平（440）

金黄色葡萄球菌中耐甲氧西林抗性基因mecC的研究进展
………………………… 陈　鑫，蒲万霞，吴　润，梁红雁，李昱辉（445）

2种甾体抗原对甘肃高山细毛羊繁殖率的影响
………………… 冯瑞林，郭　健，裴　杰，刘建斌，岳耀敬，
　　　　　　　　　　　　　　郭婷婷，牛春娥，孙晓萍，杨博辉（449）

薄层色谱法对宫衣净酊质量标准的研究
………………… 朱永刚，王　磊，王旭荣，崔东安，张景艳，
　　　　　　　　　　　　　　张　凯，张　康，李建喜，杨志强（451）

绵山羊双羔素提高细毛羊繁殖率的研究
………………… 冯瑞林，郭　健，裴　杰，刘建斌，岳耀敬，
　　　　　　　　　　　　　　郭婷婷，牛春娥，孙晓萍，杨博辉（452）

小鼠口服五氯柳胺的急性毒性研究
………………… 张吉丽，李　冰，司鸿飞，周绪正，程富胜，张继瑜（454）

绵山羊双羔素提高黑山羊繁殖率的研究
………………… 冯瑞林，郭　健，裴　杰，刘建斌，岳耀敬，
　　　　　　　　　　　　　　郭婷婷，牛春娥，孙晓萍，杨博辉（455）

紫花苜蓿草地土壤碳密度年际变化研究
………………………………………………………… 田福平，师尚礼（457）

中兽药治疗奶牛胎衣不下的系统评价
………… 崔东安，王胜义，王　磊，王　慧，妥　鑫，黄美洲，刘永明（458）

复方板黄口服液制备工艺研究
………………… 李　冰，牛建荣，周绪正，李剑勇，
　　　　　　　　　　　　　　魏小娟，杨亚军，刘希望，张继瑜（459）

金黄色葡萄球菌生物被膜研究知识图谱分析
………… 杨　峰，王旭荣，王　玲，李新圃，罗金印，张世栋，李宏胜（460）

金黄色葡萄球菌性奶牛乳房炎研究的知识图谱分析
……………… 杨　峰，王　玲，王旭荣，李新圃，罗金印，张世栋，李宏胜（461）

甘肃省某奶牛场肠杆菌性乳房炎主要病原菌的分离鉴定与耐药性分析
………… 王海瑞，王旭荣，张景艳，王　磊，曹明泽，李建喜，王学智（462）

HPLC测定宫衣净酊中葛根素含量
………………………………… 朱永刚，崔东安，王　磊，王旭荣，
　　　　　　　　　　　　　　　张景艳，林　杰，杨志强，李建喜（463）

Wnt10b、β-catenin、FGF18基因在甘肃高山细毛羊胎儿皮肤毛囊中的表达
规律研究
……… 赵　帅，岳耀敬，郭婷婷，吴瑜瑜，刘建斌，韩吉龙，
　　　　　　郭　健，牛春娥，孙晓萍，冯瑞林，王天翔，
　　　　　　　　　　　李桂英，李范文，史兆国，杨博辉（464）

不同栽培模式下甜菜和籽立苋对次生盐渍化土壤的抑盐效应
………… 代立兰，张怀山，夏曾润，王　平，王国宇，杨世柱（466）

银翘蓝芩口服液中绿原酸含量测定方法的建立
……………………………… 许春燕，刘希望，杨亚军，孔晓军，
　　　　　　　　　　　　　　李世宏，秦　哲，杨孝朴，李剑勇（468）

基于Web of Science的"阿维菌素类药物"的文献计量研究
……………… 文　豪，周绪正，李　冰，牛建荣，魏小娟，张继瑜（470）

巴尔吡尔对畜禽常见病原菌体外抑菌效果研究
……………… 张　哲，李新圃，杨　峰，罗金印，刘龙海，李宏胜（471）

抗球虫药青蒿散中青蒿素UPLC检测方法的建立
………… 黄　鑫，郭文柱，郝宝成，高旭东，梁剑平，于静怡，杜宪文（472）

绵山羊双羔素提高东北细毛羊繁殖率的研究
……………… 冯瑞林，郭　健，裴　杰，刘建斌，岳耀敬，
　　　　　　　　　　　郭婷婷，孙晓萍，牛春娥，杨博辉（473）

甘肃省售牛羊肉中雌激素残留状况分析
………………………………… 李维红，熊　琳，高雅琴，杨晓玲（474）

基于AHP、负权重和模糊数学的土壤质量评价研究
………………………………………………… 李润林，姚艳敏，董鹏程（475）

"康毒威"治疗鸡新城疫扭颈的效果
………………………………………………………… 张仁福，谢家声（477）

中药治疗种公鸡冠癣有效
………………………………… 谢家声，王贵波，李锦宇，罗超应（478）

小鼠口服五氯柳胺混悬剂的急性毒性研究
………………… 张吉丽，司鸿飞，李　冰，程富胜，周绪正，张继瑜（479）

藿芪灌注液局部刺激性试验
………………… 王东升，张世栋，苗小楼，董书伟，魏立琴，
　　　　　　　　　邝晓娇，那立冬，闫宝琪，严作廷（480）

中药治疗奶牛乳房炎的系统评价与Meta分析
………………… 杨　健，严作廷，王东升，张世栋，杨志强，董书伟（481）

多菌种协同发酵啤酒糟渣和苹果渣生产蛋白饲料的研究
………………………………… 王晓力，王　帆，孙尚琛，王永刚，
　　　　　　　　　　李　想，王春梅，朱新强，孙启忠（483）

VITEK2 Compact全自动微生物分析系统对牛乳中葡萄球菌的鉴定效果评价
………………… 林　杰，张景艳，王　磊，王海瑞，邹　璐，朱永刚，
　　　　　　　　　边亚彬，李建喜，杨志强，王旭荣（485）

固态发酵豆渣、葡萄渣和苹果渣复合蛋白饲料的研究
………………………… 朱新强，魏清伟，王永刚，冷非凡，郭陈真，
　　　　　　　　　　　庄　岩，王春梅，王晓力，孙启忠（487）

奶牛生乳中洛菲不动杆菌的分离鉴定与耐药性分析
………………………………… 林　杰，王旭荣，王　磊，王海瑞，朱永刚，
　　　　　　　　　　杨志强，李建喜，张景艳（488）

不同生长年限紫花苜蓿地下生物量的空间分布格局
………………… 周　恒，时永杰，胡　宇，陈　璐，路　远，田福平（489）

基于**Web of Science**™的"树突状细胞"研究论文产出分析
………………………………… 边亚彬，张景艳，王　磊，王旭荣，
　　　　　　　　　　杨志强，王学智，孟嘉仁，李建喜（490）

青海高原牦牛PRDM16基因克隆、生物信息学及组织差异表达分析
………………………………………………… 赵生军，张　勇，郭　宪（492）

小鼠溃疡性结肠炎模型的建立与评价
………… 曹明泽，王旭荣，王　磊，张景艳，王海瑞，李建喜，王学智（494）

紫菀不同极性段提取物对SD大鼠亚慢性毒性试验研究
……………………………………彭文静，辛蕊华，任丽花，罗永江，王贵波，
罗超应，谢家声，李锦宇，郑继方（496）

GnIH与INH表位多肽疫苗主动免疫对甘肃高山细毛羊生殖激素的影响
……………………………………张玲玲，杨博辉，岳耀敬，郭婷婷，刘建斌，袁　超，
牛春娥，冯瑞林，郭　健，孙晓萍，刘善博（498）

奶牛饲料中酿酒酵母菌的分离鉴定及其发酵液的体外抑菌活性研究
……………………………………刘龙海，李新圃，杨　峰，罗金印，张　哲，李宏胜（500）

黄花补血草总黄酮亚急性毒性试验
……………………………………刘　宇，曾豪杰，尚若锋，郝宝成，杨　珍，
郭文柱，程富胜，王学红，梁剑平（502）

抗鸡球虫药常山口服液对小鼠的急性毒性作用评价
……………………………………王　玲，郭志廷，张晓松，林春全，罗小琴，杨　峰，杨　珍（504）

INH表位多肽疫苗抗体间接ELISA测定方法的建立及优化
……………………………………张玲玲，岳耀敬，冯瑞林，李红峰，郭婷婷，袁　超，
牛春娥，刘建斌，孙晓萍，韩吉龙，刘善博，杨博辉（506）

紫菀乙醇提取物对豚鼠离体气管平滑肌收缩功能的影响
……………………………………彭文静，辛蕊华，刘　艳，罗永江，王贵波，
罗超应，谢家声，李锦宇，郑继方（508）

藏药雪山杜鹃叶挥发油成分的GC-MS分析
……………………………………郭　肖，周绪正，朱　阵，文　豪，张吉丽，张继瑜（510）

奶牛乳房炎疫苗研究进展
……………………………………刘龙海，李新圃，杨　峰，罗金印，王旭荣，
张　哲，罗增辉，李宏胜（511）

中药对牛免疫调节作用研究进展
……………………………………刘　艳，罗永江，辛蕊华，彭文静，梁　歌，郑继方（512）

黄芪多糖对"树突状细胞"形态和功能影响研究进展
……………………………………边亚彬，张景艳，王　磊，王旭荣，张　康，
王学智，孟嘉仁，李建喜（513）

奶牛子宫内膜炎病因学研究进展
……………………………………那立冬，王东升，董书伟，张世栋，
闫宝琪，杨洪早，桑梦琪，严作廷（514）

仔猪腹泻的病因及中药防治研究进展
……………………………………………杨洪早，王东升，董书伟，张世栋，
那立冬，闫宝琪，桑梦琪，严作廷（515）

可递送siRNA的非病毒纳米载体的设计
………………………………………………赵晓乐，孔晓军，李剑勇（516）

动物抗寄生虫药物的研究与应用进展
………………………………………………张吉丽，李　冰，张继瑜（517）

肝片吸虫病的研究进展
……………………………张吉丽，朱　阵，李　冰，周绪正，张继瑜（518）

纤维素酶及其在中药发酵中的运用
……………………………………………………………苏贵龙，李建喜（519）

饮食与疾病：由牛奶致癌说引发的思考
………………………………罗超应，罗磐真，李锦宇，王贵波，谢家声（520）

动物抗寄生虫药物作用机理研究进展
………………………………………………张吉丽，李　冰，周绪正，张继瑜（521）

细菌耐药性研究进展
………………朱　阵，曹明泽，张吉丽，周绪正，李　冰，张继瑜（522）

奶牛隐性子宫内膜炎诊断技术研究进展
…………闫宝琪，董书伟，王东升，那立冬，杨洪早，张世栋，严作廷（523）

羔羊腹泻细菌和病毒病原的研究进展
……………妥　鑫，刘永明，黄美州，崔东安，王　慧，王胜义，齐志明（525）

PRDM家族蛋白结构与功能研究进展
………………………………………………赵生军，张　勇，阎　萍，郭　宪（526）

牦牛瘤胃微生物降解纤维素及其资源利用的研究进展
………………………………………………苏贵龙，张景艳，王　磊，李建喜（528）

繁殖毒理学研究进展
……………赵晓乐，孔晓军，李世宏，秦　哲，刘希望，杨亚军，李剑勇（529）

植物精油在畜禽生产中的应用效果研究进展
…朱永刚，王　磊，崔东安，张景艳，林　杰，王旭荣，李建喜，杨志强（530）

基于复杂性科学探讨中药安全性评价
………………………………罗超应，罗磐真，李锦宇，王贵波，谢家声（532）

不同种植年限紫花苜蓿种植地土壤容重及含水量特征
.... 周　恒，时永杰，路　远，胡　宇，田福平，陈　璐（534）

2001—2010年张掖市甘州区土地利用景观格局演变研究
.... 李润林，董鹏程（535）

酿酒酵母菌生长特性的研究
.... 刘龙海，李新圃，杨　峰，罗金印，张　哲，李宏胜（537）

记中国农业科学院兰州畜牧与兽药研究所"中兽医药陈列馆"
.... 罗超应，王贵波，李锦宇，谢家声，王旭荣，王　磊（538）

两种植物在不同生长期控制Na^+流入的差异
.... 王春梅，张　茜，朱新强，贺洞杰，段慧荣，王晓力（539）

欧拉型藏羊生长发育曲线模型预测及趋势分析
.... 王宏博，梁春年，包鹏甲，朱新书，阎　萍（541）

酮病治疗对奶牛总抗氧化能力的影响
.... 李亚娟，王　浩，李　佳，张培军，胡俊菁，
　　　　　　　　胡国平，杜玉兰，董书伟，何宝祥（543）

丹参酮乳房注入剂中有效成分含量测定
.... 黄　鑫，梁剑平，郝宝成，高旭东，郭文柱（544）

高效液相色谱法测定二氧化氯固体消毒剂中DL-酒石酸的含量
.... 张景艳，张　宏，陈化琦，徐继英（546）

两种定值方法测定阿司匹林丁香酚酯对照品的含量
.... 杨青青，刘希望，杨亚军，李剑勇，查　飞（548）

中西兽药结合治疗安卡红种公鸡冠癣420例
.... 张仁福，王贵波，李锦宇，罗超应，谢家声（550）

宫康中水苏碱含量测定
.... 苗小楼，尚小飞，王东升，潘　虎，王　瑜，李升桦（551）

中国藏兽医药数据库系统建设与研究
.... 尚小飞，苗小楼，王　瑜，汪晓斌，张继瑜，潘　虎（552）

两种多糖对家兔眼刺激性试验
.... 张　哲，李新圃，杨　峰，罗金印，李晓强，刘龙海，李宏胜（553）

犊牛腹泻血液生化指标与主成分分析研究
.... 妥鑫，王胜义，王　慧，黄美州，崔东安，齐志明，刘永明（554）

犊牛腹泻病原调查及与临床症状相关性分析
………………… 王胜义，黄美州，王　慧，崔东安，齐志明，刘永明（555）

博物学素养与现代医药学研究
………… 杨亚军，刘希望，李世宏，孔晓军，秦　哲，杜文斌，李剑勇（556）

常山碱对小鼠巨噬细胞功能的影响
………………………………… 郭志廷，王　玲，衣云鹏，梁剑平（557）

藿芪灌注液治疗奶牛卵巢静止和持久黄体临床试验
………………… 严作廷，王东升，张世栋，董书伟，苗小楼，
　　　　　　　　那立冬，闫宝琪，朱新荣，韩积清，王雪郦（558）

肉制品中药物残留风险因子概述
………………………………………… 熊　琳，李维红，杨晓玲，高雅琴（559）

代谢组学在奶牛蹄叶炎研究中的应用前景
………………… 李亚娟，董书伟，王东升，张世栋，严作廷，
　　　　　　　　　　　　　　杨志强，杜玉兰，何宝祥（561）

奶牛乳房炎病原菌抗牛素耐药性研究进展
………………… 张亚茹，李新圃，杨　峰，罗金印，王旭荣，
　　　　　　　　　　　张　哲，刘龙海，王　丹，李宏胜（562）

苹果渣发酵生产蛋白饲料的研究进展
………………… 赵　萍，王晓力，王春梅，朱新强，张　茜（564）

奶牛健康养殖重要疾病防控关键技术研究
……………………………………………………………………… 严作廷（565）

中兽医学资源抢救和整理的重要性浅述
………………… 王贵波，罗超应，李建喜，李锦宇，谢家声，郑继方（566）

丁香酚解热作用机理研究进展
………………… 杨亚军，刘希望，赵晓乐，孔晓军，
　　　　　　　　李世宏，秦　哲，杜文斌，李剑勇（567）

疯草解毒复方中药制剂的药代动力学研究
………………… 郝宝成，权晓弟，叶永丽，高旭东，黄　鑫，
　　　　　　　　　王学红，刘建枝，王保海，梁剑平（568）

Simultaneous Determination of Diaveridine, Trimethoprim and Ormetoprim in Feed Using High Performance Liquid Chromatography Tandem Mass Spectrometry

YANG Yajun, LIU Xiwang, LI Bing, LI Shihong,
KONG Xiaojun, QIN Zhe, LI Jianyong*

(Key Laboratory of New Animal Drug Project of Gansu Province, Key Laboratory of Veterinary Pharmaceutical Development, Ministry of Agriculture, Lanzhou Institute of Husbandry and Pharmaceutical Sciences of CAAS, Lanzhou 730050, China)

Abstract: This study developed and validated a simple and reliable method for detecting and quantifying DVD, TMP and OMP in feed using dichloromethane extraction followed by HPLC-MS/MS. A matrix effect evaluation was performed using the post-extraction spiking method, and levels were less than ± 15% in all three feeds with their corresponding concentrations. LOD and LOQ, CCα and CCβ were 20 μg·kg^{-1} and 40 μg·kg^{-1}, 8.68–15.55 μg·kg^{-1} and 10.61–18.92 μg·kg^{-1} for all analytes, respectively. Calibration curves were linear for DVD, TMP and OMP with $R^2 \geq 0.990$ and $r \geq 0.995$, respectively. Recoveries of low, medium and high concentrations using the proposed method ranged from 74.4 to 105.2%. Repeatability and within-laboratory reproducibility were <7.4% (RSD). The chosen seven factors had no a significant influence on robustness. The method showed good performance when it was applied to analyze other laboratory-prepared or actual feed samples.

Key words: Diaveridine; Trimethoprim; Ormetoprim; Feed; LC-MS/MS

1 INTRODUCTION

Diaveridine (DVD, CAS: 5355-16-8), trimethoprim (TMP, CAS: 738-70-5) and ormetoprim (OMP, CAS: 6981-18-6) (chemical structures see Fig. S1), which were once called sulfonamide potentiators, inhibit bacterial dihydrofolate reductase. Sulfonamide potentiators are usually used in combination with sulfonamides, which inhibit bacterial dihydrofolate synthetase. This classic drug combination has synergistic effects, blocking folic acid metabolism in bacteria through two different mechanisms (Chen, 2009, Chap. 12 and 14). Sulfonamides potentiators can also promote the antibacterial activity of other antimicrobials (Chen, 2009, Chap.

---

* Corresponding author, E-mail: lijy1971@163.com (J.-Y. Li).
© 2016 Elsevier Ltd. All rights reserved.

12 and 14; Zander, Besier, Ackermann, & Wichelhaus, 2010; Zhou et al., 2015), such as b-lactam (Xie et al., 2005), aminoglycosides (Liu et al., 2008), fluoroquinolones (Mandal, Pal, Chowdhury, & Debmandal, 2009), and even some natural products (Wang, Cui, Shao, & Han, 2011). Thus, DVD, TMP and OMP are also considered antibacterial potentiators and have been extensively used in human and veterinary medicine and aquaculture (Declercq et al., 2013). DVD, TMP and OMP also have marked activity against coccidia (Ruff & Wilkins, 1990) and other protozoa (Cirioni, Giacometti, & Scalise, 1997; Felix et al., 2014; Lindsay, Butler, Rippey, & Blagburn, 1996). Premixes containing DVD and sulfaquinoxaline (China Agriculture Ministry, 2001) or sulfamethoxy-diazine (China Veterinary Pharmacopoeia Commission, 2011) can be added to complete feed to treat coccidiosis in poultry and intestinal infections in swine; however, these premixes are prohibited in laying hens in China (China Agriculture Ministry, 2001; China Veterinary Pharmacopoeia Commission, 2011).

In China, DVD, TMP and OMP are used as therapeutic agents to cure animal diseases after administration of appropriate preparations (China Agriculture Ministry, 2001; China Veterinary Pharmacopoeia Commission, 2011). These drugs are prohibited as common additives in commercial feeds. The physical and chemical properties of DVD, TMP and OMP are stable; thus, these drugs, which are occasionally added illicitly to commercial feed, could be concentrated in humans through the food chain. DVD is thought to be genotoxic in mammalian cells in vitro and in vivo (Ono et al., 1997; Yoshimura, 1991). Therefore, to safeguard human health from this specific risk, a maximum residue limit (MRL) was established in China, the European Union and other regulatory agencies for TMP of 50 $\mu g \cdot kg^{-1}$ in edible tissues of all food-producing species, except eggs, and 100 $\mu g \cdot kg^{-1}$ in edible tissues of horses

Veterinary drugs added illicitly to commercial feeds are an important source of residues in edible animal tissues. Therefore, monitoring DVD, TMP and OMP in feed could prevent human exposure to these drug residues in edible animal tissues. Currently, high performance liquid chromatography (HPLC)-ultraviolet (UV) detection is the only method approved for use in China to determine TMP concentrations in feed (National Standards of China, GB/T 21037—2007, 2007). However, DVD and OMP in feed have gone undetected. Therefore, the development of a simple, rapid and efficient method is urgently needed to detect DVD, TMP and OMP in feed.

Various methods for quantifying DVD, TMP or OMP have been reported, such as a gold immunochromatographic assay for TMP in milk (Wan, Feng, Zhao, Zhang, & Han, 2013), microfluidic capillary electrophoresis with UV detection at 214 nm for TMP and sulfonamides in pharmaceutical preparations (Fan, Chen, Chen, & Hu, 2003), capillary zone electrophoresis with capacitively coupled contactless conductivity detection for TMP and sulfamethoxazole in pharmaceutical tablets (da Silva, Vidal, do Lago, & Angnes, 2013), ion-pair chromatography for TMP in serum and urine (Watson, Shenkin, McIntosh, & Cohen, 1980), gas chromatography (GC) for TMP in plasma (Ernemann et al., 1990), GC-mass spectrometry (MS) for TMP in raw sewage-impacted water samples (Sichilongo, Mutsimhu, & Obuseng, 2013), laser diode thermal desorption-atmospheric pressure chemical ionization-MS for TMP in swine manure (Solliec, Massé, & Sauvé, 2014), HPLC-UV at 270 nm for TMP in feed

(Cheng, Zhang, Shen, Liu, & Wenren, 2008), LC-LTQ-Orbitrap for DVD and its metabolites in chickens (Wang et al., 2014), LC-MS/MS for TMP and other antibiotics in meat (Li, Sun, Zhang, & Pang, 2013) and honey (Economou, Petraki, Tsipi, & Botitsi, 2012), hydrophilic interaction LC-ESI⁺-MS/MS for TMP and DVD in tissues of chicken, pigs and fish (Luo et al., 2014), and HPLC-ESI⁺-MS/MS for DVD, TMP and OMP in food of animal origin (Gao et al., 2014; Jiang et al., 2009). However, no related reports have analyzed DVD, TMP and OMP simultaneously in feed.

In recent years, LC-MS/MS has become more widely used because it allows the analysis of a wide range of analytes and detection in the low $\mu g \cdot kg^{-1}$ range (Bolechová, Čáslavský, Pospíchalová, & Kosubová, 2015). Therefore, in the present study, a simple and effective multi-analyte method using liquid extraction and LC-MS/MS was developed and validated for simultaneous determination of DVD, TMP and OMP in feed. The proposed method was capable of detecting the low concentrations of these chemicals that would result from failure to comply with current Chinese regulations or from on-site contamination.

2 MATERIALS AND METHODS

2.1 Chemicals and solvents

Standard diaveridine was purchased from Sigma-Aldrich (VETRANAL® analytical standard; batch number100M1723V). Standard trimethoprim (batch number 10031-201205) and sulfanilamidum (CAS: 63-74-1, batch number 100024-201103) were purchased from the National Institutes for Food and Drug Control (Beijing, China). Standard ormetoprim was purchased from Dr. Ehrenstorfer GmbH (Augsburg, Germany; batch number 20904). Standard sulfaquinoxaline (CAS: 59-40-5, batch number H0251407) and sulfamethoxydiazine (CAS: 651-06-9, batch number H0251407) were purchased from China Institute of Veterinary Drug Control (Beijing, China). HPLC-grade methanol and formic acid were obtained from Fisher Scientific (Geel, Belgium) and Sigma-Aldrich, respectively. Deionized water (18 $M\Omega$) was prepared with a Direct-Q®3 system (Millipore, USA). Analytical grade dichloromethane, chloroform, n-hexane and sodium acetate trihydrate were furnished by Tianjin Baishi Chemical Industry Co. Ltd. (Tianjin, China), and were used without further purification.

A formic acid solution (0.2%, v/v) was prepared by pipetting 1.0 mL of formic acid into a 500 mL volumetric flask and diluting to the mark with water.

2.2 Standard solutions

Primary DVD, TMP and OMP stock solutions (approximately 1 $mg \cdot mL^{-1}$) were individually prepared in mobile phase, stored at 4℃, and were stable for 1 month (data not shown). The mixed stock solution (approximately 10 $\mu g \cdot mL^{-1}$) was prepared by pipetting 1.0 mL of the three stock solutions into a 100 mL volumetric flask and diluting to the mark with mobile phase on the day of analysis. Five-fold serial dilutions of the mixed stock solutions were also made in mobile phase to produce working stock solutions of 2 $\mu g \cdot mL^{-1}$, 400 $ng \cdot mL^{-1}$, 80 $ng \cdot mL^{-1}$, 16 $ng \cdot mL^{-1}$, 3.2 $ng \cdot mL^{-1}$ and 0.64 $ng \cdot mL^{-1}$ on the day of analysis, and these working stock solutions were used to generate the calibration curves.

2.3 Samples and extraction procedure of the optimized method

The complete feed for laying hens was obtained from Lanzhou CP Group Co., Ltd. (Lanzhou, China). The concentrated feed for growing and fattening pigs and the premix feed for the laying hens were purchased from Shaanxi Yangling Fushite Feed Co., Ltd. (Yangling, Shaanxi, China). All feed was free of DVD, TMP and OMP.

In order to ensure homogeneity prior to analysis, feed samples were ground using a laboratory mortar and pestle and filtered through a 0.45 mm standard sieve. Finely ground samples of complete feed (2 g), concentrated feed (1 g) or premix feed (1 g) were placed in a polypropylene centrifuge tube (50 mL). Sodium acetate solution (1 mol·L^{-1}, 10 mL) was added and mixed with a vortexer for 1 min (Vortex Genie 2, Scientific Industries, Portland, OR, USA). Next, 20.0 mL dichloromethane was added and the mixture was vortexed for 2 min. The mixture was extracted using an ultrasonic water bath at room temperature for 30 min, and mixed with a vortexer for 2 min every 10 min during the ultrasonic extraction. The system was centrifuged at 1 800g for 15 min (Multifuge X3R, Thermo Scientific, Rockford, IL, USA). Approximately 8 mL of the dichloromethane extract was transferred into another centrifuge tube with a 10 mL syringe. The extract was centrifuged at 1 800 g for 10 min to separate the fine feed particles. Then, 2.0 mL of the extract was pipetted to another centrifuge tube and dried under nitrogen gas at 40 ℃.

The dichloromethane extraction residue was completely dissolved with 1.0 mL mobile phase using ultrasound and mixed with a vortexer. The lipid substances were removed by liquid-liquid extraction with n-hexane (2 mL), centrifugation (1 800 g, 10 min) and removal of the n-hexane extraction layer. The procedure for removing the lipid substances was performed twice. Finally, the water phase was filtered through a 0.22 μm nylon filter for use in the following LC-MS/MS analysis. The water phase was appropriately diluted with mobile phase to ensure the concentration of analytes remained within the range of the calibration curves.

2.4 Liquid chromatography tandem mass spectrometry conditions

The experiments were performed with an Agilent 1 200 HPLC using MassHunter software version B.01.04 (Agilent Technologies, Palo Alto, CA, USA) coupled to an Agilent 6410A Triple Quadrupole mass spectrometer. The HPLC system consisted of a G1312B binary LC pump, a G1322A vacuum degasser, a temperature-controlled (G1330B) micro-well plate autosampler (G1367C) set at 4 ℃, and a G1316B thermostatted column compartment set at 40 ℃ (Agilent Technologies). The separation of DVD, TMP and OMP was accomplished with an Agilent Eclipse Plus C_{18} column (3.0 mm × 100 mm, 1.8 μm) protected with a prefilter (4 mm and 5 μm; GRACE, Williamsburg, MI, USA) at a flow rate of 300 μL·min^{-1} and using an injection volume of 5 1 L. The mobile phase was methanol–0.2% formic acid (20 + 80, v/v) with isocratic elution.

An Agilent 6410A Triple Quadrupole mass spectrometer with an electrospray ionization source interface operated in the positiveion scan mode was used for LC-MS/MS analysis. The capillary potential of the MS was +4 000 V, the gas temperature was 350 ℃, the gas flow rate was

11 L·min^{-1}, the nebulizer pressure was 35 psi, the delta EMV was 300 V, and the dwell time was 200 ms. The parameters for multiple-reaction monitoring (MRM) were optimized using the product ion scan mode, and the characteristic ions of each analyte were selected.

2.5 Method validation

The method validation was performed based on the directives from the National Standards of China GB/T 20001.4—2001 (2001), National Standards of China GB/T 27404—2008 (2008), and European Commission Decision 2002/657/EC (2002). The proposed method was validated mainly for selectivity, limits of detection (LOD) and quantification (LOQ), decision limit and detection capability, linearity range, recovery, repeatability, within-laboratory reproducibility and robustness. The recoveries experiments were performed by fortifying three different feed samples with all three analytes at three concentrations (low, medium and high) and analyzing five replicates at each concentration.

Instrumental and extraction blanks were analyzed prior to and in between fortified extracts. The instrument blank was conducted using mobile phase and analyzed at the beginning of the run. The extraction blanks were analyzed to ensure that no interference or contamination appeared during the analysis of these blanks. This experiment showed zero carryover for all the analytes, ensuring that no cross-contamination occurred.

2.5.1 Selectivity

As complex materials, feeds contain many compounds that could interfere when determining DVD, TMP and OMP. The ionization efficiency of these analytes would be influenced by compounds co-eluting with DVD, TMP and OMP, resulting in different MS responses for solvent and feed samples. Thus, a matrix effect evaluation was performed using the post-extraction spiking method (Matuszewski, Constanzer, & Chavez-Eng, 2003). Blank feed samples were extracted using the procedure described in Section 2.3. Finally, 1 mL of the water phase was added to 10 μL of a mixture of standards and filtered through a 0.22 μm nylon filter for LC-MS/MS analysis. Then, the MRM responses of matrix-matched solutions were compared to that of the standard solution with the same concentration.

In addition, the following two criteria were used for the qualitative identification of the target analytes in samples (European Commission Decision 2002/657/EC). Firstly, LC retention times of the analytes were compared with those of control compounds. The allowable retention time deviation for the analytes in samples and that of their corresponding standards was within ± 2.5%. Secondly, the allowable relative abundance deviation for the characteristic ions of the analytes and that of the characteristic ions in the corresponding standard complied with the criteria shown in Table S1.

DVD, TMP and OMP, as dihydrofolate reductase inhibitors, are often combined with sulfonamides such as sulfaquinoxaline and sulfamethoxydiazine to inhibit sensitive pathogens, such as some bacteria and protozoa. Thus, sulfonamides may potentially interfere with the determination of the analytes. Sulfaquinoxaline, sulfamethoxydiazine and sulfanilamidum, which concentrations were 5 times to the analytes, were added to the blank feed samples with DVD, TMP and OMP. The spiked samples were extracted and detected using the proposed method. Finally, LC retention times, relative abundance deviation for the characteristic ions and MRM responses of the target analytes were compared to check for interferences.

2.5.2 Limits of detection (LODs) and quantification (LOQs)

The LODs and LOQs for the three target analytes were determined as 3 and 10 times the signal to noise ratio (SNR), respectively. The samples for the determination of the LODs and LOQs were prepared as follows. For the LOD samples, 400 μL of the mixture of standards, which contained each analyte at 100 ng·mL^{-1}, was added to 2.00 g of complete feed, or 200 μL was added to 1.00 g concentrated or premix feed. For the LOQ samples, 800 μL of the mixture of standards was added to 2.00 g complete feed, or 400 μL was added into 1.00 g concentrated or premix feed. The samples were dried at 40 ℃ overnight. Thus, the concentration of each analyte in every sample for determination of the LODs and LOQs was 20 μg·kg^{-1} and 40 μg·kg^{-1}, respectively. The dried samples were processed and detected following the procedures described in Sections 2.3 (Samples and extraction procedure of the optimized method) and 2.4 (Liquid chromatography tandem mass spectrometry conditions). The SNR of each quantifier ion in each sample was calculated by comparing the peak height of quantifier ion to that of the noise.

2.5.3 Decision limit and detection capability

DVD, TMP and OMP were treated as banned substances in commercial feeds, and the permitted limits have not been established. Thus, the blank feed samples were fortified at and above the minimum required performance level in equidistant steps. Then, the samples were extracted and detected by the proposed method. After identification, Decision limit (CCα) was calculated according to European Commission Decision 2002/657/EC (2002). To obtain the detection capability (CCβ) values, 20 blank samples per feed fortified with the analytes at the obtained CCα levels and analyzed. The CCβ calculation was done using the theoretical value of CCα previously obtained plus 1.64 times the corresponding standard deviation.

2.5.4 Linearity

To construct the calibration curves, the mixed working solutions with 2 μg·mL^{-1}, 400 ng·mL^{-1}, 80 ng·mL^{-1}, 16 ng·mL^{-1}, 3.2 ng·mL^{-1} and 0.64 ng·mL^{-1} of each compound were injected into the LC-MS/MS system (three replicates per concentration). MRM determination of the working solutions and samples was conducted in a single batch. The concentration versus area of the quantifier ion was plotted. Linear least squares regression was the modeling method used. The linearity was evaluated using the coefficient of determination (R^2) and the correlation coefficient (r).

2.5.5 Accuracy and precision

The accuracy was assessed using the relative recovery results obtained from samples spiked with DVD, TMP and OMP at low, medium and high concentrations (n=5 replicates per concentration) on three different days.

Based on China Agriculture Ministry regulation No. 168 (2001), complete poultry feed is fortified for the treatment of coccidiosis with sulfaquinoxaline and DVD premix, with the concentration of DVD in the complete feed at 20 mg·kg^{-1}. Otherwise, according to the guidelines of Chinese Veterinary Pharmacopoeia (2010 Edition) for chemical medicine (China Veterinary Pharmacopoeia Commission, 2011), sulfamethoxydiazine and DVD premix is mixed in complete feed for treatment of intestinal infections in swine and coccidiosis in chickens, with the concentration of DVD in complete feed at 40 mg·kg^{-1}. Thus, the medium concentration of the three analytes in the complete feed used in the present study was 20 mg·kg^{-1}, and the low concentration was at the LOQ. In practice, the amount of the chemicals added to feed is often higher than that specified in the regulations to obtain better therapeutic efficacy. Therefore, in the present study, the high concentration of DVD, TMP and OMP present in the complete feed samples was 100 mg·kg^{-1}. In addition, the approximate proportions of concentrated feed and premix feed to complete feed were 15%–25% and 4%, respectively. Thus, the medium concentrations of analytes in concentrated feed and premix feed used in the present study were 100 mg·kg^{-1} and 500 mg·kg^{-1}, respectively.

The samples for the recovery experiment were prepared as followed. Each drug (25 mg) was mixed with 250 g complete feed for the high concentration. Each drug (60 mg) was mixed with 120 g concentrated feed for the high concentration, and 100 mg or 60 mg of each drug was mixed with 100 g or 120 g premix feed for the high or medium concentrations, respectively. These drugs were uniformly mixed with a lower amount of blank feed using a laboratory mortar and pestle. Additional blank feed was added with sufficient mixing until the calculated amount was reached to attain samples containing the highest concentrations of analytes. These samples were then diluted with blank feed to acquire samples with medium and low concentrations. The lowest concentrations of analytes in the samples were derived from the LOQs, which were obtained as described in Section 2.5.2 (Limits of detection and quantification).

The samples from the recovery experiments were also used to determine the precision of the method through the evaluation of repeatability and within-laboratory reproducibility, which was expressed as relative standard deviation (RSD). The assays were performed on three different days, and five replicates were performed with each of the three target analytes concentrations per day (n=15 per concentration).

2.5.6 Robustness

Robustness is the measure showing the performance of the method when small deliberate changes are made to the normal conditions. The robustness of the proposed method was evaluated according to European Commission Decision 2002/657/EC. In present study, the variables, which derived from extract procedure and chromatographic condition, involved sonication time, vortex time, cycles of sonication and vortex, concentration of sodium acetate solution, mobile phase, concentration of formic acid and column temperature.

3 RESULTS AND DISCUSSION

3.1 Optimization of the LC-MS/MS conditions

The presence of veterinary drug residue is confirmed when the drug is identified using both chromatographical and spectral methods. However, if this criterion cannot be satisfied because the available methods are lacking, then a new method or a better procedure should be developed for extracting and/or separating the analytes. LC-MS/MS using the MRM mode is currently considered to be the technique with the highest sensitivity and selectivity and thus was used in the present study to determine feed quality and drug residue quantity.

In order to optimize the MS/MS experimental conditions, individual standard solutions of each analyte in methanol (1 μg·mL^{-1}) were injected at a volume of 1 μL. All MS/MS parameters were then selected to obtain optimal precursor ion [M + H]$^+$ intensities for every compound. The total ion chromatogram (TIC) obtained using MRM (Fig. 1A) and the product ion spectra of DVD (Fig. 1B), TMP (Fig. 1C) and OMP (Fig. 1D) are shown in Fig. 1, and the parameters used in the MRM assay are shown in Table S2.

3.2 Validation of the analytical method

Validation was performed in terms of selectivity for matrix effects, interferences and a comparison of characteristic ions, LOD, LOQ, CCα, CCβ, linearity, range, accuracy (recovery rate), repeatability, within-laboratory reproducibility, and robustness.

3.2.1 Selectivity

Selectivity is the degree to which a method can distinguish and quantify the analyte (s) accurately in the presence of endogenous interference. To determine the selectivity of the method, matrix effects of three types of feed were analyzed. Co-eluting, undetected matrix components may reduce or enhance ion intensity of analytes and affect assay accuracy. The matrix effect during validation of an analytical method used for actual samples was best examined by comparing the MS/MS response (peak areas or peak heights) of an analyte at any given concentration spiked post-extraction into an actual sample extract to the MS/MS response of the same analyte present in the "neat" mobile phase. Detecting a matrix effect allow for an assessment of the reliability and selectivity of the HPLC-MS/MS method. If the results are not satisfactory, the calibration curve may need to be prepared using a matrix-matched solution, or the sample preparation procedures may need to be altered.

The present study used a low concentration to determine the LOQ (40 μg·kg^{-1}) and the commonly mixed concentration in complete feed (20 mg·kg^{-1}) to evaluate the matrix effects of the three kinds of feed. The resulting matrix effects of each analyte are expressed as a percentage of the MRM response (peak areas of quantifier ions) and are shown in Table 1. The results were less than ± 15% for all three kinds of feed and all concentrations. Therefore, the matrix-matched solutions were replaced with the mixed standard working solutions to prepare the calibration curves.

As experimental results, the relative intensities of ions in the DVD, TMP and OMP standard solutions at different concentrations and the samples extract were greater than 60%. The allowable deviation in this study was ± 20% according to the European Commission Decision 2002/657/EC. All sample results were ± 10% (data not shown).

The present study used a low concentration to determine the LOQ (40 μg·kg^{-1}) and the commonly mixed concentration in complete feed (20 mg·kg^{-1}) to evaluate interferences of sulfanilamides in the three kinds of feed. In general, the concentration of sulfanilamide is 5 times to sulfonamide potentiator (Chen, 2009, Chap. 12 and 14). Thus, the spiked concentrations of sulfaquinoxaline, sulfamethoxydiazine and sulfanilamidum were 200 μg·kg^{-1} and 100 mg·kg^{-1} in low concentration of LOQ and the commonly mixed concentration in complete feed, respectively.

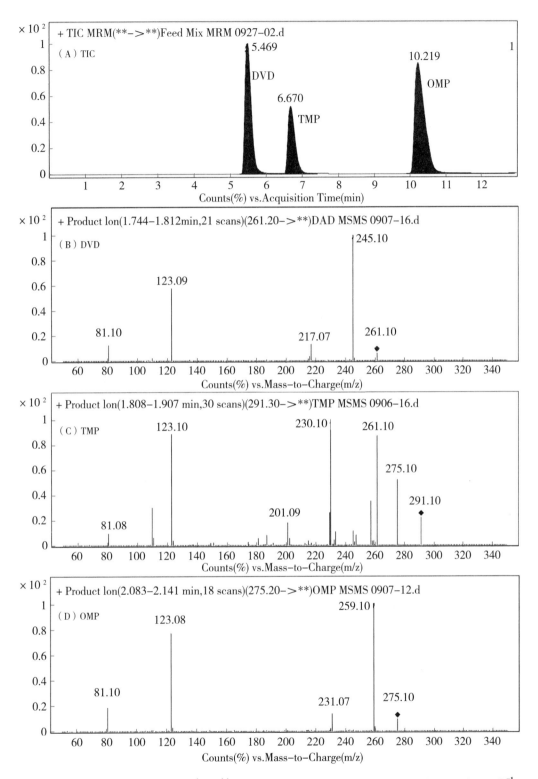

Fig. 1 Total ion chromatogram (TIC) and MS/MS spectra of three target analytes at 1 μg·mL^{-1}

The interferences results showed that there were no significant difference in the LC retention times, relative abundance deviation for the characteristic ions between all analytes without sulfonamides and all analytes, and the MRM responses of all analytes when they were combined with sulfonamides were 100.52% ± 5.68% (n=24) to the responses of all analytes without sulfonamides. Therefore, possible interferences of sulfa drugs should be negligible.

In addition, the characteristic ions of sulfonamides and other antimicrobials (Economou et al., 2012; Spielmeyer, Ahlborn, &Hamscher, 2014; Zhao et al., 2014) are different from those of DVD ([M + H]$^+$ 261.2→245.1, 261.2→123.1), TMP ([M + H]$^+$ 291.2→230.1, 291.2→275.1, 291.2→261.1, 291.2→259.1, 291.2→123.1) and OMP ([M + H]$^+$ 275.2→259.1, 275.2→123.1) in MRM with an electrospray ionization source interface operated in the positive-ion scan mode.

Table 1 Matrix effects of each feed on each analyte with corresponding concentrations.

Sample	Concentration (mg·kg^{-1})	Analyte	Matrix effect (n=5, %)	RSD (%)
Complete feed	0.04	DVD	99.32 ± 5.15	5.19
		TMP	97.25 ± 3.04	3.13
		OMP	100.44 ± 5.96	5.93
Concentrated feed	0.04	DVD	99.23 ± 3.10	3.12
		TMP	103.15 ± 3.93	3.81
		OMP	103.56 ± 5.46	5.28
Premix feed	0.04	DVD	97.36 ± 4.13	4.24
		TMP	104.94 ± 4.90	4.67
		OMP	104.05 ± 4.64	4.46
Complete feed	20	DVD	103.60 ± 3.67	3.55
		TMP	102.92 ± 3.22	3.13
		OMP	103.51 ± 1.82	1.76

3.2.2 LOD and LOQ

The LOD and LOQ are often calculated using the signal-to-noise ratio (SNR) of a chromatographic technique. The mixed concentrations of these drugs were as high as the level of mg·kg^{-1} in feed because DVD, TMP or OMP is mixed in commercial feed to cure disease or to produce better performance.

When the concentration of DVD, TMP and OMP in the samples of complete feed, concentrated feed and premix feed was 20 μg·kg^{-1} (LOD), the SNR for all the quantifier ions was markedly greater than 3 (see Fig. 2). This result indicated that the LODs for DVD, TMP and OMP using this method were less than 20 μg·kg^{-1}. Thus, this method enabled the detection of DVD, TMP and OMP in feed at very low concentrations.

Similarly, the SNR for all the quantifier ions in the test samples was markedly greater than 10 in the study examining the LOQs (chromatograms not shown). This result indicated that the LOQs for DVD, TMP and OMP using the proposed method were less than 40 μg·kg^{-1}. Therefore, the method also enabled the quantification of DVD, TMP and OMP in feed at very low concentrations.

3.2.3 Decision limit and detection capability

The decision limit (CCα) is the level used to decide whether a sample has a probability of error equal to α. The detection capability (CCβ) is the lowest content of the target species that can be detected, identified, and quantitatively determined in a sample, with an error probability equal to β (European Commission Decision 2002/657/EC). The CCα and CCβ of the proposed method were calculated based on the calibration curve of the spiked samples, according to the following equations: CCα ($\mu g \cdot kg^{-1}$) =a/b + 2.33SDa/b (α=1%) and CCβ ($\mu g \cdot kg^{-1}$) =CCα + 1.64SD (β=5%), where a is the intercept of the standard addition curve (20−40 $\mu g \cdot kg^{-1}$), b is the slope of the standard addition curve, SDa/b is the standard deviation of the within-laboratory reproducibility of a/b (n=6), and SD is the standard deviation of the within-laboratory reproducibility of the measured content (n=20) (Dasenaki & Thomaidis, 2010; European Commission Decision 2002/657/EC).

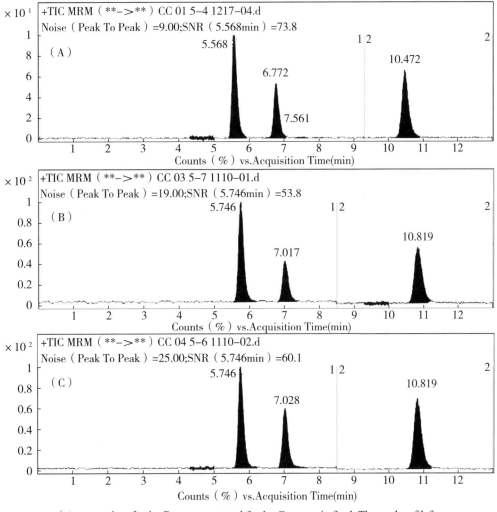

(A: complete feed, B: concentrated feed, C: premix feed. The peaks of left, middle and right are DVD, TMP and OMP, respectively.)

Fig. 2 Total ion chromatograms (TIC) of multiple-reaction monitoring (MRM) in the experiments conducted to determine the limits of detection for the three target analytes in three types feed

Table 2 The results of CCα and CCβ.

Samples	CCα (µg·kg^{-1})			CCβ (µg·kg^{-1})		
	DVD	TMP	OMP	DVD	TMP	OMP
Complete feed	8.68	12.69	10.92	10.61	15.56	12.71
Concentrated feed	12.48	15.12	12.68	13.89	18.92	14.79
Premix feed	12.39	15.55	13.90	14.63	18.59	16.89

The results are presented in Table 2. The CCα was 8.68–15.55 µg·kg^{-1}, and CCβ was 10.61–18.92 µg·kg^{-1}, which were below the LOD and LOQ set at 20 and 40 µg·kg^{-1}, respectively.

3.2.4 Linearity

To determine the method linearity, a minimum of six concentrations was analyzed with three replicates per each analyte. The matrix effect results showed that the feed matrix did not interfere with the results. So, the matrix-matched solutions were replaced with the mixed standard working solutions to construct the calibration curves. The calibration curves, which were constructed by plotting the peak area of the quantifier ions versus their concentrations, were linear for DVD, TMP and OMP over a wide range of concentrations (0.6 ng·mL^{-1} to 2 µg·mL^{-1}), with the coefficient of determination (R^2) ≥0.990 and coefficient of correlation (r) ≥0.995 (Table S3).

3.2.5 Accuracy and precision

The results for accuracy, estimated from the recovery assays, and precision are shown in Table 3. The recoveries of DVD, TMP and OMP were calculated from the quantifier ions peak areas obtained for each analyte. The repeatability and within-laboratory reproducibility values were expressed as RSD. All target analytes showed results consistent with the acceptable range as specified by the National Standards of China GB/T 27404-2008 (2008).

In order to achieve rapid, inexpensive, and simple sample processing, a sodium acetate solution (1 mol·L^{-1}) with dichloro-methane as the compound solvent was selected to extract the

Table 3 Validation results for the analytical method of the three target analytes DVD, TMP and OMP.

Samples	Concentration (mg·kg^{-1})	Analyte	Recovery (n=5, %)	Repeatability (RSD, %)	Recovery (n=3, %)	Within-laboratory reproducibility (RSD, %)
Complete feed	0.04	DVD	76.4 ± 5.1	6.6	77.2 ± 1.2	1.6
			76.6 ± 4.0	5.2		
			78.7 ± 1.2	1.5		
		TMP	75.7 ± 4.4	5.9	75.8 ± 1.6	2.1
			74.4 ± 4.8	6.4		
			77.5 ± 5.7	7.4		
		OMP	76.1 ± 5.5	7.2	77.5 ± 1.9	2.5
			76.7 ± 4.4	5.7		
			79.7 ± 5.0	6.3		

(continued)

Samples	Concentration (mg·kg^{-1})	Analyte	Recovery (n=5, %)	Repeatability (RSD, %)	Recovery (n=3, %)	Within-laboratory reproducibility (RSD, %)
Complete feed	20	DVD	94.7 ± 3.1	3.3	96.9 ± 4.2	4.3
			101.7 ± 2.8	2.8		
			94.3 ± 2.5	2.6		
		TMP	95.9 ± 2.6	2.8	95.4 ± 0.7	0.7
			95.7 ± 3.7	3.9		
			94.7 ± 3.1	3.3		
		OMP	98.3 ± 5.6	5.7	99.8 ± 4.8	4.8
			105.2 ± 1.6	1.5		
			95.8 ± 1.5	1.6		
Concentrated feed	100	DVD	96.4 ± 2.0	2.0	98.6 ± 2.3	2.3
			101.0 ± 2.2	2.2		
			98.5 ± 4.3	4.4		
		TMP	97.9 ± 3.7	3.8	98.3 ± 1.7	1.7
			100.1 ± 2.5	2.5		
			96.8 ± 2.8	2.9		
		OMP	99.6 ± 3.8	3.8	99.4 ± 2.3	2.3
			101.7 ± 2.8	2.7		
			97.1 ± 2.4	2.5		
	100	DVD	98.5 ± 2.8	2.9	98.5 ± 1.8	1.8
			100.2 ± 4.3	4.3		
			96.7 ± 1.5	1.6		
		TMP	98.0 ± 3.4	3.4	98.7 ± 2.2	2.3
			101.1 ± 4.5	4.4		
			97.2 ± 1.6	1.7		
		OMP	96.0 ± 1.4	1.4	99.0 ± 3.8	3.9
			103.4 ± 2.3	2.2		
			97.6 ± 3.2	3.3		
Premix feed	500	DVD	98.6 ± 3.1	3.1	99.6 ± 1.0	1.0
			100.5 ± 3.2	3.2		
			99.6 ± 4.1	4.1		
		TMP	95.7 ± 1.7	1.8	99.6 ± 3.4	3.4
			101.6 ± 4.1	4.0		
			101.4 ± 4.5	4.4		
		OMP	97.8 ± 3.6	3.7	98.2 ± 2.7	2.7
			95.8 ± 2.0	2.1		
			101.1 ± 3.8	3.8		

analytes in the proposed method. In a preliminary experiment, the extraction rates of DVD, TMP and OMP were investigated using dichloromethane and chloroform. The results showed that dichloromethane had the best selectivity ($P < 0.05$). The mobile phase used with isocratic elution in the proposed method was methanol-0.2% formic acid (20 + 80, v/v), and a C 18 LC column was used. Baseline separation was obtained, although the retention times of all analytes were longer than those obtained with gradient elution. The results (Table 3) showed that the proposed method had good precision (maximum 7.4%) and accuracy, approximately 70%–80% for the low concentrations of analytes and 94%–105% for the medium and high concentrations. The developed method was validated for three feed materials (complete feed, concentrated feed and premix feed) at different levels, particularly at 40 $\mu g \cdot kg^{-1}$ in all three feed types, representing the LOQ, and at 20, 100 and 500 $mg \cdot kg^{-1}$ in complete feed, concentrated feed and premix feed, respectively, corresponding to the commonly mixed concentrations in commercial feed. Finally, the within-laboratory validation results confirmed suitability of the method for analyzing DVD, TMP and OMP in feed materials.

3.2.6 Robustness

Robustness is defined as "the susceptibility of an analytical method to changes in experimental conditions which can be expressed as a list of the sample materials, analytes, storage conditions, environmental and/or sample preparation conditions under which the method can be applied as presented or with specified minor modifications. For all experimental conditions which could in practice be subject to fluctuation (e.g. stability of reagents, composition of the sample, pH, and temperature) any variations which could affect the analytical result should be indicated." Robustness and ruggedness are closely related terms often used interchangeably in the literature (Karageorgou & Samanidou, 2014). The robustness of the method was assessed according to the Youden's approach (Youden & Steiner, 1975). The basic idea is that several variations are introduced at once, instead of studying one alteration at a time. The experimental design described in European Commission Decision 2002/657/EC was applied and seven small but deliberate changes of the method (variables) were introduced. Their influence on the method results was assessed.

In present study, the method variables derived from extract procedure and chromatographic condition. Factors of extract procedure contained sonication time (5, 15 min), vortex time

Table 4 Concentrations of the spiked and detected target analytes diaveridine, trimethoprim and ormetoprim in various feed product samples.

Sample	Spiked concentration ($mg \cdot kg^{-1}$)			Detected concentration ($mg \cdot kg^{-1}$)		
	DVD	TMP	OMP	DVD	TMP	OMP
Complete feed 1	0.04	-	-	0.033	ND	ND
Complete feed 2	20	-	-	20.18	ND	ND
Complete feed 3	-	40	-	ND	39.04	ND

(continued)

Sample	Spiked concentration (mg·kg^{-1})			Detected concentration (mg·kg^{-1})		
	DVD	TMP	OMP	DVD	TMP	OMP
Complete feed 4	20	20	20	19.27	18.82	21.05
Concentrated feed 1	-	-	100	ND	ND	98.72
Concentrated feed 2	50	50	-	48.41	51.32	0.051
Premix feed 1	-	-	-	ND	ND	ND

Note: ND, not detected.

(1, 3 min), cycles of sonication and vortex (2, 4), as well as concentration of sodium acetate solution (0.5, 1.5 mol·L^{-1}). Factors of chromatographic condition included mobile phase (water 75%, 85%), concentration of formic acid (0.15%, 0.25%) and column temperature (35℃, 45℃). These seven different variables chosen within the entire analytical process were taken into account in the evaluation of method robustness.

The results showed that the factors of sonication time, cycles of sonication and vortex had very slight influence on the method, and the other chosen factors had no a significant influence on robustness. Therefore, the method proved to be fairly robust and able to withstand minor fluctuation in the operating variables that may occur during sample preparation and detection.

The results of the validation clearly demonstrated the suitability of this method for the detection and identification of all tested analytes.

3.3 Sample analysis

The proposed method was used to determine the presence and quantify the concentrations of DVD, TMP, or OMP in 11 feed samples (6 complete feed samples for poultry and swine, 2 concentrated feed samples for poultry, and 3 premix feed samples for poultry and swine) obtained from commercial vendors in several regions of the Gansu and Shaanxi provinces in China, and 7 additional samples prepared in the laboratory. The results from the analysis of the 11 samples obtained from commercial vendors showed that the target analytes were not detected. By contrast, the analytes were detected and quantified in the 7 laboratory spiked samples. The spiked concentrations and the detected concentrations are shown in Table 4. The detected concentration of each analyte was consistent with the spiked concentration of that analyte, except for OMP, which was detected in an OMP-free concentrated feed sample. This latter result was likely due to a cross-contamination that occurred during the preparation of the spiked samples.

4 CONCLUSION

In this study, a liquid extraction procedure and HPLC-MS/MS analytical method was successfully implemented and validated for detecting and quantifying DVD, TMP and OMP simultaneously in a variety of feeds. The applicability of the method for DVD, TMP and OMP were detected in spiked feed samples and actual samples. The assay was deemed suitable for regulatory monitoring of these drugs in animal feeds. This should help the regulatory agencies develop an efficient monitoring method for DVD, TMP and OMP in related commercial feeds in China.

COMPETING INTERESTS

The authors declare that no competing interests exist.

ACKNOWLEDGEMENT

This work was supported by the Project of Establishment Agricultural Standards（2013-306, 2014-350）.

APPENDIX A. SUPPLEMENTARY DATA

Supplementary data associated with this article can be found, in the online version, at http://dx.doi.org/10.1016/j.foodchem.2016.05.184.

（发表于《Food Chemistry》，院选SCI，IF：4.052. 358–366）

The Coordinated Regulation of Na$^+$ and K$^+$ in Hordeum Brevisubulatum Responding to Time of Salt Stress[☆]

WANG Chunmei[1, a], XIA Zengrun[2, a], WU Guoqiang[3], YUAN Huijun[3], WANG Xinrui[4], LI Jinhua[1], TIAN Fuping[1], ZHANG Qian[1], ZHU Xinqiang[1], HE Jiongjie[1], KUMAR Tanweer[2], WANG Xiaoli[1, *], ZHANG Jinlin[2, *]

(1. Lanzhou Institute of Husbandry and Pharmaceutical Science, Chinese Academy of Agricultural Sciences, Lanzhou 730050, People's Republic of China; 2. State Key Laboratory of Grassland Agro-ecosystem, College of Pastoral Agriculture Science and Technology, Lanzhou University, Lanzhou 730020, People's Republic of China; 3. School of Life Science and Engineering, Lanzhou University of Technology, Lanzhou 730050, People's Republic of China; 4. College of Animal Science, South China Agricultural University, Guangzhou 510642, The People's Republic of China)

Abstract: Hordeum brevisubulatum, called as wild barley, is a useful monocotyledonous halophyte for soil improvement in northern China. Although previously studied, its main salt tolerance mechanism remained controversial. The current work showed that shoot Na$^+$ concentration was increased rapidly with stress time and significantly higher than in wheat during 0–168 h of 100 mM NaCl treatment. Similar results were also found under 25 and 50 mM NaCl treatments. Even K$^+$ was increased from 0.01 to 50 mM in the cultural solution, no significant effect was found on tissue Na$^+$ concentrations. Interestingly, shoot growth was improved, and stronger root activity was maintained in H. brevisubulatum compared with wheat after 7 days treatment of 100 mM NaCl. To investigate the long-term stress impact on tissue Na$^+$, 100 mM NaCl was prolonged to 60 days. The maximum values of Na$^+$ concentrations were observed at 7th in shoot and 14th day in roots, respectively, and then decreased gradually. Micro-electrode ion flux estimation was used and it was found that increasing Na$^+$ efflux while maintaining K$^+$ influx were the major strategies to reduce the Na$^+$ concentration during long-term salt stress. Moreover, leaf Na$^+$ secretions showed little contribution to the tissue Na$^+$ decrease. Thereby, the physiological mechanism for H. brevisubulatum to survive from long-term salt stress was proposed that rapid Na$^+$ accumulation occurred

---

[☆] This research was supported by the National Natural Science Foundation of China (31201841 and 31222053), and the Agricultural Science and Technology Innovation Program of Chinese Academy of Agricultural Sciences (CAAS-ASTIP-2014-LIHPS-08).

* Corresponding authors, E-mail: 13609381223@139.com (X.-L. Wang); jlzhang@lzu.edu.cn (J.-L. Zhang).

[a] These authors have contributed equally to this work.

© 2016 Elsevier Ireland Ltd. All rights reserved.

in the shoot to respond the initial salt shock, then Na^+ efflux was triggered and K^+ influx was activated to maintain a stable K^+/Na^+ ratio in tissues.

Key words: Rapid Na^+ accumulation; Na^+ efflux; K^+ influx; Coordinated ion regulation; Na^+ secretion

1 INTRODUCTION

Salinity is a serious threat to crop yield as well as the environment protection [1]. Fortunately, halophytes have evolved various mechanisms to cope with soil salinity in long-term natural selection processes [2, 3]. Most of the halophytes are dicotyledonous. However, the majority of the economically important crops are monocotyledonous [4, 5]. Therefore, understanding of the salt-tolerance mechanism (s) of monocotyledonous halophytes, especially the wild relatives of cultivated cereals, will aid effective improvement of the salt tolerance of cereal crops [4-6].

Hordeum brevisubulatum (Trin.) Link, also known as wild barley, is a close relative of barley and wheat. It is well-known monocotyledonous halophytes and could be used as saline grass for soil improvement in north China [7]. The prime researches regarding the salt tolerance of H. brevisubulatum mainly focused on the description of the biological characteristics [8], along with the preliminary analysis of the physiological indexes during 1980—2000s [9]. Researches regarding the salt tolerance of *H. brevisubulatum* have gradually increased in recent years, but few have focused on its key physiological salt tolerance mechanism. Li et al. proposed that Na^+ compartmentation in leaves and Na^+ secretion from the leaf surface were the key adaptive mechanisms, but no obvious morphological evidence of secretory structures was found [10] except for some called "salt hairs" [11], and no succulence was observed in the genus of Hordeum. Some other researchers suggested that restricting Na^+ influx along with enhancing the osmotic adjustment by accumulating organic compounds can contribute to an overall enhancement of the salt tolerance in *H. brevisubulatum* [12, 13]. On the other hand, construction of a cDNA library, the screening of expressed gene fragments induced by salt stress [14-17], analysis of the known salt tolerance related genes, such as HbNHX1 [18], DREB1 [19], DREB2 [20], rbcS [21], HbCDPK [22] and HbCIPK2 [7], and the subcellular localization of the protein for salt stress signal transduction gene, such as CIPK [23], HbCBL1 and HbCBL2 [24], were conducted by other researchers in order to find the main genes that control the salt tolerance of *H. brevisubulatum*. However, the main genes and physiological mechanisms for salt tolerance still remain unclear.

In current work, in order to find the key physiological salt tolerance mechanisms of H. brevisubulatum, firstly, the growth response indexes of shoot and root were determined after 7 days salt treatment of 100 mM NaCl. Interestingly, significantly higher shoot Na^+ was accumulated rapidly in *H. brevisubulatum* than that in wheat. We also verified this under low concentrations (25, 50 mM). Meanwhile, treatment time was prolonged to 60 days to find the time point of peak values for Na^+ and K^+ concentration in tissues and their trends with the stress time prolonging. Surprisingly, Na^+ was declined obviously during long-term stress. So, to assess the contribution of Na^+ and K^+ fluxes to the decline of tissue Na^+ concentration, a micro-electrode ion flux estimation technique (MIFE) for net Na^+ and K^+ fluxes was used. Moreover, K^+ concentration were expanded from 0.01 to 50 mM in medium to confirm whether high Na^+ accumulation in shoot was caused by K^+ deficiency in medium. Na^+ content on the leaves surface was also measured to clarify the contribution of salt secretion to the salt tolerance of the plant species. This work would be also helpful to answer whether various mechanisms found by various researches to date are adopted during different growth stages in *H. brevisubulatum*.

2 MATERIALS AND METHODS

2.1 Plant materials and growth conditions

H. brevisubulatum seeds were collected from saline-alkaline wetlands (E100° 06′, N 39° 11′) in northwestern China. Then, *H. brevisubulatum* and wheat (*Triticum aestivum* L. cv. "Longchun 26") seeds were germinated on bibulous paper saturated with sterile water in rectangular dishes (15 cm × 8 cm × 5 cm) in the dark at 25 ℃. The germination period lasted 7 days for the *H. brevisubulatum* and 3 days for the wheat. After the leaf emergence occurred, seedlings were cultured with a modified Hoagland's nutrient solution (5 mM KNO_3, 1 mM $NH_4H_2PO_4$, 0.5 mM $Ca(NO_3)_2$, 0.5 mM $MgSO_4$, 60 μM Fe-Citrate, 92 μM H_3BO_3, 18 μM $MnCl_2 \cdot 4H_2O$, 1.6 μM $ZnSO_4 \cdot 7H_2O$, 0.6 μM $CuSO_4 \cdot 5H_2O$ and 0.7 μM $(NH_4)_6Mo_7O_{24} \cdot 4H_2O$) and the pH values were regulated as 6.0. Once the second leaf appeared, seedlings were transferred into black-painted containers with the same solution. Seedlings were grown in an environment controlled chamber at 25 ℃ in the day and 18 ℃ at night. There was a photon flux density of 2 300 $\mu mol \cdot m^{-2} \cdot s^{-1}$ in the day with a photo period of 16/8 h for the day/night cycle and a relative humidity from 65%–75%. The solution was renewed every three days to avoid ion concentration imbalance.

For the pre-experiment, the plasma membrane permeability of the roots was determined under 0–300 mM NaCl with two, three and four-leaf-old plants in order to use the appropriate NaCl concentrations and growth stages (Supplementary Fig. S1). Then, four-leaf-old plants and 0–100 mM NaCl concentrations of were used for the following experiments. Once the plants achieved four leaves, they were used for the salt treatment experiments. NaCl concentrations were increased by 25 mM/12 h in advance to achieve the final concentrations for treatments. The solutions were aerated and renewed every day to avoid oxygen and ion alterations.

2.2 Plant growth measurements

The plants of *H. brevisubulatum* and wheat were cultivated with a modified Hoagland's nutrient solution containing 25, 50, and 100 mM NaCl for 7 days. Then, the plants were harvested, gently blotted, separated into shoots and roots, weighed for fresh weights immediately and oven-dried at 80 ℃ for 3 d to obtain constant dry weights. Then, the water content and root/shoot ratio was calculated.

At the same time, root morphological parameters were determined using a root automatism scan apparatus (Perfection V700 Photo, Seiko Epson Corp, Japan) equipped with WinRHIZO software (Regent Instruments Co). The specific procedures were described in the references [25]. In each replicate, roots were placed in a transparent plastic tray filled with distilled water, and placed on the scan apparatus. Image recordings were performed at a resolution of 800 dpi, and saved as a tagged image file (TIF) format. The root phenotype traits, including the total root length, volume, surface area and number of root tips, were assayed using WinRHIZO software. The root-absorbing area was determined by a methylene-blue colorimetric method as described by Zou et al. [26]. The percentages of active absorption area and specific surface area were calculated according to Wang et al. [25].

2.3 Various terms of salt stress treatments

The plants of H. brevisubulatum and wheat were treated with a modified Hoagland nutrient solution supplemented with 25, 50 and 100 mM NaCl. Then, they were harvested for Na^+ and K^+ ion concentration measurements at 0, 6, 12, 24, 48, 96, and 168 h after salt stress for the short-term experiment and 7, 14, 28, and 60 days for the long-term experiment, respectively. The measurements of Na^+ and K^+ contents were carried out as described by Wang et al. [4]. Proline content in the tissues was tested according to Zhao [13].

2.4 Net Na^+ and K^+ fluxes measurements by MIFE

The net Na^+ and K^+ fluxes of H. brevisubulatum plants subjected to 100 mM NaCl were measured using a MIFE technique (BIO-IM, Younger USA LLC, Amherst, MA 01002, USA) as previously described [27-29] at Xuyue Science and Technology Co., Ltd., Beijing, China. In principal, prior to the flux measurement, the microelectrode was calibrated with different concentrations of the Na^+ (0.3 mM, 0.9 mM and 3 mM), and K^+ (0.1 mM, 0.5 mM and 5 mM) buffers. Only electrodes with a Nernstian slope >50 mV/decade were used in this study. The roots (with shoots retained) of the *H. brevisubulatum* were then washed gently with measuring solution and incubated in a petri dish containing a 10 ml of measuring solution (0.1 mM KCl, 0.1 mM $CaCl_2$, 0.1 mM $MgCl_2$, 0.5 mM NaCl, 0.2 mM Na_2SO_4, 0.3 mM MES, pH 6.0, adjusted with Tris) to equilibrate for 15 min. The net Na^+ flux measurements commenced from the root tip and were repeated along the root at various positions from the root tip in order to determine the optimal position (the zone that corresponded to the site where the Caparian bands developed, and the zone of the lateral root's initial development). The steady-state of ion fluxes were then recorded until the values variation amplitude was relatively stable (approximately 400 s). The micro-electrode oscillated with an excursion of 30 μm, and completed an entire cycle in 5.36 s. The net ion fluxes ($pmol \cdot cm^{-2} \cdot s^{-1}$) were calculated using Mage Flux software developed by Xuyue (http://xuyue.net/mageflux). In order to eliminate the error of free diffusion during the testing caused by the differences of ion concentrations between the plants and test solutions during MIFE measuring processes, flux values of the control were considered as the reference values.

2.5 Various K$^+$ level treatments

KNO$_3$ (5 mM) in Hoagland's nutrient solution was replaced by 0.01 mM KNO$_3$, 5 mM KNO$_3$, 5 mM KNO$_3$ + 15 mM KCl, and 5 mM KNO$_3$ + 45 mM KCl to achieved the final K$^+$ concentrations of 0.01, 5, 20 and 50 mM, respectively, while the NO$_3^-$concentration was retained, under 100 mM NaCl for 14 days. 0.01 mM replaced 0 mM there to avoid nutrition deficiency. Then, the plants were harvested for Na$^+$ and K$^+$ ion concentration measurements. The selective absorption (SA) and selective transport (ST) values were calculated according to Wang et al. [30].

2.6 Salt secretion analysis

The salt secretion from the leaves was analyzed according to Wang et al. [4]. Briefly, four-leaf-old plants of *H. brevisubulatum* were cultured in hydroponic tanks and treated with 100 mM NaCl for 60 days; while control was irrigated in the same modified Hoagland solution without NaCl. Then, the leaves were carefully separated from the shoot bases, and rinsed in 30 ml of deionized water for 1.5 min, holding the incision above the water to avoid any Na$^+$ loss from the cut end. At this point, the leaves were removed from the washing water, blotted and weighted. The roots with the remaining shoots were harvested, and after washing off any surface soil with tap water, they were immediately blotted and weighed. The Na$^+$ and K$^+$ contents in the deionized water and the entire plants were measured as described below.

2.7. Statistical analysis

The results of the growth, ion concentration and net Na$^+$ and K$^+$ fluxes of the plants were presented as a means with standard deviations (n=6). Statistical analyses, one-way analysis of variance (ANOVA), and Duncan's multiple range tests were performed using statistical software at $P < 0.05$ (Ver.16.0, SPSS Inc., Chicago, IL, USA).

3 RESULTS

3.1 Shoot growth was improved and stronger root activity was maintained in *H. brevisubulatum* compared with wheat during short-term salt stress

The growth responses in both plant species were investigated. Interestingly, it was found that fresh weight increased significantly in the shoots of *H. brevisubulatum* in 50 and 100 mM NaCl. However, no significant increases were observed in wheat (Fig. 1). The dry weights showed the similar results (Supplementary Fig. S2). Total root length, root volume, root surface area, the numbers of root tip and percentage of active absorption area of wheat were significantly reduced by 7 day treatment of 100 mM NaCl compared to the control (Table 1). However, no significant negative effect was found for *H. brevisubulatum*, furthermore, the numbers of root tips, the percentage of active absorption and specific surface areas were even enhanced by 100 mM NaCl (Table 1).

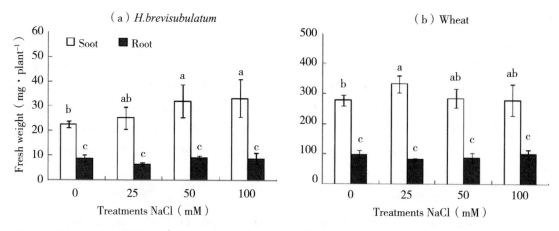

Fig. 1 Fresh weight of shoot (white columns) and root (black columns) in *H. brevisubulatum* (a) and wheat (b) under 0, 25, 50 and 100 mM NaCl (increased stepwise with 25 mM per 12 h) in modified Hoagland solution for 7 days

Note: Ten plants for H. brevisubulatum and two plants for wheat were pooled in each replicate (n=6). Values are means ± SD and bars indicate SD. The same as below.

Table 1 Root morphology and physiology indexes of *H. brevisubulatum* and wheat under control (0) and 100 mM NaCl (increased stepwise with 25 mM per 12 h) in modified Hoagland solution for 168 h

Species	Treatments NaCl (mM)	Total root length (cm·plant^{-1})	Root volume (mm^3·plant^{-1})	Root surface area (cm^2·plant^{-1})	Numbers of root tip (plant^{-1})	Percent of active absorption area (%)	Specific surface area (m^2·cm^{-3})
H. brevisubulatum	0	31.30 ± 2.54 a	9.98 ± 1.11 a	1.98 ± 0.22 a	140.68 ± 9.28 a	46.83 ± 0.26 b	1.52 ± 0.01 b
	100	26.64 ± 3.09 a	8.85 ± 0.70 a	1.71 ± 0.11 a	89.28 ± 2.79 b	47.97 ± 0.23 a	1.56 ± 0.01 a
Wheat	0	159.68 ± 9.74 a	74.58 ± 5.15 a	12.19 ± 1.84 a	525.30 ± 35.24 a	48.54 ± 0.32 a	1.49 ± 0.01 b
	100	95.29 ± 12.67 b	63.25 ± 4.92 b	8.67 ± 0.72 b	286.80 ± 19.40 b	47.12 ± 0.20 b	1.54 ± 0.02 a

3.2 Rapid Na$^+$ uptake and higher Na$^+$ levels were maintained in the shoots of *H. brevisubulatum* during short-term (within 168 h) salt stress

Shoot Na$^+$ concentration of *H. brevisubulatum* was increased gradually with prolonged time and over 53-fold after 168 h treatment of 100 mM NaCl, and it was significantly higher (2.6–6.4 folds) than that of wheat during 6–168 h (Fig. 2a). Na$^+$ concentrations in roots were relatively stable in both species (Fig. 2d). However, *H. brevisubulatum* showed an obviously lower root Na$^+$ concentration than in shoot. K$^+$ concentrations were relatively stable in shoot of both species during 0–168 h. Root K$^+$ concentrations decreased gradually from 48 h, but more stable and 29%–40% higher K$^+$ concentrations were maintained in *H. brevisubulatum* than in wheat during 96–168 h of treatments (Fig. 2b and e). These results led to a lower K$^+$/Na$^+$ ratio in shoot, but a higher

K^+/Na^+ ratio in root of *H. brevisubulatum* compared with wheat (Fig. 2c and f). Interestingly, even in the absence of NaCl, both shoot and root Na^+ concentrations were significantly higher (approximately 3 times) in *H. brevisubulatum* than those in wheat (Fig. 2a and d), leading to a significantly lower K^+/Na^+ ratio in *H. brevisubulatum* (Fig. 2c and f).

To clarify whether the high Na^+ concentration accumulation occurred under mild salt treatment in *H. brevisubulatum*, 25 and 50 mM NaCl were used. Similar result was obtained that higher Na^+ concentration was accumulated in shoot of *H. brevisubulatum* than wheat (Supplementary Figs. S3 and S4).

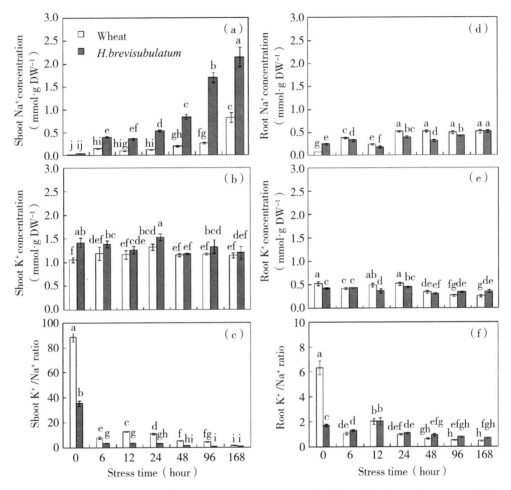

Fig. 2 Na^+ (a), K^+ (b), K^+/Na^+ ratio (c) in shoot and Na^+ (d), K^+ (e), K^+/Na^+ ratio (f) in root of *H. brevisubulatum* (black columns) and wheat (white columns) during 0–168 h treatment of 100 mM NaCl (increased stepwise with 25 mM per 12 h) in modified Hoagland solution

3.3 Na^+ concentration was reduced while K^+ was retained in *H. brevisubulatum* during long-term (7–60 days) NaCl treatments

To further assess whether the high Na^+ accumulation in *H. brevisubulatum* persisted over long-term treatments, time-courses (from 0 to 60 days) of the tissue K^+ and Na^+ concentrations were analyzed (Fig. 3a, b, c). The results indicated that shoot Na^+ concentration increased rapidly when subjected to 100 mM NaCl and reached the maximum value (2.15 mmol·gDW^{-1}) at 7th days, and then gradually reduced with the prolonged NaCl treatments (14–60 days). Although root Na^+ concentration increased not so dramatically compared with shoot during 0–7 days, it increased sharply from 7th day and reached the maximum value (2.43 mmol·gDW^{-1}) at 14th days, then decreased gradually till 60th day. It was noticeable that, although shoot Na^+ concentration increased rapidly and higher than in root during initial salt shock (0–7 days), relatively lower Na^+ levels was maintained in shoot during the following long-term stress (14–60 days) (Fig. 3a).

Although K^+ concentration had a slight decline during the shortterm (0–7 day) stress, a relatively stable level of K^+ was maintained during the following stress. Consequently, K^+/Na^+ ratio had a slight decrease during 0–168 h (Fig. 2c and f), and then increased gradually (Fig. 3c). Moreover, K^+ concentrations were higher in shoots than those in roots during all time-courses (Fig. 3b).

3.4 Na^+ and K^+ fluxes contribution to Na^+ decline during long-term (7–60 days) salt stress in *H. brevisubulatum*

In order to clarify the contribution of net Na^+ and K^+ fluxes to Na^+ decline during long-term stress, a MIFE technique at anatomically distinct zones were used. The results showed that a rapid Na^+ efflux (140.5 pmol·cm^{-2}·s^{-1}) was recorded after 7 days of 100 mM NaCl treatment compared with control, and then increased to 295.4 pmol·cm^{-2}·s^{-1} after 60 days of treatment (Fig. 4). Although, K^+ efflux (267.2 pmol·cm^{-2}·s^{-1}) was observed after 7 days of NaCl treatment, an obvious K^+ influx (−190.4 pmol·cm^{-2}·s^{-1}) was record after 60 days of treatment (Fig. 5). This was consistent with the results of tissue ion concentrations (Fig. 3) where Na^+ concentration decreased obviously from 7th day and K^+ concentration decrease at initial salt stock (0–7 days) and increased gradually in the following stages, consequently a higher K^+/Na^+ was maintained for H. brevisubulatum to adapt to salt stress.

3.5 K^+ levels in medium had small effect on Na^+ accumulation in *H. brevisubulatum*

To confirm whether high Na^+ other than K^+ accumulation in *H. brevisubulatum* was caused by K^+ deficiency in medium, 5 mM K^+ in the solution was replaced by various concentrations of K^+ (0.01–50 mM) in 100 mM NaCl treatment. Shoot Na^+ concentration did not change significantly with 0.01, 5 and 20 mM K^+, but increased significantly by 37% with 50 mM K^+. No significant differences in root Na^+ concentrations were found among 5, 20 and 50 mM K^+ treatments, even significantly decrease occurred in these three K^+ levels compared with 0.01 mM K^+. Tissue K^+ concentration increased significantly with the increase of K^+ concentrations in medium, and reached the maximum values with 50 mM K^+ (Fig. 6b). The selective absorption (SA) and selective transport (ST) values for K^+ over Na^+ were calculated and confirmed that SA values decreased sharply, while the ST values remain relatively stable with the addition of 5–50 mM K^+ (Table 2).

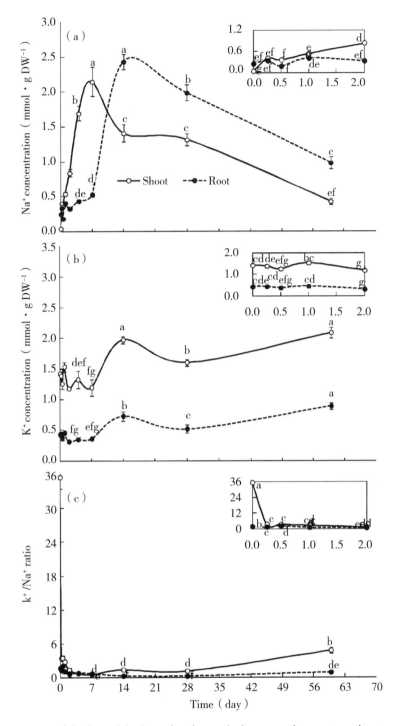

Fig. 3 Time course of Na⁺ (a), K⁺ (b), K⁺/Na⁺ ratio (c) in shoot (open circles) and root (closed circles) of *H. brevisubulatum* during 0–60 days under 100 mM NaCl (increased stepwise with 25 mM per 12 h) in modified Hoagland solution. The details of 0–2 days were showed in inserted figures

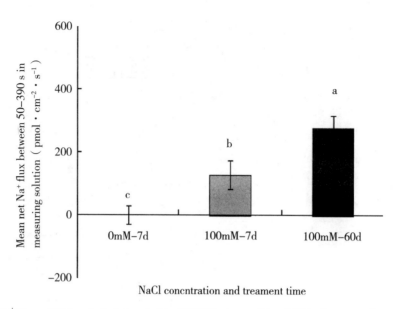

Fig. 4 Net Na⁺ flux of *H. brevisubulatum* test by MIFE between 50 and 390 s in measuring solution after 0 and 100 mM NaCl treatment (increased stepwise with 25 mM per 12 h) in modified Hoagland solution for 7 d and 100 mM NaCl for 60 d. Roots (with shoots retained) were incubated in the measuring solution to equilibrate for 15 min in advance. Steady-state ion fluxes were then recorded until the values variation amplitude is relatively stable

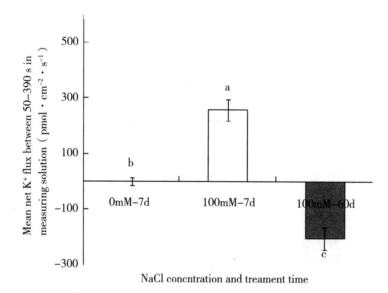

Fig. 5 Net K⁺ flux of *H. brevisubulatum* test by MIFE between 50 and 390 s in measuring solution after 0 and 100 mM NaCl treatment (increased stepwise with 25 mM per 12 h) in modified Hoagland solution for 7 d and 100 mM NaCl for 60 d. Roots (with shoots retained) were incubated in the measuring solution to equilibrate for 15 min ahead. Steady-state ion fluxes were then recorded until the values variation amplitude is relatively stable

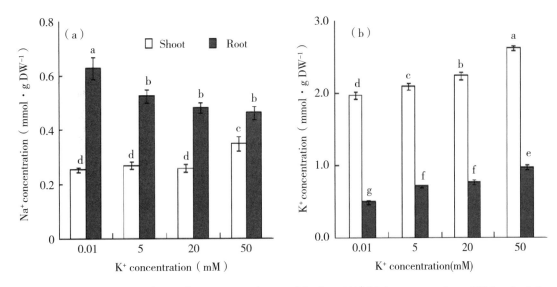

Fig. 6 The influence of K⁺ level (0.01–50 mM) on Na⁺ (a) and K⁺ (b) concentration of *H. brevisubulatum* under 100 mM NaCl (increased stepwise with 25 mM per 12 h) in modified Hoagland solution for 14 d

Table 2 Selective absorption (SA) and transport (ST) capacity for K⁺ over Na⁺ of *H. brevisubulatum* exposed to 100 mM NaCl supplied with 0.01, 5, 20 and 50 mM K⁺ (NaCl increased stepwise with 25 mM per 12 h) in modified Hoagland solution for 14 d.

K⁺ concentration (mM)	SA	ST
0.01	-	10.15 ± 1.03 a
5	119.01 ± 6.57 a	5.67 ± 0.54 b
20	33.19 ± 1.96 b	5.52 ± 0.35 b
50	12.58 ± 0.89 c	3.56 ± 0.38 c

Note: The values were calculated from Fig. 6. SA= (K⁺/Na⁺ in whole plant) / (K⁺/Na⁺ in medium); ST= (K⁺/Na⁺ in shoots) / (K⁺/Na⁺ in roots).

3.6 Loss of Na⁺ content from the leaves

To assess the contribution of salt secretion from the leaves to the decline of Na⁺ concentration in shoots during long-term salt stress, Na⁺ and K⁺ contents on leaves surface was determined under 100 mM NaCl for 60 days. Na⁺ content washed from the leaves were not significantly different between 0 and 100 mM NaCl treatments accounting for only 2.52% of the entire plant Na⁺ content with 100 mM NaCl (Fig. 7a). K⁺ content washed from the leaves accounted only about 0.1% of the entire plant K⁺ content with 0 and 100 mM NaCl.

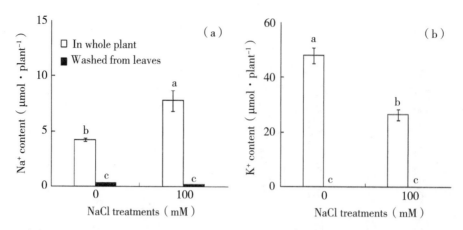

Fig. 7 Na⁺ content loss from leaves of *H. brevisubulatum* under 0 and 100 mM NaCl (increased stepwise with 25 mM per 12 h) in modified Hoagland solution for 60 d

4 DISCUSSION

4.1 Rapid and higher Na⁺ accumulation in shoot was a rapid response to short-term salt stress

Excessive Na^+ induces osmotic stress and cytosolic toxicity, resulting in growth inhibition for glycophytes, and Na^+ also acts as a competitor of K^+ with similar binding sites in major metabolic processes [31, 32]. Therefore, limiting high Na^+ concentration accu-mulation as a major salt tolerance strategy is adopted by most monocotyledonous halophytes [3-5, 33].

However, in present study, rapid and high Na^+ was accumulated in the shoots, other than in roots of *H. brevisubulatum*, even over 2-fold of wheat during short-term salt stress (Fig. 2a). Interestingly, although Na^+ levels were higher in *H. brevisubulatum*, growth rates of the shoots and root activity remained higher than in wheat (Fig. 1a, Table 1). Li et al. found that Na^+ could probably be compartmentalized into vacuole in *H. brevisubulatum* [10]. It was also found that Na^+ concentration in barley increased rapidly and sequestered into the leaves during the first several days [28], and tolerant barley genotypes accumulated significantly higher Na^+ in their leaves compared with sensitive ones [34]. Similar results were also found in Thellungiella halophila exposed to salt stress [35]. The most probable explanation is that the plants undergoing salt stress could send the amount of required Na^+ quickly to the shoot, as a "cheap" osmoticum to achieve a rapid full osmotic adjustment [36], rather than simply restricting Na^+ influx at the beginning stage of salt stress. This consequently provides an additional driving force for water uptake by roots [37] and resumes growth of shoots (Fig. 1a). In contrast, it was found that glycophytes like wheat (Fig. 2a) and pea [38] could restrict xylem Na^+ loading during the initial few days of the salt stress treatments, but failed to prevent Na^+ elevation over longer term exposure to salinity.

An improved Na^+ sequestering ability in the leaf vacuoles of *H. brevisubulatum* could be achieved by tonoplast $Na^+/H+$ exchanger, HbNHX1 [18, 39]. In the present study, it is interesting to note that rapid and higher Na^+ accumulation was also found in the lower (25 and 50 mM) NaCl treatments (Supplementary Figs. S3 and S4), consistent with the result of RT-PCR identification with HbNHX1 not induced by salt stress [18]. It was proposed that Na^+ transport under mild salinity was mediated by AtSOS1 (Salt Overly Sensitive 1), which is located across the plasma membrane of xylem parenchyma cells and could mediate Na^+ loading into the xylem, thereby controlling the long-distance Na^+ transport from roots to shoots [40, 41]. Under higher saline conditions, AKT1 (Arabidopsis K^+ transporter) regulated Na^+ uptake in the roots [31] and its expression was not influenced by outside K^+ concentrations [39, 42]. These finding were consistent with the results in current study that Na^+ concentrations in the roots were not influenced by outside K^+ levels from 5 to 50 mM (Fig. 6a). Our previous results in the salt-accumulating halophyte Suaeda maritima also supported that AKT type K^+ channels may be involved in Na^+ uptake in the roots under 150 mM NaCl [43].

4.2 Na^+ efflux was the major mechanism for *H. brevisubulatum* to maintain a low Na^+ accumulation during long-term salt stress

Although high Na^+ levels was rapidly observed in shoots of *H. brevisubulatum* (Fig. 2a), once this was achieved to the limitation of the vacuolar volume, it was better for the plants to reduce Na^+ loading rate, increase Na^+ retrieval from shoots to roots and exclude Na^+ to the soil [44, 45]. In current study, Na^+ concentration in shoots reached its maximum values at 7th day of the treatments (Fig. 3a) and then declined gradually. Consistently, Na^+ concentration in roots increased obviously from 7th day and reached the maximum values at 14th day (Fig. 3a). It was suggested that AtHKT1;1 (High-affinity K^+ Transporter) located at the plasma membrane of xylem parenchyma cells could mediated Na^+ retrieval from the xylem to the surrounding parenchyma cells [46, 47], and thereby regulated Na^+ transport from shoots to roots [48-50]. Consistently, athkt1;1 mutant accumulated more Na^+ in shoot and less Na^+ in roots compared with the wild type plants [49, 50], indicating that AtHKT1;1 could be a determinant to controlling Na^+ unloading from the xylem.

With large amount of Na^+ retrieval from shoot to root, Na^+ was efflux from roots to external environment (Fig. 4). Then, Na^+ concentrations in roots were consequently reduced from 14th day (Fig. 3a). A plasma membrane-bound Na^+/H^+ antiporter [39], encoded by SOS1 at the epidermal cells of the root tips [40], is crucial in mediating Na^+ efflux from roots to the environment [51-53]. It was noted that, regardless of halophytes or glycophytes, unidirectional Na^+ efflux accounted for large amounts of total Na^+ influx under salinity conditions [4, 54-56]. Na^+ concentrations both in shoots and roots (Fig. 3a) decreased with the increase of Na^+ efflux (Fig. 4), suggesting that Na^+ efflux from roots was the main contributor to reduced Na^+ concentration during the long-term salt stress in *H. brevisubulatum*. The similar result showed that a better ability of root cells to pump Na^+ from cytosol to external medium was also found in salt-tolerant varieties of barley [34], and accumulated approximately 30% less Na^+ than in salt-sensitive varieties after four weeks of salt stress [28].

4.3 Coordinated regulation of K^+ and Na^+ in *H. brevisubulatum* under salinity stress

Maintaining a high-cytosolic K^+/Na^+ ratio is one of the crucial and essential mechanisms of salt tolerance [31, 34, 48]. The coordinated regulation of K^+ and Na^+ (not simply maintaining a high K^+ and low Na^+) may play an important role for salt tolerance in *H. brevisubulatum*. As described above, Na^+ uptake in *H. brevisubulatum* increased rapidly and was higher than in wheat during the first 168 h of salt stress (Fig. 2a). This was probably due to the AKT1-type channels [57, 58], the subfamily 1-type HKT transporters [59-61] on root epidermal cells and SOS1 on xylem parenchyma cells. They together promoted the uptake and long distance transport of Na^+ [44, 45] into cell vacuoles of shoots to achieve a higher osmotic potential. Na^+ unloading into the xylem parenchyma cells possibly depolarizes their plasma membrane, which in turn could potentially activate the K^+ channels, such as SKOR (Stelar K^+ Outward Rectifiers), to load K^+ into xylem [45, 62, 63]. Therefore, K^+ concentrations increased during 7–14 days in both in shoot and root (Fig. 3b), following the rapid Na^+ uptake during the first 7 days (Fig. 3a). When Na^+ content achieved its limitation of vacuolar volume, the subfamily 1-type HKT transporters mediated the Na^+ retrieval from shoots to roots [44, 45]. Then, Na^+ concentration declined in the shoots and increased in the roots during 7 to 14 days of salt stress (Fig. 3a).

Once excessive Na^+ accumulated in the root cells (Fig. 3a), K^+ efflux from epidermis cells was triggered due to salt-induced membrane depolarization [64, 65] (Fig. 5), probably via the GORK (Guard cell Outward Rectifying K^+ channel) [66]. This consequently led to a decrease in K^+ concentration in both shoots and roots during 14–28 days of treatment (Fig. 3b). Then, with the increase of Na^+ efflux in roots (Fig. 4), Na^+ concentration obviously decreased during 28–60 days (Fig. 3a) and K^+ influx recovered, thereby K^+ concentration and K^+/Na^+ ratio increased progressively (Fig. 3b and c). This was consistent with the MIFE results that K^+ maintained a higher efflux at 7th day, but recovered influx at 60th day under 100 mM NaCl treatments compared with control (Fig. 5).

Accordingly, by integrating all the findings as described above, a probable pattern for elucidating salt tolerance mechanism of H. brevisubulatum was proposed, in which the roles of K^+ and Na^+ coordinated regulations during short and long-term salt stress were clarified (Fig. 8). During the first stage (0–7 days; Fig. 2a), Na^+ was absorbed quickly into the epidermal cells of roots, probably via AKT1 [57, 58] or HKT (the subfamily 1-type HKT transporters) [59, 60], then was loaded directly into the xylem via SOS1 in the plasma membranes of XPCs (Xylem Parenchyma Cell) and delivered rapidly to the shoots motivated by transpiration stream [28, 44, 45]. Then, large amount of Na^+ was rapidly sequestered into leaf vac-uoles via tonoplast Na^+/H^+ exchanger (NHX) [18, 39], in order to achieve rapid full osmotic adjustment and avoid Na^+ damage [36]. During the second stage, once Na^+ achieved its limitation in vacuole, HKT could then mediate Na^+ retrieval from xylem into XPCs to limit Na^+ accumulation in shoots [44, 45, 47, 67], and relatively larger amount of Na^+ was accumulated in roots (Fig. 3a, 7–14 days). Then in turn, this could activate K^+ channels, such as SKOR, to load the K^+ into xylem [45, 62, 63] through XPCs, and finally promote the K^+ uptake into the plants [28, 48, 50] possibly via HAK or other K^+ uptake transporters or channels. During the third stage, once Na^+ was retrieved and excessively accumulated in root cells (Fig. 3a, 7–14 days), Na^+ efflux possibly via SOS1 on root epidermal cells was triggered [40, 52, 67, 68]. At the same time, K^+ efflux via K^+ efflux channels such as GORK due to the salt-induced membrane depolarization from the cells was also triggered [66], leading to a decrease in K⁺ concentrations in the plants (Fig. 3b, 14–28 days). Then, with the increase of Na⁺ efflux (Fig. 4), Na⁺ concentration decreased and K⁺ concentration recovered gradually by its uptake, resulting in a higher K^+/Na^+ (Fig. 3a, b, c, and Fig. 5), which consequently was sustained during long-term of higher salt stress.

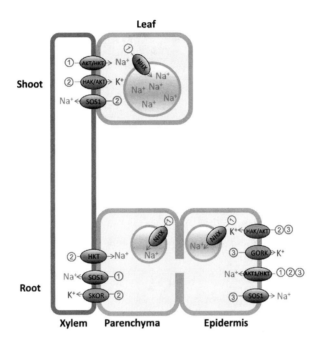

Fig. 8 A proposed schematic pattern for K⁺ and Na⁺ transporters/channels to coordinate regulation of salt tolerance of *H. brevisubulatum*

Note: ① During the first stage (0–7 days), Na⁺ was absorbed quickly into the epidermal cells of roots, probably via AKT1 or HKT, then was loaded directly into the xylem via SOS1 in the plasma membranes of XPCs and delivered rapidly to the shoots motivated by transpiration stream. Then, large amount of Na⁺ was rapidly sequestered into leaf vacuoles via NHX, in order to achieve rapid full osmotic adjustment and avoid Na⁺ damage. ② During the second stage, once Na⁺ achieved its limitation in vacuole, HKT could then mediate Na⁺ retrieval from xylem into XPCs to limit Na⁺ accumulation in shoots, and relatively larger amount of Na⁺ was accumulated in roots. Then in turn, this could activate K⁺ channels, such as SKOR, to load the K⁺ into xylem through XPCs, and finally promote the K⁺ uptake into the plants possibly via HAK or other K⁺ uptake transporters or channels. ③ During the third stage, once Na⁺ was retrieved and excessively accumulated in root cells, Na⁺ efflux possibly via SOS1 on root epidermal cells was triggered. At the same time, K⁺ efflux via K⁺ efflux channels such as GORK due to the salt-induced membrane depolarization from the cells was also triggered, leading to a decrease in K⁺ concentrations in the plants. Then, with the increase of Na⁺ efflux, Na⁺ concentration decreased and K⁺ concentration recovered gradually by its uptake, resulting in a higher K^+/Na^+, which consequently was sustained during long term of higher salt stress.

4.4 Na$^+$ loss from leaves makes a minor contribution to salt tolerance of *H. brevisubulatum*

Although high Na$^+$ level was maintained in shoots of *H. brevisubulatum*, Na$^+$ loss from leaf surface was extremely minor of Na$^+$ amount in the entire plant (2.6%, Fig. 7) compared with salt-secreting halophytes, such as Spartina anglica (60%), Limonium vulgare (33%), Glaux maritime (20%) and wild rice Porteresiacoarctata (over 50%)[69, 70]. Moreover, although some "salt hairs" were observed on the leaves [11], no salt-secreting structures was observed in *H. brevisubulatum* [10], also no bicellular glands was observed in any other species of the Pooideae family [71], indicating that Na$^+$ secreting contributes little to salt tolerance of H. brevisubulatum.

5 CONCLUSION

This study provided evidence that rapid Na$^+$ influx occurred at seedling stages with initial salt shock and Na$^+$ efflux and K$^+$ influx was enhanced to maintain a K$^+$ and Na$^+$ balance at tilling stages during long-term salt stress in the monocotyledonous halophyte *H. brevisubulatum*. A probable regulation pattern was hypothesized to elucidate its K$^+$ and Na$^+$ co-ordinated mechanism against different terms of salt stress. Although this seems a smart strategy, it is too early to speculate that this mechanism could be adopted by other wild plant species.

ACKNOWLEGEMENTS

The authors would like to thank Professor S.-M. Wang, C.-J. Li and Dr. P. Wang for their helpful discussion. We would also like to thank the anonymous reviewers for their constructive suggestions regarding the manuscript. This research was supported by the National Natural Science Foundation of China (31201841 and 31222053), and the Agricultural Science and Technology Innovation Program of Chinese Academy of Agricultural Sciences (CAAS-ASTIP-2014-LI-HPS-08).

APPENDIX A. SUPPLEMENTARY DATA

Supplementary data associated with this article can be found, in the online version, at http://dx.doi.org/10.1016/j.plantsci.2016.08.009.

（发表于《Plant Science》，院选SCI，IF：3.362. 358–366）

Genome-wide Association Study Identifies Loci for the Polled Phenotype in Yak

LIANG Chunnian[1]🌑, WANG Lizhong[2]🌑, WU Xiaoyun[1],
WANG Kun[2], DING Xuezhi[2], WANG Mingcheng[2], CHU Min[1],
XIE Xiuyue[2], QIU Qiang[2]‡*, YAN Ping[1]‡*

(1. Key Laboratory of Yak Breeding Engineering Gansu Province, Lanzhou Institute of Husbandry and Pharmaceutical Sciences, Chinese Academy of Agricultural Science, Lanzhou, China; 2. State Key Laboratory of Grassland Agro-ecosystem, College of Life Science, Lanzhou University, Lanzhou, China)

Abstract: The absence of horns, known as the polled phenotype, is an economically important trait in modern yak husbandry, but the genomic structure and genetic basis of this phenotype have yet to be discovered. Here, we conducted a genome-wide association study with a panel of 10 horned and 10 polled yaks using whole genome sequencing. We mapped the POLLED locus to a 200-kb interval, which comprises three protein-coding genes. Further characterization of the candidate region showed recent artificial selection signals resulting from the breeding process. We suggest that expressional variations rather than structural variations in protein probably contribute to the polled phenotype. Our results not only represent the first and important step in establishing the genomic structure of the polled region in yak, but also add to our understanding of the polled trait in bovid species.

1 INTRODUCTION

The yak (Bos grunniens) is endemic to the Qinghai-Tibet Plateau (QTP), the largest and harshest highland in the world [1]. More than 14 million domestic yaks are currently distributed across the QTP, providing the food, shelter, fuel and transport that enable nomadic Tibetans and other pastoralists to survive in this harsh climate; indeed, the yak has become an iconic symbol of Tibet [1, 2]. They are also strongly integrated into Tibetans' socio-cultural life. Due to its important position in Tibetan daily life, yak production and its related products are the cornerstone of Tibetan animal husbandry [3, 4]. In addition, yak live on unpolluted highland pasture where they produce green, nutritional and healthy products, much valued by modern communities [3, 4].

The bovine polled phenotype, the highly desired and favorable trait in modern husbandry

---

🌑 These authors contributed equally to this work.
‡ These authors jointly supervised this work.
* Corresponding authors, E-mail: qiuqiang@lzu.edu.cn (QQ); pingyanlz@163.com (PY).

systems, has huge practical importance for breeders and is of special interest to geneticists [5, 6] (Fig 1A shows a horned and a polled yak). Yak horns are a major cause of injuries, particularly in feedlots and during transport [7]. Nowadays, commercial beef yaks are confined to barns and fenced-in enclosures such as pastures or corrals. More hornless yak can be accommodated in the same space compared to yak with horns, and the trait would reduce economic losses due to injuries to both humans and animals under these conditions [7]. Although dehorning at a young age is routinely practiced in horned breeds of yak, this method does not eradicate the problem and there are associated animal welfare concerns. Considering an auto-somal dominant mode of inheritance for the polled trait, the approach also limits the ability to discriminate between heterozygous and homozygous polled animals [8, 9]. Thus, creating polled genetic markers to identify homozygous/heterozygous yak and breeding genetically polled yak based on non-invasive and high welfare methods is a promising alternative [7, 9], which would be valuable to modern yak husbandry in high altitude harsh environments [10]. In addition, identification of genes and causal variations associated with the polled phenotype will contribute to our knowledge and understanding of the molecular mechanisms that underlie horn differentiation and development in bovine species.

In cattle (Bos taurus), the POLLED locus has previously been mapped on the proximal end of bovine chromosome 1 (BTA1) [11]. More recent efforts to refine the polled locus and detect candidate causal mutations have included seeking additional microsatellite markers [12-14], BAC-based physical mapping [15], high-density SNP genotyping [16-19], targeted capture sequencing [17, 20, 21] and whole genome resequencing [16, 19]. Recently, at least two different alleles for polledness were reported in cattle, identifying an 80-kb duplication (BTA1:1, 909, 352-1, 989, 480 bp) in Friesian original breeds (P F allele) and a duplication of 212-bp (BTA1:1, 705, 834-1, 706, 045 bp) in place of a 10-bp deletion (1, 706, 051-1, 706, 060 bp) in Celtic original breeds (P C allele, a 202-bp insertion-deletion, InDel), respectively, suggesting the existence of allelic heterogeneity at the POLLED locus [17, 21, 22]. In addition, other sporadic mutations associated with the horn-like scurs phenotype have been described [23, 24]. Intriguingly, none of these mutations was located in known coding or regulatory regions [13, 16]. One plausible reason is that different alleles have been selected in different geographic regions or breeds, and worldwide and across-breed breeding using limited founders and artificial insemination have led to the complex inheritance pattern of the horn trait in different breeds, thus adding to the complexity of understanding the molecular basis of polledness [17]. Hence, simultaneous research in different Bovidae species needs to be undertaken to provide extra information [7, 19].

In the current breeding stage, polled yaks are developed deliberately by crossing polled cows (PP or Pp) with horned bulls (pp) by artificial insemination (S1 Fig) at the Datong Yak Breeding Farm of Qinghai Province, providing an ideal system to study the genomic structure and genetic basis of this phenotype [10]. Although our previous analysis based on an a priori candidate gene set detected associated signals, the actual location of the POLLED locus has so far not been confirmed as BTA1 in yak [10]. In the present study, we describe our efforts to determine the polled trait associated genome regions in yak based on whole genome sequencing; these were carried out independently from any recently published studies.

2 RESULTS AND DISCUSSION

2.1 Genetic variants and population structure

We sequenced 10 horned and 10 polled yaks to an average depth (raw data) of 11.2 × using an Illumina Hiseq2000 instrument, resulting in a total of 6.04 billion reads comprising approximately 595Gb of sequencing data. Using BWA-MEM software [25], reads were aligned to the B. grunniens reference genome with an average alignment rate of 91.20%, covering 99.26% of the genome [26] (S1 Table). After SNP calling using SAMtools [27], filtering the potential PCR duplicates, removing SNPs with potential errors and correcting the misalignments around

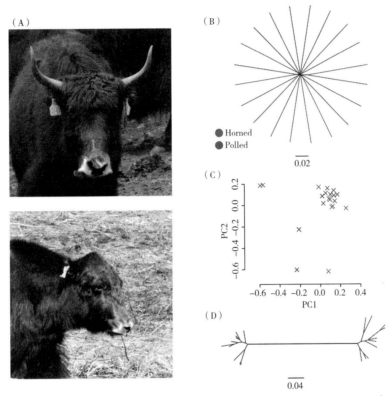

(A) Photos of horned (above) and polled (below) yak herds, taken by Chunnian Liang. (B) A neighbor-joining phylogenetic tree constructed using whole-genome SNP data. The scale bar represents level of similarity. Horned (blue) and polled (red) samples are indicated. (C) Principal component (PC) analysis plots of the first two components. (D) A neighbor-joining phylogenetic tree constructed using SNP data for the GWAS region.

Fig. 1 Phylogenetic and population structure of horned and polled yaks

doi: 10.1371/journal.pone.0158642.g001

InDels (details in Materials and Methods), approximately 8.4 million high quality SNPs were retained.

To examine the phylogenetic relationship between horned and polled yaks, we constructed a neighbor-joining tree based on our high-quality SNPs. The horned and polled yaks formed a mixed clade (Fig. 1B), indicating that pairwise distances between horned and polled yaks were not larger than those within each population. We also performed principle component analysis (PCA, Fig. 1C and S2 Table) and population structure analysis (S2 Fig.) based on the genotype data. Taken together, all of these results indicate no population genetic structure correlated with the horned/polled phenotypes, consistent with the relatively short time of breeding polled yaks. More importantly, the indistinguishable genomic background suggests that genome-wide association studies should enable high-resolution mapping of genomic regions associated with the horned/polled phenotype.

2.2 Identification of genome regions associated with the polled phenotype

Taking advantage of the yak population with no population stratification, we performed a genome-wide association study (GWAS) analysis between 10 horned and 10 polled animals using the Dominant model with PLINK [28]. This autosomal dominant Mendelian trait was mapped to a 200 kb interval between positions 1, 122, 103 and 1, 322, 666 bp on scaffold526_1 (P-values<0.000 1, which means that the probability of obtaining these frequencies by chance is very low, <0.01%) (Fig. 2A and 2B, S3 Fig.). Despite the small sample size and the consequent relatively limited statistical power [29], this exclusive signal reached genome-wide significance in the GWAS analyses and appeared to align within the polled locus of BTA1 mapped in previous studies of cattle [11-19, 21]. Further, we found that the horned and polled individuals clustered into two genetically distinct groups in a phylogenetic tree based on this region (Fig. 1D). In contrast, there was no significant differentiation between these groups when the tree was constructed for the whole genome (Fig. 1B). Moreover, most of these loci with Pvalues<0.000 1 (95%, 567 of 596) are entirely heterozygous (Pp) in polled animals, coinciding with the breeding practice of crossing polled cows (PP or Pp) with horned bulls (pp) by artificial insemination. Although all previous studies identified different polled mutations in different cattle breeds, all the clues lead to the same position on BTA1 [11-19, 21]; we therefore believe that the horned/polled trait in cattle and yak may share the same ontogenetic mechanism.

The region defined as the GWAS loci of the polled phenotype contained 3 protein-coding genes: SYNJ1, PAXBP1 and C1H21orf62 (Fig. 2C). SYNJ1 encodes synaptojanin 1, a key neural protein highly expressed in nerve terminals with essential roles in the regulation of synaptic vesicles in conventional synapses and hair cells [30, 31]. Recently, a histological analysis revealed that nervous tissue and hair follicle development have different features in horn buds and polled frontal skin during the development of the horn buds of bovine fetuses, implying that SYNJ1 maybe have an important role in horn differentiation [32]. PAXBP1 is an essential binding protein that regulates the proliferation of muscle precursor cells, which in turn, are involved in the development of normal craniofacial features and spine morphogenesis [33]. C1H21orf62 is an uncharacterized protein. Since previous transcript profiling analyses of polled and horned tissues from cattle, using the Agilent 44 k bovine array, failed to find differential expression in any of the genes located in the POLLED locus [34], we sought to re-annotate the novel protein-coding genes in this region by mapping large-scale RNA-seq data for five tissues (brain, kidney, lung, liver and heart) from a previous study [35] of domestic yak. We were, however, unable to find any new gene or open reading frame (Fig. 2D, details in Materials and Methods).

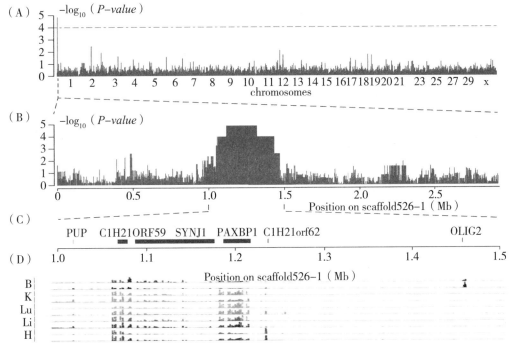

(A) Genome-wide P values (y axis) are plotted along the genome (B) and magnification of scaffold526_1. (C) All genes around the candidate GWAS region. (D) diagram of read depths (X axes) of RNA-seq data mapping of five different tissues: brain (B), kidney (K), lung (Lu), liver (Li) and heart (H), each with two replicates (Y axes).

Fig. 2 Associated mapping of the polled phenotype

doi: 10.1371/journal.pone.0158642.g002

Previous genetic and genomic research has proposed two structural variants (P_F and P_C alleles) associated with the polled phenotype based on larger scale GWAS results in many cattle breeds [17]. We therefore examined whether these two structural variants exist in the yak genome based on our whole genome sequencing data ($>10\times$), which should ensure quality and accuracy in the detection of structural variants [36]. Our results indicated that the yak genome does not include these two cattle candidate structural variants (S4 Fig. and S3 Table), contrasting with previous studies indicating that there is extensive allelic heterogeneity of the polled trait in highly mobile bovid species. We further analyzed other structural variants from this associated region and could not identify any duplications or InDels associated with the polled trait in yak (S3 Table). Due to the small sample size, short breeding history and absence of homozygous polled (PP) individuals, our results cannot be used to refine the candidate genomic region nor to detect causal variants (causal SNPs or structural variants) in yak at present. However, we are convinced that a future study based on more samples with detailed pedigree information will narrow down the candidate region of this trait [37], and this highly confidential region should be the target of focused studies to establish the functional significance of this key trait in domestic yak.

2.3 Characterization of the polled interval

Considering the breeding practice developed by crossing polled cows (PP or Pp) with horned bulls (pp) (Schematic shown in S1 Fig.), the candidate region identified could be expected to exhibit specific signatures of recent artificial selection in the polled population, including a high proportion of heterozygous, significantly differentiated nucleotide diversity levels and long-range haplotype homozygosity [38]. Based on these principles, we examined five different parameters to evaluate detailed genetic polymorphism and differentiation between horned and polled yaks: nucleotide diversity (π), the proportion of shared and private SNPs, F_{ST} (population-differentiation statistic), d_{xy} (mean pairwise comparisons of the nucleotide difference between groups) and the linkage disequilibrium (LD). Population-specific estimates of π showed that the level of nucleotide diversity is higher in polled yaks than in horned yaks in the 200 kb GWAS region (Fig. 3A), although a similar level of nucleotide diversity was observed in the rest of genome (mean pairwise nucleotide diversity of π_{horned}: 0.001 39 ± 0.000 81; π_{polled}: 0.001 36 ± 0.000 77, S5 Fig.). In addition, this region showed an elevated proportion of private

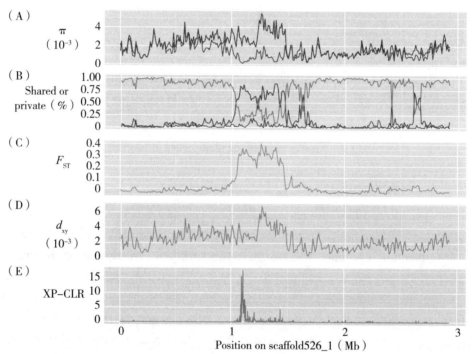

(A) the nucleotide diversity (π, blue for horned and red for polled yaks) for each population; (B) the proportion of shared polymorphisms among sites that are polymorphic in at least one population (green), the proportion of private polymorphisms among sites that are polymorphic within populations (blue for horned and red for polled yaks), private and shared polymorphisms shown in the same panel; (C) F_{ST}; (D) d_{xy}; and (E) XP-CLR of scaffold526_1.

Fig. 3 Distribution of population genomic parameters along scaffold526

doi: 10.1371/journal.pone.0158642.g003

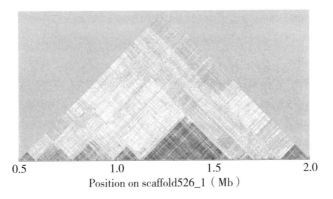

Fig. 4 Haplotype block at linkage disequilibrium along the scaffold526_1.

doi: 10.1371/journal.pone.0158642.g004

SNPs in polled yaks and a reduced level of shared SNPs between horned and polled yaks (Fig. 3B), which also accord with the breeding approach. Furthermore, we found that the GWAS region implicated in the polled phenotype showed striking genetic differentiation between the horned and polled individuals in the F_{ST} analysis (Fig. 3C, with a mean F_{ST} of only 0.000 6, S6 Fig.). The mean pairwise nucleotide difference between-group comparisons (d_{xy}) showed a divergence peak (more than 0.006) compared with the flanking regions (Fig. 3D, with whole genome level of 0.001 3 ± 0.000 6). Linkage analysis for this scaffold also revealed a higher linkage disequilibrium (LD) value with one haplotype block of 450 kb (1.14-1.59Mb, Fig. 4), which was probably defined by the causal allele and its linked neighbor variants. Taken together, these results indicate that this candidate genomic region tends to be highly diverged and exhibits clear signals of selection. As a complementary approach, we used a likelihood method (the cross-population composite likelihood ratio, XP-CLR [39]) to scan for extreme allele frequency differentiation over extended linked regions and found this region to have elevated XP-CLR values (Fig. 3E), which means that the polled trait was affected by selection during breeding activities.

2.4 Potential genetic basis and future directions

Despite the fact that the POLLED locus has been easily mapped, fine characterization of this locus and a definitive description of the molecular basis of horn ontogenesis has proved more difficult than expected [17, 22]. One important reason is that complex genomic structural variations and a possible regulatory effect rather than nonsynonymous variations in protein sequence are the major contributors to the differentiation of the horned/polled trait [16, 17, 19, 21]. A recent study reported the success of production of hornless dairy cattle using genome editing technology involving introgressing the Celtic original candidate POLLED allele (202-bp insertion-deletion), but the potential functional effect of this sequence variant remains unknown [22]. To date, none of the causative variants have been located in known coding or regulatory regions and no genes with a high probable impact have been identified in cattle [13, 16]. To test this result in yak, we scanned the SNPs within the coding regions of the POLLED locus and were unable to identify any nonsynonymous substitutions, consistent with the situation previously observed in cattle. In addition, differential expression studies among horned and polled cattle failed to reveal differences in gene expression located in the POLLED locus, but several genes outside this region showed a high level of expression divergence [19].

By combining our results with those of earlier studies, we suggest that unknown regulatory sequences and cis-regulation elements may reside in the POLLED locus, thus influencing horn development. Furthermore, the horned/polled trait is developmentally-related, and such traits are likely to be involved in the complex interaction of many genes [16, 19, 35]. This indicates that a high level of genetic heterogeneity is expected and that different species or breeds may have developed this phenotype as a result of different genomic strategies [17]. For example, sequence changes in the POLLED locus may affect the transcription factor and noncoding RNA binding, coordination of histone modifications and other chromatin remodeling activities, which can lead to transcription changes to horn-related genes in the POLLED locus and other regions of the genome [40]. Therefore, we must emphasize the important role of gene regulation in horn development. Future studies should focus on identification of novel regulated elements in the POLLED locus and involve an examination of detailed expression patterns across different horn developmental stages, which should reveal the precise mechanism responsible for the polled trait.

3 CONCLUSIONS

The importance of breeding polled yak has grown considerably due to animal welfare issues and the needs of modern yak husbandry. Herein, we report the first genome-wide association study of the POLLED phenotype in which we identified a 200-kb genomic region responsible for this economically important trait in yak. However, we need to point out that the candidate region was inevitably large because of the small sample sizes, and our current data were insufficient to detect causal variants. Further research based on larger sample sizes will be necessary to obtain more reliable estimates and refine the genomic loci that contribute to this trait. We found that this region was under artificial selection and the characterizations of the POLLED locus were concordant with the breeding process. Combined with previous results in cattle, we further suggest that expressional variations other than structural variations in protein are the major causes of the polled phenotype. The results of our study represent a critical advance towards the delimitation of a genomic region for further functional study and provide new insights into the genetic basis of the polled trait in yak and other bovine species.

4 MATERIALS AND METHODS

4.1 Sample collection and sequencing

We randomly selected 10 horned and 10 polled individuals (S1 Table) of Datong yaks from a large herd (n > 2 000, from the Datong Yak Breeding Farm of Qinghai Province (37° 15′ 35.6″ N, 101° 22′ 24.0″ E). For each yak, genomic DNA was extracted from blood samples using a standard Phenol/Chloroform method. The quality and integrity of the extracted DNA was controlled by the A260/A280 ratio and agarose gel electrophoresis. Paired-end sequencing libraries with an insert size of 500 bp were constructed according to the Illumina manufacturer's instructions, for sequencing on the Hiseq 2 000 platform, and paired-end reads were generated. Sequencing and base calling were performed according to the standard Illumina protocols. All individuals were sequenced to an average raw read sequencing depth of 11.2 × assuming a genome size of 2.66 Gb. All experimental protocols were approved by ethical committees of the Datong Yak Breeding Farm of Qinghai Province.

4.2 Sequence quality checking and mapping

We performed a per-base sequence quality checks [41] and low quality reads of the following types were filtered out: (i) Reads with ⩾10% unidentified nucleotides (N); (ii) Reads for which more than 65% of the read length had a phred quality score ⩽7; (iii) Reads with more than 10 bp aligned to the adapter, allowing 2 bp mismatches; and (iv) duplicate reads. Reads were also trimmed if they had three consecutive base pairs with a phred quality score of 13 or below, and discarded if they were shorter than 45 bp.

The pair-end sequence reads were mapped to the B. grunniens reference genome using BWA-MEM [25] (0.7.10-r789) with default parameters. The picard software (http://broadinstitute.github.io/picard/, version 1.92) was subsequently used to assign read group information containing library, lane, and sample identity. Duplicated reads were filtered and index files were built for reference as well as bam files using SAMtools (0.1.19). The Genome Analysis Toolkit (GATK, version 2.6-4-g3e5ff60) [42] was used to perform local realignment of reads to enhance the alignments in the vicinity of InDel polymorphisms. Realignment was performed with GATK in two steps. The first step used the RealignerTargetCreator to identify regions where realignment was needed, and the second step used IndelRealigner to realign the regions found in the first step, generating a realigned mapping file for each individual. The overall mapping rate of reads to the reference genome was 91.20%, with average read depths of 10.2 × (10.06 × to 10.37 ×). On average, across all samples, the reads covered 99.26% of the genome (S1 Table).

4.3 SNP calling

After alignment, we performed SNP calling using a Bayesian approach as implemented in the package SAMtools. Realigned regions were piped to SAMtools and reformatted into pileup files for SNP identification. Sequence variants from pileups were then condensed into a variant call format (VCF) file using BCFtools [27] (0.1.19). The genotype with maximum posterior probability was picked as the genotype for that locus.

The threshold of SNP calling was set to 20 for both base quality and mapping quality. SNPs were discarded based on the following conditions: (i) quality less than 20; (ii) those with too low (total depth < 2 × 20) or too high (total depth > 40 × 20) a depth (possibly bad assembly or repetitive regions); (ii) 5 bp around InDels; (iv) those occurring in a cluster (more than three SNPs with 10 bp); (v) failure in the exact test for Hardy-Weinberg equilibrium at P<0.001; or (vi) those with > 50% missing genotype data with the population.

4.4 Phasing and linkage disequilibrium

The program Beagle [43] (version: r1196) was used to infer the haplotype phase and impute missing alleles with default parameters. Linkage disequilibrium (pairwise r² statistic) was calculated using Haploview [44] (v4.2) software with the parameters '-dprime -maxDistance 1 000 -minMAF 0.05 -hwcutoff 0.001 -missingCutoff 0.5 -minGeno 0.6'.

4.5 Phylogenetic relationship and population structure analysis

To assess recent relationships between samples, we calculated pairwise estimates of Identity-By-State (IBS) scores [28]. We found no possible duplicate (IBS>0.9) that showed high pairwise genetic similarity with another sampled individual, indicating that these 20 individuals, as unrelated samples, can be used in the downstream analyses.

Neighbor-joining trees were constructed with PHYLIP (v3.696, http://evolution.genetics.washington.edu/phylip.html)) using the matrix of pairwise genetic distances ('—cluster—distance-matrix' of PLINK v1.07). The SmartPCA program from the EIGENSOFT [45] (v5.0.1) package was used to perform principle component analysis on the individuals that we sequenced with default parameters. A Tracy–Widom test was used to determine the significance level of the eigenvectors and no significant eigenvectors were found (S2 Table). In addition, ADMIXTURE [46] (v1.23, with default parameters) was used to infer the population substructure among the samples with number for population grouping parameter K set from 1 to 3.

4.6 Association analysis using PLINK

To map the poll trait loci, we performed a case-control analysis between horned and polled yaks using the Dominant model with PLINK [28] (v1.07, with parameters '—model—model-dom—fisher'). This Dominant model assumes that an effect on phenotype is only seen if you have at least one copy of the minor allele. It categorizes individuals into two groups based on whether they have at least one minor allele A (either Aa or AA) or no copies of the minor allele (aa). Fisher's exact test was used to analyze genotypic differences between the 10 horned and 10 polled samples.

4.7 Population genetic statistics

The nucleotide diversity (π) and population-differentiation statistic (F_{ST}) were calculated using VCFtools [47] (v0.1.12a) with a sliding window approach (50 kb window sliding in 10 kb steps).

d_{xy} was calculated as follows:

$$d_{xy} = \sum_{ij} x_i y_i d_{ij}$$

where, in two populations, x and y, d_{ij} measures the number of nucleotide differences between the ith haplotype from x and the jth haplotype from y.

XP-CLR values were calculated with default parameters using XP-CLR (v1.0).

4.8 Re-annotating the associated region using RNA-seq data

To find out whether there were new genes or open reading frames within our GWAS loci, we mapped previous RNA-seq data of domestic yak [32] to scaffold526_1 of the yak genome (as described in the 'Sequence quality checking and mapping' section). Read depths of RNA from five tissues (brain, kidney, lung, liver and heart, each with two duplicates) were calculated using SAMtools (parameters of 'samtools depth') and visualized as shown in Fig. 2D.

4.9 Checking P_F, P_C alleles and other structural variants

To check the existence of a P_F allele in yak, we mapped our sequencing reads to the BTA1 sequence (UMD3.1 genome build, downloaded from Ensembl release77, http://www.ensembl.org/). Sequencing depths for each sample around the genome region near the P_F allele were calculated using SAMtools. The relative depths of each sample were calculated and no P F allele (i.e. 80 kb duplication) was found in polled yaks (S4 Fig.).

Structural variants (insertions, deletions, tandem duplications and inversions) were discovered using Pindel[48] (v0.2.5a3), which uses a pattern growth approach to identify the breakpoints of these variants from paired-end short reads (with default parameters and '-c 1 -T 30 -l -k'). No P_C allele was found in polled yaks and no other structural variants were found associated with polled phenotypes.

5 SUPPORTING INFORMATION

S1 Fig. Schematic of breeding practice for Datong yaks. Sexuality is indicated by a circle (cow) or a square (bull), genotypes are indicated by different colors (PP, orange; Pp, green; pp, blue).

(TIF)

S2 Fig. Population structure plots with K = 1–3. The y axis quantifies the proportion of the individual's genome from inferred ancestral populations, and the x axis shows the different populations. The CV error of each run is given in parentheses.

(TIF)

S3 Fig. QQ-plot of GWAS P values.

(TIF)

S4 Fig. Relative depth of horned (blue) and polled (red) yaks around the P_F allele on BTA1. The green frame indicates the region of the P_F allele.

(TIF)

S5 Fig. Genome-wide distribution of π_{horned} (a) and π_{polled} (b).

(TIF)

S6 Fig. Genome-wide distribution of F_{ST}.

(TIF)

S1 Table. Overview of sample information and sequencing statistics.

(XLS)

S2 Table. Tracy-Widom (TW) statistics and P-values for the first five eigenvalues in the PCA. No significant P values.

(XLS)

S3 Table. Results of structural variants discovery.

(XLS)

ACKNOWLEDGMENTS

We thank Dr. Tao Ma, Dr. Yongzhi Yang and Dr. Bingbing Liu for their helpful comments and suggestions about this project. Special thanks to the farmers and researchers who bred the yaks from the Datong Yak Breeding Farm of Qinghai Province.

AUTHOR CONTRIBUTIONS

Conceived and designed the experiments: QQ PY. Performed the experiments: CL LW XW KW XD MW MC XX. Analyzed the data: CL LW XW KW XD MW MC XX. Contributed reagents/materials/analysis tools: CL LW XW KW XD MW MC XX. Wrote the paper: CL LW QQ PY.

（发表于《PLOS ONE》，院选SCI，IF：3.057. 1-14）

Integrated Analysis of the Roles of Long Noncoding RNA and Coding RNA Expression in Sheep (*Ovis aries*) Skin during Initiation of Secondary Hair Follicle

YUE Yaojing[●], GUO Tingting[●], YUAN Chao[●], LIU Jianbin, GUO Jian, FENG Ruilin, NIU Chune, SUN Xiaoping, YANG Bohui[*]

(Lanzhou Institute of Husbandry and Pharmaceutical Sciences, Chinese Academy of Agricultural Sciences, Jiangouyan Street, Lanzhou, China)

Abstract: Initiation of hair follicle (HF) is the first and most important stage of HF morphogenesis. However the precise molecular mechanism of initiation of hair follicle remains elusive. Meanwhile, in previous study, the more attentions had been paid to the function of genes, while the roles of non-coding RNAs (such as long noncoding RNA and microRNA) had not been described. Therefore, the roles of long noncoding RNA (LncRNA) and coding RNA in sheep skin during the initiation of sheep secondary HF were integrated and analyzed, by using strand-specific RNA sequencing (ssRNA-seq). A total of 192 significant differentially expressed genes were detected, including 67 up-regulated genes and 125 down-regulated genes between stage 0 and stage 1 of HF morphogenesis during HF initiation. Only Wnt2, FGF20 were just significant differentially expressed among Wnt, Shh, Notch and BMP signaling pathways. Further expression profile analysis of lncRNAs showed that 884 novel lncRNAs were discovered in sheep skin expression profiles. A total of 15 lncRNAs with significant differential expression were detected, 6 up-regulated and 9 down-regulated. Among of differentially expressed genes and LncRNA, XLOC002437 lncRNA and potential target gene COL6A6 were all significantly down-regulated in stage 1. Furthermore, by using RNA-hybrid, XLOC005698 may be as a competing endogenous RNA "sponges" oar-miR-3955-5p activity. Gene Ontology and KEGG pathway analyses indicated that the significantly enriched pathway was peroxisome proliferator-activated receptors (PPARs) pathway (corrected P-value < 0.05), indicating that PPAR pathway is likely to play significant roles during the initiation of secondary HF. Results suggest that the key differentially expressed genes and LncRNAs may be considered as potential candidate genes for further study on the molecular mechanisms of HF initiation, as well as supplying some potential values for understanding human hair disorders.

[●] These authors contributed equally to this work.
[*] Corresponding author, E-mail: lzyangbohui@163.com

1 INTRODUCTION

Hair follicle (HF) research is a rapidly developing area of skin biology [1]. HF morphogenesis and cycling can be successfully used for a wide range of studies into the mechanisms of morphogenesis [2], stem cell behavior [3], cell differentiation [2] and apoptosis [1, 4]. Moreover, HF investigations can provide invaluable insights into the possible causes of human hair disorders [5-8].

The morphogenesis of HF is an excellent example of mesenchymal-epithelial interactions [9]. HF formation has been divided into eight distinct developmental (stages 0-8). The stages of morphogenesis are broadly classified as follows: induction (stages 0-1), organogenesis (stages 2-5) and cytodifferentiation (stages 6-8) [10]. The first stage, stage 0 of HF morphogenesis, corresponds to the undifferentiated and single-layered epidermis with no morphological signs of HF induction. Classically, the initiation of HF is described in terms of an ordered series of mes-enchymal-epithelial interactions. At stage 1 (also described as hair placode stage), a "first dermal message" emanating from the dermis acts on an unspecified epidermis, and the formation of morphologically recognizable hair placodes occurs [11]. In sheep embryos, second HF placodes are formed at E96 [12]. These placodes then emit an epidermal signal that instructs underlying mesenchymal cells to cluster and form "dermal condensates." A "second dermal message" from the dermal condensates induces epidermal placode cells to rapidly divide downward and invade the dermis, thus enwrapping the dermal condensate, which becomes the HF germ (stage 2) [13]. Although the precise nature of the epidermal placode-inducing "first dermal message" remains poorly understood, several studies have suggested that HF initiation is an orchestrated interaction between mesenchymal and epithelial cells mediated through the secretion of stimulator and inhibitor signaling molecules, such as Wnt/β-catenin, EDA/EDAR/NF-κB, Noggin/Lef-1, Shh, BMP-2/4/7 and FGF [9, 14]. Recent studies looking beyond protein-coding genes have shown that non-coding RNA (ncRNA), such as microRNA (miRNA), natural antisense transcripts (NAT) and long non-coding RNA (lncRNA), can show higher specificity as biomarkers for some applications than protein coding genes [15-19]. Among these ncRNAs, lncRNAs are not only large in quantity but also play important roles in gene expression regulation in organisms [20-22]. However, few lncRNAs are annotated within the sheep genome. LncRNAs are generally defined as having a size greater than 200 nucleotides, and they constitute a diverse group of non-coding RNAs that are distinct from miRNAs [23]. LncRNAs have been implicated in biological, developmental and pathological processes, and they act through mechanisms such as chromatin reprogramming, cis regulation at enhancers and post-transcriptional regulation of mRNA processing [15, 21, 22, 24]. However, the roles of lncRNAs in controlling HF initiation have not been described. In this study, we used strand-specific RNA sequencing (ssRNA-seq) to identify the role of lncRNAs and mRNAs in sheep skin during the initiation of secondary HF.

2 RESULTS

2.1 Sequencing and assembly

NGS was performed on two groups, stage 0 of HF morphogenesis (n = 3) (Fig. 1A) and stage 1 of HF morphogenesis (n = 3) (Fig. 1B) , and raw reads greater than 100 million were obtained for every group (Table 1) . There was a 3' adaptor sequence in the raw data as well as small amounts of low-quality sequences and various impurities. Impurity data were removed from the raw data. For the stage 0 and stage 1 libraries, 103, 413, 896 and 95, 671, 374 clean reads were obtained, respectively.

The reads were then aligned using Top Hat [25] onto the Ovis_aries_v3.1 reference genome sequence. For the stage 0 and stage 1, 84.44% and 83.92% of the reads were aligned with the reference sheep genome, respectively (Table 1) , and 79.85% and 79.81% of the clean reads were niquely located in the reference sheep genome, respectively. Moreover, 41, 274, 699 (39.91%) and 38, 168, 756 (39.9%) clean reads of the stage 0 and stage 1 groups, respectively, were mapped to the positive strand of the reference sheep genome. Furthermore, 41, 305, 965 (39.94%) and 38, 189, 322 (39.92%) clean reads of the stage 0 and stage 1 libraries, respectively, were mapped to the negative strand of the reference sheep genome.

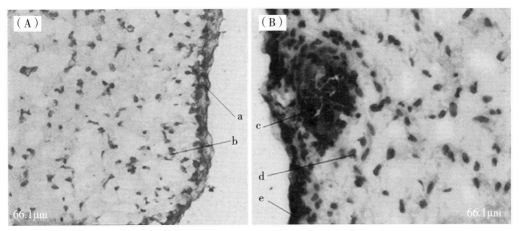

a.Stage 0 of secondary HF morphogenesis in 87E fetal ovine skin sections (200×) . hematoxylin stained section.
b. Stage 1 of secondary HF morphogenesis in 96 E fetal ovine skin sections (400×) .
(a) Epidermis; (b) eukaryotic cells; (c) hair placode; (d) eukaryotic cells; (e) Epidermis.

Fig. 1 Skin sections during the initiation of secondary hair follicle

doi: 10.1371/journal.pone.0156890.g001

Table 1 Summary of clean reads mapping to the Ovis_aries_v3.1 reference genome sequence.

Sample name	Stage 0	Stage 1
Total reads	103 413 896	95 671 374
Total mapped	87 319 755 (84.44%)	80 291 861 (83.92%)
Multiple mapped	4 739 091 (4.58%)	3 933 783 (4.11%)
Uniquely mapped	82 580 664 (79.85%)	76 358 078 (79.81%)
Reads map to '+'	41 274 699 (39.91%)	38 168 756 (39.9%)
Reads map to '-'	41 305 965 (39.94%)	38 189 322 (39.92%)

doi: 10.1371/journal.pone.0156890.t001

Out of the annotated transcripts of Stage 0 and Stage 1, 22, 572, 607 (61.92%) and 20, 955, 431 (62.51%) transcripts, respectively, were identified as protein-coding mRNAs, while 1.67% and 2.41%, respectively, were classified as different types of noncoding transcripts, such as mis-cRNA, pseudo gene, rRNA and tRNA. The remaining other types of transcripts amounted to 13, 273, 937 (36.41%) and 11, 759, 533 (35.08%) for stage 0 and stage 1, respectively, and these transcripts may include lncRNAs (Fig. 2).

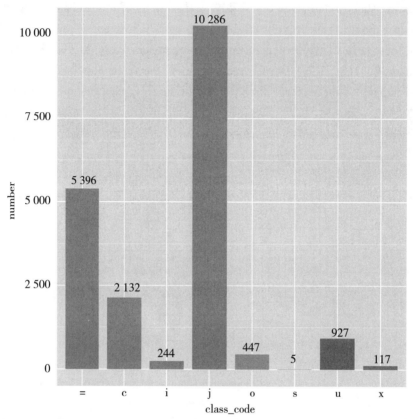

Note: =, Complete match of intron chain; c, Contained by a reference transcript; I, A transfrag falling entirely within a reference intron; j, At least one splice junction is shared with a reference transcript; o, Generic exonic overlap with a reference transcript; s, An intron of the transfrag overlaps a reference intron on the opposite strand; u, Unknown, intergenic transcript; x, Exonic overlap with reference on the opposite strand.

Fig. 2 Distribution of sheep skin transcripts by Cufflinks class code.

doi: 10.1371/journal.pone.0156890.g002

2.2 Identification of lncRNAs

After sequence assembly with Cufflinks [26] and scripture [27], the similar or identical transcripts to the known sheep non-mRNA (rRNA, tRNA, snRNA, snoRNA, pre-miRNA and pseudogenes) were filtered out from 20, 142 transcripts matching both stitching software using Cuffcompare, and the obtained transcripts were then compared with the known mRNA of the reference sheep genome. The class_code information in the results of Cuffcompare (http://cole-trapnell-lab.github.io/cufflinks/) was used to screen the candidate transcripts. The transcripts annotated by "i", "u" and "x" from class_code were used as the candidate lncRNA for lincRNA, intronic lncRNA and anti-sense lncRNA, respectively, resulting in a total of 1288 lncRNA candidate transcripts (Fig. 2). The candidate lncRNAs with coding potential were excluded using Coding-Non-Coding Index (CNCI), PhyloCSF and pfam protein domain analysis, which resulted in 884 lncRNAs (Fig. 3) including 622 lincRNAs, 188 intronic lncRNAs and 74 anti-sense lncRNAs. Analysis of the characteristics of the sheep lncRNA and the transcripts encoding a protein showed that sheep lncRNAs were significantly shorter than the mRNAs in length distribution (S1A Fig.) and ORF length (S1B Fig.). Moreover, the number of exons was also less than that of mRNAs (S1C Fig.), and the sheep lncRNAs were longer than human and mouse lncRNAs [28]. The sequence conservation of mRNA and lncRNA were conservatively scored using phastCons, resulting in the cumulative distribution curve of mRNA and LncRNA conservation scores (S1D Fig.), which indicated that the sequence conservation of lncRNA was less than that of mRNA.

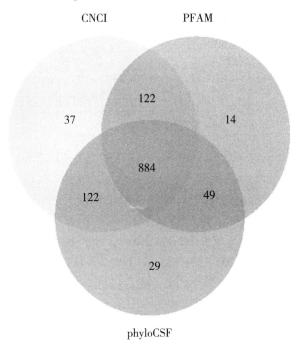

Fig. 3 Venn diagram of the number of LncRNA with coding potential analysis by CNCI, pfam and phyloCSF.

doi: 10.1371/journal.pone.0156890.g003

2.3 Differentially expressed genes and lncRNAs

The mRNA and lncRNA expression was analyzed using Cuffdiff in Cufflinks [28] software. In sheep skin during the initiation of secondary HF, the gene expression level was low, and the gene expression levels of the two groups were similar (Fig. 4a). However, the expression level of mRNA was higher than that of lncRNA (Fig. 4b).

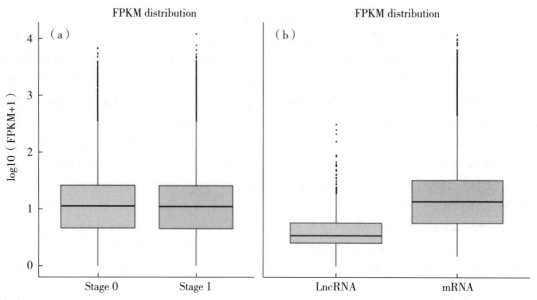

(a) The FPKM density distribution of tanscript expression in sheep skin at stage 0 and stage 1 (b) the expression level of mRNA and lncRNA in sheep skin during the initiation of secondary HF.

Fig. 4 The FPKM distribution of mRNA and lncRNA expression in sheep skin during the initiation of secondary HF

doi: 10.1371/journal.pone.0156890.g004

Using edgeR (the threshold is usually set as |log2 (Fold Change) | > 1 and q value < 0.005), the differential genes and lncRNAs between Stage 0 and Stage 1 group were screened, resulting n 209 differentially expressed genes and lncRNAs (Fig. 5). Of these genes and lncRNAs, 67 genes and six lncRNAs were upregulated, and 127 genes and nine lncRNAs were downregulated (S1 Table). In the differentially expressed genes, the downregulated gene showing the biggest difference was ADIPOQ (Gene ID ENSOARG00000020509) with FPKM value of -9.058 (q value (padj) = 1.20E-39); the upregulated gene showing the biggest difference is COL3A1 (Gene ID ENSOARG00000016476), with a multiple differential expression value of 21.90 (q value (padj) = 7.06E-107). The differentially expressed lncRNAs included 12 lincRNAs, one intronic lncRNA and one anti-sense lncRNA. Six of the differentially expressed lncRNAs were distributed on the sense strand of the sheep reference genome, and nine of these lncRNAs were distributed on the antisense strand (Table 2). XLOC_002747, XLOC_005698 and XLOC_014751 had alternative splice variants. The target gene prediction for the differential lncRNAs showed the existence of cis or trans target genes in twelve lncRNAs, but cis or trans target gene failed to be predicted in three lncRNAs, as XLOC007757, XLOC005698 and XLOC000629 (Table 3). Among of differentially expressed genes and LncRNA, lncRNA XLOC002437 and potential target gene collagen type Ⅵ alpha 6 (COL6A6, Gene ID: 101111424) were all significantly down-regulated in stage 1 of the secondary HF morphogenesis. BLAST alignment analysis was performed for the differential lncRNAs with the mature miRNA of sheep from the miRBase database, and the results showed a high consistency between XLOC005698 and oar-miR-3955-5p as well as between XLOC000629 and oar-miR-544-5p (Fig. 6A). The miRNA binding sites on LncRNAs were predicted using a web-based program RNAhybrid (version: 2.2) [29]. The minimum free energy (MFE) of XLOC005698 combined with oar—miR—3955-5 p was more lower than XLOC000629 combined with oar—miR—544-5 p, the mfe was −34.4 kcal/mol (Fig. 6B), −15.6 kcal/mol (Fig. 6C), respectively. It was suggested that XLOC005698 may be play important roles in the regulation of gene expression during the secondary HF morphogenesis by "sponges" oar-miR-3955-5p activity as a competing endogenous RNA (ceRNA). To further obtain the potential biological functions of the differential lncRNAs, cluster analysis was performed for the differentially expressed genes and lncRNAs. The 15 differential lncRNAs were first clustered into 12 small gene clusters as follows: XLOC_000629, XLOC_002747, 101123159 and 101120453 gathered in a cluster; XLOC_021775, XLOC_006635 and 101120775 gathered in a cluster; XLOC_007281, XLOC_014751 and 101121257 gathered in a cluster, and the remaining nine lncRNAs gathered respectively into eight gene clusters (S2 Table). These results preliminarily indicated that the differential lncRNAs might be involved in secondary follicle morphogenesis and the formation of placodes through a number of different metabolic processes or cellular pathways.

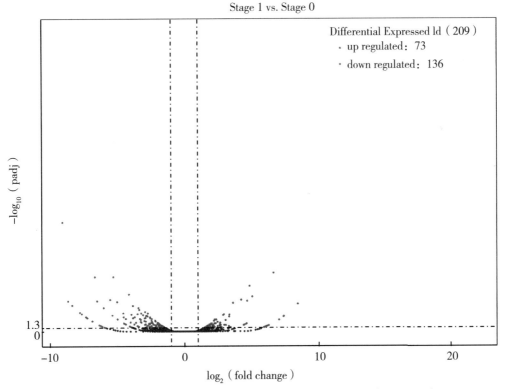

Of the 209 differentially expressed genes and LncRNA, 73 were upregulated (right, red) and 136 were downregulated (left, green) in stage 1 compared with stage 0.

Fig. 5 Differentially expressed genes and lncRNAs in sheep skin between stage 0 and stage 1 of secondary HF morphogenesis.

doi: 10.1371/journal.pone.0156890.g005

Table 2 Differentially expressed LncRNA in sheep skin between Stage 0 and Stage 1 group of secondary HF morphogenesis.

LncRNA ID	Location	Length	Strand	Exon number	Type	Stage 1 FPKM	Stage 0 FPKM	log2Fold Change	padj
XLOC016657	Chr3: 214782109-147825962	219	–	2	anti-sense lncRNA	0	30.2983	−6.120 44	0.015 537
XLOC002437	chr1: 269765163-269812147	390	–	2	lincRNA	14.636 8	61.970 6	−2.067 19	0.003 173
XLOC020709	chr9: 32640662-32645592	4 043	+	3	lincRNA	3.857 57	0.113 48	5.040 358	3.36E-13
XLOC014751	Chr3: 15940142-15943214	479	+	2, 3	lincRNA	5.921 45	0.242 519	4.407 241	0.005 568
XLOC007281	Chr15: 69763640-70034206	1 328	–	2	lincRNA	6.159 07	0.603 154	3.326 677	5.94E-06
XLOC007757	Chr17: 4435904-4441701	3 610	+	2	lincRNA	3.742 09	1.205 35	1.635 23	0.049 143
XLOC005698	chr14: 32625557-32704885	924	+	2, 3	lincRNA	7.182 29	1.244 53	2.098 051	0.014 642
XLOC014287	chr25: 33206921-33212098	5 112	–	2	lincRNA	1.803 69	0.431 517	2.059 673	0.006 547
XLOC021775	chrX: 94517001-94546262	4 043	–	6	lincRNA	3.857 57	0.113 48	5.040 358	3.36E-13
XLOC006635	Chr14: 56674446-56681460	753	–	2	lincRNA	0.7308 52	5.312 13	−2.807 51	0.010 757
XLOC018246	Chr5: 11050873-11083559	1 296	–	4	lincRNA	0.3727 59	5.133 88	−3.726 63	7.75E-06
XLOC000629	chr1: 41283122-41284685	1 460	+	2	Intronicl-ncRNA	0.270 59	2.653 64	−3.228 68	0.000 668
XLOC002747	chr10: 30088157-30091646	1 128	–	2, 3	lincRNA	0.3428 13	3.024 87	−2.967 01	0.004 374

(continued)

LncRNA ID	Location	Length Strand		Exon number	Type	Stage 1 FPKM	Stage 0 FPKM	log2Fold Change	padj
XLOC003435	chr11: 57875653-57891453	2715	+	2	lincRNA	1.225 96	8.928 3	-2.853 11	2.21 E-06
XLOC016440	chr3: 171509716-171564468	2506	-	3	lincRNA	1.484 99	5.015 01	-1.746 15	0.030 153

doi: 10.1371/journal.pone.0156890.t002

The differentially expressed genes and lncRNAs obtained from the screening were verified by strand-specific RT-PCR. Nine differentially expressed genes and lncRNAs were randomly selected from the two groups, and GAPDH was used as the internal reference. The quantitative results showed that the expression patterns of the selected differentially expressed genes in the two groups were consistent withFPKM values of these genes and lncRNAs (Fig. 7), and the sequencing results correlated with the strand-specific RT-PCR results.

2.4 Enrichment of differentially expressed genes and lncRNAs

The 160 differentially expressed genes and target genes of 12 lncRNAs containing the functional annotation information were assigned to 1,023,191GO terms, respectively. There were five, five and seven GO terms that were significantly enriched for biological process, molecular function and cellular components, respectively (corrected P-$value$ < 0.05) (Table 4). The differential genes in the induction stage of the secondary follicle morphogenesis in sheep were enriched into 136 pathways, including the PPAR signaling pathway, ECM-receptor interaction, the PI3K-Akt signaling pathway, the Wnt signaling pathway, the VEGF signaling pathway and the MAPK signaling pathway. Of these pathways, 8 genes of the PPAR signaling pathway was significantly enriched (corrected P-$value$ < 0.05) (Fig. 8, S3 Table).

Table 3 Potential cis or trans target gene, ncRNA, micRNA of differentially expressed LncRNA.

LncRNA ID	Potential target gene/ncRNA/micRNA			Cis/Trans
	Gene ID	Gene symbol	Description	
XLOC016657	101122943	TAB1	TGF-beta activated kinase 1	Cis
	105614926	SYNGR1	synaptogyrin 1	Cis
	101106836	RPL3	ribosomal protein L3	Cis
XLOC002437	101111424	COL6A6	collagen, type VI, alpha 6	Cis
	101107574	KIAA1715	KIAA1715 ortholog	Trans
XLOC020709	101111889	EFCAB1	EF-hand calcium binding domain 1	Cis
	105616018		LOC105616018 (Gene type: ncRNA)	Cis
	105609197		LOC105609197 (Gene type: ncRNA)	Cis

(continued)

LncRNA ID	Potential target gene/ncRNA/micRNA			Cis/Trans
	Gene ID	Gene symbol	Description	
XLOC014751	105607455		LOC105607455（Gene type：ncRNA）	Cis
	105607456		LOC105607456（Gene type：ncRNA）	Cis
XLOC007281	101108158	LRRC4C	leucine rich repeat containing 4C	Cis
	105602356		LOC105602356（Gene type：ncRNA）	Cis
	101115653		LOC101115653（Gene type：pseudogene）	Cis
XLOC005698	MIMAT0019245	-	oar-miR-3955-5p	Trans
XLOC014287	101113003	KLK12	kallikrein-related peptidase 12	Trans
	105605013		LOC105605013（Gene type：ncRNA）	Trans
XLOC021775	101107015	TMEM196	transmembrane protein 196	Trans
	105605574		LOC105605574（Gene type：ncRNA）	Cis
	105605574		LOC105605573（Gene type：ncRNA）	Cis
XLOC006635	101103089	ZNF577	zinc finger protein 577	Cis
	101104336		LOC101104336	Cis
	101104087	ZNF614	zinc finger protein 614	Cis
	101102585	HAS1	hyaluronan synthase 1	Cis
	101114548	ZNF613-like	zinc finger protein 613-like	Cis
XLOC018246	101102642	ZNF791	zinc finger protein 791	Cis
XLOC000629	MIMAT0019298		oar-miR-544-5p	Trans
XLOC002747	101110772	HSPH1	heat shock 105kDa/110kDa protein 1	Cis
	105616258		WD repeat-containing protein 49-like	Cis
XLOC003435	101107975	SOX9	SRY（sex determining region Y）-box 9	Cis
	101103746	PLAGL2	pleiomorphic adenoma gene-like 2	Trans
	101121742	RID5B AT	rich interactive domain 5B	Trans
XLOC016440	101121522	MCC	MCC mutated in colorectal cancers	Trans
	101105703	PLCE1	phospholipase C，epsilon 1	Trans

doi：10.1371/journal.pone.0156890.t003

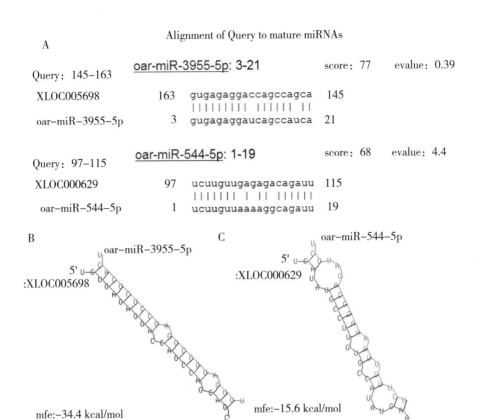

(a) Alignment between XLOC005698 and oar-miR-3955-5p as well as between XLOC000629 and oar-miR-544-5p. (b) The oar-miR-3955-5p miRNA binding sites on XLOC005698. (c) The oar-miR-544-5p miRNA binding sites on XLOC000629.

Fig. 6 Bioinformatics predicted oar-miR-3955-5p binding sites on XLOC005698, and oar-miR-544-5p binding sites on XLOC000629.

doi: 10.1371/journal.pone.0156890.g006

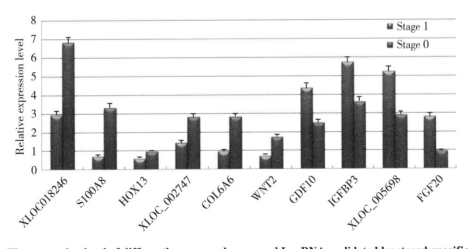

Fig. 7 The expression level of differently expressed genes and LncRNAs validated by strand-specific qPCR.

doi: 10.1371/journal.pone.0156890.g007

Table 4 Gene ontology analysis of differentially expressed genes in sheep skin between Stage 0 and Stage 1 of secondary HF morphogenesis.

GO_accession	Description	Term type	Corrected p Value
GO: 0046849	bone remodeling	biological process	1.13E-05
GO: 0048771	tissue remodeling	biological_process	1.13E-05
GO: 0044707	single-multicellular organism process	biological_process	0.032 134
GO: 0032501	multicellular organismal process	biological_process	0.032 863
GO: 0006952	defense response	biological_process	0.046 941
GO: 0005576	extracellular region	cellular_component	2.44E-05
GO: 0044421	extracellular region part	cellular_component	0.000 6
GO: 0005615	extracellular space	cellular_component	0.001 997
GO: 0005882	intermediate filament	cellular_component	0.044 882
GO: 0045111	intermediate filament cytoskeleton	cellular_component	0.044 882
GO: 0004857	enzyme inhibitor activity	molecular_function	2.44E-05
GO: 0030414	peptidase inhibitor activity	molecular_function	2.44E-05
GO: 0061134	peptidase regulator activity	molecular_function	2.44E-05
GO: 0004866	endopeptidase inhibitor activity	molecular_function	2.44E-05
GO: 0061135	endopeptidase regulator activity	molecular_function	2.44E-05
GO: 0004869	cysteine-type endopeptidase inhibitor activity	molecular_function	0.000 72
GO: 0030234	enzyme regulator activity	molecular_function	0.017 937

doi: 10.1371/journal.pone.0156890.t004

3 DISCUSSION

Initiation of HF involves a series of signaling between the epidermal cell and the dermal papilla, such as Wnt/beta-catenin, EDA/EDAR/NF-κB, Noggin/Lef-1, Ctgf/Ccn2, Shh, BMP-2/4/7, Dkk1/Dkk4 and EGF [30, 31]. This study also found that the above signaling molecules were expressed in sheep skin during the initiation of secondary HF. However, the astonishing thing is that only Wnt2, FGF0 were just significant differentially expressed. In this study, we observed that Wnt2 at stage 1 is 1.58 times less expressed than at stage 0 in sheep skin. Cadau et al also observed a maximal expression of Wnt2 at E12.5 in mouse skin and a slight decrease at E13. This decrease is more significant at E13.5, reaching a third of its initial expression. At E14.5, Wnt2 is ten times less expressed than at E12.5 [14]. In situ hybridization has revealed that Wnt2 is expressed in the epidermis and HF during the HF placode stage [32]. Therefore, Wnt2 likely acts as the secondary Wnt[32], which is a part of the placode signal and maybe is very essential for sheep HF initiation. Immediately after the formation of the placode, the dermal fibroblasts are aggregated, which is regulated by fibroblast growth factor 20 (FGF20), which is induced by epithelial Eda/Edar and Wnt/β-catenin and expressed in the HF placode [33]. FGF20 controls the aggregation of primary and secondary follicle dermis. FGF20 is also significantly upregulated at stage 1 of secondary HF, while FGFR1, FGFR2, FGFR3 and FGFR4, as potential FGF20 receptors, are not significant[33].

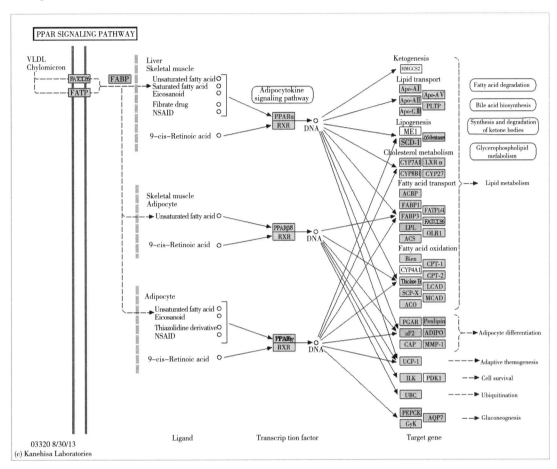

Note: The red color labels genes significantly down regulated in stage 1 compared with stage 0 (q value < 0.005). The green color labels genes were expressed in sheep skin during secondary HF initiation, but not significant difference.

Fig. 8 Differentially expressed genes between Stage 0 and Stage 1 of secondary HF morphogenesis involved in PPAR signaling pathway.

doi: 10.1371/journal.pone.0156890.g008

Recent studies have shown that the morphogenesis of HFs is not only related to the proliferation and differentiation of HF cells but also affected by other cells around the HFs, such as sebaceous gland, sweat gland[34-36]. Peroxisome proliferator-activated receptors (PPARs) are members of the nuclear hormone receptor family and have emerged as the important mediators of lipid metabolism in adipocytes and sebaceous glands [37, 38]. A set of different in vitro studies has demonstrated the important function of PPARs for cell differentiation, lipid synthesis and fatty acid uptake into cells [38]. Additionally, SCD1 and PPARs have also been implicated in the regulation of keratinocyte differentiation and the formation of a functional skin barrier [39, 40]. FATP4$^{-/-}$, Dgat1$^{-/-}$, Dgat2$^{-/-}$ and Early B cell factor 1 (Ebf1$^{-/-}$) mice have decreased intradermal adipose tissue due to defects in lipid accumulation in mature adipocytes [41-43]. Interestingly, these mice also display abnormalities in skin structure and function such as hair loss and epidermal hyperplasia. The results of this study showed that the PPAR signaling pathway was significantly enriched (corrected P-value < 0.05) and FABP4, AQP7, ADIPOQ, PEPCK, SCD, LPL and PLIN1 in PPAR signaling pathway were significantly downregulated at stage 1 of secondary HF. A previous study revealed that sebaceous gland buds form on the ental side of central primary HF during secondary HF morphogenesis on day 90 [44]. It is speculated that secondary HF morphogenesis in sheep may be promoted by reducing PPAR and then inhibiting the formation of sebaceous glands around the primary HFs.

LncRNA is a class of RNA molecules that do not encode proteins, and lncRNAs are 200 bp or more in length and have conserved secondary structures. LncRNAs may interact with proteins, DNA and RNA, and their biological function is involved in a variety of mechanisms at multiple levels of epigenetic, transcriptional and post-transcriptional levels. These mechanisms include gene imprinting, chromatin remodeling, cell cycle regulation, splicing regulation, mRNA degradation and translational regulation [45]. The databases of NONCODE mainly include human and mouse lncRNA data. In the NONCODE v4.0 and lncRNAdb database, lncRNAs have been rapidly increased from 73, 327 to 210, 831 in the last two years, respectively. This is indicated that lncRNAs are becoming a hot topic in life science research. However, only five sheep lncRNAs (antiPeg11 [46], MEG3, MEG9, Rian and Xist [47]) can be found in the above database [48]. This study obtained 884 novel sheep lncRNAs. A total of 15 lncRNAs with significant differential expression were detected, 6 up-regulated and 9 down-regulated. The cis or trans target gene prediction for lncRNAs showed LncRNA, XLOC002437 and potential target gene COL6A6 were all significantly expression. COL6A6 is expressed in a wide range of fetal and adult tissues including lung, kidney, liver, spleen, thymus, heart, skeletal muscle and skin dermis [49, 50]. Thus, defective COL6A6 results in the disorders with combined muscle and connective tissue involvement, including weakness, joint laxity and contractures, as well as abnormal skin [51-53]. The expression of collagen VI chains is highly regulated at different levels, such as gene transcription, processing of encoding RNAs, translation and post-translational modifications, and the impairment of the efficiency of each step may affect protein assembly and secretion [53, 54]. The XLOC002437 lncRNA and COL6A6 are significantly downregulated at stage 1 of secondary HF, suggesting that the interaction of COL6A6 and XLOC002437 may regulate and reduce the collagen VI α6 chain deposition in the skin by positive feedback, thereby inhibiting skin fibrosis and promoting the formation and deposition of the placode.

Studies have shown the relationship of mutual regulation among miRNAs, lncRNAs and mRNAs [55-58]. Recent studies show that lncRNA can interact with the miRNA as a competing endogenous RNA (ceRNA) to participate in the expression regulation of target genes, which exert an important role in the initiation and progression of tumor[59, 60]. For example, the LncRNA H19 as a ceRNA for miR-138 and miR-200a promotes epithelial to mesenchymal transition by functioning as miRNA sponges in colorectal cancer[59]. In this study, we found the high consistency between XLOC005698 and oar-miR-3955-5p as well as more lower MFE. XLOC005698 may be as a competing endogenous RNA "sponges" oar-miR-3955-5p activity. However, the molecular mechanism of the interaction of oar-miR-3955-5p and XLOC005698 lncRNA also need further study.

4 CONCLUSIONS

The present study applied the ssRNA-seq technique to integrated analysis of the role of LncRNA and coding RNA expression in sheep (*Ovis aries*) skin during the initiation of secondary HF.A total of 884 novel lncRNAs were discovered in sheep skin expression profiles. Differences were found in 192 expressed genes, 15 lncRNAs between the two different stages. These results laid a foundation to screen the regulatory elements or functional genes that specifically regulate the initiation of secondary HF, as well as supplying some potential values for understanding human hair disorders.

5 MATERIALS AND METHODS

5.1 Sheep skin sampling

Alpine merino sheep were obtained from a sheep stud farm located in Zhangye City, Gansu Province. All experimental and surgical procedures were approved by the Biological Studies Animal Care and Use Committee, Gansu Province, People's Republic of China. Sixty GAS ewes (2–3 years old), which had a mean fiber diameter of 18.1 ± 0.5 μm and were sourced from a single flock, were artificially inseminated with fresh sperm from a single ram (fiber diameter = 19.20 μm), and the day of insemination was designated as embryonic day (E) 0. Three fetuses were collected at E87 (stage 0 of HF morphogenesis) and E96 (stage 1 of HF morphogenesis), respectively[61]. On the day of sampling, the pregnant ewes were stunned via captive bolt and exsanguinated. The uterus was exteriorized, and the fetuses were carefully removed. The fetuses were washed in phosphate buffered saline and exsanguinated. The midside skin strips from two sides of the fetuses were snap frozen in liquid nitrogen for frozen sectioning and RNA extraction.

5.2 HF morphogenesis

Frozen skin strips from E87 and E96 were embedded in an O. C. T. compound (Sakura Finetek, USA, Inc., Torrance, CA), cut into 8-μm-thick serial sections in a cryostat, placed on Superfrost Plus glass slides (Fisher Scientific, Pittsburgh, PA, USA) and stained with hematoxylin. HF morphogenesis was studied in the longitudinal direction serial sections to observe secondary follicle morphogenesis.

5.3 Total RNA extraction, library construction and deep sequencing

Total RNA was isolated from the tissues using an RNeasy Maxi Kit (Qiagen, Hilden, Germany) according to the manufacturer's instructions. RNA quality was verified using a 2100 Bioanalyzer RNA Nano Chip (Agilent, Santa Clara, CA, USA), and the RNA Integrity Number

(RIN) value was > 8.5. The RNA was quantified using a Nano Drop ND-2000 Spectrophotometer (Nano-Drop, Wilmington, DE, USA).

The rRNA was depleted from 3 μg of total RNA using Epicentre Ribo-Zero™ rRNA Removal Kit (Epicentre, USA). The cDNA libraries were prepared from the remaining RNA without poly (A) selection using the NEBNext 1 Ultra™ Directional RNA Library Prep Kit for Illumina (NEB, USA) according to the manufacturer's instructions. The products were then purified with AMPure XP Universal PCR primers and the Index (X) Primer. The products were purified (AMPure XP system), and library quality was assessed using the Agilent Bioanalyzer 2100 system. The clustering of the index-coded samples was performed on a cBot Cluster Generation System using the TruSeq PE Cluster Kit v3-cBot-HS (Illumina). After cluster generation, the libraries were sequenced on the Illumina HiSeq 2000 platform, and 125 bp paired-end reads were generated.

Sequencing-received raw image data were transformed by base culling into sequence data, which was called raw data. Raw sequences were transformed into clean reads after removing all low quality tags, empty reads and singletons (tags that occurred only once). All paired-end clean reads were mapped to sheep reference sequences (version: Oarv3.1) by TopHat2 (version: V2.0.9) [25], and read counts for different genes and other known transcripts (misc_RNA, pseudogene, rRNA, tRNA and others) were extracted by HTSeq (version: V0.6.1) [62] with default parameters and allowing one mismatch. To monitor mapping events on both strands, both sense and complementary antisense sequences were included [63].

5.4 Identification of lncRNAs

LncRNAs were identified using the following workflow. According to the annotation of sheep reference sequences (version: Oarv3.1), the transcriptome from each dataset was assembled independently using the Cufflinks package (version: 2.1.1) [26] and Scripture (version: beta2) [27]. Transcripts smaller than 200 bp or all single-exon transcripts were excluded first. Cufflinks was then used to estimate the abundance of all transcripts based on the final transcriptome, and the transcripts with coverage less than 3 were also discarded. All transcriptomes were then pooled and merged to generate a final transcriptome using Cuffmerge, and the known protein-coding transcripts as well as rRNA, tRNA, snRNA, snoRNA, pre-miRNA and pseudogenes were identified using Cuffcompare and excluded. The remaining unknown transcripts were used to screen for putative lncRNAs. Among the different classes of class_code (http://cufflinks.cbcb.umd.edu/manual.html#class_codes), only those annotated by "u", "i", and "x" were retained, which represent potential novel intergenic, intronic and anti-sense lncRNAs, respectively. Filtering of the remaining transcripts resulted in many novel, long expressed transcripts. We first used CNCI [64] and PhyloCSF [65] to predict transcripts with coding potential. All transcripts with CNCI scores > 0, CNCI scores > 0 or PhyloCSF score > 100 were discarded. The remaining transcripts were subjected to HMMER analysis to exclude transcripts that contained any known protein domains cataloged in the Pfam database [66]. Conservation analysis of lncRNAs and mRNAs was performed using phastCons with default parameters [67]. To select bona fide lncRNAs, the lncRNAs identified using the above four methods were integrated into a comprehensive data set.

5.5 Determination of gene expression levels and detection of DEGs and lncRNAs

Gene expression FPKM values of mRNAs and lncRNAs were calculated with Cufflinks v2.0.2. Additionally, a table comprising read counts for each transcript was calculated using BEDTools version 2.17.0 [68]. We removed the low expressed transcripts (at least all of the samples had FPKM < 0.1). The set of remaining transcripts was reduced to a set of non-overlapping regions (or 'genes') by comparing all overlapping transcripts and keeping the transcript with the largest average FPKM across all samples as the representative transcript for that region. For differential expression quantification of mRNA and lncRNA genes, EdgeR version 3.0.8 [69] was used to identify differentially expressed transcripts between E87 and E96 using q value (p-adjusted) ⩽0.05 and absolute fold change ⩾1.

Differentially expressed lncRNAs were selected for target prediction via cis- or trans-regulatory effects. For the cis pathway target gene prediction, lncRNAs and potential target genes were paired and visualized using UCSC genome browser on the NCBI database. The genes transcribed within a 100 kbp window upstream or downstream of lncRNAs were considered as cis target genes [70]. For the trans pathway target gene prediction, the blast ratio (e < 1E-5) between lncRNAs and protein coding genes was calculated. RNAplex software was then used to select trans-acting target genes [71]. RNAplex parameters were set as -e -20. For the prediction of target miRNAs, lncRNAs were screened in the sense-antisense miRNA overlapping and non-overlapping regions by searching for similarity with miRBase mature miRNA sequences of sheep (Ovis aries) [72] using the BLAST program. The miRNA-binding sites on lncRNAs were then predicted using a web-based program called RNA hybrid (version: 2.2) [29].

5.6 Strand-specific real-time quantitative RT-PCR

To confirm the differentially expressed sense and antisense transcripts between super fine wool group and fine wool group, ten genes were randomly selected to verify the expression levels of genes and LncRNAs in skin by strand-specific qRT-PCR according to the protocol described in yue et al. (2015) [19]. Primers for real-time PCR were designed with Primer Express 3.0 (Applied Biosystems) (S4 Table).

5.7 GO and KEGG enrichment analysis of differentially expressed genes and LncRNA

All differentially expressed genes and the predicted target genes of the LncRNAs were mapped to GO terms in the GO database, and the gene numbers for each GO term were calculated using the GO seq R package (version: 1.18.0) [73]. The significantly enriched metabolic pathways or signal transduction pathways were identified via pathway enrichment analysis using KEGG (Kyoto Encyclopedia of Genes and Genomes), and KOBAS (version: 2.0) [74]. In above all tests, P-values were calculated using Benjamini-corrected modified Fisher's exact test, and ⩽ 0.05 was taken as a threshold of significance GO terms or pathways.

SUPPORTING INFORMATION

S1 Fig. Comparison of transcript length, ORF length, the number of exons and conservation score of mRNA and LncRNA. (A) The transcript length distribution of mRNA and LncRNA. (B) The ORF length distribution of mRNA and LncRNA. (C) the number of exons of mRNA and LncRNA. (D) The cumulative distribution curve of mRNA and LncRNA conservation scores.

(PDF)

S1 Table. Differentially expressed genes and LncRNA between Stage 0 and Stage 1 of secondary HF initiation.

(XLS)

S2 Table. Cluster genes of differentially expressed LncRNAs in sheep skin during secondary HF initiation.

(XLS)

S3 Table. Differentially expressed genes of PPAR signaling pathway in sheep skin between Stage 0 and Stage 1 of secondary HF initiation.

(XLS)

S4 Table. Relevant information of gene and primer sequences for strand-specific RT-PCR.

(XLS)

ACKNOWLEDGMENTS

This study was supported by the Central Level, Scientific Research Institutes for Basic R & D Special Fund Business (Grant No.1610322015014), the Earmarked Fund for Modern China Wool & Cashmere Technology Research System (Grant No.nycytx-40-3), and the National Natural Science Foundation for Young Scholars of China (Grant No.31402057).

AUTHOR CONTRIBUTIONS

Conceived and designed the experiments: YJY TTG BHY. Performed the experiments: YJY TTG CY JBL XPS. Analyzed the data: YJY TTG CY. Contributed reagents/materials/analysis tools: RLF JG CN. Wrote the paper: YJY CY BHY.

(发表于《PLOS ONE》，院选SCI，IF：3.057. DOI:10.1371/journal.pone.0156890 June 8, 2016)

Genetic Diversity and Phylogenetic Evolution of Tibetan Sheep Based on mtDNA D-Loop Sequences

LIU Jianbin[1,2],　DING Xuezhi[1],　ZENG Yufeng[1],　YUE Yaojing[1,2],
GUO Xian[1],　GUO Tingting[1,2],　CHU Min[1],　WANG Fan[3],　HAN Jilong[1,2],
FENG Ruilin[1,2],　SUN Xiaoping[1,2],　NIU Chune[1,2],　YANG Bohui[1,2],
GUO Jian[1,2],　YUAN Chao[1,2]

(1. Lanzhou Institute of Husbandryand Pharmaceutical Sciences of the Chinese Academy of Agricultural Sciences, Jiangouyan Street, Lanzhou, China; 2. Sheep Breeding EngineeringTechnology Research Center of ChineseAcademy of Agricultural Sciences, Jiangouyan Street, Lanzhou, China; 3. China Agricultural Veterinarian Biology Science and Technology Co., Ltd., Xujiaping, Lanzhou, China)

Abstract: The molecular and population genetic evidence of the phylogenetic status of the Tibetan sheep (*Ovis aries*) is not well understood, and little is known about this species' genetic diversity. This knowledge gap is partly due to the difficulty of sample collection. This is the first work to address this question. Here, the genetic diversity and phylogenetic relationship of 636 individual Tibetan sheep from fifteen populations were assessed using 642 complete sequences of the mitochondrial DNA D-loop. Samples were collected from the Qinghai-Tibetan Plateau area in China, and reference data were obtained from the six reference breed sequences available in GenBank. The length of the sequences varied considerably, between 1031 and 1259 bp. The haplotype diversity and nucleotide diversity were 0.992 ± 0.010 and 0.019 ± 0.001, respectively. The average number of nucleotide differences was 19.635. The mean nucleotide composition of the 350 haplotypes was 32.961% A, 29.708% T, 22.892% C, 14.439% G, 62.669% A+T, and 37.331% G+C. Phylogenetic analysis showed that all four previously defined haplogroups (A, B, C, and D) were found in the 636 individuals of the fifteen Tibetan sheep populations but that only the D haplogroup was found in Linzhou sheep. Further, the clustering analysis divided the fifteen Tibetan sheep populations into at least two clusters. The estimation of the demographic parameters from the mismatch analyses showed that haplogroups A, B, and C had at least one demographic expansion in Tibetan sheep. These results contribute to the knowledge of Tibetan sheep populations and will help inform future conservation programs about the Tibetan sheep native to the Qinghai-Tibetan Plateau.

☯ These authors contributed equally to this work.
■ These authors are co-first authors on this work.
* Corresponding author, E-mail: liujianbin@caas.cn

1 INTRODUCTION

Tibetan sheep play agricultural, economic, cultural, and even religious roles in the Qinghai- Tibetan Plateau areas in China and provide meat, wool, and pelts for the local people [1]. The Qinghai-Tibetan Plateau areas are also rich in Tibetan sheep genetic resources, with approximately 17 indigenous sheep populations [2]. Most indigenous Tibetan sheep are not only adapted to their local environment but are also considered important genetic resources and are thus one of the major components of agro-animal husbandry societies. However, most indigenous Tibetan sheep populations are composed of relatively small numbers of individuals, and many populations have been in steady decline over the last 30 years [3]. The climate and land- forms of the Qinghai-Tibetan Plateau areas are different from other areas of China. Traffic from other parts of China is blocked; thus, the Tibetan sheep are rarely influenced by external populations. These populations may now be on the verge of extinction and may ultimately be lost, given the rapid destruction of their ecological environment, the continuing introduction of modern commercial Tibetan sheep populations, and the ongoing lack of effective conservation methods [4]. To date, the genetic diversity, phylogenetic relationship, and maternal origin of the Qinghai-Tibetan Plateau populations remain uncertain and controversial.

The study of mitochondrial DNA (mtDNA) polymorphisms has proven to be tremendously useful for elucidating the molecular phylogeny of various species [5-8] due to the extremely low rate of recombination of mtDNA, its maternal lineage heredity and its relatively faster substitution rate than nuclear DNA [9]. In particular, the control region (CR), also called the displacement-loop region (D-loop) is the main noncoding regulatory region for the transcription and replication of mtDNA. One very useful approach for investigating the history and phylo- genic relationships of modern domestic animals is therefore based on mtDNA sequence analysis. The variability and structure of the mtDNA control region makes it possible to describe the genetic polymorphisms and maternal origin of Tibetan sheep, mainlybecause mtDNA displays a simple maternal inheritance without recombination and with a relatively rapid rate of evolution [10]. The even higher substitution rate in the CR, compared with the heterogeneity rate in the other parts of mtDNA, can be used to optimally characterize intraspecific and interspecific genetic diversity [11-15].

Here, we present an investigation into the mtDNA D-loop variability observed in Tibetan sheep indigenous to the Qinghai-Tibetan Plateau areas. We aimed to increase the number of Tibetan sheep samples by including six available reference genomes from GenBank for our population genetic and phylogenetic analysis of the fifteen Tibetan sheep populations based on the complete mtDNA control region. Our results provide insight into the genetic diversity, phylogenetic evolution, and maternal origin of Tibetan sheep for the conservation and improved management of sheep genetic resources.

2 MATERIALS AND METHODS

2.1 Ethic statement

We declare that we have no financial or personal relationships with other people or organizations that can inappropriately influence our work, and there is no professional or other personal interest of any nature or kind in any product, service and/or company that could be construed as influencing the position.

2.2 Sample collection

Ten milliliters of blood was collected from the jugular vein of each animal. From the 10 mL samples, 2mL samples were quickly frozen in liquid nitrogen and stored at-80℃ forgenomic DNA extraction, as described previously [16]. The total DNA was extracted from the blood using the saturated salt method [17]. The extracted DNA was quantified spectrophotometrically and adjusted to 50 ng/μL. The blood samples were collected from 636 sheep living in the Qinghai-Tibetan Plateau areas in China. The sampled individuals belonged to the fifteen Tibetan sheep populations that are distributed across Qinghai Province (Guide Black Fur sheep, n = 39; Qilian White Tibetan sheep, n = 44; Tianjun White Tibetan sheep, n = 64; Qinghai Oula sheep, n = 44), Gansu Province (Minxian Black Fur sheep, n = 67; Ganjia sheep, n = 58; Qiaoke sheep, n = 71; Gannan Oula sheep, n = 52), and the Tibet Autonomous Region (Langkazi sheep, n = 10; Jiangzi sheep, n = 46; Gangba sheep, n = 85; Huoba sheep, n = 34; Duoma sheep, n = 8; Awang sheep, n = 5; Linzhou sheep, n = 9). The sampling information (population code, sample number, altitude, longitude and latitude, accession number, sampling location, and geographical location) for the fifteen indigenous Tibetan sheep populations is shown in Table 1 and Fig. 1. This study did not involve endangered or protected Tibetan sheep populations. All experimental and sampling procedures were approved by the Institutional Animal Care and Use Committee, Lanzhou Institute of Husbandry and Pharmaceutical Sciences, Peoples Republic of China. All samples were collected with the permission of the animal owners.

2.3 Data collection

To achieve good coverage of the tested populations, a dataset of six referenced breeds was completed using the six submitted sequences containing the Ovis aries, Ovis vignei, and Ovis ammon mtDNA D-loops for the six individuals in Gen Bank (Table A in S1 File). These six breeds were from six international geographic regions and included Omusimon, Ovignei, Oammon, Oasia A, Oeurope B, and Omexic. The Gen Bank accession numbers for these reference sequences are AY091487, AY091490, AJ238300, AF039578 (haplogroup A), AF039577 (hap-logroup B), and AY582801, respectively [10, 18, 19].

2.4 Polymerase chain reaction and nucleotide sequencing

One pair of polymerase chain reaction (PCR) primers and sequencing primers was designed based on the 5' and 3' conserved flanking sequences of the complete mtDNA D-loop using the Primer Premier 5.0 software [20] and synthesized by BGI Shenzhen Technology Co., Ltd. (Shenzhen, China). The nucleotide sequences of forward primer CsumF was 5'-GGCTGGGAC CAAACCTAT-3', and the nucleotide sequence of reverse primer CsumR was 5'-GAACAACC AACCTCCCTAAG-3'. PCR was performed in a thermal cycler (Mastercycler gradient, Eppen-dorf, Germany) with a total reaction volume of approximately 30 μL, containing 2 μL genomic DNA (10 ng/μL), 3 μL (3 pM) each primer, 3 μL 10 × Ex Taq reaction buffer, 2 μL dNTP (2.5 mM), 0.2 μL Taq DNA polymerase (5 μL/U) (TaKaRa, China), and 16.8 μL ddH$_2$O. The PCR conditions were as follows: initial denaturation for 5 min at 94℃, 36 cycles of denaturation at 94℃ for 30s, annealing at 56℃ for 30s, and extension at 72℃ for 1.5min. The final extension step was followed by a 10 min extension at 72℃. The PCR amplification products were subsequently stored at 12℃ until use.

The amplified D-loop fragment was purified using a PCR gel extraction kit from Sangon Biotech Co., Ltd. (Shenzhen, China) and sequenced directly using a BigDye Terminator v3.1 cycle sequencing ready reaction kit (Applied Biosystems, Darmstadt, Germany) in an automatic sequencer (ABI-PRISM 3730 genetic analyzer, Applied Biosystems, CA, USA). PCR for the sequencing was performed in an automatic sequencer with a total reaction volume of approximately 5 μL containing 3 μL genomic DNA (10 ng/μL), 1 μL (3 pM) of each sequencing primer, 0.5 μL BigDye, and 0.5 μL ddH$_2$O. The sequencing conditions were as follows: initial denaturation for 2 min at 95℃, 25 cycles of denaturation at 95℃ for 10s, and annealing at51℃ for 10s. The final extension step was followed by a 190 s extension at 60℃. The PCR sequencing products were subsequently stored at 12℃ until use.

Table 1 Sampling information for the 15 indigenous Tibetan sheep populations.

Population	population code	Sample number	Altitude (m)	Longitude and latitude	Accession number	Sampling location
Guide Black Fur sheep	GD	39	3 100	N: 38° 61′152″ E: 103° 32′160″	KP228119-KP228157	Senduo Town, Guinan County, Hainan Tibetan Autonomous State, Qinghai Province
Qilian White Tibetan sheep	QL	44	3 540	N: 42° 20′178″ E: 116° 64′618″	KP228549-KP228592	Qilian Town, Qilian County, Delingha City, Mongolian Autonomous State, Qinghai Province
Tianjun White Tibetan sheep	TJ	64	3 217	N: 42° 18′158″ E: 116° 42′210″	KP228593-KP228656	Shengge Countryside, Tianjun County, Delingha City, Mongolian Autonomous State, Qinghai Province
Qinghai Oula sheep	QH	44	3 630	N: 34° 16′433″ E: 101° 32′141″	KP228434-KP228477	Jianke Village, Kesheng Town, Henan Mongolian Autonomous County, Qinghai Province
Minxian Black Fur sheep	MX	67	3 180	N: 36° 54′048″ E: 103° 94′107″	KP228367-KP228433	Taizi Village, Qingshui Town, Minxian County, Dingxi City, Gansu Province
Ganjia sheep	GJ	58	3 022	N: 35° 32′049″ E: 102° 40′802″	KP228158-KP228215	Xike Village, Ganjia Town, Xiahe County, Gannan Tibetan Autonomous State, Gansu Province
Qiaoke sheep	QK	71	3 410	N: 35° 42′106″ E: 102° 42′210″	KP228478-KP228548	Waeryi Village, QihamaTown, Maqu County, Gannan Tibetan Autonomous State, Gansu Province

(continued)

Population	population code	Sample number	Altitude (m)	Longitude and latitude	Accession number	Sampling location
Gannan Oula sheep	GN	52	3 616	N: 33° 51′312″ E: 101° 52′424″	KP228216-KP228267	Daerqing Administrative Village, Oula Town, Maqu County, Gannan Tibetan Autonomous State, Gansu Province
Langkazi sheep	LKZ	10	4 459	N: 28° 58′951″ E: 090° 23′757″	KP228348-KP228357	Kexi Village, Langkazi Town, Langkazi County, Shannan Territory of Tibet Autonomous Region
Jiangzi sheep	JZ	46	4 398	N: 28° 55′113″ E: 089° 47′692″	KP228302-KP228347	Reding Village, Cheren Town, Jiangzi County, Shannan Territory of Tibet Autonomous Region
Gangba sheep	GB	85	4 403	N: 28° 15′281″ E: 088° 24′787″	KP228034-KP228118	Yulie Village, GangbaTown, Gangba County, Rikaze Territory of Tibet Autonomous Region
Huoba sheep	HB	34	4 614	N: 30° 13′822″ E: 083° 00′249″	KP228268-KP228301	Rima Village, Huoba Town, Zhongba County, Rikaze Territory of Tibet Autonomous Region
Duoma sheep	DM	8	4 780	N: 29° 48′609″ E: 091° 36′191″	KP228026-KP228033	Sixth Village, MaquTown, Anduo County, Naqu Territory of Tibet Autonomous Region
Awang sheep	AW	5	4 643	N: 30° 12′101″ E: 098° 63′098″	KP228021-KP228025	Ayi Third Village, Awang Town, Gongjue County, Changdou Territory of Tibet Autonomous Region
Linzhou sheep	LZ	9	4 292	N: 29° 09′121″ E: 091° 25′063″	KP228358-KP228366	Tanggu Village, Tanggu Town, Linzhou County, Tibet Autonomous Region

doi: 10.1371/journal. pone. 0159308. t001

2.5 Data analysis

The sequences were arranged for multiple comparisons using Clustal Omega [21] and were aligned using ClustalW and BLAST [22]. These results were compared with other sequences obtained from GenBank. The reference sequences for tree construction were taken from the maternal lineages of each tree: haplogroup A (AF039578), haplogroup B (AF039577, AY582801, and AY091487), haplogroup E (AY091490, AJ238300). The diversity parameters, including the haplotype diversity, nucleotide diversity and average number of nucleotide differences, were estimated using DnaSP (Sequence Polymorphism Software) 5. 10. 01 [23]. The genetic differentiation coefficient (G_{ST}), Wright's F-statistics of subpopulation within total (F_{ST}), gene flow (N_m), molecular variance (AMOVA) test, and neutrality tests (Ewens-Watterson test, Chakraborty's test, Tajima's D test, Fu's FS test) were estimated using Arlequin version 3. 5. 1. 2 [24]. To identify differences between the geographic regions using the AMOVA program, four groups were established. The phylogenetic and molecular evolutionary relationships, average number of nucleotide substitutions per site between populations (D_{xy}), net nucleotide substitutions per site between populations (D_a), ME phylogenetic haplotype and clustering tree, and genetic distance were assessed using Molecular Evolutionary Genetics Analysis (MEGA) version 6.0 [25]. We also sketched network and mismatch distribution graphs using the median-joining method implemented in the NETWORK version 4.6.1.2 software to assess the haplotype relationships [26].

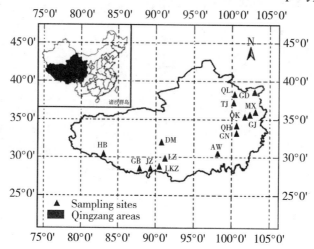

Fig. 1 Geographical locations of the 15 indigenous tibetan sheep populations sampled from the qinghai-Tibetan plateau area.

Note: The black area in the inset indicates the Qinghai-Tibetan Plateau area; the black triangles indicate the sampling sites within the plateau area (enlarged). The sampling locations of the specific populations are shown in Table 1.

doi: 10.1371/journal.pone.0159308.g001

3 RESULTS

3.1 Polymorphic site and sequencing analysis of the complete control region

Based on the reference sequences from GenBank accession numbers (AY091487, AY091490, AJ238300, AF039578, AF039577, AY582801), all of the sequences were aligned with 1274 comparative sites (707 had gaps or missing data, and 567 had no gaps or missing data), and 350 haplotypes were obtained from the 642 sequenced individuals (636 Tibetan sheep and 6 reference sequences). The length of the sequences from the fifteen Tibetan sheep populations of 636 individuals varied considerably, between 1 031 and 1 259 bp, although the majority were between 1 180 and 1 183 bp (Table B in S1 File). A total of 196 variable sites were obtained from the sequences, including 63 singleton variable sites (62 double variants and 1 triple variant) and 133 parsimony-informative variable sites (124 double variants, 7 triple variants, and 2 quadruple variants). There were 158 transitions and 38 transversions within the 196 variable sites, of which 15 sites were found to have both transitions and transversions. The most commonly observed substitution caused a transition mutation. With the exception of the insertion or deletion of several nucleotide sites, the observed variations in the length of the mtDNA D-loop sequences of the Tibetan sheep mainly resulted from variability in the number of 75bp tandem repeat motifs (between three and five repeats).

The nucleotide composition of all the haplotypes was 32.961% A, 29.708% T, 22.892% C, 14.439% G, 62.669% A+T, and 37.331% G+C. The A+T haplotype was substantially more common than the G+C haplotype, showing an AT bias (Table C in S1 File). The largest haplotype group (haplogroup A) consisted of 490 individuals and 259 haplotypes; the next largest haplo-type groups (haplogroup B and haplogroup C) consisted of 145 individuals and 43 haplotypes (64 individuals and 43 haplotypes and 81 individuals and 47 haplotypes, respectively). The smallest haplotype group (haplogroup D) consisted of 1 individual and 1 haplotype. The number of haplotypes, individuals, and frequency detected in each Tibetan sheep population of haplotype group varied from 1to49, from 0 to 62, and from 0 to 0.875, respectively (Table 2). The haplotype diversity and nucleotide diversity were calculated separately for each Tibetan sheep population (Table 2) and were estimated to be 0.992 ± 0.010 and 0.019 ± 0.001, respectively. The values for the two parameters (haplotype diversity and nucleotide diversity) ranged from 0.900 ± 0.161 to 1.000 ± 0.045 and from 0.009 ± 0.002 to 0.027 ± 0.003, respectively, thus demonstrating the high level of genetic diversity in the fifteen Tibetan sheep populations. The nucleotide diversity value of the Linzhou sheep (0.027 ± 0.003) and Jiangzi sheep (0.026 ± 0.002) populations was found to be higher than that of the other 13 Tibetan sheep populations, indicating a relatively high level of diversity. Similarly, the haplotype diversity values were highest in the Langkazi sheep (1.000 ± 0.045) and Linzhousheep (1.000 ± 0.056) populations and the lowest in the Awang sheep (0.900 ± 0.161) population.

3.2 Genetic distance and average number of nucleotide differences

Table 3 presents the genetic distance and average number of nucleotide differences between and within the fifteen Tibetan sheep populations. The genetic distance values ranged from 0.009 to 0.039 within the population diagonals, and the genetic distance values ranged from 0.014 to 0.040 among populations above the diagonals. Among the Tibetan sheep populations, the genetic distance within populations reached a maximum value in Linzhou sheep and a minimum value in Awang sheep. Similarly, the genetic distance between the populations had a maximum value for Linzhou sheep and Jiangzi sheep and a minimum value for Awang sheep and Tianjun White Tibetan sheep. The average number of nucleotide differences values ranged from 10.000 to 29.806 within populations along the digital diagonal, and the average number of nucleotide difference values ranged from 10.725 to 30.986 between the populations below the diagonals. Among the Tibetan sheep populations, the average number of nucleotide differences within the populations reached its value maximum in Linzhou sheep and its minimum value in Awang sheep. Similarly, the average number of nucleotide differences between populations reached a value maximum in the Linzhou sheep and Jiangzi sheep populations and a minimum value in the Awang sheep and Tianjun White Tibetan sheep populations.

Table 2 Genetic diversity indices of fifteen Tibetan sheep populations.

Population	Nucleotide diversity	Haplotype diversity	No. of individuals	No. of haplotypes	Haplogroup A No. of individuals	Haplogroup A Frequency or No. of haplotypes	Haplogroup B No. of individuals	Haplogroup B Frequency or No. of haplotypes	Haplogroup C No. of individuals	Haplogroup C Frequency or No. of haplotypes	Haplogroup D No. of individuals	Haplogroup D Frequency or No. of haplotypes
GD	0.019 ± 0.002	0.968 ± 0.014	39	24	30	0.750/18	3	0.125/3	6	0.125/3	0	0
QL	0.015 ± 0.002	0.998 ± 0.006	44	40	38	0.875/35	2	0.050/2	4	0.075/3	0	0
TJ	0.013 ± 0.002	0.991 ± 0.005	64	44	57	0.869/38	4	0.045/2	3	0.086/3	0	0
QH	0.021 ± 0.002	0.995 ± 0.007	44	37	32	0.703/26	7	0.162/6	5	0.135/5	0	0
MX	0.015 ± 0.002	0.932 ± 0.018	67	25	56	0.600/15	5	0.200/5	6	0.200/5	0	0
GJ	0.019 ± 0.002	0.994 ± 0.005	58	46	45	0.740/34	7	0.130/6	6	0.130/6	0	0
QK	0.021 ± 0.002	0.986 ± 0.005	71	44	53	0.705/31	9	0.159/7	9	0.136/6	0	0
GN	0.021 ± 0.002	0.986 ± 0.007	52	37	35	0.595/22	7	0.162/6	10	0.243/9	0	0
LKZ	0.023 ± 0.003	1.000 ± 0.045	10	10	8	0.800/8	1	0.100/1	1	0.100/1	0	0
JZ	0.026 ± 0.002	0.945 ± 0.024	46	27	25	0.630/17	8	0.222/6	13	0.148/4	0	0
GB	0.020 ± 0.002	0.995 ± 0.003	85	66	62	0.742/49	10	0.106/7	12	0.152/10	0	0
HB	0.016 ± 0.002	0.989 ± 0.010	34	28	29	0.821/23	1	0.036/1	4	0.143/4	0	0
DM	0.017 ± 0.003	0.964 ± 0.077	8	7	7	0.857/6	0	0	1	0.143/1	0	0
AW	0.009 ± 0.002	0.900 ± 0.161	5	4	5	1.000/4	0	0	0	0	0	0
LZ	0.027 ± 0.003	1.000 ± 0.056	9	9	8	0.778/7	0	0	1	0.111/1	1	0.111/1
Total	0.019 ± 0.001	0.992 ± 0.010	636	350	490	0.740/259	64	0.123/43	81	0.134/47	1	0.003/1

A total of 350 haplotypes; 98 haplotypes are shared.

doi: 10.1371/journai. pone. 0159308. t002

Table 3 Genetic distance (above the diagonals) and average number of nucleotide differences (below the diagonals) between and within fifteen Tibetan sheep populations.

Population	GD	QL	TJ	QH	MX	GJ	QK	GN	LKZ	JZ	GB	HB	DM	AW	LZ
GD	0.026/21.579	0.023	0.022	0.027	0.023	0.026	0.028	0.029	0.027	0.033	0.027	0.024	0.023	0.020	0.032
QL	17.798	0.020/16.499	0.019	0.025	0.020	0.023	0.025	0.027	0.023	0.032	0.025	0.021	0.020	0.016	0.029
TJ	16.934	14.965	0.017/13.658	0.024	0.019	0.022	0.024	0.026	0.022	0.032	0.024	0.020	0.019	0.014	0.028
QH	18.492	17.325	16.504	0.029/19.902	0.025	0.027	0.029	0.030	0.028	0.034	0.028	0.026	0.025	0.021	0.033
MX	19.649	16.300	15.141	17.257	0.020/18.043	0.023	0.025	0.026	0.024	0.032	0.025	0.021	0.020	0.016	0.029
GJ	19.696	18.182	17.170	18.649	18.325	0.026/20.149	0.027	0.029	0.026	0.034	0.027	0.024	0.023	0.019	0.032
QK	18.884	17.717	16.691	17.376	18.033	18.937	0.029/20.542	0.030	0.028	0.035	0.029	0.026	0.025	0.021	0.034
GN	20.697	20.079	19.199	20.221	20.092	21.245	19.986	0.031/23.154	0.030	0.035	0.030	0.027	0.027	0.024	0.035
LKZ	19.374	15.409	14.528	18.091	17.825	17.034	15.986	18.638	0.029/21.444	0.035	0.028	0.025	0.024	0.019	0.032
JZ	25.452	25.394	24.866	23.535	25.398	26.043	24.275	25.783	21.957	0.037/29.009	0.034	0.033	0.033	0.031	0.040
GB	20.334	19.307	18.359	19.930	19.488	20.700	19.701	22.497	18.198	26.131	0.029/22.250	0.026	0.026	0.021	0.034
HB	17.951	16.383	15.484	17.757	16.548	18.608	17.936	20.147	15.865	25.246	19.446	0.022/17.118	0.021	0.017	0.030
DM	17.619	13.765	12.941	17.045	15.715	15.897	14.900	17.601	17.550	21.582	17.051	14.353	0.021/16.393	0.016	0.029
AW	17.831	12.055	10.725	14.586	15.442	14.417	14.332	17.208	16.02	23.209	15.892	12.665	14.275	0.009/10.000	0.026
LZ	23.966	22.663	21.528	22.876	22.919	24.602	23.203	25.808	20.411	30.986	25.421	22.944	19.361 39	19.044	0.039/29.806

doi: 10.1371/journal.pone.0159308.t003

3.3 Genetic differentiation

To examine the genetic differentiation between the fifteen Tibetan sheep populations, we calculated Wright's F-statistics of subpopulation within total (F_{ST}) and genetic differentiation coefficient (G_{ST}) (Table 4). We also calculated the gene flow (N_m) (Table D in S1 File), the average number of nucleotide substitutions per site (D_{xy}), and the number of net nucleotide substitutions per site (D_a) among the fifteen studied Tibetan sheep populations (Table E in S1 File). Estimates for the pairwise F_{ST} values (above diagonals) are given in Table 4. The F_{ST} values ranged from −0.046 to 0.237. Duoma sheep and Langkazi sheep had the closest pairwise F_{ST} value (F_{ST} = −0.046) among the fifteen Tibetan sheep populations. Awang sheep were more distantly related to Jiangzi sheep than they were to the other Tibetan sheep populations. All F_{ST} values were smaller than 0.25, indicating that significant genetic differentiation has not occurred among the fifteen Tibetan sheep populations. The results show that the F_{ST} values between Tibetan sheep in decreasing order were 14 (Minxian Black Fur sheep), 13 (Guide Black Fur sheep and Jiangzi sheep), 12 (Qiaoke sheep), 10 (Gangba sheep, Gannan Oula sheep and Tian-jun White Tibetan sheep), 9 (Gangjia sheep, Huoba sheep, and Qinghai Oula sheep), 7 (Qilian White Tibetan sheep), 4 (Langkazi sheep and Linzhou sheep), 3 (Duoma sheep), and 1 (Awang sheep). The distribution of the fifteen Tibetan sheep populations varied according to their F_{ST} values (P<0.05, or P<0.01). The G_{ST} values ranged from 0.001 to 0.047 (Table 4). The G_{ST} value between the Langkazi sheep and Linzhou sheep was the smallest (G_{ST} = 0.001), and the G_{ST} value was the largest (G_{ST} = 0.047) (Jiangzi sheep and Awang sheep, Minxian Black Fur sheep and Awang sheep, respectively). The mean G_{ST} was 0.018, which indicates that most of the genetic diversity occurred within populations and that 1.762% of the total population differentiation came from intrapopulation, whereas the remaining 98.238% came from differences among individuals in each population. Thus, the gene divergence between the populations was very low. The result of the variation observed among and within the 15 Tibetan sheep populations was not differentiation.

Table D in S1 File presents the N_m of the sequence values and haplotype values between the fifteen Tibetan sheep populations. The N_m of sequences values ranged from −731.043 to 495.657, demonstrating that gene exchange was either extremely frequent or extremely rare between the fifteen Tibetan sheep populations. The N_m of the sequence values between the Gannan Oula sheep and Qiaoke sheep was the smallest (N_m = −731.043), and the N_m of the sequence values between the Minxian Black Fur sheep and Qilian White Tibetan sheep was the largest (N_m = 495.657). The mean N_m of the sequences was −9.593, implying a relatively distant relationship. The N_m of the haplotype values ranged from 5.041 to 177.660. Notably, the N_m between Qilian White Tibetan sheep and Ganjia sheep was 35.24 times greater than the Nm between Jiangzi sheep and Awang sheep. The N_m of the haplotype values between the Jiangzi sheep and Awang sheep was the smallest (N_m = 5.041), and the N_m of the haplotype values between the Ganjia sheep and Qilian White Tibetan sheep was the largest (N_m = 177.660). The mean haplotype N_m was 22.594, indicating that gene flow did not occur between the populations in the past.

Table 4 Estimates of pairwise F_{ST} values (above the diagonals) and G_{ST} values (below the diagonals) between fifteen Tibetan sheep populations.

Population	GD	QL	TJ	QH	MX	GJ	QK	GN	LKZ	JZ	GB	HB	DM	AW	LZ
GD	—	0.005*	0.027*	−0.001*	0.007*	−0.002*	0.003*	0.003*	−0.023*	0.056**	0.001**	−0.009*	−0.013*	0.115	−0.012*
QL	0.006	—	0.002*	0.011	0.001*	−0.001	0.009*	0.028**	−0.037	0.104**	0.010*	−0.016	−0.023	0.050	−0.022
TJ	0.011	0.004	—	0.037*	0.016*	0.016*	0.034**	0.057**	−0.021	0.148**	0.038*	0.006*	−0.014	0.065	−0.004
QH	0.006	0.002	0.004	—	0.020*	−0.008*	−0.009*	−0.009*	−0.028	0.032**	−0.009*	0.010*	0.011	0.098	−0.001
MX	0.024	0.018	0.015	0.016	—	0.010*	0.021*	0.033**	−0.022*	0.108**	0.022*	−0.005*	−0.008**	0.092*	−0.006*
GJ	0.010	0.001	0.006	0.003	0.022	—	−0.005*	0.002*	−0.032	0.062**	−0.003*	−0.001*	−0.005	0.079	−0.015
QK	0.009	0.005	0.006	0.003	0.022	0.005	—	−0.001*	−0.034*	0.044**	−0.006*	0.008*	0.007	0.102	−0.013*
GN	0.009	0.003	0.004	0.002	0.018	0.005	0.005	—	−0.020	0.014*	−0.003*	0.022*	0.016	0.141	−0.006
LKZ	0.015	0.011	0.015	0.013	0.025	0.015	0.017	0.013	—	0.039*	−0.031	−0.034	−0.046	0.019	−0.045
JZ	0.021	0.014	0.017	0.014	0.031	0.016	0.018	0.017	0.021	—	0.037**	0.093**	0.093**	0.237	0.051*
GB	0.008	0.002	0.004	0.002	0.017	0.004	0.003	0.003	0.017	0.015	—	0.007*	0.004	0.099	−0.008
HB	0.010	0.002	0.007	0.004	0.022	0.003	0.006	0.006	0.010	0.014	0.005	—	−0.029	0.072	−0.023
DM	0.023	0.022	0.025	0.020	0.029	0.024	0.025	0.022	0.010	0.028	0.025	0.020	—	0.076	−0.020
AW	0.043	0.042	0.044	0.040	0.047	0.044	0.045	0.042	0.030	0.047	0.044	0.042	0.025	—	0.050
LZ	0.016	0.014	0.018	0.013	0.023	0.018	0.019	0.015	0.001	0.022	0.019	0.013	0.002	0.019	—

F_{ST}=Wright's F−statisticsofasubpopulationwithintotalpopulation; G_{ST} =genetic differentiation coefficient. The F_{ST} and G_{ST} values were each calculated with 1023 permutations.

*Significant at P value < 0.05 ** Significant at P value = 0.000.

doi: 10. 1371/journal. pone. 0159308. t004

Table E in S1 File shows the D_{xy} and D_a values among the fifteen Tibetan sheep populations. The D_{xy} values ranged from −0.001 1 to 0.005 0. The D_{xy} value between Langkazi sheep and Linzhou sheep was the smallest (D_{xy} = −0.001 1), and the D_{xy} value between the Jiangzi sheep and Awang sheep was the largest (D_{xy} = 0.005 0). The mean D_{xy} was 0.001, indicating that a low average number of nucleotide substitutions occurred per site between the fifteen Tibetan sheep populations. The D_a values were in the range of 0.010-0.028. The mean D_a was 0.019. Similarly, the number of net nucleotide substitutions per site between populations of the fifteen Tibetan sheep populations was highest in the Jiangzi sheep and Linzhou sheep (D_a = 0.28) and lowest in the Tianjun White Tibetan sheep and Awang sheep (D_a = 0.010).

3.4 Phylogenetic relationship

To extend our knowledge of the phylogenetic relationship of the fifteen Tibetan sheep populations, a phylogenetic tree was constructed using minimum evolution (ME), neighbor joining using the Maximum Composite Likelihood method (Fig. 2) and an unweighted pair-group method with arithmetic means (UPGMA) dendrogram based on the complete mtDNA D-loop sequences of 642 individuals (Fig. A, Fig. B, and Fig. C in S1 File) and 350 haplotypes (Fig. 3) from fifteen Tibetan sheep populations and six reference breeds. The six methods produced nearly consistent topological structures and similar support levels; therefore, only the ME tree is presented (Figs. 2 and 3). According to the ME tree, NJ tree, UPGMA tree, and median-joining network dendrogram (Fig. D in S1 File), we determined four distinct cluster haplogroups: A (Fig. A in S1 File), B (Fig. B in S1 File), C (Fig. C in S1 File), and D. Of the 350 haplotypes, there was no common haplotype identified in all of the Tibetan sheep populations; 98 haplotypes were shared, and 252 haplotypes were singletons, including 38 in Gangba sheep, 33 in Ganjia sheep, 28 in Tianjun White Tibetan sheep, and 24 in Qinghai Oula sheep. The leading haplotype (Hap 39) was found in 39 individuals. The next most common haplotype was Hap 42, composed of 19 individuals, and the remaining nine haplotypes were composed of seven to 10 individuals. Haplotype 42 was composed of Jiangzi sheep, Minxian Black Fur sheep, Qilian White Tibetan sheep and Tianjun White Tibetan sheep. Haplotype 4 was composed of fourteen of the Tibetan sheep populations, excluding Langkazi sheep, and showed close clustering. The majorities of the 490 individuals were grouped in haplogroup A (Fig. A in S1 File), followed by haplogroups B (Fig. B in S1 File) (64) and C (Fig. C in S1 File) (81); however, only one animal from the Linzhou sheep (LZ03) belonged to haplogroup D. The Duoma sheep were composed of two haplogroups, the Awang sheep were composed of one haplogroup, and the remaining 13 Tibetan sheep populations were composed of three haplogroups (Table 2). The four references breeds—Oasia A, Oeurope B, Omusimon, and Omexic—belonged to haplogroups A and B. The other two reference breeds—Omusimon and Ovignei—clustered within a group (Fig. 3 and Fig. D in S1 File). Further, the genetic distance between populations was analyzed using the Maximum Composite Likelihood method and are in the units of the number of base substitutions per site (Fig. 2). More specifically, the neighbor-joining phylogenetic tree of the 642 sequences of the mtDNA D-loop based on units of the number of base substitutions per site effectively divided the 15 indigenous Tibetan sheep populations and six reference breeds into four groups. Oammon and Ovignei were genetically distinct and were the first to separate. The 15 indigenous Tibetan sheep populations and four reference breeds were then divided into three sub-clusters. The first cluster included Jiangzi sheep, Qilian White Tibetan sheep, Qinghai Oula sheep, Gannan Oula sheep, Qiaoke sheep, Minxian Black Fur sheep, and Guide Black Fur sheep. The second cluster included OasiaA, Awang sheep, Tianjun White Tibetan sheep, Ganjia sheep, Langkazi sheep, Duoma sheep, Gangba sheep, Huoba sheep, and Linzhou sheep. The third cluster included Omexic, Oeurore B, and Omusimon. An analysis of molecular variance (AMOVA) was conducted, and the results are shown in Table F in S1 File. The AMOVA revealed a variation of 4.46% among the populations and of 95.54% within the populations; this finding was significant at P<0.05. The F_{ST} was 0.045, which indicated that 4.5% of the total genetic variation was due to population differences, and the remaining 95.5% came from differences among individuals in each population.

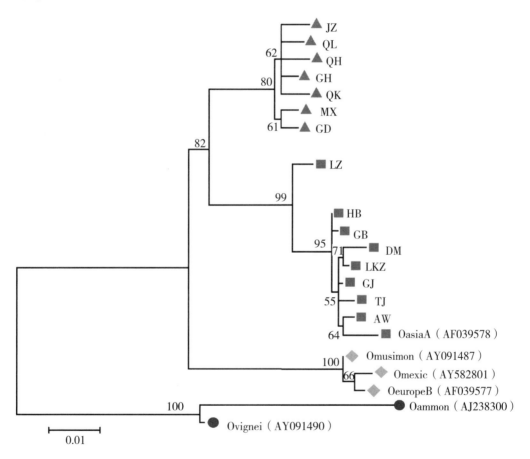

Fig. 2 A neighbor-joining phylogenetic tree of the 21 populations based on 642 Sequences of mtDNA D-loop.

Note: The distances were computed using the Maximum Composite Likelihood method and are in the units of the number of base substitutions per site.

doi: 10.1371/journal.pone.0159308.g002

Fig. 3 The ME phylogenetic tree of 350 haplotypes.

Note: The ME phylogenetic tree show that the 350 haplotypes and 636 sequences of Tibetan sheep populations and six reference breeds fall into five distinct clusters: haplogroup A, haplogroup B, haplogroup C, haplogroup D (Hap 259 of LZ 03) and haplogroup E (Omusimon and Ovignei), respectively. Haplogroups for individuals defined bytheentire haplotypes are shaded in blue (haplogroup A), green (haplogroup B), and red (haplogroup C).

doi: 10. 1371/journal. pone. 0159308. g003

3.5 Population expansions

Because the sample size for most of the populations was more than 30 individuals, the detection of population expansion was performed at the individual population level (data not shown) and in all haplotype sequences. The mismatch distribution analysis of the complete dataset (lineages A, B, C, D, and fifteen Tibetan sheep populations of mtDNA D-loop) is shown in Fig. 4 and Fig. E in S1 File. Neutrality tests (Ewens-Watterson test, Chakraborty's test, Tajima's D test, Fu's FS test) were used to detect population expansion (Table G in S1 File). The charts of the mismatch distribution for the samples of the fifteen Tibetan sheep populations and the total samples were multimodal. However, the mismatch distribution for Linzhou sheep was a unimodal function. The mismatch distribution of the complete dataset showed that there were two major peaks, with maximum values at 4 and 27 pairwise differences and two smaller peaks at 45 and 51 pairwise differences (Fig. E in S1 File). These results suggest that at least two expansion events occurred during the population demographic history of the Tibetan sheep population. The mismatch distribution analysis revealed a unimodal bell-shaped distribution of pairwise sequence differences in lineages A, B and C, but that of the lineage D was a sampling function. Mismatch analysis of lineages A, B and C suggested that a single population expansion event occurred in the demographic history of Tibetan sheep populations. The complete dataset of fifteen Tibetan sheep populations did not produce a significantly negative Ewens-Watterson test, whereas Chakraborty's neutrality test of Jiangzi sheep was significant (12.629, $P = 0.034$), and Tajima's D neutrality of Tianjun White Tibetan sheep test was also significant (-0.466, $P = 0.020$). Fu's F_S value was -7.484 for the fifteen Tibetan sheep populations, of which Ganan Oula sheep, Qiaoke sheep, Huoba sheep, Gangba sheep, Ganjia sheep, Qinghai Oula sheep, Qilian White Tibetan sheep, and Tian-jun White Tibetan sheep were highly significant ($P<0.01$ or $P<0.001$). This finding suggests the occurrence of two expansion events in the demographic history of the fifteen Tibetan sheep populations. This result is consistent with a demographic model showing two large and sudden expansions, as inferred from the mismatch distribution.

4 DISCUSSION

4.1 High mtDNA D-loop diversity of tibetan sheep populations

The haplotype diversity and nucleotide diversity of the total individuals were 0.992 ± 0.010 and 0.019 ± 0.001, respectively. The fifteen Tibetan sheep populations in our study showed a high level of haplotype and nucleotide diversity. This finding is consistent with archeological data and other genetic diversity studies [15, 27-29], but the haplotype diversity found here was higher than that found in a previous study [30], and the nucleotide diversity found here was lower than that found in a previous study [4]. These results indicate a relatively higher level of genetic diversity in the fifteen Tibetan sheep populations compared with other sheep populations [1, 4, 31]. For example, the haplotype diversity and nucleotide diversity values of Turkish sheep breeds distributed in a Turkish population were 0.950 ± 0.011 and 0.014 ± 0.001[31]. However, according to Walsh's work on the required sample size for the diagnosis of conservation units [32], a sample of 59 individuals is necessary to reject the hypothesis that individuals with unstamped ("hidden") character states exist in the population size. Thus, the sample size necessary to reject a hidden state frequency of 0.05 is 56 when sampling from a finite population of 500 individuals. Our genetic diversity estimation is therefore a precise reflection of Tibetan sheep due to the large sample size used in this study. For the Linzhou sheep, Langkazi sheep, Huoba sheep, Qinghai Oula sheep, Guide Black Fur sheep, Tianjun White Tibetan sheep, Ganjia sheep, Qiaoke sheep, Gangba sheep, and Gannan Oula sheep with broad distribution, a high genetic diversity could only be observed with such a large sample size and wide collection area. However, an even higher diversity may be found if even more samples were used, and a further investigation of the genetic diversity of these fifteen Tibetan sheep populations is still worth further research. These Tibetan sheep populations experienced a genetic bottleneck during the 20th century and are classified as the most rare populations of sheep [33]. In addition, the positive Ewens-Watterson and Chakraborty's values were significantly different among the fifteen Tibetan sheep populations, suggesting a previous decline in the population size of the mtDNA D-loop diversity. This finding was consistent with the results of a previous study [33]. Such genetic diversity may be caused by an increased mutation rate in the mtDNA D-loop, the maternal effects of multiple wild ancestors, overlapping generations, the mixing of populations from different geographical locations, natural selection favoring heterozygosis or subdivision accompanied by genetic drift [30].

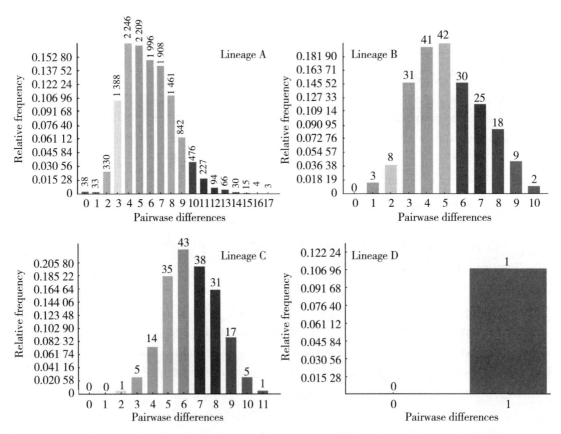

Fig. 4 The mismatch distribution of complete dataset of four lineages of the fifteen tibetan sheep populations.

Note: The results were summarized in fourlineages of the mtDNAtypes of the fifteen Tibetan sheep populations on the Qinghai-Tibetan Plateau areasshowed thatthere was at leastone demographic expansion.

doi: 10.1371/journal.pone.0159308.g004

4.2 Maternal origins of the tibetan sheep populations

The sequence motifs from the 1 180 bp to the 1 183 bp region of the mtDNA D-loop form the basis for the four major haplogroups (A-D) in the Tibetan sheep mtDNA haplotypes. Of these groups, haplogroup D is quite rare. The Tibetan sheep haplotypes were found to belong to all four major haplogroups, although only 0.157% belonged to haplogroup D, and these sheep were exclusively from the Linzhou sheep haplotype. This finding demonstrated that Tibetan sheep populations possess abundant mtDNA diversity and therefore a widespread origin of their maternal lineages. This study revealed a significant biogeographical association of the Asian Ovis mtDNA haplotypes with haplogroup A. Furthermore, the thoroughbred Tibetan sheep has been proposed to be shared in the haplogroup A, and the contribution of Asian sheep breeds to this population has also been reported. In this study, the overall sequences of all fifteen Tibetan sheep populations, including the fourteen Tibetan sheep populations respectively other than Duoma sheep and Awang sheep, were found in the common haplogroups B and C. It is generally believed that domestic sheep have two maternal lineages (haplogroup A and haplogroup B) based on earlier mtDNA analysis [4, 10, 18, 34]. Recently, a new maternal lineage (haplogroup C) was found in Chinese domestic sheep [27, 30]. The ME phylogenetic tree and median-joining analyses in our study revealed the presence of four mtDNA hap- logroups in the Tibetan sheep populations. Of these groups, the haplogroup of lineage A was predominant, and the haplogroups of lineage B and lineage C were the second most common. In this paper, the proportion of haplotypes of lineage D was 0.157%, further demonstrating that lineage D is the most rare of the mtDNA lineages. Our findings were consistent with the results of previous studies on domestic sheep breeds in China [35, 36]. Previous studies identified three mtDNA haplogroups in both China [29, 36, 37] and other countries [38, 39, 40]. The four mtDNA haplogroups of lineages A, B, C, and D found in the Tibetan sheep populations in the Qinghai-Tibetan plateau areas further supported the hypothesis of multiple maternal origins in Chinese domestic sheep.

4.3 Genetic differentiation of tibetan sheep populations

The F_{ST} value represents the level of genetic differentiation within a given population. Thus, there is "little differentiation" at a value of 0.05, "moderate differentiation" at values of 0.05-0.25, and "great differentiation" at values >0.25. In this study, the AMOVA analysis also revealed the distinct population of Qinghai-Tibetan Plateau areas among other Tibetan sheep populations with a significant positive variance. Gene flow (N_m), also known as gene migration, refers to the transfer of alleles from one population to another. N_m haplotype values >1 and N_m sequences <1 indicate a poor gene exchange, such that genetic drift will result in substantial local differentiation [41, 42]. The low G_{ST} value, combined with the low N_m of sequences used in this study, indicate that the great differentiation mainly resulted from the independent evolution of each isolated population and substantial local differentiation caused by the genetic drift [43]. An important factor leading to this result is likely the lower effective population sizes, as the Gannan Oula sheep, Qiaoke sheep, Ganjia sheep, and Qianlian White Tibetan sheep live in canyons and valleys and therefore have a limited ability to migrate and correspondingly lower population sizes relative to the other Tibetan sheep populations. As the effective population size declines, the nucleotide substitutions have a greater probability of reaching fixation [44, 45]. In addition, the estimated divergence time (data not shown) among the fifteen Tibetan sheep populations was consistent with the Pleistocene climate fluctuations and the uplift of the Qinghai-Tibetan Plateau, indicating that known paleogeographic factors might have played important roles in the speciation of Tibetan sheep.

4.4 Genetic relationships among the tibetan sheep populations

Our study showed that the fifteen Tibetan sheep populations native to the Qinghai-Tibetan Plateau are clustered into four groups: 490 Tibetan sheep represent the maternal origin of the haplogroup of lineage A, 64 Tibetan sheep represent the maternal origin of the haplogroup of lineage B, 81 Tibetan sheep represent the maternal origin of the haplogroup of lineage C, and 1 Tibetan sheep represents the maternal origin of the haplogroup of lineage D. This genetic relationship displayed a high consistency with traditional classification schemes and the results of previous studies [27, 46-49]. All fifteen Tibetan sheep populations belong to four maternal origins. The genetic differentiation of the fifteen Tibetan sheep populations was mainly the result of geographic isolation, natural selection, different living conditions, and breeding history. Because Tibetan sheep are a portable food and wool resource, the commercial trade and extensive transport of sheep along human migratory paths might help account for the observed pattern by promoting genetic exchange. Other study methods, such as genetic approaches, including the degree method and the phylogenetic relationship clustering method, also indicated that indigenous sheep were the maternal origin of haplogroups A, B, C, and D [46, 48].

4.5 Population expansion of tibetan sheep populations

Because the sample sizes of most of the populations were less than 34 individuals, the detection of population expansion was performed at the level of the individual populations (data not shown). The mismatch distribution analysis of the complete dataset, haplogroups A, B, C, D, and fifteen Tibetan sheep populations of the mtDNA D-loop, is presented in Fig. 4 and Fig. E in S1 File. Neutralitytests (Ewens-Watterson test, Chakraborty's test, Tajima's D test, Fu's FS test) were used to detect population expansion (Table G in S1 File). The complete dataset of all Tibetan sheep populations had a significantly large negative Tajima's D value and F_s value (Tajima's D = -0.466, P = 0.020; F_S = -7.484, P = 0.001). This result was consistent with a demographic model showing two large and sudden expansions, as inferred from the mismatch distribution. The mismatch distribution of the complete dataset suggested that there were two major peaks with maximum values at 4 and 27 pairwise differences and two smaller peaks at 45 and 51 differences. These results suggest that at least two expansion events occurred in the population demographic history of the Tibetan sheep living on the Qinghai-Tibetan Plateau. The mismatch distribution analysis revealed a unimodal bell-shaped distribution of the pairwise sequence differences in haplogroups A, B, and C. However, the distribution of lineage D was a sambong function. Mismatch analysis of haplogroups A, B, and C suggested that single population expansion events occurred in the demographic history of the Tibetan sheep populations. This finding was similar to the previously reported results [29].

5 CONCLUSION

China holds abundant populations of Tibetan sheep, with significant mtDNA haplotype diversity observed in the sheep of the Qinghai-Tibetan Plateau areas. Here, the large-scale mtDNA D-loop sequences analysis of fifteen Tibetan sheep populations has provided evidence for four maternal haplogroups with high diversity. Phylogenetic analysis showed that all four previously defined haplogroups (A, B, C, and D) could be identified in the 636 tested individuals of the fifteen Tibetan sheep populations, although the D haplogroup was only found in the Linzhou sheep. The estimation of demographic parameters from the mismatch analyses shows that haplogroups A, B, and C had at least one demographic expansion in the Tibetan sheep of the Qinghai-Tibetan Plateau areas.

SUPPORTING INFORMATION

S1 File. The UPGMA phylogenetic tree show that the 490 sequences of 15 Tibetan sheep populations (Fig. A). The UPGMA phylogenetic tree show that the 64 sequences of 12 Tibetan sheep populations (Fig. B). The UPGMA phylogenetic tree show that the 81 sequences of 14 Tibetan sheep populations (Fig. C). Median-joining networks for the mtDNA D-loop in the control region show that 636 sequences of 15 Tibetan sheep populations and six reference breeds fall into five distinct cluster haplogroup A, haplogroup B, haplogroup C, haplogroup D and haplogroup E, respectively. The majorities of the 490 individuals were grouped in haplogroup A, followedby haplogroup B (64) and C (81); however, only one animal from LZ03 belonged to haplogroup D. The AW population was composed of one haplogroup, the DM population was composed of two haplogroups, and the remaining 13 Tibetan sheep populations were composed of three haplogroups (Fig. D). The mismatch distribution of the complete dataset of the mtDNA types of Tibetan sheep of the four lineages on the Qinghai-Tibetan Plateau areas showed that there were two major peaks, with maximum values at 4 and 27 pairwise differences and two smaller peaks at 45 and 51 pairwise differences (Fig. E). Mitochondrial genomes of the 6 reference breeds included in the phylogenetic analyses in this study (Table A). The length of the complete mtDNA D-loop sequence in fifteen Tibetan sheep populations (Table B). Base pair composition of mtDNA D-loop of fifteen Tibetan sheep populations (Table C). Gene flow (Nm) of the sequence (above the diagonals) and Nm of the haplotype (below the diagonals) between fifteen Tibetan sheep populations (Table D). Dxy (the average number of nuc. subs. per site between populations) (above the diagonals) and Da (the number of net nuc. subs. per site between populations) (below the diagonals) of the difference in the number of nucleotides per site differences between fifteen Tibetan sheep populations (Table E). Hierarchical analysis of the molecular variance (AMOVA) of the D-loop region of mtDNA for fifteen Tibetan sheep populations (Table F). Neutrality tests for fifteen Tibetan sheep populations (Table G).

(XLSX)

ACKNOWLEDGMENTS

This work was supported by special fund from the Major International（Regional）Joint Research Project（NSFC-CGIAR 31461143020）, Gansu Provincial Funds for Distinguished Young Scientists（1308RJDA015）, Gansu Provincial Natural Science Foundation（145RJZA061）, and Gansu Provincial Agricultural biotechnology research and application projects（GNSW-2014-21）, and the Central Level, Scientific Research Institutes for Basic R&D Special Fund Business（1610322012006, 1610322015002）.

AUTHOR CONTRIBUTIONS

Conceived and designed the experiments: JBL XZD YFZ YJY XG TTG MC. Performed the experiments: FW JLH RLF XPS CEN JG. Analyzed the data: XG CY. Contributed reagents/materials/analysis tools: JBL XZD YFZ. Wrote the paper: JBL XZD YFZ BHY.

（发表于《PLOS ONE》，院选SCI，IF：3.057. DOI: 10.1371/journal.pone.0159308 July 27, 2016）

Evaluation of Crossbreeding of Australian Superfine Merinos with Gansu Alpine Finewool Sheep to Improve Wool Characteristics

LI Wenhui[1,*], GUO Jian[2], LI Fanwen[1], NIU Chune[2]

(1. Gansu Provincial Sheep Breeding Technology Extension Station, Huang Cheng town, Sunan, Gansu, China; 2. Lanzhou Institute of Husbandry and Pharmaceutical Sciences, CAAS, Jiangouyan Street, Lanzhou, China)

Abstract: Crossbreeding of Australian Superfine Merinos (ASMs) with Gansu Alpine Finewool (GAF) sheep and an evaluation of the potential benefits of this genetic cross has not been previously conducted. 13 ASMs were crossbred with GAF sheep over a five year period with backcrossing designed to assess heterosis. Data from 11,178 lambs sired by 189 rams were used in the study. Genotype, birth year, birth type, dam age, sex and/or management group, and record age were fitted as fixed effects and within-genotype sire fitted as a random effect. Crossbreeds of 1/2 ASM expressed the most desirable effects for improving average fiber diameter (AFD), clean fleece weight (CFW), yield, coefficient of variation of AFD (CVAFD), yearling staple length (YSL) to AFD ratio (YSL/AFD), and CFW to metabolic yearling bodyweight ($YWT^{0.75}$) ratio ($CFW/YWT^{0.75}$) but showed the least post-weaning average daily gain (powADG) and YWT. Genotype of backcrossing with 1/4 ASM obtained moderate improvements in AFD, CFW, CVAFD, and YSL/AFD but the highest YSL, WWT, and prwADG. Except for yield (−1.42%) and $CFW/YWT^{0.75}$ (−1%), heterosis estimates were generally low and positive, and ranged from 0.1% for CVAFD to 4% for powADG, which indicates the potential to improve relevant traits through exploiting heterosis to a varying extent. The ASMs sampled in this study were found to be superior to GAFs for AFD, CFW, yield, and CVAFD by 19.82%, 11.68%, 14.47%, and 6.99%, respectively, but inferior for YSL, PowADG, and YWT by 4.36%, 50.97%, and 16.93%, respectively. ASMs also appeared to be more efficient than GAFs in clean wool production (25.34%) and staple length growth (16.17%). The results of our study strongly suggest that an infusion of ASM genes via cross-breeding is an effective and appropriate approach to improve wool microns and wool production from GAF sheep, and we make recommendations to tackle the undesirable traits of YWT and YSL from ASM introduction.

---

[1] These authors contributed equally to this work.
[*] Corresponding author, E-mail: whleee2004@163.com

1 INTRODUCTION

Because of consumer demand for lighter weight fabrics and the severe challenge of competition from other synthetic fabrics, selective breeding of superfine wool sheep has been one of the most prevalent practices in the global fine wool sheep industry over the past 25 years. The significant achievements made in super and ultra-fine wool sheep breeding resulting from breeding programs like 'T13' [1] have put Australia in firm first place in the global fine wool sheep industry, and has resulted in the Merino enterprise producing wool in a more efficient manner. Although the total wool production figure in Australia decreased from 817, 454 tons in 1991/92 to 365, 561 tons in 2014/15, the production of superfine (≤18.5μm) increased 192%, and the proportion of superfine (≤18.5μm) wool production increased from 3.96% in 1991/92 to 25.87% in 2014/15 [2]. Adequate superfine/ultrafine wool sheep genetic resources have been made available due to these achievements in Australia. Meanwhile, there is practical and imperative need to develop superfine wool sheep breeds in China while maintaining adaptability to the environment. However, the challenge comes from the dearth of domestic superfine wool sheep genetic resources, and genetic progress in within-breed selection for reduced wool microns is known to have been slow.

Introducing Australian Merino genes into fine wool breeds is one of the dominant practices in fine wool sheep breeding in China. Notable numbers of Australian Merinos were imported in 1972, which was supposed to contribute to the development of Xingjiang Finewool sheep, and between 1984 and 1986, when 416 head (including 8 rams donated by the Australian government as part of the Australian Centre for International Agricultural Research project and used in the station of this study) were imported into China [3]. These genetic resources from Australia have made a significant contribution to the fine wool sheep gene pool in China, and previous researches on this issue have demonstrated that the introduction of the Australian Merino into different fine wool sheep breeds in China [4-5, 7] and in America [6] has improved fleece weight, staple length, and clean fleece yield. During that period of time, imported Australian Merino rams were mainly of medium and strong wool type, and the goal of the crossbreeding them was to improve wool production while less attention was paid to improving wool fineness [4-5, 7].

There is potential for Australian Superfine Merino (ASM) genetic resources to be used to improve the fine wool sheep breeds in other countries through crossbreeding. In a previous study, the Australian Merino involving rotational crosses were used to produce superfine wool [8]. Limited reports have been made available on the impact of introducing ASM genes into the local fine wool sheep breeds gene pool in China. For the Merino industry, the likely benefits of crossbreeding are in the expression of heterosis and a wider use of genetic resources [9]. However, there needs to be caution when introducing new breeds since, while they may excel in a desirable trait, they are often inferior in other characteristics that contribute to overall merit of a sheep enterprise [10], and experiments designed to show which breeds and what methods of utilizing the better ones is one of the key requirements as to develop a more efficient animal industry while most effectively using world breed resources [11].

The Gansu Alpine Finewool (GAF) sheep is a dual purpose sheep breed that was developed in the high and cold Qilian mountainous pasture, Gansu province, China, where the altitude is between 2 600m to 3 500m above sea level. It was formally identified by the Gansu Provincial government in 1980. There are approximately 1.2 million head of this breed in China, mainly distributed in the Sunan and Tianzhu counties of Gansu province [12]. It is one of the most important fine wool sheep breeds in China and has underpinned the progressive development of that sector of the sheep industry in China for the past over three decades.

A program aimed at selecting superfine strain in GAF sheep commenced in the late 1990s. A hypothesis was made that introducing ASM genes into the GAF sheep gene pool would improve fleece quality, especially wool characteristics by affecting wool fiber diameter more effectively than within-breed selection. When carrying out a crossbreeding program, the expectation is, firstly, the infusion of new genes will change the genetic make-up of the existing population to the desirable phenotypic characteristics in the long term, and secondly, to utilize heterosis in the short term. It is necessary to differentiate the contribution of heterosis from that of the true breed genetic difference. Few articles in the literature have covered the knowledge in this field and previous studies have only evaluated the effects of crossbreeding of strong and medium Australian Merinos with Chinese local finewool sheep by way of comparing the performance of the crossbred progeny with the purebred local ones [4, 5]. Nevertheless, estimates on heterosis in the crossbred progeny in these studies were not made for some reason.

13ASM rams were imported from Australia for an attempt to improve wool fineness and other characteristics by crossbreeding with GAF sheep, which commenced in 2005. The objective of this study is to evaluate the comprehensive effects of introducing ASM genes into the GAF sheep gene pool by crossbreeding. We test the hypothesis that the introduction of ASMs is an alternative to remedy the dearth of superfine Merino genetic resources in the Chinese domestic fine wool sheep population gene pool and give possible recommendations on the way that ASMs should be used in China.

2 MATERIALS AND METHODS

2.1 Project site

The project was carried out at Gansu Provincial Sheep Breeding Technology Extension Station located on the northern slopes of the Lenglong summit of the Eastern segment of the Qilian mountains, approximately 37°53′N and 101°45′E, Gansu province. It is a typically cold semi-arid alpine environment at an altitude of 2 600 to 3 500m. The station possesses the nucleus flock of 12 000 GAF sheep (relative to the whole 1.2 million population of the breed). Breeding, selection, and dissemination of high genetic merit animals from this nucleus is the core role the organization plays in the sheep industry in the Northern sector of China.

GAF sheep were developed by grading local Mongolian and Tibetan ewes with Xingjiang fine wool and Caucasian fine wool rams followed by self-breeding within F2 and/or F3 progeny flocks and then many years of selection. Since 1980, when the breed was formally recognized by the government of the province, continuous efforts have been made to improve growth, wool production, and quality while maintaining their adaptability to the high-and-cold environment in that area, mainly through within-breed selection and introduction of exotic breeds. On a number of occasions during 1980 to 2000, Merino-type rams were introduced into the nucleus population raised at the station: Australian medium and strong-wool Merino rams in 1986, New Zealand Merino rams in 1989, Xinjiang fine wool rams in 1992, and Chinese Merino rams from Inner Mongolia in 1995. These Merino gene infusions have played a singularly important role in the comprehensive genetic improvement of GAF sheep regarding wool production and quality performance.

2.2 Crossbreeding program

The crossbreeding program was conducted during 2005 to 2011 at Gansu Provincial Sheep Breeding Technology Extension Station. Thirteen ASM rams selected from two stud farms in Victoria were imported in 2005. Their average wool fiber diameter (AFD) was 14.88 μm (13.6 μm–16.2 μm) when they were one year old. The average clean fleece weight (CFW) recorded in 2006 at the first shearing in China was 5.42 kg (3.8–7.33 kg), the average clean fleece yield (yield) was 57.21% and the AFD was 18.36 μm; the above mentioned traits were superior to GAF rams of the same age, although their average body weight was 84.31 kg, which was 13.23 kg lower. Starting in 2005, crossbreeding was carried out between pure ASM rams and GAF ewes for four consecutive years and F1 crossbreeding progeny (AG) were obtained from four drops between 2006 and 2009. Ram lambs from F1 AG were selected from these drops and reared for breeding purposes from which 46 rams were backcrossed with purebred GAF ewes so that backcrossing progeny (AGG) were obtained during years 2008 to 2010. Meanwhile, GAF purebred breeding (GG) were routinely carried out during 2005 up to 2009 as control group. Information on a total of 11,178 progeny sired by 189 rams were involved in the study are shown in Table 1, which shows the investigated traits together with the associated data structure and the number of animals involved.

2.3 Sheep management and traits measurement

The experimental ewes were from six ewe breeding flocks (BFs), the size of which were approximately 750 except for one which had only 500. The mobs of ewes were grazed on similar pastures and the same supplementary feeding and management system was applied. Trans-cervical fresh semen artificial insemination (AI) was applied once a year between November 20th and December 10th, after which the back-up rams routinely joined with the ewes for another 20 days. At commencement of the AI program, vasectomized teaser rams were used twice daily (morning and afternoon) for estrus detection. The estrus ewes were drafted out and allocated to mating groups of 15–20 ewes. AI was conducted each morning and individual ewes were inseminated once per day on two consecutive days.

The corresponding lambing period took place from mid-April and continued until the end of May. Lambs were weighed at birth, ear-tagged and the pedigree recorded within 24 h of birth. Birth year (BY), sex, birth type (BT, single or twin) and dam age (DA, adult or maiden) were also recorded. Weaning and weaning assessment was conducted at an average age of 115 days, when weaning weight (WWT) was recorded. Pre-weaning average daily gain (prwADG) was calculated as the product of the difference between WWT and BWT divided by the number of weaning days. After weaning, animals were managed under post-weaning sex and/or management groups (SG: male1, male2, and female).

From approximately 4 500 male weaners, 70-90 ram lambs were selected each year as breeding rams and formed a special rearing flock (male1). These rams were supplementary fed from October to the following May inclusively (approximately 210kg of concentrate and 120kg of oat hay per head). A further 300-400 ram lambs were selected to form a common rearing flock (male2) and were also supplementary fed but at a lower rate than the special rearing group (approximately 46kg of concentrate and 50kg of oat hay per head from January to May). Approximately 2,100 out of 4,500 ewe lambs were selected each year as replacement ewes and were managed in two or three groups (female) with the same supplementary feeding regime as the common rearing ram lambs. Selection of both ewe and ram weaners were mainly based on weaning weight and visual scores of wool characteristics; surplus lambs were sold.

Yearling bodyweight (YWT) and yearling mid-side wool staple length (YSL, measured with a steel ruler to the nearest 0.5cm) were obtained on retained animals in the middle of June in the next year at yearling assessment when the animals were 13-14 months old. Post-weaning average daily gain (powADG) was calculated as the product of the difference between YWT and WWT divided by rearing days between weaning and yearling assessment. Mid-side wool samples were also taken randomly from all the yearling animals classed as 'top' grade and 'first' grade with clear identification in an attempt to keep similar numbers of animals in different subgroups at this time to test the yield, AFD, and coefficient of variation of mean fiberBWT, birth weight; WWT, weaning weight; prwADG, pre-weaning average daily gain; powADG, post-weaning average daily gain; YWT, yearling body weight; GFW, greasy fleece weight; CVAFD, coefficient of variation of average fiber diameter; YSL, yearling staple length; YSL/AFD, yearling staple length to average fiber diameter ratio; $CFW/YWT^{0.75}$, clean fleece weight to metabolic yearling body weight ratio; AG, ASMx GAF; AGG, AG x GAF; GG, GAF purebred.

Table 1 Number of animals and data structure for the AG, AGG, and GG Genotypes involved in the investigation over five years.

Genotype	AG				GG					AGG			Total
Birth Year	2006	2007	2008	2009	2006	2007	2008	2009	2010	2008	2009	2010	
BWT	1 371	1 290	1 438	594	1 510	1 649	1 241	485	221	222	397	760	11 178
WWT	1 176	1 137	1 095	478	846	1 360	971	361	150	155	305	491	8 525
prwADG	1 131	1 137	1 071	478	835	1 360	949	361	150	155	305	491	8 423
YWT	569	514	369	155	390	613	229	48	44	52	115	174	3 272
powADG	542	495	345	146	327	582	213	45	39	48	107	161	3 050
GFW	420	459	289	140	267	531	213	38	35	40	90	137	2 659
CFW	111	143	69	40	55	103	44	12	13	14	18	31	653

(continued)

Genotype	AG				GG					AGG			Total
Birth Year	2006	2007	2008	2009	2006	2007	2008	2009	2010	2008	2009	2010	
Yield	153	181	89	45	95	150	59	16	18	16	24	40	886
AFD	341	272	185	64	249	379	74	24	38	22	25	127	1 800
CVAFD	341	272	185	64	249	378	74	24	38	22	25	127	1 799
YSL	569	514	369	155	390	613	229	48	44	52	115	174	3 272
YSUAFD	328	247	174	56	231	360	71	24	36	20	22	118	1 687
CFW/YWT^{075}	104	133	65	35	51	98	43	12	12	14	17	29	613

doi: 10.1371/journal.pone.0166374.t001

diameter (CVAFD). Approximately 20 days after assessment, all the sheep were shorn and individual greasy fleece weight (GFW) recorded. CFW was calculated as the product of GFW multiplied by the yield. Two secondary traits were used in the investigation, they are respectively the ratio of clean fleece weight to metabolic yearling body weight (CFW/YWT$^{0.75}$) and staple length to average fiber diameter (YSL/AFD) ratio. Since CFW and YWT were recorded at different ages, they were initially adjusted to the same record age 435 days before calculating the final CFW/YWT$^{0.75}$ ratio.

All experimental and surgical procedures were approved by the Institutional Animal Care and Use committee, Lanzhou Institute of Husbandry and Pharmaceutical Sciences, People's Republic of China. All efforts were made in animal handling to minimize suffering during AI and other activities.

2.4 Statistical analysis

The data were analyzed using general linear model realized with ASReml-3 [13]. The statistical model included genotype (cross AG, backcross AGG, and purebred GG), SG, BY, BT, DA, and BF as fixed effects. BF was included in pre-weaning but excluded from post-weaning traits. Age at trait recording was fitted as a covariate for WWT, YWT, YSL, GFW, CFW, and YSL/AFD. The within-genotype sire effect was treated as a random effect. Sire effects were preliminarily included in the model and then excluded when the effects were found to be non-significant. The significance of the sire effect was analyzed by log-likelihood ratio testing with χ^2-twice the log-likelihood difference between the model without and the model with the sire effect. The significance of the random effect was tested at $P<0.05$ by comparing the differences in log-likelihood with values for a χ^2 distribution with three degrees of freedom. Two-way interactions among BT, SG, BY, DA, and BF were preliminarily fitted in the linear model and then those found to be non-significant were excluded. Under this general linear model, each individual effect was expected to function independently of the others.

Estimates of heterosis and true breeding value comparison between two breeds. Backcrossing was designed to estimate the heterosis of the two breeds. 46 F1 AG sires were used to mate with purebred GAF sheep. In the backcross, only half of the gene pairs involved a difference in breed of origin. Thus, heterosis expression is 50%. Accordingly, Heterosis (%) = 50 (2 x AGG/ (AG+GG) −1).

Based on the theory that if there is no heterosis (H) between the two breeds, the difference in the predicted breeding value between the two breeds is simply double the difference between the crossbred and purebred GAF sheep; in the current case, Breeding value difference=2 (AG/(H+1) -GG)

3 RESULTS

Predicted means and standard errors of five growth traits: BWT, WWT, prwADG, powADG, and YWT are presented in Table 2, while that of the traits representing wool characteristics: GFW, AFD, CFW, yield, YSL, CVAFD, and two derived relative traits YSL/AFD and CFW/YWT$^{0.75}$ are presented in Table 3. Overall means were adjusted to be single-born from adult ewes for the traits which were significantly influenced by BT and DA, and powADG and wool traits were additionally adjusted to be reared in the female post-weaning management group.

3.1 Non-genetic fixed effect

All growth traits were significantly influenced by BY, SG, BT, and DA ($P<0.01$) except that YWT was not significantly influenced by either BT or DA. BF had a significant influence on BWT, WWT, and PrwADG. BY and SG significantly influenced all five wool traits and two derived relative traits with the exception that SG had no significant effect on YSL/AFD in this investigation. BT had no significant effect on all wool traits except GFW. DA only had a significant effect on yield but not on other wool traits. As shown in Tables 2 and 3, age at the time of recording had a significant influence on WWT, YWT, GFW, CFW, and YSL/AFD. Results of all two-way interactions with their significant test are respectively presented in Table 2 for growth traits and in Table 3 for wool traits and two derived relative traits.

3.2 Effect of genotypes

Comparisons in Tables 2 and Table 3 of the traits in the progeny from three genotypes: ASM x GAF (AG), GAF purebred (GG), and AG x GAF (AGG) show the effects of crossbreeding. Except for BWT and GFW, all growth, wool, and relative traits were influenced significantly by genotype. Within-genotype sire, as a random effect, significantly affected all growth traits and wool traits involved in this study but not the two relative traits. We observed the highest WWT (24.99 ± 0.165 kg) and PreADG (184.08 ± 1.48 g/d) for the AGG group, which were significantly higher than the GG group for WWT and the AG and GG groups for PreADG. No significant differences between the AG and GG groups were observed for the WWT and PreADG traits. The lowest post-weaning traits: PowADG (32.11 ± 0.72 g/d) and YWT (34.86 ± 0.23 kg) were observed for the AG group, both of which were significantly lower than the GG and AGG groups. No significant differences were found between GG and AGG groups for PowADG and YWT.

The lowest AFD was recorded for the crossbreed AG group (16.00 ± 0.13μm) and purebred GG was the highest, while the backcross AGG group was in the middle. The three genotypes were significantly different from one another for AFD. For CVAFD, the AG group was observed to be the lowest and so the most desirable, while the GG group showed the highest BWT, birth weight; WWT, weaning weight; prwADG, pre-weaning average daily gain; powADG, post-weaning average daily gain; YWT, yearling body weight; AG, ASMx GAF; AGG, AG x GAF; GG, GAF purebred. Predicted overall means and means of fixed effects were adjusted to single-born from adult ewe for pre-weaning growth traits, while they were additionally adjusted to manage under female sex and/or management group for post-weaning growth traits. In addition to the above adjustments, weaning weight was adjusted to the average of 116.7 days of age.

Table 2 Predicted means (±standard error) for fixed effects plus the significance of any two-way interactions.

Source of variation	BWT(kg)	WWT(kg)	prwADG(g/d)	powADG(g/d)	YWT
Overall Mean	3.81(0.01)	24.67(0.09)	180.05(0.74)	37.90(0.46)	36.57(0.14)
Genotype(GT)	n.s.	***	***	***	***
AG	3.78(0.03)	24.60(0.12)ab	178.77(0.97)a	32.11(0.72)a	34.86(0.23)a
GG	3.82(0.02)	24.46(0.15)a	177.71(1.21)a	41.44(0.55)b	37.44(0.20)b
AGG	3.83(0.02)	24.990(0.165)b	184.08(1.48)b	39.72(1.15)b	37.38(0.29)b
Birth Year(BY)	***	***		***	***
2006	3.66(0.02)c	23.60(0.14)d	165.33(0.98)d	34.14(0.55)c	35.13(0.18)b
2007	3.89(0.02)a	25.33(0.11)a	184.10(0.89)ab	27.69(0.52)d	33.92(0.17)a
2008	3.88(0.02)a	24.97(0.15)b	184.74(1.22)a	42.86(0.82)b	38.47(0.24)d
2009	3.90(0.02)a	24.99(0.16)ab	181.38(1.39)bc	34.64(0.97)c	36.00(0.29)c
2010	3.70(0.03)c	24.29(0.22)c	180.26(1.80)c	49.30(1.17)a	38.64(0.34)d
Sex(Sex Group, SG)	***	***	***	***	***
male1	3.88(0.01)	25.26(0.11)	184.30(0.88)	118.83(0.91)a	66.41(0.28)a
male2				51.60(0.85)b	44.41(0.25)b
female	3.74(0.01)	24.08(0.10)	175.75(0.85)	37.90(0.46)c	36.57(0.14)c
BirthType(BT)	***	***	***	**	n.s.
single	3.82(0.01)	24.67(0.09)	180.05(0.74)	37.90(0.46)	
twins	2.92(0.03)	21.94(0.27)	163.17(2.23)	41.36(0.16)	
Dam Age(DA)	***	***	***	***	n.s.
maiden	3.36(0.02)	21.71(0.17)	157.68(1.29)	42.38(1.00)	
adult	3.82(0.01)	24.67(0.09)	180.05(0.74)	37.90(0.46)	
Birth Flock(BF)	***	***	***	—	—
Record Age	NA	116.7***	NA	NA	n.s.
Sire	***	***	***	***	***
BT × DA	**	n.s.	n.s.	n.s.	n.s.

(continued)

Source of variation	BWT (kg)	WWT (kg)	prwADG (g/d)	powADG (g/d)	YWT
BT × BF	**	***	***	—	—
BT × GT	n.s.	*	*	n.s.	n.s.
BT × BY	n.s.	***	***	n.s.	n.s.
DA × SG	n.s.	*	n.s.	n.s.	n.s.
DA × BY	n.s.	**	n.s.	n.s.	n.s.
SG × BY	**	*	n.s.	***	***
SG × BF	n.s.	*	*	—	—
BF × GT		***	***	—	—
BF × BY		n.s.	n.s.	—	—
GT × BY	n.s.	***	***	***	**

***P<0.001; **P<0.01; *P<0.05; n.s. not significant. Means with different levels within effects followed by the same letters are not significantly different (at P = 0.05). The same as Table 3.

doi: 10.1371/journal.pone.0166374.t002

Table 3 Predicted means (± standard error) for wool traits plus significance of two-way interactions of fixed effects.

Source of variation	GFW (kg)	AFD (Mm)	CFW (kg)	Yield (%)	YSL (cm)	CVAFD (%)	YSL/AFD (cm/pm)	CFW/YWT$^{0.75}$ (g/kg)
Overall Mean	3.79 (0.03)	16.95 (0.12)	2.03 (0.03)	54.94 (0.33)	10.16 (0.04)	21.13 (0.13)	0.618 (0.003)	130.71 (1.29)
Genotype (GT)	n.s.	***	*	***	*	*	***	***
AG	3.72 (0.04)	16.00 (0.13) a	2.09 (0.03) a	57.02 (0.44) a	10.04 (0.05) a	20.94 (0.19) a	0.644 (0.003) a	139.17 (1.50) a
GG	3.73 (0.04)	17.72 (0.14) c	1.95 (0.05) b	53.93 (0.40) b	10.13 (0.06) a	21.68 (0.22) c	0.592 (0.004) c	124.77 (1.86) b
AGG	3.78 (0.07)	16.93 (0.19) b	2.09 (0.08) ab	53.91 (1.05) b	10.37 (0.07) b	21.35 (0.33) ab	0.626 (0.007) b	129.32 (3.07) b
Birth Year (BY)	***	***	***	***	***	***	***	***
2006	3.33 (0.04) d	17.69 (0.13) ab	1.86 (0.04) b	56.90 (0.43) a	9.89 (0.05) a	23.33 (0.17) a	0.580 (0.004) de	125.68 (1.92) b
2007	3.46 (0.03) c	16.32 (0.13) d	1.92 (0.04) b	55.33 (0.36) b	9.93 (0.04) a	22.92 (0.17) b	0.620 (0.003) be	130.88 (1.70) b
2008	4.26 (0.04) a	16.11 (0.15) d	2.18 (0.05) a	51.72 (0.55) c	10.25 (0.06) b	19.14 (0.24) d	0.657 (0.006) a	134.30 (2.26) ab

(continued)

Source of variation	GFW (kg)	AFD (Mm)	CFW (kg)	Yield (%)	YSL (cm)	CVAFD (%)	YSL/AFD (cm/pm)	CFW/YWT$^{0.75}$ (g/kg)
2009	3.90 (0.06)b	17.11 (0.18)c	2.14 (0.07)a	58.05 (0.74)a	10.38 (0.08)b	20.88 (0.34)c	0.624 (0.009)b	138.63 (2.97)a
2010	3.74 (0.07)b	17.87 (0.17)a	1.92 (0.09)b	52.77 (1.00)c	10.20 (0.09)b	20.51 (0.29)c	0.586 (0.007)d	118.29 (3.71)c
Sex (Sex Group, SG)	***	***	***	***	***	***	n.s.	***
male1	6.76 (0.05)a	17.68 (0.33)a	3.20 (0.05)a	50.26 (0.61)a	10.63 (0.07)a	18.81 (0.34)a	0.584 (0.007)	125.50 (2.16)b
male2	3.98 (0.08)b	16.07 (0.37)c	2.03 (0.09)b	55.11 (0.52)b	10.15 (0.07)b	22.07 (0.67)b	0.583 (0.016)	112.84 (3.66)c
female	3.79 (0.03)c	16.95 (0.12)b	2.03 (0.03)b	54.94 (0.33)b	10.16 (0.04)b	21.13 (0.13)b	0.618 (0.003)	130.71 (1.29)a
Birth Type (BT)	**	n.s.	n.s.	n.s.	n.s.	n.s.	n.s.	n.s.
single	3.79 (0.03)	16.65 (0.06)	—	54.94 (0.33)				
twins	3.46 (0.07)	17.25 (0.21)	—	57.19 (1.20)				
Damage (DA)	n.s.	n.s.	n.s.	*	n.s.	n.s.	n.s.	n.s.
maiden	—	—	—	56.98 (0.81)				
adult	—	—	—	54.94 (0.33)				
Record Age	434.8***	NA	435.3**	NA	415.4***	—	415.0***	—
Sire	***	***	**	**	***	***	n.s.	n.s.
BT × GT	n.s.	n.s.	n.s.	**	n.s.	n.s.	n.s.	n.s.
BT × SG	n.s.	**	n.s.	n.s.	n.s.	n.s.	n.s.	n.s.
SG × BV	***	n.s.	**	***	***	**	n.s.	n.s.
SG × GT	*	n.s.	n.s.	*	n.s.	n.s.	***	n.s.
GT × BY	n.s.	n.s.	n.s.	n.s.	*	**	n.s.	n.s.

doi: 10.1371/journal.pone.0166374.t003

GFW, greasy fleece weight; CVAFD, coefficient of variation of average fiber diameter; YSL, yearling staple length; YSL/AFD, yearling staple length to average fiber diameter ratio; CFW/YWT 0.75, clean fleece weight to metabolic yearling body weight ratio; AG, ASMx GAF; AGG, AG x GAF; GG, GAF purebred. Predicted overall means and means of fixed effects were adjusted to manage under female sex and/or management group for all wool traits, adjusted to be single-born for GFW, and were additionally adjusted to be born from adult ewe for yield. In addition to the above adjustments, GFW, CFW, YSL, and YSL/AFD were respectively adjusted to averages of 434.8, 435.3, 415.4, and 415 days of age.

and the AGG group was in the middle. The two crossbred genotypes showed significantly improved CVAFD over the purebred GG group, and no significant differences were found between themselves. Although no significant differences were found between the three genotypes for GFW, the AG group produced 0.14kg (P<0.05) more clean fleece weight than the GG group. No significant differences were found either between AG and AGG, or between GG and AGG, for CFW. Highest yield (57.02 ± 0.44%) was observed for AG compared to GG and AGG while the latter two groups showed a similar yield. AGG realized the highest YSL (10.37cm ± 0.07cm, P<0.05) compared to AG and GG with no significant difference found between the latter two groups.

AG was observed to have the highest YSL/AFD, with AGG in the middle and GG the least, and were all significantly different from one another between the genotypes. The highest $CFW/YWT^{0.75}$ was recorded for AG, GG was the lowest, and AGG in the middle. AG showed significant higher $CFW/YWT^{0.75}$ than the other two groups, between which no significant difference was found for the trait.

The within-genotype sire group had a significant influence on all growth traits and wool traits involved in the study except for the two derived relative traits (shown in Tables 2 and 3). Genotype x BF significantly influenced BWT, WWT, and prwADG at the 0.1% significance level, and genotype x BY interaction had a significant effect on WWT (P<0.001), prwADG (P<0.001), powADG (P<0.001), YWT (P<0.01), YSL (P<0.05), and CVAFD (P<0.01). Genotype x BT had a significant effect on WWT (P<0.05), PreADG (P<0.05), and yield (P<0.01). The SG x genotype interaction was observed to have a significant influence on GFW (P<0.05), yield (P<0.05), and YSL/AFD (P<0.001).

3.3 Estimates of heterosis and comparison between breeds

From Table 4, we can see that desirable positive estimates of heterosis were observed for CFW (1.72%) and GFW (0.71%), while undesirable negative heterosis for yield (-1.42%) and positive but low heterosis for AFD (0.21%) and CVAFD (0.1%) were recorded in this study. Desirable positive estimates of heterosis for growth traits were observed, the magnitude of which ranged from 0.42% for BWT (the lowest) to 4% for powADG (the highest). Heterosis estimates were desirable and positive for trait YSL/AFD but undesirable and negative for trait $CFW/YWT^{0.75}$. Means of the purebred ASM population were estimated by excluding the extent of heterosis Heterosis (%) =50 (2 x AGG/ (AG+GG) -1), estimated means of purebred ASM=2 x AG/ (H+1) -GG, and, therefore, the true breed difference (ASM-GAF) =2 (AG/ (H+1) -GG). BWT, birth weight; WWT, weaning weight; prwADG, pre-weaning average daily gain; powADG, post-weaning average daily gain; YWT, yearling body weight; GFW, greasy fleece weight; CVAFD, coefficient of variation of average fiber diameter; YSL, yearling staple length; YSL/AFD, yearling staple length to average fiber diameter ratio; CFW/YWT 0.75, clean fleece weight to metabolic yearling body weight ratio; AG, ASMx GAF; AGG, AG x GAF; GG, GAF purebred.

Table 4 Estimation of heterosis and comparison of true breed difference.

Traits	Average of AG+GG	Means of AGG	Heterosis (%)	Estimated means of Purebred ASM	Difference (ASM-GAF)	
					Absolute	Relative (%)
Wool Traits						
AFD (μm)	16.86	16.93	0.21	14.211	−3.51	−19.82
GFW (kg)	3.72	3.78	0.71	3.661	−0.07	−1.77
Yield (%)	55.48	53.91	−1.42	61.738	7.81	14.47
CFW (kg)	2.02	2.09	1.72	2.172	0.23	11.68
YSL (cm)	10.09	10.37	1.39	9.685	−0.44	−4.36
CVAFD (%)	21.31	21.35	0.10	20.163	−1.51	−6.99
Growth Traits						
BWT (kg)	3.80	3.83	0.42	3.704	−0.11	−2.93
WWT (kg)	24.53	24.99	0.94	24.283	−0.18	−0.72
prwADG (g/d)	178.24	184.08	1.64	174.055	−3.65	−2.06
powADG (g/d)	36.77	39.72	4.00	20.315	−21.12	−50.97
VWT (kg)	36.15	37.38	1.70	31.106	−6.34	−16.93
Relative traits						
YSL/AFD (cm/pm)	0.618	0.626	0.647	0.688	0.096	16.17
CFW/YWT$^{0.75}$ (g/kg)	131.97	129.32	−1.004	156.39	31.62	25.34

doi: 10.1371/journal.pone.0166374.t004

from F1 crossbred performances. Hence, this allowed us to compare the true genetic differences between the two breeds (shown in Table 4). ASMs sampled in this study were superior to GAF sheep for AFD (−3.5μm : −19.82%), yield (7.81% : 14.47%), CFW (0.23kg : 11.68%), CVAFD (−1.51% : −6.99%), YSL/AFD (0.096 cm/μm : 16.17%), and CFW/YWT 0.75 (31.62 g/kg : 25.34%). However, ASMs appeared to have less YSL (−0.44cm : −4.36%) and GFW (−0.07kg : −1.77%) than GAF sheep. They also showed inferior BWT (−0.11kg : −2.93%), WWT (−0.18kg : −0.72%), prwADG (−3.65g/d : −2.06%), powADG (−21.12g/d : −50.97%), and YWT (−6.34kg : −16.93%) to GAF sheep.

4 DISCUSSION

4.1 The impact of ASM genotype level

The paucity of superfine Merino genetic resources is one of the key obstacles that hinders the transformation of the fine wool industry in China into a more sustainable and efficient state. Our study gave testimony to the hypothesis that introducing ASMs and crossing them with Chinese Merino-type fine wool sheep is an option to resolve the issue, and we have provided useful information on the way to introduce ASM genes into the GAF sheep gene pool. The best performance displayed by the crossbred AG genotype in the investigation of wool traits (AFD, CVAFD, Yield, CFW, YSL/AFD, and CFW/YWT$^{0.75}$) showed that this crossbreeding was mainly a success. However, undesirable results were observed for the PowADG and YWT traits. Previous research has reported that the crossbreeding of Australian Merinos with GAF sheep resulted in desirable improvement of CFW and yield [7], and similar improvements were also observed for CFW and yield with Xinjiang Finewool sheep [4] and Inner Mongolian Finewool sheep [5]. These researches also reported improvement for YSL with no significant decrease of YWT. An apparent reason for this disagreement may be that the Australian Merino rams used in the current study were superfine Merino stock while those used in the previous studies were of strong and medium wool-type [4, 5, 7]. On the other hand, the backcross genotype (AGG) with 1/4 ASM was generally observed to have the best performance in YSL among the three genotypes, and to be superior in WWT, PrwADG, AFD, and CVAFD over purebred GAF sheep, and similar PowADG, YWT, CFW, yield, and CFW/YWT$^{0.75}$ with GG genotype. We found an improving trend for AFD, CVAFD, YSL/AFD, and CFW/YWT$^{0.75}$ with increasing Australian Merino inheritance, which was also found in terms of yield and AFD when crossing Australian Merino with Rambouilleit sheep in a previous research [14].

The improvement of AFD brought about by the infusion of ASM genes is generally consistent with what was expected in the planning of the crossbreeding program. If we assume that there was no heterosis present in the crossbred progeny, the decrease of AFD will generally bring about a negative change in CFW and body weight of the animal because it is widely accepted that AFD is positively genetically correlated with CFW and body weight. Weighted means of genetic correlation between AFD and post-weaning body weight and adult body weight are 0.20 and 0.15, respectively, and that between AFD and CFW is 0.28 [15].

The positive genetic correlation between AFD and YWT could be a reasonable explanation for the inferior expression of growth traits in the crossbred genotype. As the inheritance of ASM decreases in the backcross AGG genotype, growth performance was considerably improved, which implies that the inferior performance of the growth traits in the crossbred genotype may, firstly, partially come from the additive genetic contribution of ASM genes, and, secondly, partially come from the adaptive inferiority of ASMs to the local environment or from the breed x environment interaction. However, the positive genetic correlation between AFD and CFW cannot explain the results found in this investigation that the crossbreeding not only improved AFD but also simultaneously improved CFW, especially in cases where YWT decreased.

Wool production is generally believed to depend on wool fiber density; skin area, which is closely related to body weight; and wool staple length. As there was a decrease in body weight and not much difference in staple length between the crossbred and purebred GAF sheep, the superior CFW realized in the AG genotype may be a consequence of the possibly increased fiber density, direct information on which we did not record in the current study.

In a previous study, researchers argued that the decrease in body weight (and hence the surface area) in animals with lower fiber diameter is too small to account for the increases observed in follicle density [16]. They concluded that selection for reduced fiber diameter may decrease live weight, and finer diameter may affect nutrient metabolism through two adaptations that tend to maintain fleece weight: an increase in follicle density and/or an increase in relative fiber length. In this investigation, with an increase of inheritance from the ASMs, the AFD decreased but relative fiber length increased, which is in conformity with previous findings [16].

The superior CFW/YWT$^{0.75}$ ratio observed in genotypes with ASM genes demonstrated that the Australian Merino is a more efficient wool producer than GAF, as has been similarly reported when Australian Merinos were crossed with the Polwarth breed [17]. However, what we have to keep in mind is that the economic advantage of a breed depends largely on the price advantage of finer wool compared to the advantage of the price of larger hoggets [17]. In these dry, cold, and harsh highland pastures, animals with relatively large body weight are more welcomed than small sized animals.

Except for two relative traits, significant within-genotype sire effects for most of the growth and wool traits implies that there is potential for improving growth and wool traits through exploiting variation amongst sire progeny groups while carrying out crossbreeding programs. These variations may come from the genetic differences between the individual sires, so more emphasis should be put on selecting breeding rams in a specific Merino production system.

Inclusion of the various fixed effects with their interactions in the model was mainly to allow for the assumption on the effect of genotype function independent of others. The significant year effect found in the study for all traits was apparently a reflection of varying climate and pasture conditions during the years of the study. The effect of SG mainly resulted from different retention rates at weaning and different rearing conditions during the post-weaning period. The results showed that the superior rearing conditions in breeding rams (male1) resulted in a notable decrease of variation of average wool fiber diameter since it may reduce the variation of fiber diameter along wool staple, especially when the rams received more supplementary feeding during dry and cold winter periods. On the other hand, significantly lower yields observed in the breeding ram group mainly resulted from the fact that these animals spent more time staying in or around the shed to get more supplementary feed, and so there was more chance of their wool becoming contaminated than the two other groups. The effect of DA and BT indicates that lambs born as twins and born to maiden dams were restricted in pre-weaning growth, but they can display compensatory growth post-weaning. Similar fixed effects were also reported in research based on the same population of GAF sheep [12].

4.2 Impact of heterosis and true breed difference

There is limited information on heterosis estimates for the crossing of Australian Merinos with Chinese Merino-type finewool sheep in the literature. The major challenge for estimating heterosis has been mainly because of the fact that the performance of the exotic Australian Merino purebred could not be obtained in the imported environment, and the backcrossing has not specifically designed to deliberately obtain heterosis between the two breeds. Attempts were made to evaluatethe true breed difference of Australian Merino and Xinjiang fine wool sheep assuming that the heterosis was similar with the estimates between strains or bloodline of Australian merino, which was obviously inadequate to give a precise evaluation result [4]. The heterosis estimates in our study generally agreed with the conclusion that heterosis for growth traits and wool production is usually in the range of 1% to 10% in crossbreeding Australian Merino strains or bloodlines [18]. The current estimates of heterosis are within the range of estimates between different strains of Australian Merino for traits YWT, yield, and CFW except that the estimates for AFD was much lower than in other studies (0.8% [19], 1.2% [20]). The negative direction of heterosis for yield (-1.42%) was also consistent with the results of these studies (-0.3% [19], -4.8% [20]). Our estimates of heterosis for GFW (0.71%) and WWT (0.94%) were much lower than that of the results obtained from crossbreeding Australian Merino with Polwarth sheep [17], which were respectively 2%-3% for GFW and 10% for WWT. Interestingly, the direction of heterosis estimates for the CFW/YWT$^{0.75}$ ratio in our study (-1%) is similar with that (-2%) previously reported [17].

Except for the post-weaning growth rate (4%), the heterosis estimates of all other traits are relatively low (-2%-2%). One of the reasons for this is that, before this crossbreeding program commenced in 2005, Australian Merino genes had already been infused into the gene pool of this nucleus population of GAF sheep on a number of occasions. For example, purebred strong and medium wool Australian Merino rams were introduced in 1986, and some Xinjiang Finewool rams with Australian Merino blood were introduced in 1992, and so on. These Australian Merino gene infusion into the GAF sheep gene pool may have brought the genetic distance between the two breeds closer. This argument is supported by the research results of crossbreeding Australian Merino with a number of finewool sheep breeds in Inner Mongolia where heterosis was found to be lower in genotypes to which Australian Merino genes had been previously introduced than in those to which the genes were being introduced for the first time [5].

As is well-known, crossbreeding strategy has been widely adopted to exploit heterosis in the short term for the improvement of the targeted sheep population. Our study implies that there is some potential in exploiting heterosis through crossbreeding ASMs with GAF sheep. More importantly, heterosis estimation provide information on which the true breed difference resulted from additive genetic effect can be calculated. A better knowledge of both superiority and inferiority of the exotic breed to the local breed will help design a more effective and reliable crossbreeding program. As heterosis cannot be passed on to the next generation by interbreeding, what a sheep breeder expects from the introduction of exotic breed genes in the long run is to benefit from the additive genetic effect of the exotic breed. The true breed differences will surely provide valuable information for designing crossbreeding programs between GAF and ASM sheep.

Wool fiber diameter and its coefficient of variation. The estimated breed superiority of ASM over GAF sheep sampled for AFD (-19.82%) and CVAFD (-6.99%) in this study implies that the improvements of the two traits in the genotypes with ASM inheritance were mainly resulted from the additive genetic effects of the infused ASM gene. The heterosis of the two traits were undesirable but very low, hence can be ignored. Previous research has demonstrated the coefficient of variation of fiber diameter is negatively genetically related to staple strength (-0.46 to -0.86 [21]; -0.52 [15]). Researchers have previously reported that the correlation between sire estimated breeding values for coefficient of variation of fiber diameter (CVAFD) measured at a hogget shearing and staple strength (SS) measured a year later was −0.61 [22]; CVAFD is generally viewed as an indirect indicator of SS, which is one of the most important wool quality traits but expensive to measure. The results suggest that infusion of ASM genes is an effective approach to improve wool fiber diameter and its coefficient of variation.

Wool production traits. CFW is the most important wool production trait to be improved in any Merino breeding strategy. In this study, both heterosis and the true breed differences contributed to changes in CFW and yield. However, the magnitude of superiority presented by ASMs for CFW (11.68%) and yield (14.47%) over GAF sheep were much higher than that of the heterosis expressed (1.72% and −1.42%, respectively, for CFW and yield). The superiority of ASMs for CFW is obviously attributed to its advantage expressed in yield compared with the GAF sheep, which was observed to have higher GFW (1.77%) than the ASMs.

Wool staple length. The ASMs sampled in the study were estimated to be shorter in wool staple length (−0.44cm, −4.36%) than purebred GAF sheep. Both breed genetic difference and heterosis (1.39%) contributed to the variation of YSL. The superiority of ASMs for AFD may be the cause to its inferiority in YSL, as AFD is known to have a positive genetic correlation with staple length (0.19[15]). Nevertheless, keeping a 1/4 level of ASM inheritance in a breeding strategy will allow optimal growth of wool staple length, as showed in the AGG genotype. On the other hand, the YSL/AFD ratio is widely used to express the relative growth of wool staple length in a breeding program focused on improving AFD. The estimated 16.17% superiority of ASMs for YSL/AFD ratio over GAF sheep implies that the former were more capable of keeping a relative higher growth in wool staple length while decreasing wool fiber diameter than pure bred GAF sheep.

Growth traits. The ASMs sampled in this study were estimated to be genetically inferior to purebred GAF sheep for growth traits. The magnitude of their true breed differences were low (−0.72% to −2.93%) for pre-weaning growth traits and relatively high for YWT (−16.93%) and post-weaning growth rate (−50.97%). Apparently, the low YWT estimates for ASMs in this study were the result of low post-weaning growth. The desirable heterosis estimated in the study can be a remedy to the undesirable breed effect when introducing ASM genes into the GAF sheep gene pool. Inferiority of ASMs for post-weaning growth estimated in this study may generally be due to the rule that the wool fiber diameter is positively genetically correlated with body weight-associated growth traits. The weighted means of estimates in the literature for genetic correlation between fiber diameter and post-weaning body weight and adult body weight are 0.2 and 0.15, respectively [15]. In our study, it may due to the low body weight of the original ASM rams imported. The significant influence of the within-genotype sire effect on the growth traits support the above argument. In the high-and-cold GAF sheep benefiting area, animals with higher body weight and growth are more welcomed than animals with smaller body weight. Accordingly, more emphasis should be placed on body weight selection when importing ASM rams for crossbreeding programs with GAF sheep.

Clean fleece weight to metabolic body weight ratio. $CFW/YWT^{0.75}$ has previously been used as an indicator of wool production efficiency [17]. In our study, ASMs were estimated to be 25.34%

more efficient in wool production than GAF sheep. The implication of the result is that the ASMs have the genetic makeup to produce finer, hence more valuable, wool with the same feed consumption for maintenance and production as GAF sheep. This characteristic is mostly required for the establishment of an economically efficient and eco-friendly Merino breeding and production system in China.

Precautions have to be taken while using the results derived from our study, since the ASM rams introduced might not be truly representative of Australian Superfine Merinos, and the imported rams were crossed with the GAF nucleus population raised in the environmental and management conditions in the Gansu Provincial Sheep Breeding Technology Extension Station. Therefore, further studies and evaluations are warranted to address this issue.

5 CONCLUSIONS

The results of this study provide useful information on the potential as well as the method of benefiting from the introduction of ASM genes into the domestic finewool breed gene pool in China. Crossbreeds of 1/2 ASM expressed the most desirable effect for improving AFD, CFW, yield, CVAFD, YSL/AFD, and CFW/YWT$^{0.75}$ but showed the least post-weaning growth rate and YWT. A backcross genotype with 1/4 ASM obtained a moderate improvement for AFD, CFW, CVAFD, YSL/AFD, and CFW/YWT$^{0.75}$ but the highest YSL, prwADG among the three genotypes, and similar YWT and powADG with purebred GAF sheep. There appeared to be a clear improving trend for AFD, CVAFD, YSL/AFD, and CFW/YWT$^{0.75}$ with the increase of ASM inheritance. Except for yield (-1.42%) and CFW/YWT$^{0.75}$ (-1%), the heterosis estimates were generally low and positive and ranged from 0.1% for CVAFD to 4% for powADG. There is potential to improve the relevant traits through exploiting heterosis to a varying extent.

Utilizing heterosis provides alternative remedies to traits like YSL, PowADG, and YWT to which the infusion of ASM genes resulted in undesirable changes. Heterosis estimates provide detailed information for an appropriate crossbreeding program design for the fine wool sheep selection context and allow us to balance the potential benefits from exploiting heterosis and genetic variation between the breeds.

ASMs sampled in this study were demonstrated to be superior to GAF in terms of AFD, CVAFD, yield, and CFW but inferior for YSL, PowADG, and YWT. We also demonstrated that they were more efficient than GAF in wool production and staple length growth. Our study strongly suggests that an infusion of ASM genes is an effective and appropriate approach to improve wool microns as well as wool production in GAF sheep. However, as large body size is more sought after in the finewool sheep industry in the semi-arid, high-and-cold environment in Gansu province, more emphasis should be placed on body weight selection when importing ASM rams and carrying out crossbreeding. In addition, we recommend that the first crossbreeding be carried out in the nucleus population of GAF sheep, and, therefore, replacement breeding rams with 1/2 ASM genotypes be intensively selected and provided for the multiplier flocks and commercial flocks in the whole industry.

Our study identified that the introduction of ASM rams is potentially an effective and viable approach to remedy the paucity of superfine Merino genetic resources while developing superfine Merino breeds in China. Because of the small number, hence the inadequate representation of rams sampled in this study, further research is warranted to verify the conclusions of this study.

ACKNOWLEDGMENTS

Contributions of the technical staff at Gansu Provincial Sheep Breeding Technology Extension Station on performing AI and data collection are acknowledged.

AUTHOR CONTRIBUTIONS

Conceptualization: WL JG.
Data curation: WL.
Formal analysis: WL JG CN.
Funding acquisition: WL JG FL CN.
Investigation: WL JG FL CN.
Methodology: WL JG.
Project administration: WL JG FL CN.
Resources: WL JG FL CN.
Software: WL.
Supervision: WL JG FL.
Validation: WL JG FL CN.
Visualization: WL JG FL CN.
Writing – original draft: WL JG.
Writing – review & editing: WL JG.

（发表于《PLOS ONE》，SCI，IF：3.057.）

Microwave-assisted Extraction of Three Bioactive Alkaloids from *Peganum harmala* L. and Their Acaricidal Activity against *Psoroptes cuniculi* in Vitro

SHANG Xiaofei[1], GUO Xiao[1], LI Bing[1], PAN Hu[1],
ZHANG Jiyu[1], ZHANG Yu[2, *], MIAO Xiaolou[1, *]

(1. Key Laboratory of New Animal Drug Project, Gansu Province, Key Laboratory of Veterinary Pharmaceutical Development of Ministry of Agriculture, Lanzhou Institute of Husbandry and Pharmaceutical Sciences of Chinese Academy of Agricultural Science, Lanzhou 730050, China;
2. Department of Emergency, Lanzhou General Hospital of PLA, Lanzhou 730050, China)

Abstract: Ethnopharmacological relevance: *Peganum harmala* L. is a perennial herbaceous, glabrous plant that grows in semi-arid conditions, steppe areas and sandy soils. It is used to treat fever, diarrhoea, subcutaneous tumours, arthralgia, rheumatism, cough, amnesia and parasitic diseases in folk medicines. In this paper, we aimed to develop a simpler and faster method for the extraction of three alkaloids from *Peganum harmala* L. than other conventional methods by optimizing the parameters of a microwave-assisted extraction (MAE) method, and to investigate the acaricidal activities of three compounds against *Psoroptes cuniculi*.

Materials and methods: After optimizing the operating parameters with the single factor experiment and a Box-Behnken design combined with a response-surface methodology, a MAE method was developed for extracting the alkaloids from the seeds, and a high-performance liquid chromatography was used to quantify these compounds. An in vitro experiments were used to study the acaricidal activities.

Results: The optimal conditions of MAE method were as follows: liquid-to-solid ratio 31.3 : 1 mL/g, ethanol concentration 75.5%, extraction time 10.1 min, temperature 80.7 ℃, and microwave power 600 W. Compared to the heat reflux extraction (HRE, 60 min) and the ultrasonic-assisted extraction (UAE, 30 min) methods, MAE method require the shortest time (10 min) and obtain the highest yield of three compounds (61.9 mg/g). Meanwhile, the LT 50 values for the vasicine (1.25 and 2.5 mg/mL), harmaline (1.25 and 2.5 mg/mL), harmine (1.25 and 2.5 mg/mL) and MAE extract (100 mg/mL) against Psoroptes cuniculi were 12.188 h, 9.791 h, 11.994 h, 10.095 h, 11.293 h, 9.273 h and 17.322 h, respectively.

Conclusions: The MAE method developed exhibited the highest extraction yield within the

---

* Corresponding authors, E-mail addresses: shangxf928@163.com (X. Miao).
© 2016 Elsevier Ireland Ltd. All rights reserved.

shortest time and thus could be used to extract the active compounds from *Peganum harmala* L. on an industrial basis. As the active compounds of *Peganum harmala* L., vasicine, harmalin and harmine presented the marked acaricidal activities against Psoroptes cuniculi, and could be widely applied for the treatments of acariasis in animals.

Key words: Microwave-assisted extraction; *Peganum harmala* L.; Alkaloids; Acaricidal activity

1 INTRODUCTION

Peganum harmala L. (Zygophyllaceae family) commonly known as 'Harmal, Harmel and Syrian rue', is a perennial herbaceous, glabrous plant that grows in semi-arid conditions, steppe areas and sandy soils. This plant is native to the eastern Mediterranean region and is widely distributed in Middle and East Asia and North Africa (Fig. 1). As a traditional medicine, seeds of *Peganum harmala* L. were used as powder, decoction, maceration or infusion for fever, diarrhoea, abortion and subcutaneous tumours and are widely used as a remedy for dolorous events in North Africa and the Middle East (Bellakhdar, 1997; Monsef et al., 2004). Bremner et al. (2009) also reported that this plant largely used in Spanish traditional medicine as anti-inflammatory agents. In China, the seeds and aerial parts of *Peganum harmala* L. (called as 'Luo Tuo Peng') were used to treat arthralgia and rheumatism, cough and excessive phlegm, amnesia, menoxenia, enteritis and dysentery by the Uygur and the Mongolian people. Meanwhile, due to the hypotoxicity of the seeds, after decocting with water or soaking in wine it also was used as antihelmintic and insecticide to control and treat the endoparasitic diseases and ectoparasitic diseases of both human and animals by local people, such as roundworm and mites (Chinese Pharmacopoeia Committee, 1998; Commission of Flora Reipublicae Popularis Sinicas, 2004; Mu and Azi, 2004; Li, 2005; Baolidebate et al., 2011). According to the traditional uses, it has been applied in the ointments with other medicines, which was used to treat rheumatic arthritis, mange, and has been approved by government (Chinese Pharmacopoeia Committee., 1998). In 1999, it officially adopted in the Uygur Medicine Volume of the Ministerial Drug Standards of the People's Republic of China (Wen et al., 2013).

Up to now, a series of alkaloids with the core frameworks of β-carboline and quinazolone have been obtained that were demonstrated to be the active compounds of this plant, such as harmine, harmaline and vasicine (Fig. 2). The modern pharmacology proved that these compounds have marked analgesic activity, anticancer, antiproliferative, anti-parasitological and antifungal activities, hypothermic properties and other activities (Zhao et al., 2010; Farouk et al., 2008; Sarpeleh et al., 2009; Nenaah, 2010). Meanwhile, the total alkaloid also presented the toxicity, the LD 50 and 95% confident limit of total alkaloid by i.p, i.v and p.o in mice are 144 mg/kg (131.7-157.9), 56 mg/kg (48.3-64.9 mg/kg) and 289 mg/kg (218.5-382.3 mg/kg) (Yang et al., 1998). Recently, people have developed the various methods to extract the alkaloids from Peganum harmala L., such as heat-reflux extraction, chloroform extraction, cation exchange resin extraction, enzyme extraction, mixing solvent extraction and ultrasound-assisted extraction. Due to the rigid tissue of seeds, the conventional extraction methods reported require more times and chemical reagents to extract the alkaloids from *Peganum harmala* L., and the extraction efficiency was low (Zhang et al., 1993; Liao, 2014; Li et al., 2005; Wang et al., 2002; Cheng et al., 2008; Liu et al., 2010). Therefore, to find a more efficient, better environmental protection and faster procession method with higher alkaloids yields still should be developed.

Fig. 1 A photograph of *Peganum harmala* L.

Now, MAE has been successfully applied to extract the active compounds from different plants (Omirou et al., 2009; Beejmohun et al., 2007). But up to now, no paper has been reported for the simultaneous extraction of the alkaloids with MAE method from *Peganum harmala* L. Moreover, these alkaloids have many marked anti-parasitological activities were reported, such as the insecticidal and antimalarial effects, antileishmanial (Goel et al., 2009; Mirzaie et al., 2007; Khaliq et al., 2009; Pragya et al., 2008). But the acaricidal activities against Psoroptes cuniculi for controlling animal acariasis have not studied. Considering the traditional uses on controlling parasitic diseases of *Peganum harmala* L., in this study we aimed to develop an efficient MAE method for the simultaneous extraction and determination of three bioactive alkaloids, harmine, harmaline and vasicine, and evaluate the acaricidal activities of these three compounds against Psoroptes cuniculi in vitro.

2 MATERIALS AND METHODS

2.1 Material, standards and reagents

The seed samples were collected in Sep. 2014 from Alxa Left Banner of the Inner Mongolia Autonomous region of China and were dried to constant weight at 60℃. The raw material was identified by Prof. Zhigang Ma, Pharmacy College of Lanzhou University, China. A voucher specimen of *Peganum harmala* L. with accession number 492 was submitted to the Herbarium of

Lanzhou Institute of Animal and Veterinary Pharmaceutical Sciences, Chinese Academy of Agricultural Science (Lanzhou, China). Analytical standard grade vasicine (purity ≥95%, No. 92951) was purchased from Sigma-Aldrich Company (St. Louis, USA). Harma-line (purity ≥98%) was obtained from Shanghai Tauto Biotech Company, and harmine (purity ≥98%) was obtained from Shanghai R&D Center for Standardization of Chinese Medicines (China). Ethanol (analytical grade) was purchased from Tianjin Guangfu Chemical Reagent Company (China) and acetonitrile (HPLC grade) were purchased from Fisher Scientific (England).

2.2 Microwave-assisted extraction

2.2.1 MAE experiment

The MAE experiment was carried out using a WX-4000 Microwave Workstation (Shanghai PreeKem Microwave Chemical Technology Co.) equipped with a temperature sensor, a pressure sensor, a digital control panel, six closed vessels. It operates at atmospheric pressure with a microwave frequency of 2 450 MHz. During extraction, the temperature, time and power can be controlled. Seeds of *Peganum harmala* L. (0.2 g) were mixed with 80% ethanol (v/v, 6 mL) in a vessel and irradiated with microwave power (600 W) at 80℃ for 8 min. After cooling, the solutions were transferred to a 10 mL volumetric flask and diluted to volume. Then, the supernatant (10 μL) was injected into the high-performance liquid chromatography (HPLC) system for quantitative analysis after filtering through a 0.45 mm filter. Finally, after drying the solution, the weight of dry total extract was measured.

2.2.1.1 Single factor experimental design. The single-factor experimental design was employed for the selection of the variable factors. Seeds of Peganum harmala L. (0.2 g) were treated under the different MAE conditions (ethanol concentration, liquid-to-solid ratio, extraction time, temperature and microwave power) (Table 1), and each experiment was performed in triplicate.

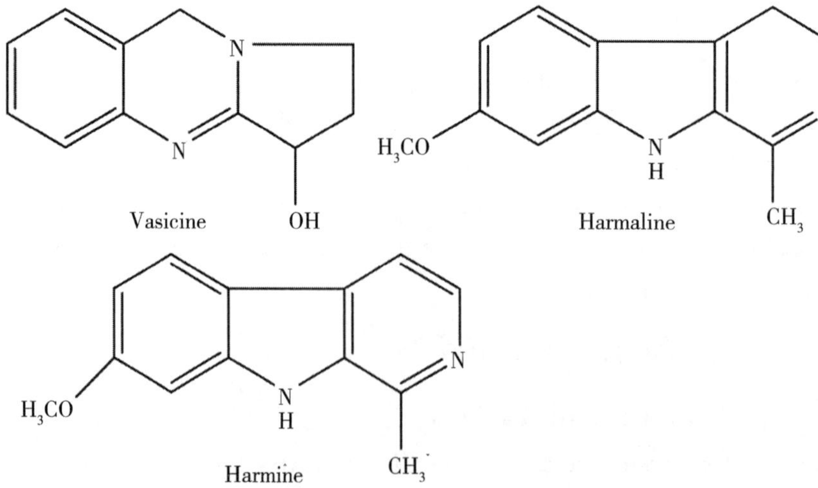

Fig. 2 Chemical structures of vasicine, harmaline and harmine from *Peganum harmala* L.

Table 1 Factors and levels of single factor experiment of MAE.

Factors	Levels				
Ethanol concentration (%)	20	40	60	80[a]	95
Liquid-to-solid ratio (mL/g)	10	20	30[a]	40	50
Extraction time (min)	4	6	8[a]	10	12
Temperature (℃)	[b]	40	60	80[a]	100
Microwave power (W)	200	400	600[a]	800	1 000

[a] The levels kept constant when other factors were investigated.

[b] 20℃ was not applied in the test due to the indoor temperature more than 20℃.

2.2.1.2 Box-Behnken design. In the present study, the Box-Behnken design was adopted for the response surface methodology (RSM) by Design Expert Version 9.0 software (Stat-Ease, Inc., U.S.A). Four main effective factors were selected as independent variables, and each had three levels, respectively. A total of 29 experiments were run in triplicate, and the average extraction yields of alkaloids were taken for statistical analysis.

2.2.2 Ultrasound-assisted extraction

Seeds of *Peganum harmala* L. (0.5 g) were mixed with 75% ethanol (v/v, 18 mL) in a flask for 30 min at 40 kHz in an ultrasonic apparatus (KQ-250 model, Kunshan Ultrasonic Equipment Co., Ltd., China). After extracting, the solution was filtered and transferred to a 25 mL volumetric flask to volume. Then, the supernatant (10 μL) was injected into the HPLC system for quantitative analysis after filtering through a 0.45 mm filter. Finally, after drying the solution, the weight of dry total extract was measured.

2.2.3 Heat reflux extraction

The sample powder (1.0 g) was placed in the reflux apparatus and extracted with 75% ethanol (v/v, 30 mL) on a water bath at 80℃ for 1 h. After extracting, the solution was filtered, transferred to a 50 mL volumetric flask and diluted to volume. Then, the supernatant (10 μL) was injected into the HPLC system for quantitative analysis after filtering through a 0.45 mm filter. Finally, after drying the solution, the weight of dry total extract was measured.

2.2.4 HPLC analysis

RP-HPLC analysis of the extract was performed on a Waters apparatus (two solvent delivery systems, model 600, and a Photodiode Array detector, model 996) using a gradient solvent system comprised of CH_3CN (A) and 0.1 M ammonium acetate buffer solution (B). The gradient condition was initially 4% A; a 15 min linear change to 8% A, a 30 min linear change to 10% A, and an 84 min linear change to 30% A at 0.7 mL/min. The on-line UV spectra were recorded at 280 nm. Data acquisition and quantification were performed using millennium 2.10 version software (Waters). A T3C-18 column (250 mm × 4.6 mm, 5μm, Waters, Ireland) was maintained at ambient temperature (30.0℃). The mobile phase was filtered through a Millipore 0.45 mm filter and degassed prior to use. The peaks were detected, and vasicine, harmaline and harmine were detected by comparison with chemical standards (Fig. 3).

2.3 Acaricidal activity in vitro

2.3.1 Collection of mites

Psoroptes cuniculi mites were isolated from the infested ear cerumen of naturally infected rabbits, and the adult mites (Psor-optes cuniculi) that were in good condition based on stereo-microscopic examination were collected and counted with a needle in Petri dishes for testing (Walton and Currie, 2007). After the collection of the materials, the rabbits were immediately treated with ivermectin. The experiments complied with the rulings of the Gansu Experimental Animal Center (Gansu, China) and were officially approved by the Ministry of Health, P.R. China in accordance with the NIH guidelines.

2.3.2 In vitro test

In this test, the in vitro acaricidal activities of MAE extract (MAEE, 100 mg/mL, w/v), vasicine, harmaline and harmine (2.5 and 1.25 mg/mL, w/v, respectively) were investigated. All drugs were diluted in 10% glycerin, and this solution was applied to the untreated group. The experiments were performed according to the methods of Macchioni et al. (2004). Specifically, one hundred μL of each drug were added to the culture plate, and the liquid excess was absorbed with filter paper. Next, 10 adult mites were collected from the infested ear cerumen with a needle and placed in each plate. All plates were incubated at 25℃ in 75% relative humidity. At 1 h, 3 h, 6 h, 9 h, 12 h, 18 h, and 24 h, the reactions of the mites were observed. Five replicates were performed for each extract concentration.

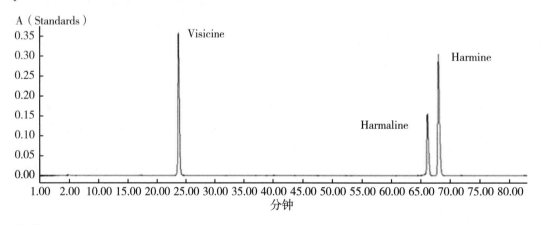

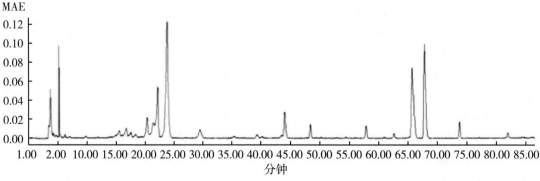

Fig. 3 The HPLC chromatograms of the standards (A), MAE, UAE and HRE.

2.4 Statistical analysis

The obtained data were analyzed with the SPSS software program version 18.0 and are expressed as the means ± SDs. The significance of each coefficient was determined in the regression model and the analysis of various (ANOVA) for the evaluation of the second-order model in the Box-Behnken Design. In the acaricidal tests, the data were analyzed with one-way ANOVAs followed by Student's two-tailed t-tests to compare the test and control groups. Tukey's tests were used for comparisons of three or more groups. P-values below 0.05 (Po0.05) were considered indicative of significance. The median lethal time (LT_{50}) was calculated using the complementary log-log (CLL) model.

3 RESULTS

3.1 Single factor experimental design

The effect of ethanol concentrations on the extraction yields of alkaloids is shown in Fig. 4A. When the ethanol concentration increases from 20% to 80%, the yields of the three compounds also increase and reach a maximum at 80%. Then, the yields decline at a concentration of 95%. Thus, an 80% ethanol concentration was selected for the following experiments.

As seen from Fig. 4B, the yields of three compounds increased with the extension of extraction time from 4 min, reaching a maximum between 8 min and 10 min. Further study showed that except for harmine, the yields at 8 min were greater than at 10 min, so for the other two compounds maximum yields were obtained at an extraction time of less than 10 min. Thus, an extraction time of 8 min was selected for the following experiments.

Meanwhile, as shown in Fig. 4C, the temperature had a strong effect on the yields of the three compounds, and the extraction yield of the alkaloids increased significantly with the increases of the temperature from 20 to 80℃. Although higher temperatures had a positive effect on the extraction yields, temperature cannot be increased indefinitely. Hence, a temperature of 80℃ was chosen for further optimization.

As we can see in Fig. 4D, as the liquid-to-solid ratio increases from 10 : 1 to 30 : 1 (mL/g), the yields of the alkaloids are significantly enhanced. However, when the liquid-to-solid ratio exceeds 30 : 1, the yields are not markedly increased. Therefore, considering water consumption, a liquid-to-solid ratio of 30 : 1 was selected for further investigation.

Finally, Fig. 4E illustrated the effect of the microwave power on the yields of alkaloids. The results showed that the along with the increase of the power, the yields of three compounds are somewhat improved, but without significant differences. When the microwave was 1 000 W, the components may be decomposing because of longer heating time at high temperature or other reasons. In some cases, high power levels have negative effects on extraction (Qiu et al., 2012; Eskilsson et al., 2000). Considering the yields of the three compounds and energy consumption, 600 W was selected for further investigation.

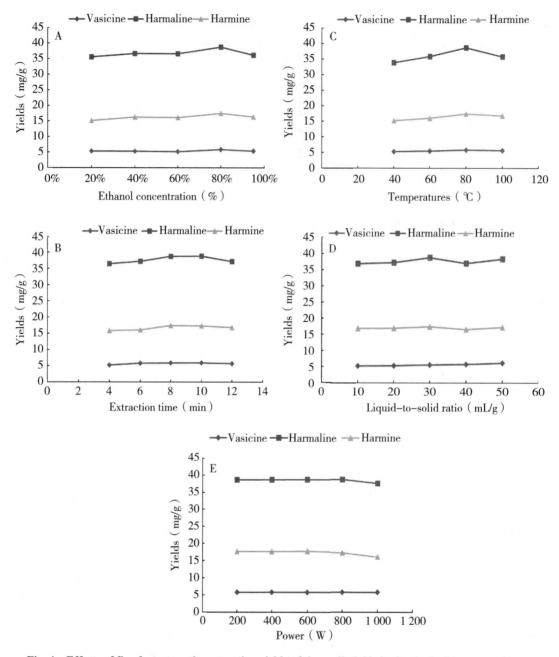

Fig. 4 Effects of five factors on the extraction yields of three alkaloids in the single factor experiment.

3.2 Analysis of the Box-Behnken design

According to the results of the single-factor experiment, the liquid-to-solid ratio (X_1), ethanol concentration (X_2), temperature (X_3) and extraction time (X_4) were selected as the independent variables for the Box-Behnken design. The design matrix of 29 runs and the responses were shown in Table 2. A second-order polynomial function was fitted to correlate the relationship of each independent variable to the response. The final equation in terms of actual factors is:

Vasicine=+0.491 34−0.041 723X_1+0.135 48X_2+0.001 173 33X_3+0.173 42X_4+0.000 561 667X_1X2−

0.000 033 75X_1X_3+0.007 087 5X_1X_4+0.000 048 333 3X_2X_3+0.001 341 67X_2X_4+0.001 631 25X_3X_4−0.001 130 17X_1^2−0.001 106 19X_2^2−0.000 140 354X_3^2−0.031 067X_4^2.

Harmaline=−39.492 99+0.809 07X_1+0.447 07X_2+0.754 63X_3+3.879 58X_4+0.00028X_1X_2−0.005 742 5X_1X_3+0.008 875X_1X_4+0.000 83X_2X_3−0.000 583 33X_2X_4+0.002 431 25X_3X_4−0.007 321 37X_1^2−0.003 620 85X_2^2−0.004 131 1X_3^2−0.222 39X_4^2.

Harmine=3.766 7+0.215 56X_1+0.078 734X_2+0.124 82X_3+0.338 67X_4−0.000 303 33X_1X_2−0.001 635X_1X_3+0.008 45X_1X_4+0.000 631 6X_2X_3+0.001 958 33X_2X_4−0.000 668 75X_3X_4−0.002 337 67X_1^2−0.000 918 407X_2^2−0.000 674 104X_3^2−0.032 879X_4^2.

As shown in Table 3, the regression models of three compounds are marked significant (P<0.01), and the lack of fit is not significant (P>0.05), which suggests that the three models are reliable. All of the R^2 values of the models were above 0.9, except for harmine, for which R^2 =0.891 9. The values for visicine and harmaline were 0.941 9 and 0.934 1, respectively.

3.3 Response surface analysis

To understand the interaction of the four factors as they affect the yields of the three compounds, the three-dimensional response surfaces of the model graphs and contour plots were calculated and were shown in Fig. 5, Fig. 6 and Fig. 7.

Table 2 Experiments conditions and the extraction yields for the Box-Behnken design.

Run	Factors				Extraction yield (mg/g)		
	X_1 (mL/g)	X_2 (%)	X_3 (℃)	X_4 (min)	Vasicine	Harmaline	Harmine
1	40	80	60	10	5.763	37.372	16.989
2	30	80	80	10	5.854	38.178	17.214
3	20	95	80	10	5.237	36.166	16.624
4	30	65	80	12	5.618	37.471	17.066
5	40	80	80	12	5.806	37.718	17.199
6	20	80	100	10	5.685	37.091	16.876
7	30	65	80	8	5.684	37.831	17.045
8	30	65	100	10	5.589	36.98	17.042
9	40	80	80	8	5.602	37.633	16.922
10	30	95	80	12	5.423	35.731	16.914
11	30	65	60	10	5.685	36.610	16.875
12	20	80	80	8	5.782	36.647	16.947
13	40	65	80	10	5.630	36.670	16.989
14	20	80	80	12	5.419	36.021	16.658
15	40	95	80	10	5.515	36.153	16.641
16	40	80	100	10	5.673	34.843	16.585

(continued)

Run	Factors				Extraction yield (mg/g)		
	X_1 (mL/g)	X_2 (%)	X_3 (℃)	X_4 (min)	Vasicine	Harmaline	Harmine
17	30	80	100	12	5.757	36.384	17.264
18	30	80	60	12	5.737	35.398	16.702
19	30	80	80	10	5.900	38.791	17.251
20	30	95	100	10	5.513	36.207	17.200
21	30	80	80	10	5.856	38.483	17.179
22	30	95	60	10	5.551	34.841	16.275
23	30	80	100	8	5.531	36.339	17.052
24	30	80	80	10	5.924	38.784	17.357
25	20	80	60	10	5.748	35.026	16.272
26	30	80	60	8	5.772	34.739	16.383
27	20	65	80	10	5.689	37.851	16.890
28	30	95	80	8	5.328	36.021	16.658
29	30	80	80	10	5.892	38.433	17.290

X_1: Liquid-to-solid ratio; X_2: Ethanol concentration; X_3: Temperature; X_4: Extraction time

Table 3　ANOVA of response surface quadratic model for extraction yields of vasicine, harmaline and harmine.

Source	F value	P value (Prob > F)
	Vasicine	
Model	16.62	<0.000 1**
X_1	4.46	0.053 0
X_2	42.78	<0.000 1
X_3	626	0.025 4
X_4	0.090	0.768 3
X_1X_2	8.27	0.012 2
X_1X_3	0.053	0.821 2
X_1X_4	23.40	0.000 3

(continued)

Source	F value	P value (Prob > F)
Vasicine		
X_2X_3	0.24	0.628 4
X_2X_4	1.89	0.191 2
X_3X_4	4.96	0.042 9
X_1^2	24.12	0.000 2
X_2^2	116.97	<0.000 1
X_3^2	5.95	0.028 6
X_4^2	29.16	<0.000 1
Lack of fit	4.95	0.068 5
R^2	0.9419	
Harmaline		
Model	14.18	<0.0001**
X^1	3.08	0.101 1
X^2	39.76	<0.000 1
X^3	3.76	0.072 9
X^4	1.02	0.330 1
X^1X^2	0.039	0.846 3
X^1X^3	29.15	<0.000 1
X^1X^4	0.70	0.417 4
X^2X^3	1.37	0.261 3
X^2X^4	0.006 767	0.935 6
X^3X^4	021	0.657 1
X_1^2	19.22	0.000 6
X_2^2	23.78	0.000 2
X_3^2	97.84	<0.000 1
X_4^2	28.36	0.000 1
Lack of fit	3.38	0.125 8
R^2	0.934 1	

(continued)

Source	F value	P value (Prob > F)
Harmine		
Model	8.25	0.000 2**
X^1	4.77	0.046 5
X^2	10.84	0.005 3
X^3	27.12	0.000 1
X^4	2.70	0.122 7
X^1X^2	0.086	0.773 7
X^1X^3	12.99	0.002 9
X^1X^4	4.09	0.062 6
X^2X^3	7.34	0.016 9
X^2X^4	0.71	0.415 0
X^3X^4	0.15	0.707 8
X_1^2	23.51	0.000 3
X_2^2	11.71	0.004 1
X_3^2	25.24	0.000 2
X_4^2	4.31	0.056 8
Lack of fit	5.35	0.060 1
R^2	0.891 9	

X_1: Liquid-to-solid ratio; X_2: Ethanol concentration; X_3: Temperature; X_4: Extraction time; ** Represents P<0.01.

Fig. 5 shows the response surface and contour plots for visicine. The yield of visicine was markedly affected by the ethanol concentration (P<0.01), followed by the factors of temperature, liquid-to-solid ratio and extraction time. From Fig. 5, we can see that the yield of visicine increased significantly when the ethanol concentration increased from 65% to 80%. However, the yield decreased when the concentration increased from 80% to 95%. Meanwhile, when the temperature was increased from 60℃ to 75℃, the yield of visicine was also increased. In addition, the yields reached their maxima when the liquid-to-solid ratio and extraction time approached 30 mL/g and 10 min, respectively. These results indicated that the optimized conditions for extracting visicine are a liquid-to-solid ratio of 30.1 mL/g, an ethanol concentration of 76.4%, and a temperature of 71.8℃ and an extraction time of 9.9 min

Fig. 6 displays the response surface and contour plots for harmaline. The yield of harmaline was significantly affected by the ethanol concentration (P<0.01), followed by the factors of temperature, liquid to solid ratio and extraction time. Fig. 6 shows that the yield of harmaline was a minimum when the temperature was 60 ℃ and the liquid-to-solid ratio was 20 mL/g. As the temperature and liquid to solid ratio increased, the yields were also markedly increased to more than 38 mg/g. At the same time, when the ethanol concentration was between 77% and 83%, the yield reached its maximum. Thus, the optimized conditions for the maximum yield of harmaline were a liquid-to-solid ratio of 31.3 mL/g, an ethanol concentration of 72.6%, and a temperature of 80.3 ℃ and an extraction time of 10.0 min

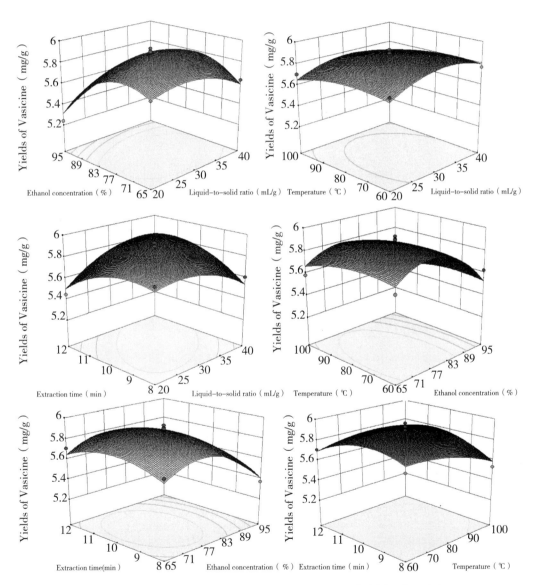

Fig. 5 A. 3D surface of model graphs of the yields of visicine (mg/g)

Fig. 7 shows the response surface and contour plots for harmine as affected by the four factors noted. The yield of harmine was markedly affected by the temperature and the ethanol concentration (P<0.01), followed by the liquid-to-solid ratio and extraction time. As shown in Fig. 7, the yield of harmine increased significantly when the ethanol concentration increased from 65% to 85%. However, when the concentration increased over 85% the yield was decreased. These findings indicated that an intermediate ethanol concentration provided the highest yield. In addition, Fig. 7 also showed that when the extraction time approached 10 min, the yield reached a maximum. However, with the extraction time was extended, the yield of harmine was decreased. It could be concluded that the optimized extraction conditions for harmine were at a liquid-to-solid ratio of 31.4 mL/g, an ethanol concentration of 77.1%, a temperature of 84.7℃ and an extraction time of 10.6 min

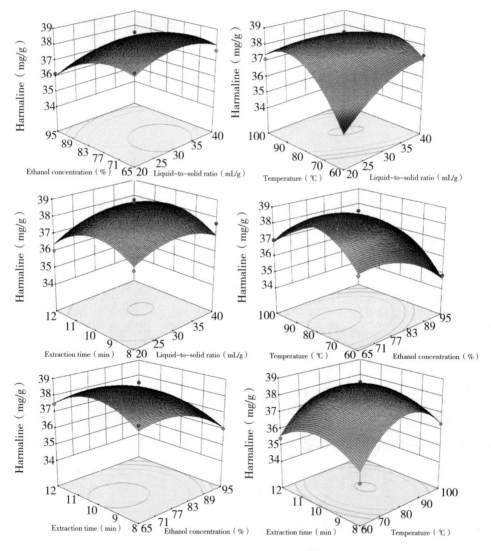

Fig. 6 3D surface of model graphs of the yields of harmaline (mg/g)

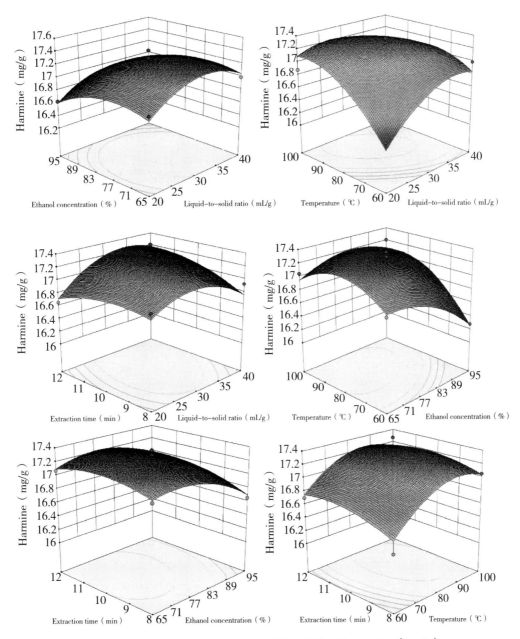

Fig. 7   3D surface of model graphs of the yields of harmaline (mg/g)

3.4 Optimization and verification experiment

In order to develop an optimum extraction condition for the three alkaloids simultaneously, the desirability function was employed using response surface analysis. The results indicated that the predicted optimal conditions for all three compounds are as follows: a liquid-to-solid ratio of 31.3 (mL/g), an ethanol concentration of 75.5%, a temperature of 80.7 ℃ and an extraction time of 10.1 min with the microwave power at 600 W as the single factor experiment. Next, the predicted optimal conditions were used to evaluate the yields of all three compounds. The results showed that the predicted values for the extraction yields of vasicine, harmaline and harmine were

5.8 mg/g, 38.8 mg/g and 17.2 mg/g, which were close to the experimental results (5.9 mg/g, 38.7 mg/g and 17.3 mg/g) and did not show significant differences. Thus, we thought that the predictive validity of the established models is confirmed.

3.5 Comparison of MAE with other two extraction methods

In the test, to evaluate the extraction efficiency of the MAE method that we developed, we compared and investigated the heat reflux extraction method and the ultrasound-assisted extraction method for extracting three compounds. Because microwave energy corresponds to electromagnetic waves could enhance the penetration of solvent into matrix and increase the release of bioactive compounds by destroying the cell wall of seeds within a short time (Chemat et al., 2013; Fabiano-Tixier et al., 2011; Liu et al., 2005), compared to the yields of vasicine (5.9 mg/g), harmaline (38.7 mg/g) and harmine (17.3 mg/g) obtained by the MAE method within 10 min, the yields of three compounds for the HRE method within 60 min and the UAE within 30 min were 5.3 mg/g, 32.2 mg/g, 14.7 mg/g; and 4.8 mg/g, 29.1 mg/g and 13.1 mg/g, respectively. And the total yield of three compounds was 61.9 mg/g, which were 1.2- and 1.3- fold higher than HRE and UAE ($P<0.01$). The latter two methods not only provided the lower extraction yields but also required much longer extraction times than did the MAE. Meanwhile, the HRE method could extract the most substance (542.5 mg/g), follows are MAE (525.8 mg/g) and UAE (493.2 mg/g). But compared to UAE and HRE, the contents of vasicine (1.1%), harmaline (7.4%) and harmine (3.3%) in the MAE extract are highest (Table 4). Taken together, the MAE method we developed is the best method for the extraction of alkaloids from Peganum harmala L. with the shortest extraction times and the highest yields.

3.6 Method validation

In our work, the RP-HPLC method was used to analyze the content of the three alkaloids in the plant. The results showed that using CH_3CN −0.1 M ammonium acetate buffer solution as a gradient solvent mobile phase, the three alkaloids are well-separated without apparent interference from other components in the sample within 80 min. The standard curves of three compounds are constructed by linear regression of peak areas versus nominal concentrations, and in the concentration range used; the chromatographic method presents good linearity (Table 5). Next, the intra-day variation and inter-day precision were investigated. After analyzing the same solution in triplicate at five different times within one day, the RSD of intra-day variation was in the range of 0.87%−2.55%. Meanwhile, after analyzing the same solution in triplicate for three consecutive days, RSD of the inter-day precision was in the range of 1.58%−3.45%. Finally, the recovery test showed that the recovery of the three compounds ranged from 94.78% to 104.83%. According to the above results, the HPLC chromatographic condition was accurate and applicable to the determination of the three alkaloids in the seeds of *Peganum harmala* L..

Table 4 Comparison of MAE with other extraction methods.

Extraction methods	Extraction Time (min)	Solvent	Temperature (°C)	Liquid-to-solid ratio (mL/g)	Yields of three alkaloids (mg/g)				Weight of total extract (mg/g)	Purity of three alkaloids in total extract (%)			
					Vasicine	Harmaline	Harmine	Total		Vasicine	Harmaline	Harmine	Total
MAE	10 min	75% ethanol	80 °C	30 mL/g	5.859 ± 0.019	38.708 ± 0.123	17.325 ± 0.111	61.892 ± 0.249	525.791	1.114%	7.362%	3.295%	11.771%
UAE	30 min	75% ethanol	25 °C[a]	36 mL/g	4.816 ± 0.020**	29.062 ± 0.113**	13.132 ± 0.061**	47.010 ± 0.139**	493.242	0.976%	5.892%	2.662%	9.531%
HRE	60 min	75% ethanol	80 °C	30 mL/g	5.327 ± 0.024**	32.240 ± 0.117**	14.725 ± 0.128**	52.292 ± 0.267**	542.532	0.982%	5.943%	2.714%	9.639%

[a] Room temperature, 25 °C; ** $P<0.01$ versus MAE.

Table 5 Retention times, regression equations, correlation coefficients and linear range for the three alkaloids.

Compounds	Retention times (min)	Regression equation	R^2	Linear range (ng)
Vasicine	23.392	y=3 348 479.181x−321 383.542	0.999	0.427−8.530
Harmaline	66.234	y=291 302.916x−21 106.292t	0.999	1.264−15.800
Harmine	68.044	y=641 665.382x+45 222.909	0.999	2.140−26.750

3.7 Acaricidal activity in vitro

The results revealed that three compounds and MAEE demonstrated a significant acaricidal effect against Psoroptes cuniculi. Specifically, compared to control group (8.0%), the mean mortalities at 24 h in the vasicine, harmaline and harmine groups of 2.5 mg/mL doses and MAEE of 100 mg/mL were 94.00%, 90.00%, 96.00% and 68% with significant difference, respectively. Meanwhile, after culturing 9 h with ivermectin, the mean mortality of positive drug group was 100% (Table 6).

Table 6 The acaricidal activity of MAEE, vasicine, harmaline and harmine against *Psoroptes cuniculi* in vitro.

Group(s)		Time (h, Mean mortality (%) ± SD)						
Drug(s)	Concentration(s) (mg/mL)	1 h	3 h	6 h	9 h	12 h	18 h	24 h
Vasicine	1.25	0.00 ± 0.00**	2.00 ± 4.47**	12.00 ± 4.47**#	32.00 ± 8.37**	48.00 ± 8.37**	76.00 ± 8.94**	84.00 ± 5.48*
	2.5	0.00 ± 0.00**	6.00 ± 5.48**	18.00 ± 8.37**	46.00 ± 5.48**	58.00 ± 8.37**	84.00 ± 5.48**	94.00 ± 5.48
Harmaline	1.25	0.00 ± 0.00**	2.00 ± 4.47**	6.00 ± 5.48**	36.00 ± 5.48**	48.00 ± 8.37**	80.00 ± 10.00**	86.00 ± 8.94
	2.5	0.00 ± 0.00**	8.00 ± 4.47**	12.00 ± 4.47**	46.00 ± 5.48**	56.00 ± 5.48**	86.00 ± 11.40	90.00 ± 7.07
Harmine	1.25	0.00 ± 0.00**	4.00 ± 5.48**	6.00 ± 5.48**	38.00 ± 8.37**	50.00 ± 10.00**	84.00 ± 5.48	90.00 ± 7.07
	2.5	0.00 ± 0.00**	10.00 ± 7.07**	12.00 ± 4.47**#	46.00 ± 8.94**	62.00 ± 8.37**	92.00 ± 8.37	96.00 ± 5.48
MAEE	100	0.00 ± 0.00**	2.00 ± 4.47**	6.00 ± 5.48**	8.00 ± 3.74**	30.00 ± 7.07**	58.00 ± 8.37**	68.00 ± 8.37**
Ivermectin	10	42.00 ± 8.37	72.00 ± 8.37	96.00 ± 5.48	100.0 ± 0.00	100.0 ± 0.00	100.0 ± 0.00	100.0 ± 0.00
Control		0.00 ± 0.00	0.00 ± 0.00	0.00 ± 0.00	0.00 ± 0.00	2.00 ± 4.47	6.00 ± 5.48	8.00 ± 8.37

* $P<0.05$. ** $P<0.01$ compared with mean mortality in positive drug group (ivermectin) at each corresponding time. # $P<0.05$. ## $P<0.01$ compared with mean mortality in control group at each corresponding group.

Then, further analyses indicated that they exhibited the great toxicity, and the LT 50 values for the vasicine (1.25 and 2.5 mg/mL), harmaline (1.25 and 2.5 mg/mL), harmine (1.25 and 2.5 mg/mL) and MAEE (100 mg/mL) were 12.188 h, 9.791 h, 11.994 h, 10.095 h, 11.293 h, 9.273 h and 17.322 h, respectively (Table 7).

Table 7 The LT_{50} values of MAEE, vasicine, harmaline and harmine against
Psoroptes cuniculi in vitro by CLL model.

Group(s)		Regression	LT_{50}(h) (95%FL)	Pearsonχ^2
Drug(s)	Concentration(s)			
Vasicine	1.25 mg/mL	Y=3.594x−3.903	12.188 (10.928−13.644)	0.565
	2.5 mg/mL	Y=3.581x−3.548	9.791 (8.741−10.922)	2.504
Harmaline	1.25 mg/mL	Y=3.984x−4.299	11.994 (10.830−13.297)	3.793
	2.5 mg/mL	Y=3.440−3.454	10.095 (8.993−11.303)	7.323
Harmine	1.25 mg/mL	Y=4.005x−4.217	11.293 (10.193−12.506)	7.926
	2.5 mg/mL	Y=3.789x−3.665	9.273 (7.122−11.782)	13.640
MAEE	100 mg/mL	Y=3.458x−4.283	17.322 (15.354−20.128)	6.169

4 DISCUSSION

Due to the unique heating mechanism, moderate cost, good performance and high efficiency, microwave-assisted extraction (MAE) has become a promising technique and has drawn significant attention in medicinal plant research for over ten years (Chan et al., 2011). In the present studies, an efficient MAE method for the simultaneous extraction and determination of three bioactive alkaloids, harmine, harmaline and vasicine were developed, which demonstrated the marked bioactivities and had the higher content than other alkaloids of Peganum harmala L.. The single factor experiment and the response surface methodological (RSM) test were used to optimize the extraction parameters, and the yields of the MAE method were compared to those of the conventional approaches. At the same time, HPLC method was also established for the simultaneous determination of the three bioactive compounds.

Because microwave energy corresponds to electromagnetic waves converted into heat within the material depending on its dielectric constant, and could enhance the penetration of solvent into matrix and increase the release of bioactive compounds by destroying the cell wall of seeds within a short time (Omirou et al., 2009; Beejmohun et al., 2007). Compared to the HRE (60 min), the UAE (30 min) methods and other conventional methods reported, the MAE methods only require 10 min and the simpler extraction procedure, and could decrease the decompose of the components caused by the longer heating time at high temperature. Then, 75% ethanol was proved to be the best choice to extract three compounds. And compared to chloroform and other chemical reagent with high toxicity applied in the conventional extraction methods, this solvent of MAE is relatively low toxicity and low cost, and is environmental-friendly. Of course, the yields of vasicine (5.859 mg/g), harmalin (38.708 mg/g) and harmine (17.325 mg/g) of MAE were highest compared to UAE and HRE ($P<0.01$). And total yield of three compounds was 61.892 mg/g, which were 1.184- and 1.317-fold higher than HRE and UAE ($P<0.01$), respectively. Meanwhile, the contents and purity of vasicine (1.114%), harmaline (7.362%) and harmine (3.295%) in the MAE extract are highest among of UAE and HRE, although the weight of the total extract from seeds is less than HRE and more than UAE. These results proved that compared to other extract methods, three alkaloids are easier to extract and purify by adopting MAE from *Peganum harmala* L.. Taken together, the MAE process we developed was validated to be fast and reliable with the highest yield, and the HPLC chromatographic conditions were suitable for determining the three alkaloids in the seeds of *Peganum harmala* L.

Today, along with the increasing harm for the productivity and the quality of animal products induced by animal acariasis, which is one of important veterinary skin diseases, and the increase resistance in target species, toxicity and environmental hazards induced by the overuse of chemical control drugs, to find new acaricidal medicine from natural plant has become to a main new way (Dagleish et al., 2007; Halley et al., 1993; O'Brien, 1999; Currie et al., 2004). And the literatures reported that herbal medicines and their active compounds presented the marked acaricidal activity and could be used as potential acaricidal agents, such as Eugenia caryophyllata, Eupatorium adenophorum, etc. (Ruan, 2005; Nong et al., 2012). In this paper, we firstly studied the acaricidal activities of the MAE extract and its active compounds, vasicine, harmalin and harmine. The results showed that compared to control group, three compounds and MAEE demonstrated the acaricidal effect against Psoroptes cuniculi. Among of them, due to the high concentration, the acaricidal activity of MAEE is little; harmine presented the best acaricidal activity with LT_{50} 9.273 h, followed by vasicine (9.791 h), harmaline (10.095 h) and MAEE (17.322 h) (Tables 6 and 7). These results indicated that as the potential acaricidal agents, vasicine, harmalin and harmine exhibited the marked toxicity against *Psoroptes cuniculi* and could be widely applied for the treatments of acariasis in animals. Meanwhile, due to effortless degraded in the environment, not remain in livestock and not as prone to resistance, and relatively safe for humans, animals, and the environment and other advantageous features, as an alternative medicine and adjuvant therapy with ivermectin *Peganum harmala* L could be used to treat animal's mange topically, and is worthy to study and develop. Then, due to the different acaricidal mechanism of both *Peganum harmala* L and chemical drugs, it could be used to kill the chemical drugs resistance of mites. The further studies should be conducted to study the structure-activity relationships associated with acaricidal activity and the acaricidal mechanism.

5 CONCLUSION

In a word, compared to the conventional extraction methods, the developed MAE showed the highest extraction yield within the shortest time. And the optimal conditions for the extraction yields of the alkaloids were as follows: liquid-to-solid ratio 31.316 : 1 mL/g, ethanol concentration 75.466%, extraction time 10.111 min, temperature 80.692℃, and microwave power 600 W. Meanwhile, the procedure of MAE is simpler and more economic, which could be used to extract the active compounds from *Peganum harmala* L. in industry. Microwave-assisted extraction method is better than other current arts for extracting the alkaloids of *Peganum harmala* L.. Moreover, MAE extract and vasicine, harmalin and harmine presented the marked acaricidal activities against Psoroptes cuni-culi, and as the potential acaricidal agents, they could be widely applied for the treatments of acariasis in animals.

AUTHORS' CONTRIBUTIONS

XS and XM conceived the study, XS, JZ and XG determined the index, XS, BL and YZ wrote the manuscript, and HP performed the statistical analyses. All authors read and approved the final version of the manuscript.

ACKNOWLEDGEMENTS

This work was financed by the National Natural Science Foundation of China (31302136), the Special Fund of the Chinese Central Government for Basic Scientific Research Operations in Commonwealth Research Institutes (No. 1610322014011). The authors would also like to express their gratitude to Lanzhou University PhD English writing foreign teacher Allan Grey who thoroughly corrected the English in the paper.

（发表于《Journal of Ethnopharmacology》，院选SCI，IF：3.055. 350–361）

Synthesis and Pharmacological Evaluation of Novel Pleuromutilin Derivatives with Substituted Benzimidazole Moieties

AI Xin[1], PU Xiuying[2], YI Yunpeng[1], LIU Yu[1], XU Shuijin[3], LIANG Jianping[1], SHANG Ruofeng[1, *]

(1. Key Laboratory of New Animal Drug Project of Gansu Province, Key Laboratory of Veterinary Pharmaceutical Development, Ministry of Agriculture, Lanzhou Institute of Husbandry and Pharmaceutical Sciences of CAAS, Lanzhou 730050, China; 2. College of Life Science and Engineering, Lanzhou University of Technology, Lanzhou 730050, China)

Abstract: A series of novel pleuromutilin derivatives with substituted benzimidazole moieties were designed and synthesized from pleuromutilin and 5-amino-2-mercaptobenzimidazole through sequential reactions. All the newly synthesized compounds were characterized by IR, NMR, and HRMS. Each of the derivatives was evaluated in vitro for their antibacterial activity against Escherichia coli (E. coli) and five Gram (+) inoculums. 14-O-((5-amino-benzimidazole-2-yl) thioacetyl) mutilin (3) was the most active compound and showed highest antibacterial activities. Furthermore, we evaluated the inhibition activities of compound 3 on short-term S. aureus and MRSA growth and cytochrome P450 (CYP). The bioassay results indicate that compound 3 could be considered potential antibacterial agents but with intermediate inhibition of CYP3A4.

Key words: Pleuromutilin derivatives; Synthesis; Antibacterial activity; Inhibition; CYP3A4

1 INTRODUCTION

The emergence of multidrug-resistant bacteria has led to the reduction or loss of the curative effects of many available drugs and remains a global human threat with the potential for catastrophic consequences in the future [1]. Infections caused by drug-resistant Gram-positive organisms, such as methicillin-resistant Staphylococcus aureus (MRSA) and vancomycin-resistant enterococci (VRE), remain a critical issue because of not only the morbidity and mortality caused by these infections but also the associated economic cost [2]. To solve the drastic increase of pathogenic

---

*Corresponding author, E-mail: shangrf1974@163.com

Sample Availability: Samples of the compounds 2, 3 and 4a–i are available from the authors.

© 2016 by the authors; licensee MDPI, Basel, Switzerland. This article is an open access article distributed under the terms and conditions of the Creative Commons Attribution (CC-BY) license (http://creativecommons.org/licenses/by/4.0/).

bacterial resistance, especially multiresistant bacteria, there is a pressing need to develop novel antibiotics against these dreadful pathogens.

The natural compound pleuromutilin (Figure 1) was first discovered and isolated in a crystalline form from cultures of two species of basidiomycetes, Pleurotus mutilus and P. passeckerianus in 1951 [3]. This compound has a modest antibacterial activity [4]. However, modification of the C-14 position of pleuromutilin may improve its biological activity and thus has led to three drugs: tiamulin, valnemulin, and retapamulin (Figure 1) [5-7].

Pleuromutilin derivatives selectively inhibited bacterial protein synthesis by binding to the ribosomes at the acceptor (A) and donor (P) site [8, 9]. Further studies demonstrated that the interactions of tricyclic core of the tiamulin are mediated through hydrophobic interactions and hydrogen bond, which are formed mainly by the nucleotides of domain V [10, 11]. The C-11 hydroxyl groups and the C-21 keto groups of pleuromutilin derivatives were located in a position suitable for hydrogen bonding to G-2505 phosphate and G-2061, respectively. Meanwhile, C-14 side chains pointed toward the P-site and minor hydrophobic contacts were formed with ribosomal components [11].

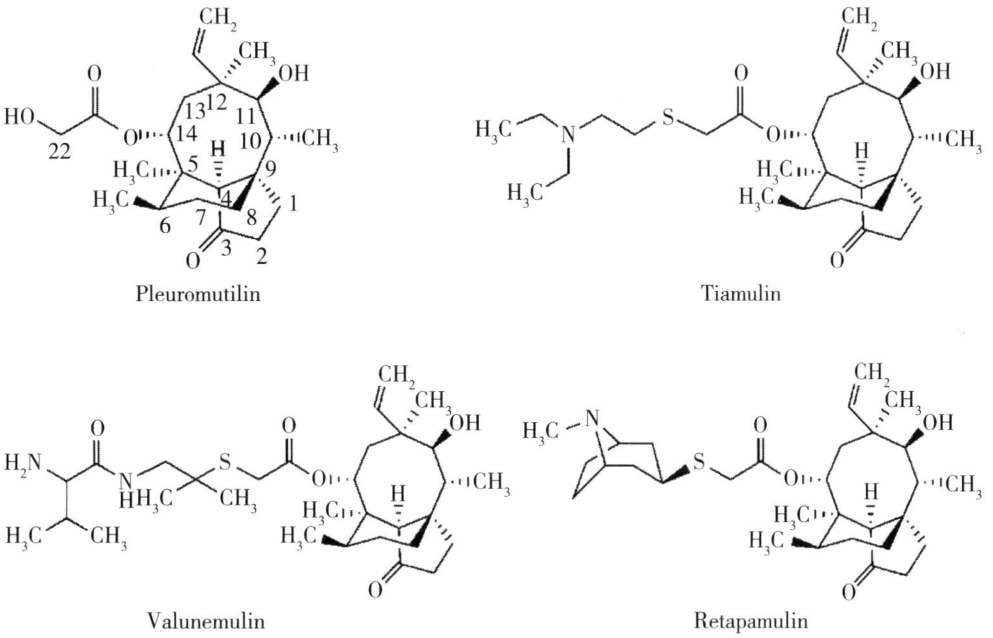

Figure 1 Structural formulas of pleuromutilin, tiamulin, valnemulin, and retapamulin.

Ling [12] and the present authors proposed that heterocyclic rings bearing polar groups at the C14 side chain of pleuromutilin derivatives improved their antibacterial activity [13-15]. Two classes of pleuromutilin derivatives with 1, 3, 4-thiadiazole and pyrimidine ring were reported and selected as a few compounds with excellent in vitro antibacterial activity against both sensitive and resistant Gram-positive bacterial strains. In this work, we describe the design, synthesis, and antibacterial studies for a novel pleuromutilin derivatives with a substituted benzimidazole moiety. This new series bears the fused heterocyclic ring at the C14 side chain of pleuromutilin derivatives instead of a single heterocyclic structure and thus are distinct from previous reported compounds.

2 RESULTS AND DISCUSSION

2.1 Chemistry

The general synthetic route to building the pleuromutilin derivatives is illustrated in Scheme 1. The lead compound 14-O-((5-amino-benzimidazole-2-yl) thioacetyl) mutilin (3) was prepared by nucleophilic substitution of 22-O-tosylpleuromutilin (2) with 5-amino-2-mercaptobenzimidazole under basic conditions with a 66% yield. Compounds 4a–l were directly obtained with a 51%–94% yield via condensation reactions between the amino group of compound 3 and the carboxyl group of carboxylic acids. The reactions were performed at room temperature in the presence of 1-ethyl-3-(3-dimethyllaminopropyl) carbodiimide hydrochloride (EDCI) and 1-hydroxybenzotriazole (HOBt), which was used to suppress racemization and improve the efficiency of the amide synthesis [13]. The synthesis and the IR, 1 H-NMR, and 13 C-NMR spectra of all the new compounds are reported in the Supplementary data.

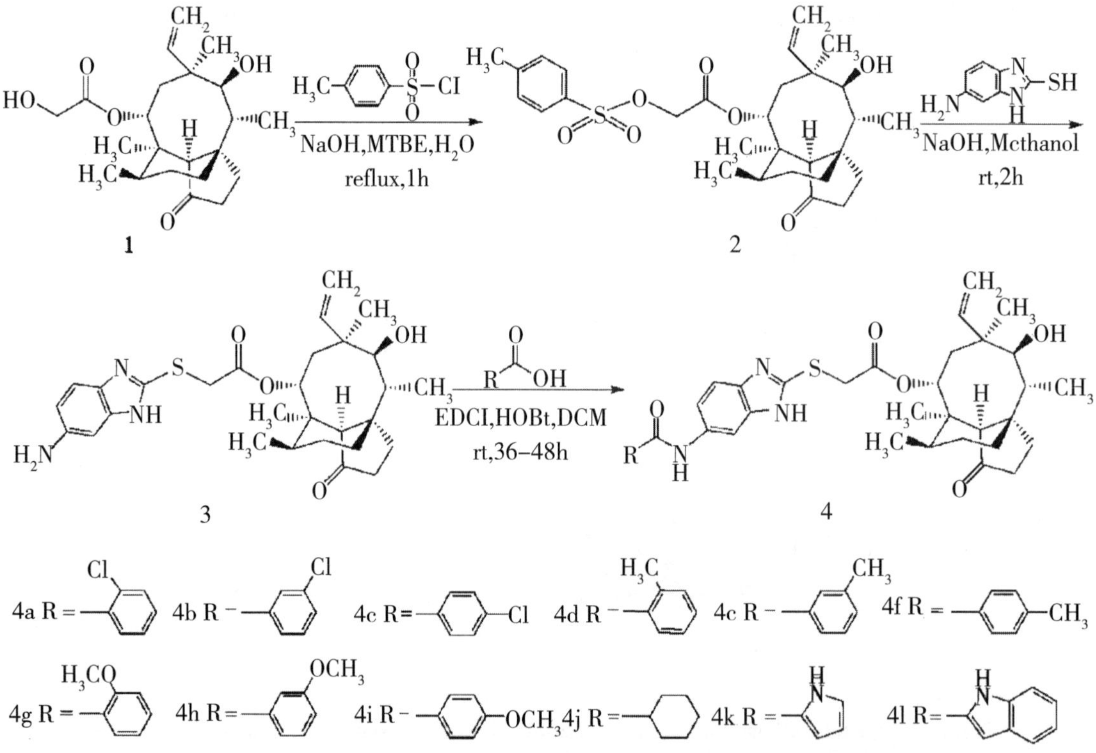

Scheme 1 General synthetic scheme for the pleuromutilin derivatives.

2.2 Antibacterial activity

The synthesized pleuromutilin derivatives 3 and 4a–l, with tiamulin fumarate used a as reference drug, were screened for their in vitro antibacterial activity against Escherichia coli (E. coli), *Staphylococcus aureus* (*S. aureus*), Methicillin-resistant Staphylococcus aureus (MRSA), *Staphylococcus vitulinus* (*S. vitulinus*), *Staphylococcus warneri* (*S. warneri*), and *Staphylococcus haemolyticus* (*S. haemolyticus*). The antibacterial activities are reported in Table 1 as the minimum inhibitory concentration (MIC) using the agar dilution method. The MICs of the 13 new pleuromutilin derivatives in vitro against *E. coli* ranged from 10 to 80 μg/mL; against *S. aureus*, MRSA, *S. vitulinus*, *S. warneri*, and *S. haemolyticus* ranged from 0.156 to 40 μg/mL, respectively. It can be observed that compound 3 showed the highest antibacterial activities than the other synthesized compounds and was comparable to the reference drug. Compounds 4j-l exhibited relatively moderate inhibitory activities against *S. aureus*, MRSA, *S. vitulinus*, *S. warneri*, and *S. haemolyticus* and weak activity against *E. coli*. However, the other compounds showed lower antibacterial activities.

Table 1 In vitro antibacterial activity (MIC) of the synthesized pleuromutilin derivatives.

Compound No.	MIC (μg/mL)					
	E. coli	*S. aureus*	MRSA	*S. vitulinus*	*S. warneri*	*S. haemolyticus*
3	10	0.156	0.156	0.156	0.156	0.156
4a	40	40	40	40	40	5
4b	80	40	20	10	20	20
4c	40	40	40	40	40	40
4d	80	10	40	40	40	40
4e	80	40	40	40	40	40
4f	80	40	40	40	20	40
4g	80	40	40	40	40	10
4h	40	40	40	40	40	10
4i	40	40	20	40	20	5
4j	20	5	1.25	1.25	2.5	10
4k	20	1.25	1.25	1.25	5	1.25
4l	20	10	10	5	2.5	10
Tiamulin	1.25	0.156	0.156	0.156	0.156	0.156

The Oxford cup assays were carried out to evaluate the antibacterial activities against the above-mentioned six bacterial strains. The zones of inhibition for two concentrations (320 and 160 μg/mL) of the synthetic compounds and tiamulin fumarate were measured. Data are reported as diameters of growth inhibition (mm), and the results are given in Table 2. The results of Oxford cup assay correspond with that obtained by agar dilution method (MICs) as a whole. Compound 3 showed the best growth inhibition against the pathogens particularly haemolyticus and comparable to tiamulin, whereas compounds 4j-l showed moderate growth inhibition except against E. coli.

Table 2 Zone of inhibition for *E. coli*, *S. aureus*, MRSA, *S. vitulinus*, *S. warneri*, and *S. haemolyticus* in mm.

Compound	E. coli		S. aureus		MRSA		S. vitulinus		S. warneri		S. haemolyticus	
	320	160	320	160	320	160	320	160	320	160	320	160
3	25.52	21.38	29.56	27.28	30.56	25.42	29.56	25.48	32.28	28.02	33.04	30.28
4a	11.28	0.00	12.44	9.02	20.16	15.88	19.22	13.78	22.38	21.58	24.92	20.38
4b	0.00	10.38	15.84	12.28	21.68	17.26	23.82	20.48	19.48	16.46	20.38	20.30
4c	17.66	17.16	16.46	14.30	15.46	14.58	21.76	20.54	22.24	19.18	20.95	16.95
4d	0.00	0.00	17.84	14.72	19.70	16.30	19.12	18.18	22.28	20.60	18.36	17.38
4e	0.00	0.00	19.36	17.88	27.76	25.58	20.26	18.10	21.46	20.82	20.80	18.32
4f	8.64	0.00	17.84	16.72	17.32	14.06	15.08	14.00	22.94	22.86	20.72	16.20
4g	9.38	0.00	18.34	13.74	21.54	19.12	18.00	16.24	20.10	18.54	25.58	24.30
4h	18.02	15.32	18.72	15.88	19.70	16.30	21.66	19.78	19.78	17.92	28.41	25.41
4i	17.72	14.22	21.48	18.62	25.70	21.32	18.20	17.62	21.78	18.58	32.86	23.04
4j	19.78	17.44	27.25	15.42	31.90	28.00	29.10	25.98	27.69	24.72	26.36	23.00
4k	20.06	18.54	28.86	26.62	30.68	26.54	27.00	24.68	25.54	21.22	29.88	25.52
4l	20.48	17.36	24.72	20.88	21.58	19.76	26.54	22.06	29.56	25.36	25.12	20.80
Tiamulin	27.10	25.84	28.16	25.12	30.48	28.14	30.94	18.58	30.52	22.42	31.48	28.80

Compound 3, being the best antibacterial derivative, was chosen for further evaluation on its antibacterial effect to inhibit short-term S. aureus and MRSA growth (Figure 2). This includes the GI_{50} of tiamulin fumarate for S. aureus and MRSA. The inhibition study revealed that compound 3 and tiamulin fumarate showed higher inhibition efficacy against *S. aureus* than that against MRSA, respectively. Compared with tiamulin fumarate (GI_{50} was 0.2535 for MRSA), compound 3 displayed higher growth inhibition with a GI_{50} of 0.1748 μM against MRSA. However, there was no significant difference between compound 3 (GI_{50} was 0.0987 for *S. aureus*) and tiamulin fumarate (GI_{50} was 0.0983 for *S. aureus*) when they inhibited *S. aureus* growth.

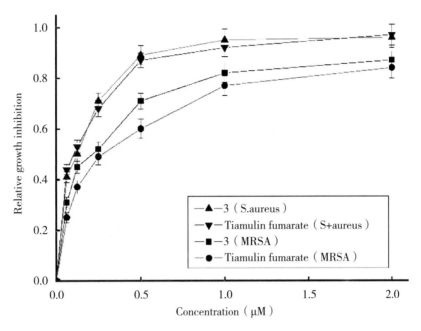

Figure 2 Inhibition of the growth of S. aureus and MRSA by compound 3 and tiamulin fumarate.

Note: Data points are the mean from at least three independent experiments carried out in duplicate. The standard deviation is shown as error bars.

2.3 Inhibitory Effects on Cytochrome P450

The cytochrome P450 (CYP) proteins form a large family of heme proteins that catalyze many reactions involved xenobiotic metabolism and biosynthesis of cholesterol, steroids, and other lipid components. Some chemicals could affect the disposition of conventional pharmaceuticals through the inhibition of CYP450 enzymes. The in vitro inhibitory effects of compound 3 on the activities of five common major human CYP 450 were evaluated, as some pleuromutilin derivatives showed potent inhibition of CYP3A4, especially azamulin, a synthetic azole pleuromutilin [16]. The selective probe substrate concentrations were prepared near to its respective Km value, and the compound 3 concentration range is from 0.05 to 100 mM. The CYP 450 inhibition was analyzed by determining IC_{50} values, and the results are summarized in Table 3 and Figure 3.

Table 3 Some experiment parameters and cytochrome P450 inhibition.

P450 Isozyme	Probe	Final Concentrations (mM)[a]	Metabolite	IC_{50} (μM)
CYP1A2	Phenacetin	45	Acetaminophen	>50
CYP2C19	Mephenytoin	55	4-OH-S-mephenytoin	>50
CYP2D6	Dextromethorphan	10	Dextrophan	>100
CYP2C9	Diclofenac	10	4-OH-diclofenac	10.70
CYP3A4	Midazolam	5	1-OH-midazolam	1.69

[a] The final concentrations of probe were set according to [17].

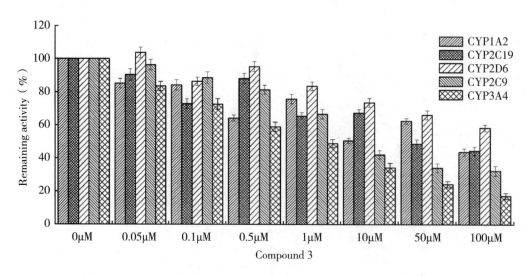

Figure 3　Inhibitory effects of compound 3 on CYP-catalyzed reactions.

Note：Each data point represents the mean of triplicate experiments and the standard deviation is shown as error bars.

Compound 3 showed intermediate inhibition of CYP3A4 and CYP2C9, with IC_{50} values of 1.69 and 10.70 μM, respectively. In contrast, compound 3 displayed low to no significant inhibitory effects on CYP1A2 ($IC_{50} > 50$), CYP2C19 ($IC_{50} > 50$), and CYP2D6 ($IC_{50} > 100$). It was reported that the selective inhibition for CYP3A4 may be primarily a function of the pleuromutilin portion of the molecule [16, 18]. However, side chain structures of pleuromutilin derivatives have some influence on CYP3A4 inhibition profiles [12]. Our studies show that compound 3 is potent inhibitor of CYP3A4, a member of the CYP450 family with the most common and the most versatile function. Further research on design and synthesis of new pleuromutilin derivatives should be considered to improve CYP450 inhibition.

3　EXPERIMENTAL SECTION

3.1　Synthesis

3.1.1　General

Starting materials, reagents, and solvents were of analytical grade and obtained from commercial sources. All reactions were monitored by thin-layer chromatography (TLC) analysis on silica gel GF254 plates and visualized under UV illumination at 254 nm for UV active materials after elution. Further visualization was achieved by staining with a 0.05% $KMnO_4$ aqueous solution. All column chromatography purifications were carried out on silica gel (200–300 mesh, Qingdao Haiyang Chemical Co., Ltd., Qingdao, China) through conventional methods. IR spectra were obtained on a NEXUS-670 spectrometer (Nicolet Thermo, Edina, MN, USA) using KBr thin films, and the absorptions are reported in cm^{-1}. NMR spectra were recorded in appropriate solvents using a Bruker-400 MHz spectrometer (Bruker BioSpin, Zürich, Switzerland). The chemical shifts (δ) were expressed in parts per million (ppm) relative to tetramethylsilane. The multiplicities of the NMR signals were designated as s (singlet), d (doublet), t (triplet), q (quartet), m (multiplet), br (broad), etc. High-resolution mass spectra (HRMS) were obtained on a Bruker Daltonics APEX II 47e mass spectrometer (Billerica, MA, USA) equipped with an electrospray ion source.

3.1.2 14-O- ((5-Amino-benzimidazole-2-yl) thioacetyl) Mutilin (3)

A variety of 5-amino-2-mercaptobenzimidazole (1.65 g, 10 mmol) and 10 M NaOH (1.1 mL, 11 mmol) in 50 mL of methanol was stirred for 20 min at room temperature. Compound 2 (0.52 g, 10 mmol) dissolved in 7.5 mL of dichloromethane was added by dropwise. After stirring for 24 h at room temperature, the reaction mixture was evaporated under reduced pressure to dryness. The crude product was extracted three times with a solution of ethyl acetate (60 mL) and water (20 mL). The organic phase was treated with saturated NaHCO 3 and the target compound 3 was then precipitated without further purification by chromatography to yield 3.47 g (66%). IR (KBr): 3405, 3331, 2935, 1729, 1633, 1445, 1420, 1384, 1359, 1269, 1154, 1118, 1018, 981, 916 cm^{-1}; ^{1}H-NMR (400 MHz, DMSO) δ 7.33 (s, 1H), 6.74 (s, 1H), 6.62 (d, J = 8.5 Hz, 1H), 6.43 (dd, J = 17.4 Hz, 11.0, 1H), 5.78 (d, J = 8.5 Hz, 1H), 5.21 (dd, J = 49.7 Hz, 14.2, 2H), 3.82 (s, 2H), 3.49 (s, 1H), 3.35 (d, J = 6.2 Hz, 1H), 2.37-2.28 (m, 1H), 2.27-2.11 (m, 2H), 2.08 (d, J = 5.7 Hz, 1H), 2.01 (dd, J = 19.6 Hz, 11.0, 1H), 1.77 (d, J = 12.0 Hz, 1H), 1.69-1.59 (m, 2H), 1.56-1.21 (m, 9H), 1.17-1.04 (m, 4H), 0.94-0.79 (m, 4H), 0.72 (t, J = 7.5 Hz, 3H); ^{13}C-NMR (101 MHz, DMSO) δ 216.11, 167.45, 158.40, 137.99, 131.75, 124.81, 121.52, 116.18, 109.05, 73.48, 69.59, 56.99, 52.30, 44.32, 43.37, 42.84, 40.73, 35.60, 34.89, 34.36, 33.36, 29.26, 25.72, 25.29, 23.71, 15.65, 13.69, 10.38; HRMS (ES) calcd [M + H]$^+$ for $C_{29}H_{39}N_3O_4S$ 526.2934, found 526.2780.

3.1.3 General Procedure for the Synthesis of Compounds 4a-l

A mixture of the carboxylic acids derivative (2.4 mmol), 1.05 g of compound 3 (2.0 mmol), 0.46 g of 1-ethyl-3- (3-dimethylaminopropyl) carbodiimide hydrochloride (2.4 mmol), and 0.34 g of 1-hydroxybenzotriazole (2.4 mmol) were dissolved in 50 mL of dichloromethane and stirred for 36-48 h at room temperature. The mixture was washed with saturated aqueous NaHCO$_3$ and brine, and then dried with Na$_2$SO$_4$ overnight. The crude residue obtained was purified by silica gel column chromatography (petroleum ether: ethyl acetate 1 : 1-1 : 4 v/v) to afford the desired compounds.

14-O- ((2-Chlorobenzamide-5-aminobenzimidazole-2-yl) thioacetyl) mutilin (4a). Synthesized according to the general procedure for 36 h. White solid; yield 94% (1.25 g). IR (KBr): 3422, 2933, 1726, 1655, 1603, 1544, 1445, 1406, 1353, 1272, 1153, 1117, 1016, 980, 749 cm^{-1}; ^{1}H-NMR (400 MHz, CDCl$_3$) δ 10.98 (s, 1H), 8.20 (s, 2H), 7.67 (d, J = 6.0 Hz, 1H), 7.34 (dd, J = 38.3 Hz, 15.1, 4H), 6.41-6.25 (m, 1H), 5.70 (d, J = 7.8 Hz, 1H), 5.12 (dd, J = 70.7 Hz, 14.0, 2H), 4.10-4.01 (m, 1H), 3.85 (s, 2H), 3.26 (s, 1H), 2.24 (s, 1H), 2.20-2.05 (m, 2H), 2.00 (s, 1H), 1.97 (s, 1H), 1.93 (dd, J = 15.9 Hz, 8.4, 1H), 1.68 (d, J = 13.9 Hz, 1H), 1.57-1.48 (m, 2H), 1.45-1.16 (m, 9H), 1.11-0.95 (m, 4H), 0.80 (d, J = 2.7 Hz,

3H), 0.65 (d, J = 2.9 Hz, 3H); ^{13}C-NMR (101 MHz, CDCl$_3$) δ 216.00, 170.15, 167.49, 163.84, 148.04, 137.76, 134.32, 131.63, 130.63, 129.73, 129.38, 129.14, 126.29, 116.30, 73.61, 69.65, 59.39, 57.11, 44.43, 43.46, 42.98, 40.86, 35.72, 35.04, 34.78, 34.10, 33.44, 29.40, 25.86, 25.45, 23.83, 20.02, 15.81, 13.82, 13.18, 10.48; HRMS (ES) calcd [M + H]$^+$ for C$_{36}$H$_{42}$N$_3$O$_5$SCl 664.2600, found 664.2598.

14-O- ((3-Chlorobenzamide-5-aminobenzimidazole-2-yl) thioacetyl) mutilin (4b). Synthesized according to the general procedure for 40 h. White solid; yield 73% (0.97 g). IR (KBr): 3422, 2934, 1725, 1655, 1543, 1445, 1406, 1284, 1153, 1117, 1017, 980, 916 cm^{-1}; ^{1}H-NMR (400 MHz, CDCl3) δ 8.40 (s, 1H), 8.03 (s, 1H), 7.82 (s, 1H), 7.70 (d, J = 6.3 Hz, 1H), 7.40 (d, J = 7.0 Hz, 1H), 7.29 (s, 2H), 6.37-6.24 (m, 1H), 5.68 (d, J = 7.5 Hz, 1H), 5.08 (dd, J = 51.3 Hz, 14.0, 2H), 4.14-3.96 (m, 1H), 3.87 (q, J = 16.1 Hz, 2H), 3.26 (s, 1H), 2.20 (d, J = 17.1 Hz, 1H), 2.17-2.04 (m, 2H), 2.00 (s, 1H), 1.97 (s, 1H), 1.93 (dd, J = 15.4 Hz, 7.9, 1H), 1.67 (d, J = 14.2 Hz, 1H), 1.54 (d, J = 10.4 Hz, 1H), 1.50-1.11 (m, 9H), 1.03 (d, J = 20.1 Hz, 4H), 0.79 (s, 3H), 0.65 (s, 3H); ^{13}C-NMR (101 MHz, CDCl$_3$) δ 216.08, 170.17, 167.54, 163.89, 148.04, 137.76, 135.75, 133.86, 131.64, 130.76, 129.02, 126.50, 124.33, 116.28, 73.60, 69.76, 59.40, 57.11, 44.44, 43.49, 43.00, 40.87, 35.73, 35.03, 34.09, 33.46, 29.39, 29.10, 25.86, 25.47, 23.83, 20.03, 15.84, 13.82, 13.19, 10.48; HRMS (ES) calcd [M + H]$^+$ for C$_{36}$H$_{42}$N$_3$O$_5$SCl 664.2600, found 664.2599.

14-O- ((4-Chlorobenzamide-5-aminobenzimidazole-2-yl) thioacetyl) mutilin (4c). Synthesized according to the general procedure for 36 h. White solid; yield 89% (1.18 g). IR (KBr): 3373, 2934, 1727, 1654, 1598, 1544, 1486, 1445, 1285, 1153, 1116, 1099, 980, 938 cm^{-1}; ^{1}H-NMR (400 MHz, CDCl$_3$) δ 7.97 (dd, J = 28.4 Hz, 21.8, 3H), 7.63 (dd, J = 52.2 Hz, 16.8, 2H), 7.40-7.21 (m, 2H), 6.11-5.97 (m, 1H), 5.51 (d, J = 6.8 Hz, 1H), 4.99 (dd, J = 74.7 Hz, 14.4, 2H), 4.45 (s, 1H), 4.14-4.02 (m, 2H), 3.37 (s, 1H), 2.35 (s, 1H), 2.20-2.10 (m, 1H), 2.04 (dd, J = 20.3 Hz, 9.6, 2H), 1.97 (s, 1H), 1.95-1.87 (m, 1H), 1.69-1.54 (m, 2H), 1.45 (s, 1H), 1.40-1.10 (m, 9H), 1.03-0.87 (m, 4H), 0.78 (s, 3H), 0.60 (s, 3H); ^{13}C-NMR (101 MHz, CDCl$_3$) δ 217.54, 167.31, 164.62, 149.95, 149.21, 141.07, 140.69, 136.66, 135.89, 134.34, 133.99, 132.77, 129.97, 128.84, 117.44, 115.78, 110.25, 73.06, 70.59, 60.19, 57.69, 45.40, 44.60, 41.99, 36.81, 34.45, 30.56, 29.15, 27.06, 24.91, 21.21, 16.52, 14.97, 11.95; HRMS (ES) calcd [M + H]$^+$ for C$_{36}$H$_{42}$N$_3$O$_5$SCl 664.2600, found 664.2601.

14-O- ((2-Methylbenzamide-5-aminobenzimidazole-2-yl) thioacetyl) mutilin (4d). Synthesized according to the general procedure for 40 h. White solid; yield 71% (0.91 g). IR (KBr): 3423, 2932, 1726, 1655, 1601, 1542, 1447, 1406, 1278, 1153, 1117, 1018, 981, 939, 917, 741 cm^{-1}; ^{1}H-NMR (400 MHz, CDCl$_3$) δ 8.32 (s, 1H), 7.96-7.72 (m, 1H), 7.35 (dd, J = 57.9 Hz, 6.5, 3H), 7.18 (s, 1H), 6.80 (d, J = 70.7 Hz, 1H), 6.35 (ddd, J = 15.6 Hz, 10.2, 8.3, 1H), 5.69 (d, J = 7.0 Hz, 1H), 5.12

(dd, J = 49.2 Hz, 14.2, 2H), 3.82 (s, 2H), 3.26 (s, 1H), 2.44 (s, 3H), 2.17 (dd, J = 43.0 Hz, 5.2, 3H), 1.99 (s, 1H), 1.89 (dd, J = 22.6 Hz, 14.6, 1H), 1.66 (s, 1H), 1.55 (d, J = 9.6 Hz, 1H), 1.31 (ddt, J = 19.6 Hz, 14.3, 9.5, 10H), 1.01 (d, J = 10.8 Hz, 4H), 0.80 (d, J = 6.8 Hz, 3H), 0.64 (d, J = 6.8 Hz, 3H); ^{13}C-NMR (101 MHz, CDCl$_3$) δ 216.06, 167.50, 147.94, 137.77, 135.53, 135.22, 132.02, 130.26, 129.29, 127.83, 125.71, 124.95, 116.31, 73.61, 69.61, 64.57, 59.40, 57.11, 44.43, 43.44, 42.97, 40.85, 35.72, 35.03, 34.02, 33.44, 29.56, 29.39, 25.85, 25.44, 23.83, 20.03, 18.84, 15.80, 13.82, 13.18, 10.48; HRMS (ES) calcd [M + H]$^+$ for C$_{37}$H$_{45}$N$_3$O$_5$S 644.3162, found 644.3160.

14-O-((3-Methylbenzamide-5-aminobenzimidazole-2-yl) thioacetyl) mutilin (4e). Synthesized according to the general procedure for 36 h. White solid; yield 76% (0.98 g). IR (KBr): 3395, 2932, 1726, 1649, 1602, 1541, 1446, 1406, 1284, 1153, 1117, 1018, 981, 938, 917, 740 cm^{-1}; ^{1}H-NMR (400 MHz, CDCl$_3$) δ 8.28 (d, J = 5.4 Hz, 1H), 8.12 (s, 1H), 7.63 (dd, J = 10.1 Hz, 6.6, 2H), 7.40-7.26 (m, 2H), 7.07 (s, 1H), 6.32 (dd, J = 17.4 Hz, 11.1, 1H), 5.67 (d, J = 8.4 Hz, 1H), 5.29-4.93 (m, 2H), 3.88 (s, 2H), 3.24 (d, J = 6.1 Hz, 1H), 2.32 (s, 3H), 2.20 (d, J = 6.8 Hz, 1H), 2.11 (d, J = 7.8 Hz, 2H), 1.97 (d, J = 3.0 Hz, 1H), 1.90 (dd, J = 16.0 Hz, 8.4, 1H), 1.64 (s, 1H), 1.53 (d, J = 10.2 Hz, 2H), 1.50-1.10 (m, 9H), 1.00 (s, 4H), 0.78 (d, J = 6.8 Hz, 3H), 0.63 (d, J = 6.8, 3H); ^{13}C-NMR (101 MHz, CDCl$_3$) δ 216.08, 167.31, 165.55, 147.94, 137.79, 137.70, 133.94, 132.00, 131.62, 129.90, 127.67, 126.87, 123.15, 116.27, 114.94, 73.60, 69.61, 64.57, 57.11, 44.43, 43.44, 42.97, 40.85, 35.71, 35.03, 34.15, 33.44, 29.57, 29.39, 25.85, 25.46, 23.82, 20.39, 18.17, 15.78, 13.80, 10.47; HRMS (ES) calcd [M + H]$^+$ for C$_{37}$H$_{45}$N$_3$O$_5$S 644.3162, found 644.3154.

14-O-((4-Methylbenzamide-5-aminobenzimidazole-2-yl) thioacetyl) mutilin (4f). Synthesized according to the general procedure for 40 h. White solid; yield 61% (0.79 g). IR (KBr): 3422, 2928, 1726, 1635, 1609, 1541, 1446, 1425, 1297, 1180, 1117, 1019, 980, 940, 912, 747 cm^{-1}; ^{1}H-NMR (400 MHz, CDCl$_3$) δ 8.43 (d, J = 43.5 Hz, 1H), 8.34 (d, J = 7.7 Hz, 1H), 7.73-7.42 (m, 3H), 7.17 (t, J = 7.5 Hz, 1H), 7.05 (d, J = 8.3 Hz, 1H), 6.40 (dd, J = 17.4 Hz, 11.0, 1H), 5.75 (d, J = 8.3 Hz, 1H), 5.18 (dd, J = 51.6 Hz, 14.2, 2H), 4.09 (d, J = 10.7 Hz, 3H), 3.94 (d, J = 4.8 Hz, 2H), 3.31 (s, 1H), 2.41-2.23 (m, 2H), 2.23-2.12 (m, 2H), 2.05 (s, 1H), 2.04 (s, 1H), 1.96 (dd, J = 15.8 Hz, 8.4, 1H), 1.73 (t, J = 13.6 Hz, 1H), 1.61 (d, J = 9.9 Hz, 2H), 1.55-1.20 (m, 9H), 1.06 (s, 4H), 0.86 (d, J = 6.9 Hz, 3H), 0.70 (d, J = 6.8 Hz, 3H); ^{13}C-NMR (101 MHz, CDCl$_3$) δ 215.94, 167.41, 167.32, 162.45, 156.29, 147.67, 137.80, 132.32, 131.53, 120.79, 116.26, 110.62, 73.61, 69.51, 59.39, 57.11, 55.30, 44.43, 43.44, 42.97, 40.85, 35.73, 35.04, 34.17, 33.44, 29.41, 25.86, 25.43, 23.83, 20.02, 18.17, 15.76, 13.79, 13.18, 10.46; HRMS (ES) calcd [M + H]$^+$ for C$_{37}$H$_{45}$N$_3$O$_5$S 644.3162, found 644.3183.

14-O-((2-Methoxybenzamide-5-aminobenzimidazole-2-yl) thioacetyl) mutilin (4g).

Synthesized according to the general procedure for 36 h. White solid; yield 68% (0.90 g). IR (KBr): 3357, 2934, 1729, 1653, 1600, 1551, 1457, 1406, 1292, 1237, 1162, 1117, 1019, 980, 938, 916, 756 cm^{-1}; ^{1}H-NMR (400 MHz, CDCl$_3$) δ 8.04 (s, 1H), 7.88 (d, J = 6.7 Hz, 2H), 7.57–7.20 (m, 4H), 6.11–5.98 (m, 1H), 5.53 (d, J = 7.1 Hz, 1H), 5.01 (dd, J = 74.0 Hz, 14.3, 2H), 4.47 (s, 1H), 4.09 (s, 2H), 3.38 (s, 1H), 2.39 (s, 3H), 2.36 (s, 1H), 2.16 (d, J = 11.1 Hz, 1H), 2.11–1.87 (m, 4H), 1.62 (dd, J = 25.4 Hz, 12.7, 2H), 1.47 (s, 1H), 1.42–1.11 (m, 9H), 0.99 (d, J = 28.2 Hz, 4H), 0.80 (d, J = 5.3 Hz, 3H), 0.62 (d, J = 3.2 Hz, 3H); ^{13}C-NMR (101 MHz, CDCl$_3$) δ 217.68, 167.48, 165.71, 149.15, 141.90, 141.22, 140.67, 136.04, 134.44, 132.93, 129.44, 128.21, 117.52, 115.93, 115.52, 102.88, 73.21, 70.74, 60.34, 57.84, 45.55, 44.75, 43.71, 42.14, 36.96, 34.59, 30.71, 29.28, 27.21, 25.06, 21.60, 16.68, 15.12, 14.69, 12.10; HRMS (ES) calcd [M + H]$^+$ for C$_{37}$H$_{45}$N$_3$O$_6$S 660.3145, found 660.3149.

14-O-((3-Methoxybenzamide-5-aminobenzimidazole-2-yl)thioacetyl)mutilin (4h). Synthesized according to the general procedure for 40 h. White solid; yield 57% (0.75 g). IR (KBr): 3385, 2936, 1727, 1654, 1598, 1541, 1447, 1349, 1287, 1154, 1117, 1039, 980, 937, 918, 804 cm^{-1}; ^{1}H-NMR (400 MHz, CDCl$_3$) δ 8.30 (s, 1H), 8.10 (s, 1H), 7.38 (s, 2H), 7.36 (s, 1H), 7.27 (t, J = 7.8, 1H), 6.98 (d, J = 6.0, 1H), 6.32 (dd, J = 17.4 Hz, 11.1, 1H), 5.68 (d, J = 8.3 Hz, 1H), 5.10 (dd, J = 43.3 Hz, 14.2, 2H), 4.05 (dt, J = 7.1 Hz, 6.2, 1H), 3.84 (d, J = 15.7 Hz, 2H), 3.76 (s, 3H), 3.24 (s, 1H), 2.29–2.17 (m, 1H), 2.16–2.04 (m, 2H), 1.97 (d, J = 1.0 Hz, 1H), 1.94–1.85 (m, 1H), 1.66 (d, J = 14.4 Hz, 1H), 1.53 (d, J = 9.8 Hz, 1H), 1.44–1.17 (m, 8H), 1.00 (s, 4H), 0.79 (d, J = 6.7 Hz, 3H), 0.64 (d, J = 6.7 Hz, 4H); ^{13}C-NMR (101 MHz, CDCl$_3$) δ 216.18, 170.22, 167.42, 165.11, 158.89, 147.98, 137.72, 135.42, 131.85, 128.78, 117.97, 116.89, 116.30, 111.55, 76.31, 73.56, 69.58, 59.42, 57.08, 54.47, 44.41, 43.37, 42.94, 40.82, 35.69, 34.99, 34.07, 33.44, 29.36, 25.82, 25.42, 23.80, 20.06, 15.80, 13.80, 13.19, 10.51; HRMS (ES) calcd [M + H]$^+$ for C$_{37}$H$_{45}$N$_3$O$_6$S 660.3145, found 660.3098.

14-O-((4-Methoxybenzamide-5-aminobenzimidazole-2-yl)thioacetyl)mutilin (4i). Synthesized according to the general procedure for 36 h. White solid; yield 58% (0.76 g). IR (KBr): 3386, 2932, 1725, 1647, 1606, 1544, 1508, 1458, 1406, 1351, 1286, 1254, 1179, 1119, 1076, 980, 939, 917, 762 cm^{-1}; ^{1}H-NMR (400 MHz, CDCl$_3$) δ 8.21 (s, 1H), 8.06 (d, J = 24.1 Hz, 1H), 7.81 (d, J = 8.8 Hz, 2H), 7.24 (s, 1H), 6.86 (d, J = 8.7 Hz, 2H), 6.32 (dd, J = 17.4 Hz, 11.0, 1H), 5.67 (d, J = 8.4 Hz, 1H), 5.09 (dd, J = 43.1 Hz, 14.2, 2H), 3.87 (d, J = 3.7 Hz, 2H), 3.78 (s, 3H), 3.25 (s, 1H), 2.20 (dd, J = 13.7 Hz, 7.1, 1H), 2.10 (dt, J = 17.5 Hz, 7.3, 2H), 1.97 (s, 1H), 1.97 (s, 1H), 1.90 (dd, J = 16.0 Hz, 8.5, 1H), 1.66 (d, J = 14.0 Hz, 1H), 1.54 (d, J = 7.1 Hz, 1H), 1.49–1.12 (m, 9H), 0.99 (s, 4H), 0.78 (d, J = 6.9 Hz, 3H), 0.64 (d, J = 6.9 Hz, 3H); ^{13}C-NMR (101 MHz, CDCl$_3$) δ 216.05, 167.41, 164.80, 161.49, 147.85, 137.78, 132.08, 128.05, 126.11, 116.27, 112.99,

76.23, 73.60, 69.61, 59.39, 57.11, 54.45, 44.43, 43.45, 42.98, 40.95, 40.85, 35.72, 35.03, 34.15, 33.45, 29.39, 25.85, 25.45, 23.82, 20.02, 15.79, 13.80, 13.18, 10.47; HRMS (ES) calcd [M + H]$^+$ for C$_{37}$H$_{45}$N$_3$O$_6$S 660.3145, found 660.3143.

14-O-((Cyclohexanecarboxamide-5-aminobenzimidazole-2-yl) thioacetyl) mutilin (4j). Synthesized according to the general procedure for 36 h. White solid; yield 81% (1.03 g). IR (KBr): 3365, 2931, 1728, 1664, 1601, 1541, 1448, 1407, 1345, 1277, 1208, 1152, 1117, 1018, 980, 938, 917, 807 cm^{-1}; ^{1}H-NMR (400 MHz, CDCl$_3$) δ 8.16 (s, 1H), 7.61 (s, 1H), 7.39 (d, J = 7.9 Hz, 1H), 6.93 (s, 1H), 6.39 (dd, J = 17.4 Hz, 11.0, 1H), 5.74 (d, J = 8.4 Hz, 1H), 5.33-5.00 (m, 3H), 3.96 (d, J = 6.1 Hz, 1H), 3.30 (d, J = 4.6 Hz, 1H), 2.39-2.10 (m, 4H), 2.08-1.95 (m, 4H), 1.69 (dd, J = 72.2 Hz, 26.0, 10H), 1.41-1.15 (m, 9H), 1.05 (s, 4H), 0.86 (d, J = 6.9 Hz, 3H), 0.71 (d, J = 6.9 Hz, 3H); ^{13}C-NMR (101 MHz, CDCl$_3$) δ 216.03, 173.89, 167.29, 147.69, 137.80, 131.96, 129.91, 127.84, 116.25, 73.61, 69.53, 59.40, 57.12, 52.41, 45.65, 44.44, 43.44, 42.97, 40.86, 35.73, 35.04, 34.17, 33.46, 29.41, 28.82, 25.86, 25.44, 24.70, 23.83, 20.03, 18.18, 15.76, 13.80, 10.46; HRMS (ES) calcd [M + H]$^+$ for C$_{36}$H$_{49}$N$_3$O$_5$S 636.3495, found 636.3497.

14-O-(((1H-Pyrrole-2-carboxamide)-5-aminobenzimidazole-2-yl) thioacetyl) mutilin (4k). Synthesized according to the general procedure for 48 h. White solid; yield 76% (0.94 g). IR (KBr): 3374, 2933, 1724, 1638, 1601, 1554, 1444, 1408, 1352, 1271, 1191, 1151, 1117, 1017, 980, 937, 917, 746 cm^{-1}; ^{1}H-NMR (400 MHz, CDCl$_3$) δ 9.95 (s, 1H), 8.09 (s, 1H), 7.85 (s, 1H), 7.30 (s, 1H), 7.09-6.85 (m, 2H), 6.75 (s, 1H), 6.29 (dd, J = 17.3 Hz, 11.1, 1H), 6.19 (s, 1H), 5.67 (d, J = 8.0 Hz, 1H), 5.05 (dd, J = 41.0 Hz, 14.2, 2H), 3.87 (s, 2H), 3.25 (s, 1H), 2.19 (d, J = 6.7 Hz, 1H), 2.16-2.02 (m, 2H), 1.98 (d, J = 2.7 Hz, 1H), 1.89 (dd, J = 15.7 Hz, 7.9, 1H), 1.65 (d, J = 13.8 Hz, 1H), 1.54 (s, 1H), 1.53-1.09 (m, 9H), 1.06-0.92 (m, 4H), 0.77 (d, J = 6.7 Hz, 3H), 0.62 (d, J = 6.7 Hz, 3H); ^{13}C-NMR (101 MHz, CDCl$_3$) δ 216.21, 167.56, 158.57, 158.51, 147.60, 147.23, 137.70, 131.83, 124.91, 121.63, 116.29, 109.28, 109.15, 76.21, 73.59, 69.69, 57.10, 52.41, 49.03, 44.43, 43.48, 42.95, 40.84, 35.71, 35.00, 34.47, 33.47, 29.36, 25.83, 25.40, 23.82, 15.76, 13.80, 10.49; HRMS (ES) calcd [M + H]$^+$ for C$_{34}$H$_{43}$N$_3$O$_5$S 619.2951, found 619.2953.

14-O-(((1H-Indole-2-carboxamide)-5-aminobenzimidazole-2-yl) thioacetyl) mutilin (4l). Synthesized according to the general procedure for 48 h. White solid; yield 81% (1.08 g). IR (KBr): 3380, 2934, 1725, 1648, 1601, 1545, 1445, 1418, 1342, 1308, 1276, 1240, 1149, 1117, 1017, 980, 937, 917, 812, 747, 628 cm^{-1}; ^{1}H-NMR (400 MHz, CDCl$_3$) δ 12.55 (s, 1H), 11.73 (d, J = 11.3 Hz, 1H), 10.16 (d, J = 22.0 Hz, 1H), 8.05 (s, 1H), 7.68 (d, J = 8.0 Hz, 1H), 7.53-7.46 (m, 1H), 7.43 (d, J = 1.7 Hz, 2H), 7.22 (t, J = 7.6 Hz, 1H), 7.07 (t, J = 7.5 Hz, 1H), 6.07 (dd, J = 17.1 Hz, 11.9, 1H), 5.53 (d, J = 8.1 Hz, 1H), 5.02 (dd, J = 48.6 Hz, 14.5, 2H), 4.48 (t, J = 5.9 Hz, 1H), 4.12 (d, J = 10.1 Hz, 1H), 3.38 (s, 1H), 2.37 (s, 1H), 2.17 (dd, J = 18.6 Hz, 10.9, 1H), 2.04 (d, J = 9.4 Hz, 1H), 1.99 (s, 1H), 1.93 (dd,

J = 16.0 Hz, 8.3, 1H), 1.62 (s, 2H), 1.46 (d, J = 7.4 Hz, 1H), 1.43–1.11 (m, 9H), 0.97 (d, J = 6.2 Hz, 4H), 0.80 (d, J = 6.8 Hz, 3H), 0.63 (t, J = 6.3 Hz, 3H); HRMS (ES) calcd [M + H]$^+$ for $C_{38}H_{45}N_4O_5S$ 669.3107, found 669.3109.

3.2 Biological evaluation

3.2.1 MIC Testing

The MIC for the synthesized compounds (3 and 4a–l) and tiamulin fumarate used as a reference drug were determined using the agar dilution method according to the National Committee for Clinical Laboratory Standards (NCCLS). Tested compounds were dissolved in 25%–40% DMSO to a solution with a concentration of 1 280 μg/mL, whereas tiamulin fumarate was directly dissolved in 10 mL of distilled water at the same concentration as the tested compounds. All the solutions were then diluted twofold with sterile water to provide 11 dilutions (final concentration is 0.625 μg/mL). A 2 mL volume of the 2-fold serial dilution of each test compound/drug was incorporated into 18 mL of hot Muellere-Hinton agar medium. Inoculums, including five Gram (+), S. aureus, MRSA, S. vitulinus, S. warneri, S. haemolyticus, and one Gram (−), E. coli were prepared from blood slants and adjusted to approximately 10^5–10^6 CFU/mL with sterile saline (0.90% NaCl). A 10 μL amount of bacterial suspension was spotted onto Muellere-Hinton agar plates containing serial dilutions of the compounds/drug. The plates were incubated at 36.5 ℃ for 24–48 h. The same procedure was repeated in triplicate.

3.2.2 Oxford Cup Assays

The procedure for Oxford cup assays was the same as that previously reported [13]. Briefly, inoculums were prepared in 0.9% saline using McFarland standard and spread uniformly on nutrient agar plates. The 320 and 160 μg/mL of all the tested compounds were added individually into the Oxford cups that were placed at equal distances above the agar surfaces. The zone of inhibition for each concentration was measured after a 24–36 h incubation at 37 ℃. The same procedure was repeated in triplicate.

3.2.3 Inhibition of the Bacterial Growth

The inhibition activities of test compounds were evaluated by measuring the absorbance of the bacterial suspension. In brief, inoculums were prepared from blood slants and adjusted to approximately 10^6 CFU/mL in Muellere-Hinton broth. The 0.062 5, 0.125, 0.25, 0.5, 1, and 2 μM of tested compounds were prepared and added to the prepared inoculums. A sample of inoculum with no compound was used as control. The absorbance of bacterial suspension at 450 nm was measured using an UV spectrophotometer after 3.5 h of incubation, and the inhibition was calculated as $1 - (A^P/A^0)$, where A^P and A^0 are the absorbance of bacterial suspension in the presence and absence of compounds, respectively. The same procedure was independently repeated in triplicate.

3.2.4 Cytochrome P450 Inhibition Assay

Compound 3 was screened for its ability to inhibit cytochrome P450 using a cocktail assay described previously [19, 20]. Briefly, 50 μL of compound 3 was added to 96-well plates, with final concentrations of 0.05, 0.1, 0.5, 1, 10, 50, and 100 μM, respectively. Twenty microliters of human liver microsomes (final concentration of 0.3 mg/mL) and 20 μL of probe substrates (Table 3) in 0.1 M Tris were added. After preincubation at 37℃ for 10 min, 10 μL of NADPH was added (final concentration of 1 mM), and the reaction started to incubate at the same temperature for 15 min. The reactions were then quenched by the addition of 100 μL of acetonitrile with a mixture of propranolol and nadolol (50 nM) used as an internal standard. After reactions were finished, the plates were centrifuged and the supernatants were analyzed for the five metabolites (Table 3) by LC/MS/MS.

4 CONCLUSIONS

Novel antibacterial pleuromutilin derivatives were synthesized by introduction of the substituted benzimidazole moieties to its C22 side. These derivatives were initially evaluated for their in vitro antibacterial activities against E. coli, S. aureus, MRSA, S. vitulinus, S. warneri, and S. haemolyticus. Compounds 3 and 4j – l exhibited promising in vitro antibacterial effects against all the pathogens except E. coli. The further evaluation of compound 3 displayed higher growth inhibition with GI_{50} values of 0.098 7 and 0.174 8 μM against S. aureus and MRSA, respectively. The CYP450 inhibition assay of compound 3 showed intermediate in vitro inhibitory potency for CYP3A4, a member of the CYP450 family with the most common and the most versatile function. This study indicates that further designing new pleuromutilin derivatives should be considered to improve the CYP450 inhibition profile.

Supplementary Materials: Supplementary materials can be accessed at: http://www.mdpi.com/1420-3049/21/11/1488/s1.

Acknowledgments: The authors are grateful for financial support from the National Key Technology Support Program (No. 2015BAD11B02) and the Agricultural Science and Technology Innovation Program (ASTIP, No. CAAS-ASTIP-2014-LIHPS-04).

Author Contributions: X. A., J. L., and R. S. designed research; X. A., X. P., Y. Y., Y. L., and S. X. performed research and analyzed the data; R. S., and J. L. wrote the paper. All authors read and approved the final manuscript.

Conflicts of Interest: The authors declare no conflict of interest.

（发表于《molecules》, SCI, IF: 2.465.）

Short Communication: N-Acetylcysteine-mediated Modulation of Antibiotic Susceptibility of Bovine Mastitis Pathogens

YANG F., LIU L. H., LI X. P., LUO J. Y., ZHANG Z.,
YAN Z. T., ZHANG S. D., and LI H. S.*

Key Lab of New Animal Drug Project, Key Laboratory of Veterinary Pharmaceutical Development, Ministry of Agriculture, Lanzhou Institute of Husbandry and Pharmaceutical Sciences of Chinese Academy of Agricultural Science, Lanzhou 730050, China

Abstract: The aim of this study was to investigate the effects of N-acetylcysteine (NAC) on antibiotic susceptibility of bovine mastitis pathogens including *Staphylococcus aureus*, *Streptococcus dysgalactiae*, *Escherichia coli*, and *Streptococcus agalactiae*. Minimum inhibitory concentrations (MIC) were tested by the agar-based E-test method. The presence of 10 mM NAC reduced the MIC of penicillin and ampicillin but enhanced the MIC of erythromycin and ciprofloxacin for all of the strains. In addition, NAC-mediated modulation of MIC of kanamycin, tetracycline, and vancomycin was diverse, depending on the target bacterial pathogen and antibiotic being used. The results suggest that NAC is an important modulator of antibiotic activity against the major bovine mastitis pathogens.

Key words: N-acetylcysteine; Bovine mastitis; Pathogen; Antibiotic susceptibility

SHORT COMMUNICATION

Bovine mastitis is one of the most costly diseases affecting the dairy industry worldwide (Perreten et al., 2013). Although mastitis can be caused by 137 different microorganisms (Watts, 1988), Staphylococcus aureus, Streptococcus dysgalactiae, Escherichia coli, and Streptococcus agalactiae are the main etiological agents commonly associated with the disease (Nair et al., 2005). To date, antibiotic therapy is the standard treatment of mastitis. However, the results of this therapy have been disappointing due to the misuse of antibiotics (Pereira et al., 2011; Barrero et al., 2014).

N-Acetylcysteine (NAC) is a mucolytic agent that disrupts disulfide bonds in mucus and reduces the viscosity of secretions (El-Feky et al., 2009). Based on these characteristics, NAC has been widely used as an adjuvant in combination with antibiotics during medical treatment of bacterial infectious diseases including chronic bronchitis, vascular catheter-related infection, and urinary tract infection (Marchese et al., 2003; Olofsson et al., 2003; Aslam et al., 2007).

---

*Corresponding author, E-mail: lihsheng@sina.com

However, the effect of NAC on bovine mastitis pathogens has not been studied. The aim of this study was to investigate the effects of NAC on antibiotic susceptibility of Staph. aureus, Strep. dysgalactiae, E. coli, and Strep. agalactiae isolated from bovine mastitis cases.

The Staph. aureus, Strep. dysgalactiae, E. coli, and Strep. agalactiae strains were isolated from subclinical bovine mastitis in Gansu province in China during 2015. Mastitis infection was confirmed by the California mastitis test. Identification was performed by morphological characterization and biochemical testing as previously described (Cressier and Bissonnette, 2011). Minimum inhibitory concentrations of penicillin, ampicillin, erythromycin, kanamycin, tetracycline, ciprofloxacin, and vancomycin were determined by the E-test (BioMerieux, Marseille, France) method (Liu et al., 2014). Antimicrobial agent concentrations ranged from 0.002 to 32 μg/mL for penicillin and ciprofloxacin, and 0.016 to 256 μg/mL for ampicillin, erythromycin, kanamycin, tetracycline, and vancomycin.

The effects of NAC (Sigma-Aldrich, Lyon, France) on antibiotic susceptibility of pathogens to 7 antibiotics belonging to different groups were studied by measuring their MIC in the presence and absence of 10 mM NAC in the medium, respectively. Antibiotic susceptibility of 2 Staph. aureus strains (LZ 0215, LZ 84184), 2 Strep. dysgalactiae strains (LZ 717, LZ 211), 2 E. coli strains (LZ 2552, LZ 282), and 2 Strep. agalactiae strains (LZ 17, LZ 21) were determined in this study. The presence of 10 mM NAC did not affect the growth of these strains. The experiments were carried out at least twice, and the representative results are mentioned here.

The effects of NAC on antibiotic susceptibility of Staph. aureus and Strep. dysgalactiae are shown in Table 1, and E. coli and Strep. agalactiae are shown in Table 2. In the case of β-lactam antibiotics, the MIC of both penicillin and ampicillin decreased for all of the strains in the presence of NAC. Conversely, NAC increased the MIC of erythromycin and ciprofloxacin for all of the strains. It also led to reduction in MIC of tetracycline for all of the strains except that of E. coli. Similarly, the MIC of kanamycin decreased for Strep. agalactiae and Strep. dysgalactiae strains but increased for Staph. aureus and E. coli strains in the presence of NAC. In addition, changes in vancomycin MIC were not observed in Staph. aureus, Strep. dysgalactiae, and E. coli strains at the presence of NAC, although it resulted in a reduction in MIC of vancomycin against Strep. agalactiae.

Table 1 Effect of N-acetylcysteine (NAC[1]) on susceptibility of *Staphylococcus aureus* and *Streptococcus dysgalactiae* to different antibiotics, as measured by MIC (μg/mL)

Antibiotic	Staph. aureus				Strep. dysgalactiae			
	LZ 0215		LZ 84184		LZ 717		LZ 211	
	Control	+NAC	Control	+NAC	Control	+NAC	Control	+NAC
Penicillin	1.5	0.38	0.5	0.25	1.5	0.75	0.016	0.004
Ampicillin	2	0.25	0.75	0.5	0.75	0.25	0.032	<0.016
Erythromycin	0.094	3	0.064	8	0.75	3	0.032	0.064
Kanamycin	3	12	1.5	8	32	2	128	48
Tetracycline	0.75	0.5	0.5	0.25	>256	96	0.25	0.047

(continued)

Antibiotic	Staph. aureus				Strep. dysgalactiae			
	LZ 0215		LZ 84184		LZ 717		LZ 211	
	Control	+NAC	Control	+NAC	Control	+NAC	Control	+NAC
Ciprofloxacin	0.19	0.25	1	1.5	0.5	1.5	0.75	1
Vancomycin	1.5	1.5	1.5	1.5	2	2	0.75	0.75

[1] In each case, the final NAC concentration was 10 mM.

The presence of NAC enhances the efficacy of β-lactams antibiotics against all of the strains, while it gives protection against erythromycin and ciprofloxacin. These data are similar to those reported by Goswami and Jawali (2010). However, in our study, it is interesting that NAC reduced the antibacterial activity of kanamycin against pathogens belonging to Streptococcus spp. alone. And for tetracycline, the presence of NAC decreased the antibacterial activity against grampositive bacteria but enhanced that against gram-negative bacteria. In addition, NAC gives protection against vancomycin for Strep. agalactiae alone. These findings suggest that the effects of NAC on bacterial antibiotic susceptibility are significantly associated with bacterial species, shape, and structure. It was reported that NAC in combination with fosfomycin and tigecycline displayed opposite effects on bacterial biofilm formation (Marchese et al., 2003; Aslam et al., 2007). The bacteria enclosed in biofilm became 10 to 1000 times more tolerant to antibiotics than equivalent planktonic cultures (Mah and O'Toole, 2001). In this study, it is possible that NAC-mediated modulation of antibiotic susceptibility against the pathogens may be mediated by biofilm-forming ability, and the specific mechanisms are being explored in our laboratory.

Table 2 Effect of N-acetylcysteine (NAC[1]) on susceptibility of *Escherichia coli* and *Streptococcus agalactiae* to different antibiotics, as measured by MIC (μg/mL).

Antibiotic	E. coli				Strep. agalactiae			
	LZ 2552		LZ 282		LZ 17		LZ 21	
	Control	+NAC	Control	+NAC	Control	+NAC	Control	+NAC
Penicillin	>32	32	>32	12	0.047	0.016	0.032	0.016
Ampicillin	2	1.5	3	1	0.094	<0.016	0.064	<0.016
Erythromycin	16	>256	48	>256	0.032	0.125	0.047	0.25
Kanamycin	0.75	1.5	3	25	>256	24	>256	32
Tetracycline	3	8	3	8	48	32	96	32
Ciprofloxacin	0.004	0.012	0.004	0.012	0.5	1.5	0.5	1.5
Vancomycin	>256	>256	>256	>256	1.5	0.75	1	0.5

[1] In each case, the final NAC concentration was 10 mM.

In conclusion, the present study indicates that NAC is an important modulator of antibiotic activity against the major bovine mastitis pathogens. A combination of β-lactam antibiotics and NAC is recommended during antibiotic therapy of bovine mastitis caused by these major pathogens.

ACKNOWLEDGMENTS

This study was supported by the Special Fund of Chinese Central Government for Basic Scientific Research Operations in Commonwealth Research Institutes (No. 1610322015007) and the Natural Science Foundation of Gansu Province (No. 145RJYA311, China).

(发表于《Journal of Dairy Science》,院选SCI, IF: 2.408. 4 300–4 302.)

Influences of Season, Parity, Lactation, udder Area, Milk Yield, and Clinical Symptoms on Intramammary Infection in Dairy Cows

ZHANG Z., LI X. P., YANG F., LUO J. Y., WANG X. R., LIU L. H., LI H. S.[*]

(Key Lab of New Animal Drug Project, Key Laboratory of Veterinary Pharmaceutical Development, Ministry of Agriculture, Lanzhou Institute of Husbandry and Pharmaceutical Sciences of Chinese Academy of Agricultural Science, Lanzhou 730050, China)

Abstract: The aim of this study was to evaluate the influences of season, parity, lactation, udder area, milk yield, and clinical symptoms on bacterial intramammary infection (IMI) in dairy cows. A total of 2,106 mastitis pathogens in 12 species were isolated from 125 dairy farms distributed in 30 different cities in China, and the information about these factors was recorded at the same time. Mastitis pathogens were isolated from 63.43% of the milk samples, whereas Streptococcus agalactiae accounted for 38.61% of all pathogens, followed by Str. dysgalactiae (28.16%), Staphylococcus aureus (19.10%), Escherichia coli (6.90%), and other pathogens (7.23%). According to our investigation, IMI was more common in spring with the isolation rate of pathogens at 81.04%, and lowest in winter (52.34%). Cows were more likely to be infected by environmental pathogens (E. coli or Str. uberis) in summer, in rear quarters and in cows with higher daily milk yield or lower somatic cell count. In addition, Str. dysgalactiae exhibited a higher prevalence with increased parity. Different clinical symptoms of quarters with bacterial IMI were seen in this study, and mastitis pathogens were isolated from healthy quarters.

Key words: Intramammary infection; Pathogen; Relationship; Bovine mastitis

1 INTRODUCTION

Dairy cow mastitis is a serious disease associated with both high incidence (van den Borne et al., 2010) and economic losses (Holland et al., 2015), posing a major challenge to the dairy industry (Boboš et al., 2013; Leelahapongsathon et al., 2014). There are approximately 220 million dairy cattle worldwide. The incidence of clinical mastitis (CM) is estimated to range between 16 and 48 cases per 100 cows (Kvapilik et al., 2014), and the prevalence of subclinical mastitis (SCM) is reported to be 20% to 80% globally (Kivaria, 2006; Contreras and Rodríguez, 2011). Research has demonstrated a wide range in the cost of mastitis, ranging from

---

[*] Corresponding author, E-mail: lihsheng@sina.com

$16.43 to $572.19 per cow (Holland et al., 2015). Beside the financial implications of mastitis, the importance of mastitis with respect to public health should not be overlooked (Bradley, 2002). Milk from cows with mastitis accidentally mixed into bulk milk enters the food chain and has the potential to transmit pathogenic organisms and antibiotic residues to humans (Hameed et al., 2007). In addition, the extensive use of antibiotics to both prevent and treat mastitis likely contributes to the rise in antimicrobial resistance in the management of human infectious diseases (Leblanc et al., 2006).

Mastitis is complex, developing as a result of the interaction between various factors associated with the host, specific pathogens, environment, and management (Demme and Abegaz, 2015; Rashad et al., 2016). Over 200 different organisms have been recorded to cause bovine mastitis (Blowey and Edmondson, 2010). However, IMI is mostly caused by a much smaller range of pathogens. A better understanding of the prevalence and distribution of the major mastitis pathogens is important for the dairy industry to help guide specific control measures (Piessens et al., 2011).

According to the data from National Bureau of Statistics of the People's Republic of China, approximately 15 million dairy cattle are present in China (Li et al., 2015). The incidence of mastitis was estimated to range between 16 and 75 cases per 100 cows. Annual losses associated with bovine mastitis were estimated to be 15 to 45 billion yuan, accounting for 38% of total direct costs related to dairy cattle health (Song and Yang, 2010; Memon, 2013). Although a great deal of research has been performed concerning various aspects of bovine mastitis in China, studies on the correlation between IMI and factors such as season, parity, lactation, and so on have not been reported. This study aimed to eluci date the prevalence of mastitis pathogens, identify the relationship between IMI and these factors and SCC, test for pathogens in samples from healthy cows, and provide information that can be used in bovine mastitis control programs.

2 MATERIALS AND METHODS

2.1 Herds and cows

Data from 125 dairy farms distributed in 30 different cities in China were included in this study (Figure 1). The cows were primarily major China Holstein (hybrid of Holstein and Yellow cattle), and were housed in either a free stall or a tie stall with straw, sawdust, or something others as bedding. Most farms used the bucket-type milking machines for milking cows twice daily and postmilking teat disinfection and selective dry cow therapy based on udder health status. The mean herd size of the study herds was 553 cows, with an average production of 5, 862 kg of milk/cow per year.

2.2 Sampling and processing

Before sampling, the first streams of milk were discarded, and teat ends were disinfected with cotton swabs soaked in 75% alcohol and allowed to dry. A total of 3, 134 milk samples were collected from 3 072 dairy cattle and samples (n = 2 493) without evidence of CM were performed by Lanzhou Mastitis Test (LMT; scored at −, ±, +, ++, and +++, corresponding to negative, suspicious, weak positive, positive, and strong positive, respectively; Liu et al., 1983). The LMT is a diagnostic method of SCM, which is similar to the California Mastitis Test.

Samples of CM (n = 641), SCM (n = 1 808, LMT score at +, ++, or +++), and healthy cows (n = 685, LMT score at − or ±) were placed on ice and transported to the laboratory within 6 h of collection for bacteriological studies. Information regarding season, parity, lactation, udder area, milk yield, and clinical symptoms were recorded at the same time. Not all samples contain information as mentioned above because no information was recorded in some small farms and the information on some samples collected by farmers and their veterinary surgeons was not recorded.

2.3 Bacterial identification

Culturing and identification of the microorganisms to species level was carried out according to standard procedures described by Yuan et al. (1991) within 24 h after sampling. Briefly, a bacteriological loop was used to spread approximately 0.01 to 0.02 mL of each milk sample on blood agar, Chapman agar, Edward agar, MacConkey agar, and nutrient broth. The plates were incubated at 37℃ and examined after 24 h. Suspected colonies were purified and transferred to nutrient broth and agar slant culture medium. When slow-growing or unusual bacteria were suspected, longer incubation periods or on incubator environment of 10% CO_2 were used. If no growth occurred within 7 d, samples were considered negative.

All isolates were identified based on colony appearance, conventional Gram staining, Christie, Atkins, Munch-Petersen (CAMP) reaction, and other biochemical methods as described by Yuan et al. (1991). The API Test (BioMérieux SA, Marcy l'Etoile, France) was also used if necessary.

2.4 Statistical analysis

First, databases containing the results of bacterial identification and factors (season, parity, lactation, udder area, milk yield, and clinical symptoms) were built in a Microsoft Excel (Microsoft Corp., Redmond, WA) worksheet for analysis. These data were then used to estimate isolation rate of bacterium/pathogens on quarters level (number of bacterium/pathogens/IMI cases per 100 quarters) and distribution of pathogens (the proportion of specific pathogen in the total pathogens in each group). Second, chi-square analysis, as applied in the SPSS software version 22.0 (SPSS Inc., Chicago, IL), was used to assess the association between IMI and all above-mentioned factors. Statistical significance in this step was assessed at $P < 0.05$.

3 RESULTS

3.1 Distribution of mastitis pathogens

Out of 3 134 milk samples collected and processed, 86.85% (n = 2 722) were culture positive for bacteria. A total of 3 410 bacterial isolates of 24 species were obtained, and 2 106 isolates (63.43%) of 12 species were identified as mastitis pathogens. *Streptococcus agalactiae* (38.61%), *Str. dysgalactiae* (28.16%), *Staphylococcus aureus* (19.10%), and *Escherichia coli* (6.90%) were the predominant isolates, accounting for 92.77% of all mastitis pathogens

(n = 2 106) , and most of them were contagious bacteria (85.87%) .

3.2 Relationships between mastitis pathogens and seasons

As shown in Table 1, IMI was more common ($P < 0.01$) in spring with the isolation rate of pathogens at 81.04%, and lowest in winter (52.34%) . However, the isolation rate of *S. aureus* was highest in winter (41.07%) and lowest in spring (16.23%) . A higher isolation rate of environmental pathogens (*E. coli* and *Str. uberis*) pathogens was seen in summer compared with other seasons ($P < 0.05$) . However, the isolation rate of *Str. agalactiae* was significantly lower in summer compared with other seasons ($P < 0.05$) .

3.3 Relationships between mastitis pathogens and parity of cows

The data demonstrated in Table 2 revealed that mastitis pathogens were isolated with approximately equal frequencies between parities, and the distribution was similar to the general trend, of which *Str. agalactiae* was the dominant pathogen, followed by *Str. dysgalactiae*, *S. aureus*, and other pathogens. However, the isolation rate of *Str. agalactiae* increased gradually with increasing parity.

3.4 Relationships between mastitis pathogens and lactation

Among 2 493 milk samples from lactating cows, 1 650 (66.19%) were collected in the early lactation period (1–300 d in milk) and the rest were from the late lactation period (7–10 d before dry period) . As recorded in Table 3, the distribution of mastitis pathogens was similar to the general trend. However, the isolation rate of pathogens was higher ($P < 0.01$) in the early lactation stage (60.18%) than the late lactation stage (50.95%) .

Table 1 Relationships between pathogenic species and season

Season	No. of samples	Isolation rate of pathogens (%)	Distribution of pathogens[1] (%)											
			S.agal.	S.dys.	S.aur.	E.coli	S.uber.	Prot.	Kleb.	P.aer.	N.ast.	S.pyo.s	C.alb	C.alb
Spring	571	81.04	55.85	17.18	16.23	6.92	2.39	0.72	0.95	0.95	1.43			
Summer	716	59.50	23.47	28.17	21.36	15.73	3.05	0.47	0.70	2.11	3.17	0.70	0.70	0.23
Fall	392	77.04	40.73	28.48	23.84	5.96	1.99		1.99	1.32	1.32			
Winter	107	52.34	28.57	17.86	41.07	7.14					1.79			
Total	1 786	69.76	41.41	23.11	20.39	9.47	2.3	0.08	0.10	1.36	1.20	0.24	0.24	0.08

[1] *S. agal.* = *Streptococcus agalactiae*; *S. dys.* = *Streptococcus dysgalactiae*; *S. aur.* = *Staphylococcus aureus*; *E. coli* = *Escherichia coli*; *S. uber.* = *Streptococcus uberis*; *Prot.* = *Proteus*; *Kleb.* = *Klebsiella*; *P. aer.* = *Pseudomonas aeruginosa*; *N. ast.* = *Nocardia asteroides*; *S. pyo.* = *Streptococcus pyogenes*; *C. pyo.* = *Corynebacterium pyogenes*; *C. alb.* = *Candida albicans*. The same as below.

Table 2 Relationships between pathogenic species and parity.

Parity	No. of samples	Isolation rate of pathogens (%)	Distribution of pathogens[1] (%)											
			S. agal.	S. dys.	S. aur.	E. coli	S. uber.	Prot.	Kith.	P. aer.	N. ast.	S. pyo.	C. pyo.	C. alb.
1	258	65.00	36.54	25.64	17.11	14.10			0.64	3.85	1.28			0.64
2	281	61.80	35.42	17.36	24.31	11.81	6.94	1.39		1.39	0.69		0.69	
3	308	68.21	42.93	28.27	13.09	12.04	1.05		0.52	1.05	0.52	0.52		
4	212	65.56	47.46	11.86	15.25	16.10	0.85	1.69	1.69	4.26	0.82			
5	161	68.46	52.80	12.36	19.10	13.48		1.12			1.12			
>5	100	61.90	50.64	19.49	11.69	13.42	3.04				1.72			
Total	1 410	65.15	44.30	19.16	16.76	13.49	2.82	1.40	0.48	1.76	1.01	0.82	0.69	0.64

3.5 Relationships between mastitis pathogens and LMT

As shown in Table 3, high LMT score was positively correlated with IMI. However, in late lactation, *S. aureus* and *E. coli* IMI was negatively correlated with LMT score. Furthermore, 241 mastitis pathogens were isolated from LMT negative samples (n = 574).

Table 3 Relationships between pathogenic species and lactation stage.

Lactation stage	Type	LMT[2] score	No. of samples	Isolation rate of pathogens (%)	Distribution of pathogens[1] (%)											
					S. agal.	S. dys.	S. aur.	E. coli	S. uber.	Prot.	Kleb.	P. aer.	N. ast.	S. pyo.	C. pyo.	C. alb.
Early lactation period	Health	−	127	35.37	37.93	20.69	34.48		6.90							
		±	30	40.00	75.00		25.00									
	Sub-health	+	80	51.69	52.17	15.21	23.91		6.52	2.17						
		++	1 060	62.39	34.81	34.56	17.71	2.95	5.29	2.21	1.48	0.37	0.37	0.12		0.12
		+++	353	66.13	31.01	36.23	19.16	4.14	5.57	0.70	0.70	0.35	1.39		0.35	0.35
	Total		1 650	60.18	35.27	33.10	19.16	2.94	5.41	1.73	1.15	0.33	0.57	0.08	0.08	0.16

(continued)

Lactation stage	Type	LMT[2] score	No. of samples	Isolation rate of pathogens (%)	Distribution of pathogens[1] (%)											
					S. agal.	S. dys.	S. aur.	E. coli	S. uber.	Prot.	Kleb.	P. aer.	N. ast.	S. pyo.	C. pyo.	C. alb.
Late lactation period	Health	−	447	43.16	19.72	29.58	18.31	26.76	4.93	0.70						
		+	81	50	43.60	23.08	15.38	17.95								
	Sub-health	+	115	53.39	42.86	25.40	19.05	9.52	1.59							
		++	122	64.18	56.98	19.77	10.47	6.98	4.65							
		+++	78	72.29	70.00	3.33	15.00	6.67						5.00		
	Total		843	50.95	38.67	23.14	16.28	17.22	3.29	0.31				0.66		
Total			2 493	57.96	36.30	30.09	18.29	7.25	4.77	1.30	0.80	0.23	0.60	0.06	0.06	0.11

[2] LMT = Lanzhou Mastitis Test. −, ±, +, ++, and +++ correspond to negative, suspicious, weak positive, positive, and strong positive, respectively.

3.6 Relationships between mastitis pathogens and udder area

As shown in Table 4, a nonsignificant association was found between different udder areas. The distribution of pathogens was similar to the general trend, but the level of IMI of front quarters was lower than posterior quarters. However, *E. coli* IMI was more common in posterior quarters ($P < 0.01$), and *Str. agalactiae* IMI was more common in front quarters ($P = 0.05$).

3.7 Relationships between mastitis pathogens and milk yield

From the results presented in Table 5, the isolation rate of pathogens increased with the increasing daily milk yield (DMY) except for cows producing <30 kg per d (this might be due in part to only 14 samples in this group) indicated that increased DMY was a risk for IMI ($P > 0.05$). However, cows with higher DMY tend to have a lower LMT positive rate. The IMI level of *E. coli* and *Str. uberis* was higher in cows with high DMY ($P < 0.01$).

Table 4 Relationships between pathogenic species and udder area.

Udder area	No. of samples	Isolation rate of pathogens (%)	Distribution of pathogens[1] (%)										
			S. agal.	S. dys.	S. aur.	E. coli	S. uber.	Prot.	Kleb.	P. aer.	N. ast.	S. pyo.	C. pyo.
Front left	699	64.19	42.23	29.28	15.14	3.37	3.98	0.59	0.19	0.19	0.39	0.19	0.39
Front right	699	60.00	41.77	26.58	17.30	6.12	3.38	1.90	1.90	0.42	0.63		
Rear left	744	63.82	38.42	20.45	20.51	6.78	4.90	1.51	0.94	1.13	0.56	0.18	

(continued)

Udder area	No. of samples	Isolation rate of pathogens (%)	Distribution of pathogens[1] (%)										
			S. agal.	S. dys.	S. aur.	E. coli	S. uber.	Prot.	Kleb.	P. aer.	N. ast.	S. pyo.	C. pyo.
Rear right	746	64.90	32.96	31.30	19.26	7.78	3.33	0.74	0.92	1.85	1.11		
Total	2 888	63.26	38.68	26.90	18.12	6.06	3.90	1.17	0.97	0.93	0.68	0.09	0.09

Table 5 Relationships between pathogenic species and milk yield

Milk Yield (kg)	No. of samples	Isolation rate of pathogens (%)	LMT[2] positive rate (%)	Distribution of pathogens[1] (%)										
				S. agal.	S. dys.	S. aur.	E. coli	S. uber.	P. aer.	Kleb.	S. pyo.	Prot.	N. ast.	c. pyo.	c. alb.
<10	347	59.65	71.48	53.62	22.22	15.46	4.83	0.97	0.97				0.48	1.45	
10-20	480	60.42	54.16	44.48	22.41	9.66	16.55	2.41	1.72	0.34	0.34	0.34	1.03	0.34	0.34
20-30	314	61.98	38.98	26.29	22.69	20.62	19.59	3.61	3.61	1.55	0.52	1.03	0.52		
>30	14	28.57	35.71	50.00	50.00										
Total	1 155	60.23	55.01	42.14	22.59	14.40	13.82	2.30	2.02	0.58	0.29	0.57	1.01	0.14	0.14

[1] LMT = Lanzhou Mastitis Test.

3.8 Relationships between mastitis pathogens and clinical symptoms of udder

As shown in Table 6, clinical symptoms (including redness, swelling, increased heat, and pain) of the udder can be caused by most pathogens, but the features vary depending on the type of bacteria. For example, udder will become larger and harder if it was infected with either *Pseudomonas aeruginosa* or *S. aureus* ($P < 0.01$). Fever caused by *E. coli* IMI was significantly lower than other pathogens ($P < 0.01$). Redness and swelling induced by *Str. agalactiae* IMI was significantly lower than others ($P < 0.01$). Bloody milk (bleeding inside the udder) caused by *Str. agalactiae* IMI and nonbacteria (possibly virus, mycoplasma, mechanical damage, and so on) was significantly higher than other pathogens ($P < 0.05$).

Table 6 Relationships between pathogenic species and clinical symptom of udder (%)

Pathogenic species	Total cases	Udder size		Hardness level of udder		Udder temperature		Pain of udder		Red swollen of udder		Color of milk				Floes in milk	
		En-large	Nor-mal	Hard	Nor-mal	Hot	Nor-mal	Pain	Nor-mal	Obvi-ous	Nor-mal	Yel-low	Red	Cream color	Hoar	Present	None
Streptococcus agalactiae	143	80.42	19.58	95.80	5.59	93.91	6.03	64.83	35.17	69.67	30.33	73.70	13.70	12.33	12.33	93.18	6.81
Streptococcus dysgalactiae	40	72.50	27.50	92.50	7.50	82.50	17.50	65.00	35.00	85.92	14.71	77.50	7.50	10.00	10.00	88.50	11.43
Staphylococcus aureus	66	92.42	7.46	96.97	4.55	83.33	16.67	80.60	19.40	89.47	11.76	73.13	7.46	19.40	19.40	87.30	12.70
Escherichia coli	46	76.09	23.91	89.13	10.87	68.89	31.11	71.74	28.26	86.05	13.95	84.78	2.17	13.04	13.04	80.95	19.05
Pseudomonas aeruginosa	14	100	0	100	0	92.86	7.14	71.73	28.57	91.67	8.33	85.70	0	14.29	14.29	100	0
Nonbacteria	12	83.33	16.66	91.67	8.33	100	0	58.33	41.67	83.33	16.67	66.67	16.67	8.33	8.33	88.89	11.11
Total	321	82.24	17.73	94.70	6.23	86.91	13.06	69.14	30.87	79.58	20.75	75.90	9.51	13.53	13.53	89.77	10.21

4 DISCUSSION

A database containing information about bacteriological findings and factors recorded on individual cow based on our study was established. The results showed that contagious agent (*Str. agalactiae*, *Str. dysgalactiae*, and *S. aureus*) accounted for 85.87% of all IMI. This is in agreement with previous studies (Olde Riekerink et al., 2007a; Abdel-Rady and Sayed, 2009; Demme and Abegaz, 2015) and may be because these organisms can be spread easily from infected quarters to healthy ones through contaminated milkers' hands, cloth towels, or milking equipment (Fox and Gay, 1993). Rapid changes in dairy industry (from management style to mastitis control measures) have occurred during last few years, which may induce hard conditions for environmental pathogens (Zadoks and Fitzpatrick, 2009). A study carried out in Estonia found that contagious pathogens were more frequently in SCM (Kalmus et al., 2011), and only 641 samples were collected from CM compared with 1 808 from SCM in our study.

Streptococcus agalactiae, also known as group B *Streptococcus*, is one of the leading causes of bovine mastitis and an important human pathogen (Keefe, 1997). With increased use of udder health technologies and mastitis control programs, the prevalence of *Str. agalactiae* IMI has decreased during the last few decades. In Denmark, for example, *Str. agalactiae* IMI was reported to decrease from 40% in 1950 to about 2% in 1992 (Mweu et al., 2012). In Massachusetts, an intraherd prevalence of 44.7% was found between 1976 and 1982, and only 10% of samples was positive for *Str. agalactiae* in 1992 (Keefe, 1997). In Ragusa, *Str. agalactiae* was isolated in fewer than 2% of 18, 711 milk samples (Ferguson et al., 2007). In China, *Str. agalactiae* was still a major pathogen of bovine mastitis (Xu et al., 2005). However, in some other reports in China, the isolation rate of Str. agalactiae was only 3.48%. The prevalence of dominant mastitis pathogens differs considerably among countries. Even in the same country, it is always different among regions due to different management level and climate.

According to our results, mastitis pathogens were most prevalent in spring, which was in agreement with the study reported by Penev et al. (2014). This mainly due to the increased air temperature and high air humidity in spring after cold winter, which was beneficial for bacteria and accompanied by increased cases of IMI. In China, spring is a rainy season, which is a risk factor associated with CM (Oliveira et al., 2015). In addition, the weather in spring is highly variable, which may result in lower immunity of cattle. Opposite to results of Olde Riekerink et al. (2007b), who established the highest proportion of IMI during the winter, we found the lowest proportion of IMI in winter.

Taking each pathogen into consideration, environmental pathogens (*E. coli* and *Str. uberis*) were more common in summer. During the summer, the humidity and temperature are favorable for the development of coliform in the bedding. Costa et al. (1998) reported an increased occurrence of environmental IMI in Brazil mainly caused by *Enterobacteriaceae* and *Str. Uberis* during hot and wet weather. *Streptococcus uberis* and *E. coli* were found to be more common during summer (Osterås et al., 2006; Koivula et al., 2007), which was in agreement with our findings. In addition, *S. aureus* was found more common in winter, which was in agreement with early research (Makovec and Ruegg, 2003).

In the present study, the influence of parity on IMI was not obvious, ranging from 61.80 to 68.46%. Only *Str. dysgalactiae* exhibited a higher prevalence with increased parity, which is in agreement with an early study reported by Osterås et al. (2006). Laevens et al. (1997) were also unable to find the effect of parity for mastitis, but Sharma et al. (2013) revealed that the risk of an increase in SSC with higher parity.

A statistically significant higher incidence of IMI was found in the early lactation stage (60.18%) than the late lactation stage (50.95%). This was in line with many previous studies on cows (Nyman et al., 2007; Penev et al., 2014), and possibly due to impaired immune function in early lactation (Persson Waller et al., 2009). A negative correlation was found between LMT score (SCC) and the prevalence of *E. coli*. This was in agreement with Erskine et al. (1988), who found *E. coli* was more common in milk samples with low SCC (43.5%) than high SCC samples (8.0%). At cow level, there is a tendency to believe that low SCC fails to protect the udder from environmental pathogens (Burvenich et al., 2003), and cows may suffer even higher prevalence (Suriyasathaporn et al., 2000). Further research is required to better understand this phenomenon because SCC is not as simple as the number of cells present in milk; the type/sub-type of cell, its ability to function, and the speed of its recruitment to the udder will all almost certainly play a role (Bradley, 2002). Of note, mastitis pathogens were also isolated from healthy quarters (LMT score at -) as shown in Table 4. A similar result was obtained by Özenç et al. (2008) in Turkey. Hence, bacteriological examinations should be carried out together with LMT for mastitis diagnosis.

According to our results as shown in Table 5, a higher prevalence of IMI was found in the rear quarters than front quarters, but it was not statistically significant. However, *E. coli* exhibited a significant difference between quarters. A recent study (Dimitar and Metodija, 2012) reported that the rear quarters with IMI (49.39%) in relation to the front one (33.04%). During a 4-yr retrospective study, Shpigel et al. (1998) also found the rear quarters had a higher incidence risk than the front quarters. The higher prevalence rate of IMI in the rear udder quarters may be due to their lower position compared with the front one, which make them more easily injured, and also greater milk yield produced by the rear quarters (Kocak, 2006). *Streptococcus agalactiae* was found mainly in front quarters ($P = 0.05$), and further research is needed to explain this phenomenon.

From the results presented in Table 6, increased DMY was a risk for IMI. This result was in agreement with Gröhn et al. (2004), who found IMI occurred more frequently in cows with higher milk production. Another report made in dairy goats indicated that the effect of IMI on milk yield was different for various pathogen groups (Koop et al., 2010). A negative relation between milk yield and IMI was reported by Gan et al. (2013), who also found the degree of IMI had a significant effect ($P < 0.05$) on SCC.

The prevalence of *Str. agalactiae* was higher ($P < 0.05$) in cows with lower DMY, whereas *E. coli* and *Str. uberis* IMI were common in cows with higher milk production ($P < 0.01$). Most mastitis pathogens damage secretory tissue in the mammary gland, which is subsequently replaced by nonsecretory tissue (Zhao and Lacasse, 2008; Hertl et al., 2014). However, in moderate cases of E. coli IMI, the main changes were superficial and confined to the tissue, without serious involvement of the secretory tissue (Frost and Brooker, 1986).

The variation of clinical signs between udders infected with pathogens was obtained in the present study. Clinically, Bacillus species IMI is often presented as a hard quarter with white clots (Blowey and Edmondson, 2010). This was proved by the result of *Pseudomonas aeruginosa* in the present work. A survey with larger samples is needed in future work due to the limited samples used in this study. All of these findings may be helpful in identifying and treating bovine mastitis.

5 CONCLUSIONS

Various factors (season, parity, lactation, udder area, milk yield, and clinical symptoms) were proven to be associated with IMI. Contagious organisms were major pathogens for IMI in China, accounting for 85.87% of all isolates. Cows were more likely to be infected by environmental pathogens (E. coli or Str. uberis) in summer, in the rear quarters, and in cows with higher DMY or lower SCC. *Streptococcus dysgalactiae* exhibited a higher prevalence with increased parity. In addition, a variation of clinical signs between quarters with bacterial IMI was seen in this study, and mastitis pathogens were isolated from healthy quarters.

ACKNOWLEDGMENTS

This research is supported by The National Science & Technology Pillar Program during the 12th Five-year Plan Period (2012BAD12B03), The Science & Technology Pillar Program of Gansu (144NKCA240), Agricultural Science & Technology Innovation Project of Gansu (GNCX-2013-59), and the Innovation Project of the Chinese Academy of Agricultural Sciences (research team of cow diseases). No conflict of interest exists in the submission of this manuscript.

（发表于《Journal of Dairy Science》，院选SCI，IF：2.408. 6 484-6 493.）

Acaricidal Activity of Oregano Oil and its Major Component, Carvacrol, Thymol and p-cymene Against *Psoroptes cuniculi* in Vitro and in Vivo

SHANG Xiaofei[1], WANG Yu[1], ZHOU Xuzheng[1], GUO Xiao[1], DONG Shuwei[1], WANG Dongsheng[1], ZHANG Jiyu[1], PAN Hu[1], ZHANG Yu[2, *], MIAO Xiaolou[1, *]

(1. Key Laboratory of New Animal Drug Project of Gansu Province, Key Laboratory of Veterinary Pharmaceutical Development of Ministry of Agriculture, Lanzhou Institute of Husbandry and Pharmaceutical Sciences of Chinese Academy of Agricultural Science, Lanzhou 730050, China;
2. Department of Emergency, Lanzhou General Hospital of PLA, Lanzhou 730050, China)

Abstract: Oregano oil possesses marked antioxidant and antimicrobial activity and is widely applied in animal husbandry. In the present study, we aimed to investigate the acaricidal activities of oregano oil and its major component, carvacrol, thymol and p-cymene against Psoroptes cuniculi in vitro and in vivo. The results revealed that oregano oil exhibited significant acaricidal effects against P. cuniculi that were dose- and time-dependent response. In in vitro test, concentrations of 0.05% and 0.02% (v/v) killed all of the mites within 1 h and 6 h, respectively. Moreover, 0.1 mg/mL (w/v) carvacrol, 0.2 mg/mL (w/v) thymol and 1% pcymene (v/v) also possessed marked acaricidal activities, and compared with the control group, elicited mean mortalities of 84.00%, 96.00% and 66% at 24 h, respectively. The median lethal times (LT 50) against P. cuniculi of the concentrations of 0.02%, 0.01% and 0.005% (v/v) of oregano oil, thymol, carvacrol and p-cymene were 2.171 h, 11.396 h, 26.102 h, and 4.424 h, 8.957 h and 15.201 h, respectively. Meanwhile, twenty naturaly infested rabbits were used to four homogeneity groups: negative control (without treatment), positive control (treated with ivermectin), group treated with 1% of oregano oil and other group with 5% of oregano oil. All the treatments were topically. After the treatment of 1% and 5% oregano oil, the P. cuniculi were completely eliminated in the rabbits, and at the end of the test (day 20), the rabbits of all treatment groups exhibited favorable mental and physical statuses. These results indicated that oregano oil could be widely applied as a potential acaricidal agent in the treatment of animal acariasis in the future.

Key words: Oregano oil; Carvacrol; Thymol; Acaricidal activity; Psoroptes cuniculi

---

*Corresponding authors, E-mail: shangxf928@163.com (X. Miao).
© 2016 Elsevier B.V. All rights reserved.

1 INTRODUCTION

Animal acariasis is a parasitic infection of the body surfaces or epidermises of animals; it is a critical veterinary skin disease that reduces the yield and the quality of animal products and is often fatal (Dagleish et al., 2007). Additionally, psoroptosis is a highly contagious disease. Particularly in rabbits, *Psoroptes cuniculi* infestation can cause intense pruritus and the formation of crusts and scabs that can completely cover the external ear canal and the internal surfaces of the pinna (Bates, 1999; Nong et al., 2013). Today, psoroptic acariasis is a global disease that causes substantial losses in the United States, Turkey, South Korea, India and other countries (Fichi et al., 2007; Eo and Oh-Deog, 2010).

To treat and control sarcoptic mange, chemical drugs have been widely used in the veterinary clinic, and these drugs exhibit relatively satisfactory treatment effectiveness. However, due to the negative public safety effects and environmental hazards induced by the overuse of some agents, the clinical applications of several chemical acaricides are limited by the governments (O'Brien, 1999). Thus, the identification of alternative medicines, particularly natural products for mite control, has attracted the interest of many people (Samish and Rehacek, 1999).

Oregano (Labiateae family) has been widely used as a flavoring agent for meat and traditional medicines in some European and Asian countries since ancient times (Zhu et al., 2007). Oregano oil (OR) is primarily composed of monoterpenoids and monoterpenes, which demonstrated marked antioxidant, antimicrobial and acaricidal effects, such as carvacrol, thymol, p-cymene and others (Zhu et al., 2007; Nechita et al., 2015). However, the acaricidal activities of oregano oil against P. cuniculi in vitro and in vivo have not previously been investigated.

Therefore, in the present study, we aimed to investigate the acaricidal activities of OR and its major components, carvacrol, thymol and p-cymene against *P. cuniculi* in vitro and to evaluate the clinical acaricidal efficacy of oregano oil in rabbits in vivo.

2 MATERIALS AND METHODS

2.1 Oregano oil and compounds

Oregano oil was purchased from Sigma-Aldrich, Co. (lot: MKBR7813V, USA) Carvacrol (lot: B21365, LC ≥ 98%), thymol (lot: YY91875, GC > 99%) and p-cymene (lot: B20334, LC ≥ 98%) were purchased from Shanghai Yuanye Biotech. Co., Ltd. (China).

2.2 Collection of mites

P. cuniculi mites were isolated from the ear cerumen of naturally infested New Zealand White rabbits, and the adult mites that were in good condition based on stereomicroscopic examination were collected and counted with a needle in Petri dishes for testing (Walton and Currie, 2007). After the collection of the materials, the rabbits were immediately treated with ivermectin. The experiments complied with the rulings of the Gansu Experimental Animal Center (Gansu, China) and were officially approved by the Ministry of Health, P.R. China in accordance with the NIH guidelines.

2.3 Acaricidal activity in vitro

In this test, the in vitro acaricidal activities of OR (0.02%, 0.01%, and 0.005%, v/v), carvacrol (0.2 mg/mL, w/v), thymol (0.1 mg/mL, w/v) and p-cymene (1%, v/v) were investigated. All drugs were diluted in 10% glycerin, and this solution was applied to the untreated group. The experiments were performed according to the methods of Macchioni et al. (2004). Specifically, one hundred μl of each drug were added to the culture plate, and the liquid excess was absorbed with filter paper. Next, 10 adult mites were collected from the infested ear cerumen with a needle and placed in each plate in order to complete contact the liquid. All plates were incubated at 25 ℃ in 75% relative humidity. At 1 h, 3 h, 6 h, 9 h, 12 h, 18 h, and 24 h, the reactions of the mites were observed, such as mite's body or foot movement. Five replicates were performed for each extract concentration.

2.4 Acaricidal activity in vivo

Oregano oil was used to treat the naturally infested rabbits. OR was diluted to 1% and 5% (v/v) with 50% glycerol, and the in vivo acaricidal activities were evaluated. The experimental procedures were performed as previously described by Guillot and Wright (1981), Fichi et al. (2007) and Nong et al. (2013). Twenty naturally infested rabbits with similar ages, weights and clinical scores were divided randomly into four groups (groups A, B, C and D), and the extent of the scabbing of the external ear canal and the present or absence of mites were evaluated clinically and with a stereomicroscope according to Fichi et al. (2007). There were no significant differences in the levels of infection in the right ears of all four groups. Prior to treatment, none of the rabbits had been treated with any anti-acariasis drugs or other agents, and no other complicating diseases were observed in the rabbits.

The right ears of the rabbits of group A were treated with 2 mL 5% OR, and those of group B were treated with 2 ml of 1% OR. Group C was treated with ivermectin (1%, positve control), and the animals in the negative control group were treated with 2 mL 50% glycerin. All the treatments were topically. These treatments were applied three times at 0, 5 and 10 days. Subsequently, on days 0, 8, 15 and 20 after the beginning of treatment, the right ears of all rabbits were examined with a stereomicroscope to evaluate the presence of scabs and mites.

2.5 Statistical analysis

The obtained data were analyzed with the SPSS software program version 18.0 and are expressed as the means ± the SDs. The data were analyzed with one-way ANOVAs followed by Student's two-tailed t-tests to compare the test and control groups. Tukey's tests were used for comparisons of three or more groups. P-values below 0.05 ($P < 0.05$) were considered indicative of significance. At the same time, a non parametric test as Kruskal-Wallis followed the Wilcoxon rank-sum test was used to analyze the data. The median lethal time (LT 50) was calculated using the complementary log-log (CLL) model.

3 RESULTS

3.1 Acaricidal activity in vitro

The results revealed that compared with the control, OR demonstrated a significant acaricidal effect against P. cuniculi with a dose- and time-dependent response. Specifically, the 0.05% and 0.02% concentrations of OR killed all of the mites within 1 h and 6 h, respectively. Moreover, the mean mortalities at 24 h following the 0.01% and 0.005% OR treatments were 84.00% and 48.00%, respectively. Further analyses indicated that the LT 50 values for the OR concentrations of 0.02%, 0.01% and 0.005% (v/v) were 2.171 h, 11.396 h and 26.102 h, respectively (Tables 1 and 2).

The major compounds of OR (carvacrol, thymol and p-cymene), possessed marked acaricidal activities against P. cuniculi, and the mean mortalities at 24 h in the carvacrol, thymol and p-cymene groups were 84.00%, 96.00% and 66.00%, respectively. Of these compounds, 0.2 mg/mL thymol exhibited the greatest toxicity with an LT_{50} value of 4.424 h. The LT_{50} values for 0.1 mg/mL carvacrol and 1% p-cymene were 8.957 h and 15.201 h, respectively. And as the positive drug, the mean mortality at 24 h of ivermectin was 100% against P. cuniculi (Tables 1 and 2).

3.2 Acaricidal activity in vivo

The results related to the treatment effects of topically applied 1% and 5% OR on the infested rabbits were as follows. Before treatment, no differences were observed between the four selected groups (day 0). After two treatments with OR or ivermectin, the numbers of scabs and mites on the right ears were substantially decreased. At day 8, there were significant differences between treatment groups and the untreatment control group. After three treatments (day 15), the right ears of the rabbits that were treated with 5% OR or ivermectin were free of scabs and/or mites, those treated with 1% OR exhibited only small scabs or minimal secretions in the ear canals and no mites. The untreated control group remained infested, and their statuses become poor, and they exhibited marasmus. At the end of the test (day 20), the rabbits in all of the treatment groups exhibited favorable mental and physical statuses, and no mites or scabs were found in the right ears of these rabbits (Table 3).

Table 1 The acaricidal activities of oregano oil, carvacrol, thymol and p-cymene against *Psoroptes cuniculi* in vitro.

Group(s) Treatment(s)	1 h Mean mortality (%) ± SD	1 h Mean rank	3 h Mean mortality (%) ± SD	3 h Mean rank	6 h Mean mortality (%) ± SD	6 h Mean rank	9 h Mean mortality (%) ± SD	9 h Mean rank	12 h Mean mortality (%) ± SD	12 h Mean rank	18 h Mean mortality (%) ± SD	18 h Mean rank	24 h Mean mortality (%) ± SD	24 h Mean rank
Oregano oil 0.05% (v/v)	100.0 ± 0.00A	43.0	100.0 ± 0.00A	43.0	100.0 ± 0.00A	39.0	100.0 ± 0.00A	38.0	100.0 ± 0.00A	38.0	100.0 ± 0.00A	37.0	100.0 ± 0.00Aa	35.5
Oregano oil 0.02% (v/v)	14.00 ± 5.48Ba	30.4	58.00 ± 8.37Ba	32.6	100.0 ± 0.00A	39.0	100.0 ± 0.00A	38.0	100.0 ± 0.00A	38.0	100.0 ± 0.00A	37.0	100.0 ± 0.00Aa	35.5
Oregano oil 0.01% (v/v)	0.00 ± 0.00c	12.0	4.00 ± 5.48C	13.2	20.00 ± 7.07Ba	18.6	28.00 ± 8.37B	17.3	56.00 ± 11.40B	21.2	78.00 ± 8.37B	22.0	84.00 ± 11.40Ab	22.6
Oregano oil 0.005% (v/v)	0.00 ± 0.00c	12.0	2.00 ± 4.47C	11.1	4.00 ± 5.48C	8.7	4.00 ± 5.48Ca	6.9	8.00 ± 8.37C	6.80	38.00 ± 8.37C	9.80	48.00 ± 8.37B	8.2
Carvacrol 0.1 mg/mL (w/v)	10.00 ± 7.07B	26.0	24.00 ± 8.94D	22.7	30.00 ± 10.00B	22.1	46.00 ± 8.94D	22.8	52.00 ± 8.37B	19.8	72.00 ± 8.37B	19.80	84.00 ± 11.40Ab	22.6
Thymol 0.2 mg/mL (w/v)	6.00 ± 5.48BCb	21.6	44.00 ± 8.94Bb	28.6	56.00 ± 11.40D	27.8	76.00 ± 11.40E	27.9	78.00 ± 8.37D	27.8	92.00 ± 8.37AB	30.2	96.00 ± 5.48A	30.7
p-Cymene 1% (V/V)	0.00 ± 0.00C	12.0	0.00 ± 0.00C	9.0	6.00 ± 5.48Cb	10.3	18.00 ± 8.37Cb	13.6	28.00 ± 8.37E	13.0	42.00 ± 8.37C	11.2	66.00 ± 5.48C	13.4
Ivermectin 1% (w/v)	42.00 ± 8.37E	38.0	76.00 ± 5.48E	37.8	96.00 ± 5.48A	36.0	100.0 ± 0.00A	38.0	100.0 ± 0.00A	38.0	100.0 ± 0.00A	37.0	100.0 ± 0.00Aa	35.5
Control	0.00 ± 0.00C	12.0	0.00 ± 0.00C	9.0	0.00 ± 0.00C	5.5	0.00 ± 0.00C	4.5	2.00 ± 4.47C	4.4	4.00 ± 5.48D	3.0	4.00 ± 5.48D	3.0

The difference between data with the different capital letter within a column is used to represent $P < 0.01$, and the difference between data with the different small letters within a column is used to represent $P < 0.05$.

Table 2 The LT$_{50}$ values of oregano oil, carvacrol, thymol and p-cymene against Psoroptes cuniculi in vitro by CLL model.

Group(s) Drug(s)	Concentration(s)	Regression line	LT$_{50}$(h) (95%FL)	Pearson Chi-square
Oregano Oil	0.02%	Y = 3.906x−1.315	2.171 (1.841−32.521)	7.923
	0.01%	Y = 3.249x−3.434	11.396 (10.128−12.857)	2.684
	0.005%	Y = 2.987x−4.232	26.102 (18.935−66.282)	10.939
Carvacrol	0.1 mg/mL	Y = 1.613x−1.536	8.957 (7.291−11.117)	6.305
Thymol	0.2 mg/mL	Y = 2.175x−1.405	4.424 (3.615−5.254)	3.497
p-Cymene	1%	Y = 2.954x−3.481	15.201 (7.702−773.245)	44.188

Table 3 Acaricidal activity of two different concentrations of oregano oil against Psoroptes cuniculi in rabbits, measured by clinical score of infestation.

Groups	Day(s)			
	0	8	15	20
Group A (5%)	2.8 ± 0.84 A	1.1 ± 0.55 A	0.0 ± 0.00 A	0.0 ± 0.00
Group B (1%)	2.4 ± 0.55 A	1.3 ± 0.67 A	0.2 ± 0.27 A	0.0 ± 0.00
Group C (Ivermectin)	2.6 ± 0.55 A	1.0 ± 0.61 A	0.0 ± 0.00 A	0.0 ± 0.00
Group D (Negative control)	2.8 ± 0.84 A	2.8 ± 0.84 B	3.0 ± 0.71 B	−[a]

The difference between data with the different capital letter within a column is very significant ($P < 0.01$).

[a] After the observation at day 15, rabbits of group D were treated.

4 DISCUSSION

In the present study, the results revealed that OR exhibited a significant dose- and time-dependent acaricidal effect against P. cuniculi. Indeed, the 0.05% and 0.02% concentrations of OR killed all mites within 1 h and 6 h, respectively. Meanwhile, compared to control group and p-cymene, carvacrol and thymol also possessed the marked acaricidal activity with the mean mortality 84.00% and 96.00% at 24 h. Moreover, the result indicated that the LT 50 values against P. cuniculi of the concentrations of 0.02%, 0.01% and 0.005% (v/v) OR, 0.2 mg/mL (w/v) thymol, 0.1 mg/mL (w/v) carvacrol and 1% p-cymene (v/v) were 2.171 h, 11.396 h, 26.102 h, and 4.424 h, 8.957 h and 15.201 h, respectively (Tables 1 and 2). These results indicated that the three compounds all demonstrated acaricidal activities, but their activities were low compared with OR at the same or similar concentrations. Therefore, we suggest that these compounds may exhibit synergistic acaricidal effects.

Specifically, the synergetic effects of thymol and carvacrol against Spodoptera littoralis larvae, Amblyomma sculptum and Dermacentor nitens larvae and other organisms have been reported (Pavela, 2010; Novato et al., 2015). Meanwhile, considering that other monoterpene compounds, such as eugenol, share the same phenolic hydroxyl group in their structures and all exhibit strong acaricidal activity; we suggest that this phenolic hydroxyl group may play an important role in the killing of ticks and mites via action on the GABA receptor and the octopamine receptor. Additionally, because they share the same target or mechanism of action against mites and ticks, the acaricidal effects of each compound may be synergistically combined and enhanced. Therefore, additional studies of the structure-activity relationships associated with acaricidal activity should be conducted.

Finally, in contrast to the rabbits of the negative control group, after the three treatments topically, the P. cuniculi were completely eliminated in the rabbits in the 1% and 5% oregano oil groups (Table 3). Moreover, at the end of the test (day 20), the rabbits in all the treatment groups exhibited favorable mental and physical statuses. We believe that after further in-depth study, the potential acaricidal agent OR could be widely applied for the treatments of acariasis in animals.

5 CONCLUSION

The results of the present study demonstrated that oregano oil and its three major compounds exhibited significant acaricidal activities against *P. cuniculi* in vitro. Additionally, oregano oil exhibited good clinical efficacy in vivo. Moreover, the acaricidal mechanism of OR should be studied, and other active compounds of OR and the structure-activity relationship associated with the acaricidal activities of active compounds should also be further investigated.

ACKNOWLEDGEMENTS

This work was financed by the National Natural Science Foundation of China (31302136), the Special Fund for Agro-scientific Research in the Public Interest (201303040-14) and National Science and Technology Infrastructure Program of China (2015BAD11B01).

（发表于《Veterinary Parasitology》，院选SCI，IF：2.242. 93-96.）

Lowering Effects of Aspirin Eugenol Ester on Blood Lipids in Rats with High Fat Diet

KARAM Isam[†], MA Ning[†], LIU Xiwang, KONG Xiaojun,
ZHAO Xiaole, YANG Yajun[*], LI Jianyong[*]

(Key Lab of New Animal Drug Project of Gansu Province, Key Lab of Veterinary Pharmaceutical Development, Ministry of Agriculture, Lanzhou Institute of Husbandry and Pharmaceutical Science of CAAS, Lanzhou 730050, China)

Abstract: (Background): Aspirin and eugenol were esterified to synthesize aspirin eugenol ester (AEE). As a pale yellow and odourless crystal, AEE reduced the gastrointestinal damage of aspirin and vulnerability of eugenol. The study was conducted to evaluate the preventive effects of AEE on blood lipids in rats with high fat diet (HFD). (Methods): Suspensions of AEE and simvastatin were prepared in 5% carboxymethyl cellulose sodium (CMC-Na). In order to observe the intervention effects, the drugs and HFD were administrated at the same time. Based on individual weekly body weight (BW), AEE was intragastrically administered at the dosage of 18, 36 and 54 mg/kg. Simvastatin (10 mg/kg) and CMC-Na (20 mg/kg) were used as control drug. After 6 weeks of administration, the changes of BW and blood lipid indices including triglyceride (TG), low density lipoprotein (LDL), high density lipoprotein (HDL) and total cholesterol (TCH) were determined in the experiment. (Results): The rat blood lipids profile in model group was remarkably different after feeding 6-weeks HFD. TG, TCH and LDL indexes in model group were increased significantly compared with those in control group ($P < 0.01$). AEE at the dosage of 54 mg/kg significantly decreased levels of TG, TCH and LDL ($P < 0.01$), and slowed the rate of BW gain in comparison with model group ($P < 0.05$). Moreover, high dose AEE showed better effects than simvastatin on reducing TCH level and similar effects on TG, HDL and LDL. (Conclusion): AEE could remarkably reduce levels of TG, TCH and LDL in rats with high fat diet, and slow the rate of body weight gain. It was conducted that AEE was a potential candidate on reducing blood lipids level. The mechanism of action of AEE should be investigated in further studies.

Key words: Aspirin eugenol ester (AEE); Blood lipids; Body weight; High fat diet; Rats

---

[*] Corresponding authors, E-mail: yangyue10224@163.com; lijy1971@163.com.
[†] Equal contributors.

1 BACKGROUND

Hyperlipidemia is a heterogeneous group of disorders characterized by an excess of lipids in blood stream such as the increased serum levels of triglycerides (TG), total cholesterol (TCH), low-density lipoprotein (LDL) as well as decreased levels of high-density lipoprotein (HDL) [1, 2]. Hyperlipidemia, as the major risk factor for the development of cardiovascular diseases, is becoming a major health problem in the world [3].

Aspirin could ameliorate hyperlipidemia induced by high fat diet and hyperinsulinemia in rats [4]. Moreover, aspirin could diminish hypertriglyceridemia in obese rodents and has potential in hyperlipidemia prevention [5]. Eugenol as volatile oil is extracted from dry alabastrum of Eugenia caryophyllata. Therapeutic effects of eugenol on hyperlipidemia had been proved in previous study [6, 7]. Four-week administration of eugenol could significantly decrease the serum lipid profile in normal albino rabbits [8].

Based on prodrug principle and therapeutic effects of aspirin and eugenol on hyperlipidemia, aspirin eugenol ester (AEE) as a new drug was synthesized [9]. The metabolites of AEE had been confirmed in beagle dog and liver microsomes. AEE could be metabolized into aspirin and eugenol in vitro and in vivo, which could show their original activities and act synergistically [10]. AEE also reduced the side effects of its precursor such as the gastrointestinal damage of aspirin and irritation of eugenol [11]. The acute toxicity of AEE was less than its precursor, which was 0.02 times of aspirin and 0.27 times of eugenol [9, 11]. The teratogenicity and mutagenicity of AEE have been investigated. AEE did not show any mutagenesis in Ames test and the mouse bone marrow micronucleus assay [11, 12]. Moreover, the effects of AEE had been evaluated in animal model. The results showed that AEE had positive effects on antithrombosis, anti-inflammatory, analgesia and antipyretic [9, 11, 13].

Therapeutic strategies for hyperlipidemia treatment depend on reducing blood lipids. There are many chemical drugs that lower cholesterol level in the body such as statins, fibrates, ezetimibe and nicotinic acid. However, most of them are expensive and have undesirable effects [14]. There is an obvious need for more efficacious and alternative treatment options for hyperlipidemia. In our previous study, AEE (50 mg/kg and 160 mg/kg) could reduce TCH and TG in rats with standard diet [11]. Moreover, five-week administration of AEE (54 mg/kg) could normalize blood lipids profile in hyperlipidemic rats [15]. So there are increasing interest to evaluate the intervention effects of AEE on blood lipids in rat with high fat diet. This study will increase the understanding of AEE and provide impetus for further studies.

2 METHODS

2.1 Chemicals and reagents

Aspirin eugenol ester (AEE), transparent crystal with the purity of 99.5% by RE-HPLC, was prepared in Key Lab of New Animal Drug Project of Gansu Province, Key Lab of Veterinary Pharmaceutical Development of AgriculturaL·ministry, Lanzhou Institute of Husbandry and Pharmaceutical Sciences of CAAS. CMC-Na and simvastatin was supplied by Tianjin Chemical

Reagent Company (Tianjin, China). Standard compressed rat feed and high diet food were supplied by Keao Xieli Co., Ltd (Beijing, China). Standard rat diet consisted of 12.3% lipids, 63.3% carbohydrates, and 24.4% proteins (kcal) and high fat diet (standard rat diet 77.8%, yolk power 10%, lard 10%, cholesterol 2%, bile salts 0.2%) consisted of 41.5% lipids, 40.2% carbohydrates, and 18.3% proteins (kcal). The TG, TCH, LDL and HDL kits were provided by Ningbo Medical System Biotechnology Co., Ltd (Ningbo, China). Erba XL-640 analyzer (German) was used to measure the blood lipid indices.

2.2 Animals

Seventy Sprague-Dawley (SD) male rats were purchased from the animal breeding facilities of Gansu University of Chinese Medicine (Lanzhou, China). The rats were housed in plastic cages of appropriate size (50 cm × 35 cm × 20 cm, ten rats per cage) with stainless steel wire cover and chopped bedding. Rat feed and drinking water were supplied ad libitum. Light/dark regime was 12/12 h and living temperature was 22 ± 2℃ with relative humidity of 55% ± 10%. Animals were acclimatized for 2 weeks before study initiation.

2.3 Serum sampling

At the end of the experiment, rats were fasted for 10-12 h and then anaesthetized with 1% pentobarbital sodium. The blood samples were collected from the heart with vacuum tube. The sera were obtained by centrifuging for 15 mins at the speed of 4 000 g at 4℃. Serum samples were stored at -80℃ until the day of analysis.

2.4 Drug preparation

AEE and simvastatin liquid suspensions were prepared in 0.5% of CMC-Na.

2.5 Study design

Rats were randomly divided into seven groups including control, model, CMC-Na, simvastatin and three AEE groups. The detailed design of the experiment is shown in Table 1. Simvastatin was used as a positive control (10 mg/kg). The administrations of food and drugs were started at the same time. The volumes of CMC-Na and drug suspensions were nearly equal. After 6-weeks administration, the blood lipid levels were analyzed.

2.6 Statistics

The statistical analyses were carried out using IBM SPSS 19.0 (USA). All data obtained from the experiment are expressed as mean ± standard deviation (SD). Statistical differences were evaluated by using one-way ANOVA with Tukey's multiple comparison tests. P-values less than 0.05 were considered statistically significant.

Table 1 Study design of the experiment.

Groups	Food	Drug	Dosage (mg/kg)	Concentration (mg/mL)
Control	SRD	—	—	—
Model	HFD	—	—	—

(continued)

Groups	Food	Drug	Dosage (mg/kg)	Concentration (mg/mL)
CMC-Na	HFD	CMC-Na	20	5
Statin	HFD	Simvastatin	10	2.5
AEE Low	HFD	AEE	18	4.5
AEE Mid	HFD	AEE	36	9
AEE High	HFD	AEE	54	13.5

Rats were divided into seven groups. Control group was received SRD and the rest groups were received HFD, respectively. Based on the individual weekly body weight, rats were given with different drug suspension volume. CMC-Na as a vehicle was used in control group. Simvastatin was designed as positive drug. Different dose of three levels of AEE were administrated in the study. The administration period was 6 weeks and then blood lipids were analyzed AEE aspirin eugenol ester, SRD standard rat diet, HFD high fat diet.

3 RESULTS

3.1 Intervention effects of AEE

After feeding rats with HFD for 6 weeks, the blood lipid profile was notably different among the groups. There were significant differences between control and model groups (Table 2). In model group, TG, TCH and LDL were increased significantly in comparison with the control group ($P < 0.01$). However, no change was observed in HDL between control and model groups. There was no significant difference between CMC-Na and model groups. In comparing with model group, low and intermediate dose of AEE had no effect on blood lipid indexes except TCH index in AEE low group ($P < 0.05$). AEE high dose significantly decreased TG, TCH and LDL compared to model group ($P < 0.01$). AEE in different dosages had no influence on HDL. Simvastatin as a positive drug control could significantly decrease TCH and LDL ($P < 0.01$). With regard to TG and HDL, no changes were observed between simvastatin and model groups.

The three groups were used to compare with control and statin groups (Fig. 1). When compared with control group, TG, TCH and LDL in AEE groups were significant increased ($P < 0.01$ or $P < 0.05$). Notably, the mean values of TG, TCH and LDL in AEE high group were less than the values in low and intermediate AEE groups. Simvastatin and AEE made similar effects on TG, HDL and LDL indexes. However, simvastatin showed more positive effects on TCH than AEE L and AEE M but not AEE H ($P < 0.01$ or $P < 0.05$, Fig. 1d). In order to find out the relationship between drug effects and the dosage, intermediate and high AEE groups were compared with low dose AEE group. The results showed that there was only significant difference on TCH index in AEE high group ($P < 0.01$, Fig. 1d). Interestingly, the mean values of TG and LDL in AEE high group were the smallest in all AEE groups.

Table 2 The blood lipids levels after drugs administration for six weeks (n = 10) .

Variables	Control	Model	CMC-Na	Statin	AEE Low	AEE Mid	AEE High
TG (mmol/L)	$0.53 \pm 0.12^{**}$	0.89 ± 0.07	0.79 ± 0.19	0.83 ± 0.05	0.79 ± 0.23	0.77 ± 0.10	$0.71 \pm 0.07^{**}$
HDL (mmol/L)	0.72 ± 0.04	0.74 ± 0.08	0.69 ± 0.09	0.66 ± 0.13	0.73 ± 0.09	079 ± 0.10	0.65 ± 0.15
LDL (mmol/L)	$0.25 \pm 0.02^{**}$	0.50 ± 0.09	0.46 ± 0.06	$0.39 \pm 0.13^{**}$	0.45 ± 0.05	0.46 ± 0.10	$0.36 \pm 0.08^{**}$
TCH (mmol/L)	$1.4 \pm 0.10^{**}$	2.29 ± 0.26	2.13 ± 0.22	1.80 ± 0.1	$2.10 \pm 0.22^{*}$	2.15 ± 0.15	$1.66 \pm 0.15^{**}$

Note: $^{*}P < 0.05$, $^{**}P < 0.01$ significant difference compared to model group. TG triglyceride, HDL high density lipoprotein, LDL low density lipoprotein, TCH total cholesterol. Blood lipids indices were increased in model group. The blood lipid levels were reduced in simvastatin and high dose AEE groups in varying degrees.

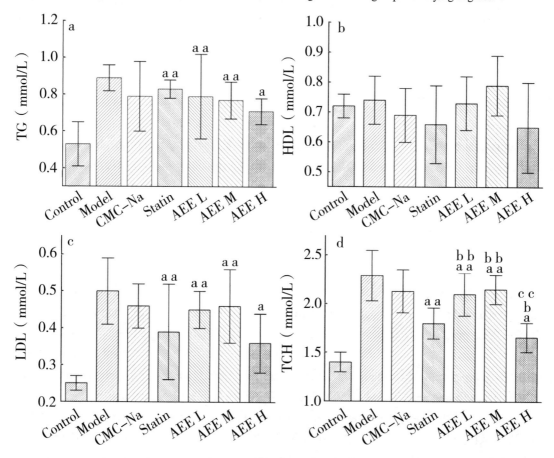

Fig. 1 Comparative effects of AEE and simvastatin on blood lipid indexes (n = 10) .

Note: $^{a}P < 0.05$, $^{aa}P < 0.01$ significant difference compared to control group; $^{b}P < 0.05$, $^{bb}P < 0.01$ significant difference compared to statin group; $^{c}P < 0.05$, $^{cc}P < 0.01$ significant difference compared to AEE L. CMC-Na: carboxymethylcellulose sodium, AEE L: low dose of AEE (18 mg/kg) , AEE M: intermediate dose of AEE (36 mg/kg) , AEE H: high dose of AEE (54 mg/kg) . a, b, c and d were TG, HDL, LDL and TCH, respectively.

3.2 Body weight

The body weights of all groups were recorded at the beginning and ending of the experiment (Table 3). At the beginning of the experiment, there was no significant difference between all groups. At week 6, the body weights in the group supplemented with AEE high dose were significantly lower than those in model group ($P < 0.05$). The results showed that high dose of AEE could reduce the rate of body weight gain. The body weights in AEE intermediate and high groups were remarkably lower than control group ($P < 0.05$ or $P < 0.01$). No difference was observed among statin and AEE groups. The gains in body weights in AEE high group were significantly lower than those in AEE low group ($P < 0.05$).

Table 3 Body weights of rats at the beginning and the end of the experiment (n = 10).

Groups	Week 1	Week 6
Control	209 ± 15	385 ± 20
Model	216 ± 11	368 ± 25
CMC-Na	209 ± 10	365 ± 20
Statin	220 ± 22	355 ± 20^a
AEE Low	220 ± 17	366 ± 31
AEE Mid	211 ± 15	356 ± 20^a
AEE High	213 ± 14	340 ± 15^{*aab}

Note: $^*P < 0.05$ significant difference compared to model group; $^aP < 0.05$, $^{aa}P < 0.01$ significant difference compared to control group; $^bP < 0.05$ significant difference compared to AEE low group; Week 1: beginning of the experiment; Week 6: the end of the experiment.

3.3 Discussion

Based on prodrug principle, aspirin and eugenol, as starting precursors, were esterified to synthesize aspirin eugenol ester (AEE). The positive effects of AEE on the symptoms of inflammation, fever, pain and thrombosis had been confirmed in animal disease model [9, 11, 13]. The blood chemistry results in 15-day oral dose toxicity study showed AEE could significantly reduce the values of TG and TCH both in male and female rat, which indicated that this compound was potential for curing hyperlipidemia [11]. In subsequent experiment, the regulation effects of AEE on blood lipid profile in hyperlipidemic rats had been confirmed [15]. Based on this fact, it is essential to evaluate the intervention effects of AEE on blood lipids in rats with high fat diet fed.

The intervention effect of AEE was evaluated by comparing AEE groups with model group. AEE low and intermediate dose showed no significant effect on blood lipid indices except reducing TCH at AEE low dose. It is noteworthy that high dose of AEE was highly effective in reducing blood lipid indices such as TG and TCH. These meant that the activity of AEE at high dose was better than low and intermediate doses. So there was a positive relationship between drug efficacy and dosage. Under the present experimental conditions, the optimal dose was considered to be

54 mg/kg for normalizing blood lipid profile in high fat diet fed rat.

HDL is responsible for the transportation of cholesterol to liver, which is essential for cholesterol removal [16]. HDL levels of rat fed with high fat diet for six weeks showed no changes in this study. There are several possible reasons for this result. First of all, HFD compositions such as the lack of saturated fat may be insufficient to make substantial change on HDL. Second, the constant values of HDL may be attributable to the short duration of the experiment (only 6 weeks). Finally, animal model used in the experiment may be a potential cause. Literature indicated that most cholesterol in rat blood circulation existed in HDL and this contributed to reducing susceptibility of HDL level to HFD [17, 18]. HFD is a commonly used material to induce animal disease such as hyperlipidemia, diabetes, obesity and arteriosclerosis [19, 20]. In this experiment, the results of body weight were not consistent with those in our previous study. There was no difference in body weight between control and model group. However, HFD with same components significantly increased body weight at the end of eighth week in Wistar rat [15]. The shortage of HFD consumption time may be the reason for the results of no significant difference in body weight. Metabolism and liver function could be substantially changed with HFD feeding [21]. It was necessary to give the rat time to adjust for HFD gradually, which may slow the rate of the body weight gain in certain period. It was found that AEE dosage had different influence on body weight. AEE high and middle dose made an impact on body weight and there was significant difference between low and high AEE groups. The results of AEE on body weight were also the reason why 54 mg/kg was suggested for optimal dosage.

Simvastatin is an inhibiter of 3-hydroxy-3-methylglutaryl coenzyme A (HMG-CoA) which plays a crucial role in cholesterol synthesis in liver cells. Simvastatin was used as a positive drug control in this study and it decreased the levels of TCH and LDL. The effects of 10 mg/kg simvastatin on TCH and LDL were similar with 54 mg/kg AEE. However, simvastatin had no influence in TG index and showed significant difference from high dose AEE. From these results, it could be concluded that AEE was more effective than simvastatin on reducing blood lipid level. CMC-Na was widely used as a reliable drug vehicle in pharmaceutical industry [22, 23]. The effect of CMC-Na on blood lipid indexes was eliminated by administrating equal volume CMC-Na to control group. Therefore, it manifested that the effects of AEE was not related to CMC-Na.

Therefore, intragastrical administration of AEE had a significant intervention effects in HFD fed rat. AEE was decomposed into salicylic acid and eugenol after absorption. The effects of AEE on reducing blood lipid indexes may be mainly from synergetic action of aspirin and eugenol. More studies are necessary to investigate the action mechanism of AEE such as evaluation of inhibitory effect on digestive enzymes and influence on metabolic targets.

4 CONCLUSIONS

Under the present study condition, AEE at daily dosage of 54 mg/kg BW for six weeks could remarkably reduce levels of TG, TCH and LDL in rats with high fat diet, and slow the rate of body weight gain. Moreover, AEE at 54 mg/kg showed better effects than simvastatin at 10 mg/kg on reducing TCH level and similar effects on TG, HDL and LDL.

Abbreviations

AEE: Aspirin eugenol ester; BW: Body weight; CMC-Na: Sodium carboxymethyl cellulose; HDL: High density lipoprotein; HFD: High fat diet

LDL: Low density lipoprotein; SRD: Standard rat diet; TCH: Total cholesterol

TG: Triglycerides

Acknowledgements

The work was supported by the National Natural Science Foundation of China (No.31402254) and the Natural Science Foundation for Youth of Gansu Province (1506RJYA148).

Funding

The funders had no role in the design of the study and collection, analysis, and interpretation of data and in writing the manuscript.

Availability of data and materials

All data generated or analyzed during the current study are available from the corresponding author on reasonable request.

Authors' contributions

Conceived and designed the experiments: NM JYL. Performed the experiments: NM IM ZXL. Analyzed the data: NM YJY XWL IM. Contributed reagents/materials/analysis tools: IM NM XWL XJK. Wrote the paper: NM YJY JYL. All authors read and approved the final manuscript.

Competing interests

The authors have declared that no competing interests existed.

Consent for publication

Not applicable.

Ethics approval

The study protocol was approved by the Institutional Animal Care and Use Committee of Lanzhou Institute of Husbandry and Pharmaceutical Science of the Chinese Academy of Agricultural Sciences. Animal welfare and experimental procedures were performed strictly in accordance with the Guidelines for the Care and Use of Laboratory Animals issued by the US National Institutes of Health.

(发表于《Lipids in Health and Disease》, SCI, IF: 2.137.196.)

Characterization of the Complete Mitochondrial Genome Sequence of Wild Yak (*Bos mutus*)

LIANG Chunnian, WU Xiaoyun, DING Xuezhi, WANG Hongbo,
GUO Xian, CHU Min, BAO Pengjia, YAN Ping

(Key Laboratory for Yak Breeding Engineering of Gansu Province, Lanzhou Institute of Husbandry and Pharmaceutical Sciences, Chinese Academy of Agricultural Sciences, Lanzhou 730050, China)

Abstract: Wild yak is a special breed in China and it is regarded as an important genetic resource for sustainably developing the animal husbandry in Tibetan area and enriching region's biodiversity. The complete mitochondrial genome of wild yak (16,322bp in length) displayed 37 typical animal mitochondrial genes and A+T-rich (61.01%), with an overall G+C content of only 38.99%. It contained a non-coding control region (D-loop), 13 protein-coding genes, two rRNA genes, and 22 tRNA genes. Most of the genes have ATG initiation codons, whereas ND2, ND3, and ND5 genes start with ATA and were encoded on H-strand. The gene order of wild yak mitogenome is identical to that observed in most other vertebrates. The complete mitochondrial genome sequence of wild yak reported here could provide valuable information for developing genetic markers and phylogenetic analysis in yak.

Key words: Genome; Mitochondrion; Wild yak (Bos mutus)

Wild yak (*Bos mutus*) is a unique bovine distributing in Qinghai-Tibet Plateau area in China. It is an important genetic resource as the base for sustainably developing the animal husbandry in Tibetan area and as the gene bank for persistence of the region's biodiversity. The wild yak has the strong body conformation, high-altitude adaptation and cold-resistant ability (Wiener et al., 2003). However, wild yak has suffered from population shrinking and genetic characterization degenerating. In this study, mitochondrial DNA (mtDNA) were extracted from the individual blood samples by the genomic DNA isolation kit (TaKaRa, Dalian, China) in accordance with the manufacturer's protocol and used as templates for PCRs. In total, six pairs of PCR primers were designed to cover the mitochondrial genomes. PCR products were purified and then sequenced by Life Technologies (Shanghai, China). The sequenced mtDNA fragments were assembled by DNAStar 5.1 software (DNASTAR, Inc., Madison, WI). The protein-coding gene sequences were found using the ORF finder tool (Biostar Microtech USA Corporation,

---

Corresponding author, E-mail: pingyanlz@163.com.

Doral, FL) at NCBI (http://www.ncbi.nlm.nih.gov/gorf/gorf.html). Putative tRNA genes were identified using the program tRNAscan-SE (Biostar Microtech USA Corporation, Doral, FL) (Schattner et al., 2005) and non-coding regions and rRNA genes were identified using BLAST search. Gene boundaries were confirmed based on comparison and alignment with other published mt genomes of species in Bos taurus and Bos grunniens (Anderson et al., 1982; Guo et al., 2006; Wang et al., 2010).

This study showed that the complete mtDNA of wild yak is a 16, 322bp circular DNA molecule (GenBank accession no. KR106993). The structural organization of mitogenome is identical to that observed in most other vertebrates (Broughton et al., 2001; Guo et al., 2006; Tu et al., 2014). The mtDNA sequence was A+T-rich (61.01%), with an overall G+C content of only 38.99%. The whole wild yak mtDNA contains 37 typical genes: 13 protein-coding genes, 22 tRNA genes, two rRNA genes, and one non-coding region (D-loop region). Ten protein-coding genes used ATG as a start codon, with the exception of ND2, ND3, and ND5 with ATA. These genes have four types of stop codon, including eight genes (ND1, COX1, COX2, ATP8, ATP6, ND4L, ND5, and ND6) ended with TAA. ND2 was terminated by TAG and CYTB by AGA. Another three genes (COX3, ND3, and ND4) ended with the incomplete stop codon "T-" which is the 5' 0 terminal of the adjacent gene. Twenty-two tRNA genes (ranging from 60 to 74bp) are interspersed in mtDNA. Most of the ORFs and tRNAs were located on the heavy-strand (H), with the exception of the eight tRNA and ND6 genes. The D-loop region located between tRNA^{Pro} and tRNA^{Phe} with a length of 892bp, which is responsible for transcription and replication of the mitochondrial genome. The small non-coding region, a putative origin of the light-strand replication, was located between tRNA^{Asn} and tRNA^{Cys} genes in the length of 31bp. The 16S rRNA gene located between the tRNA^{Phe} and tRNA^{Val} had a length of 1571bp, whereas the 12S rRNA gene located between tRNA^{Val} and tRNA^{Leu} had a length of 957bp.

The complete wild yak mtDNA sequence, together with gene annotations, was submitted to the GenBank with the accession number KR106993. Based on the published mitochondrial genome sequences of 11 closely species, a molecular phylogenetic tree was constructed (Figure 1). As shown in Figure 1, the wild yak mitochondrial genome identified in this study is closely related to genome of domestic yak.

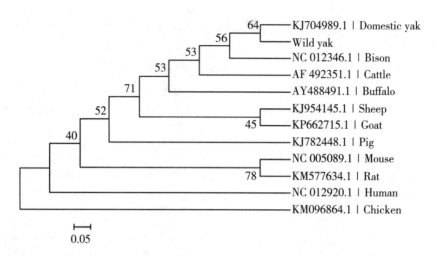

Figure 1 The phylogenetic tree of wild yak (Bos mutus) in this study and other 11 close species which have reported mitogenome sequences using UPGMA algorithm. The boldface values above the branch are bootstrap values obtained by 1000 step.

DECLARATION OF INTEREST

This project was funded by the Special Fund for Agro-scientific Research in the Public Interest (Grant no. 201203008), "Five-twelfth" National Science and Technology Support Program (No. 2012BAD13B05) and Agricultural Science and Technology Innovation Program (No. CAAS-ASTIP-2014-LIHPS-01). The authors report no conflicts of interest.

（发表于《MITOCHONDR DNA》，SCI，IF：1.76. 1–2.）

Determination of Antibacterial agent Tilmicosin in Pig Plasma by LC/MS/MS and its Application to Pharmacokinetics

LI Bing[1,2,3], GONG Shiyue[4], ZHOU Xuzheng[1,2,3], YANG Yajun[1,2,3], LI Jianyong[1,2,3], WEI Xiaojuan[1,2,3], CHENG Fusheng[1,2,3], NIU Jianrong[1,2,3], LIU Xiwang[1,2,3], ZHANG Jiyu[1,2,3*]

(1. Key Laboratory of Veterinary Pharmaceutical Development, Ministry of Agriculture, Lanzhou, China; 2. Key Laboratory of New Animal Drug Project of Gansu Province, Lanzhou, China; 3. Lanzhou Institute of Husbandry and Pharmaceutical Sciences of CAAS, Lanzhou, China; 4. China Agricultural Vet. (Tianjin) Biological Medicine Co. Ltd, Tianjin, China)

Abstract: A rapid and sensitive high-performance liquid chromatography-tandem mass spectrometry (LC-MS/MS) method was developed and validated to quantify tilmicosin in pig plasma. Plasma samples were prepared by liquid-liquid extraction. Chromatographic separation was achieved on a C 18 column (2.1 mm × 30 mm, 3.5 μm) using acetonitrile-water (90 : 10, v/v; water included 0.1% formic acid) as the mobile phase. Mass detection was carried out using positive electrospray ionization in multiple reaction monitoring mode. The calibration curve was linear from 0.5 to 2 000 ng/mL ($r^2=0.9998$). The intra- and inter-day accuracy and precision were within the acceptable limits of ± 10% for all tilmicosin concentrations. The recoveries ranged from 95 to 99% for the three tested concentrations. The LC-MS/MS method described herein was simple, fast and less laborious than other methods, achieved high sensitivity using a small sample volume, and was successfully applied to pharmacokinetic studies of tilmicosin enteric granules after oral delivery to pigs. In comparison with tilmicosin premix, tilmicosin enteric granules slowed the elimination rate of tilmicosin, prolonged its period of action and significantly improved its bioavailability.

Key words: LC/MS/MS; Pharmacokinetics; Pig plasma; Tilmicosin; Tilmicosin enteric granules

1 INTRODUCTION

Pig pleuropneumonia caused by Actinobacillus pleuropneumoniae (APP) is a contagious disease that causes significant economic losses worldwide (Gottschalk, 2015). Following the first report of APP in the UK in 1957, the disease occurred in North America, Germany and Canada successively, after which it developed into a worldwide disease. APP is especially prevalent

---

Corresponding author, E-mail: infzjy@sina.com.

in Europe and America, where it results in significant economic losses to the pig industry (Liu et al., 2011). Recently, an APP epidemic has emerged in China. Acute APP infection can induce severe clinical signs, including lung lesions, and is characterized by pleuritis, respiratory distress and mortality (Tobias et al., 2014). APP infection may become chronic or subclinical without previous signs of the disease; therefore, outbreaks may suddenly appear, while subclinical infections may remain silent (Marois et al., 2009).

The virulence of APP varies among biotypes: biotype I has been divided into 13 serotypes, whereas biotype II has been divided into two serotypes, for a total of 15 (Zimmerman, D'Allaire & Taylor, 2006; Zhang et al., 2015). Serotypes 2 and 9, two virulent serotypes, are prevalent in outbreaks in European countries, particularly in France (Gottschalk, Morvan, Broes, Desrosier, & Kobisch, 2005). In addition, drug treatment rapidly induces drug resistance in APP, making disease treatment and control difficult (Zhang et al., 2015).

Currently, clinical APP outbreaks are mainly controlled by giving antibiotics to all pigs in an infected facility, whether or not they are clinically affected (Hoflack, Maes, Mateusen, Verdonck, & de Kruif, 2001; Gottschalk, 2012). Control and prevention of APP are currently achieved by vaccination (Tumamao et al., 2004) or the use of antimicrobials such as enrofloxacin, amoxicillin and ceftiofur (Mengelers, Kuiper, Pijpers, Verheijden, & van Miert, 2000; Hoflack et al., 2001; Klinkenbeg et al., 2014). APP serotypes show large differences in virulence and weak cross-protection capacity among serotypes (Chen, Xiao, & Wen, 2006). Indeed, vaccination does not always prevent infection of susceptible pigs by APP (Velthuis, De Jong, Kamp, Stockhofe, & Verheijden, 2003). Widespread use of antibiotics contributes to the problem of drug resistance in APP, which inhibits successful treatment. Therefore, despite the large number of studies on this disease, new drugs with increased efficacy are needed.

Tilmicosin is a new semisynthetic macrolide antibiotic with a wide range of veterinary uses, including treatment of bacterial and myco-plasma infections. Tilmicosin has antimicrobial activity stronger than that of tylosin against Pasteurella haemolyticus, Pasteurella multocida and Mycoplasma spp. (Jiang, Shen, & Hu, 2002). In Australia, Italy and the USA, tilmicosin has been approved to prevent and treat infection by susceptible bacteria in bovines, goats, sheep, swine and chickens. In particular, tilmicosin is used to treat livestock diseases related to respiratory tract infection, such as pig pleuropneumonia, pasteurellosis, chicken mycoplasmosis and mastitis. Tilmicosin has been approved in China to treat Pasteurella infection in swine and bovines, as well as to treat Mycoplasma infection in chickens. Currently, tilmicosin is usually applied clinically in the form of premixes and injectable tilmicosin formulations. Tilmicosin premixes are orally administered and absorbed rapidly, but incompletely. In addition, tilmicosin premixes taste bitter and are unpalatable to pigs. Tilmicosin injection is not safe, because it can lead to tachycardia and decreased myocardial contractility through cardiovascular toxicity. Tachypnea, emesis and convulsion occur in pigs after intramuscular injection of tilmicosin at a dose of 10 mg/kg, whereas most pigs die following intramuscular injection of 20 mg/kg tilmicosin (China Veterinary Pharmacopoeia, 2005). The narrow therapeutic window of tilmicosin greatly limits its clinical application. Therefore, new preparations of tilmicosin derivatives with better pharmacokinetic profiles are needed to overcome the limitations of the drug.

In a previous study, Hubei Longxiang Pharmaceutical Co. Ltd (Wuxue City, Hubei, China) successfully prepared tilmicosin enteric granules as an oral preparation. Tilmicosin enteric granules are prepared by microencapsulation, which increases the bioavailability of tilmicosin and improves its palatability to pigs, in comparison with tilmicosin premixes. Tilmicosin enteric granules show significant treatment efficacy in pigs suffering from pleuropneumonia, as well as good thermodynamic stability and storage stability.

In order to define the pharmacokinetic profile of tilmicosin enteric granules, a method for the determination of tilmicosin in pig plasma must be established. High-performance liquid chromatography-tandem mass spectrometry (LC-MS/MS) was chosen for the determination of tilmicosin enteric granules because it is accurate and sensitive. LC-MS/MS was successfully applied to characterize the pharmacokinetics of tilmicosin enteric granules in pigs following oral administration at a dose of 100 mg/kg.

2 MATERIALS AND METHODS

2.1 Reagents and chemicals

Tilmicosin and internal standard (IS) roxithromycin (Fig. 1) were provided by Dr Ehrenstorfer GmbH (Germany) (batch numbers 31 102 and 30 822, respectively). Tilmicosin and the IS had purity > 99%. Acetonitrile and methanol (MS grade) were purchased from Fisher Chemical (Waltham, MA, USA). Formic acid (HPLC grade) was purchased from Fisher Chemical (Waltham, MA, USA). NaOH and NaCl (analytical grade) were purchased from China Pharmaceutical Group Corporation. Water was purified through a Milli-Q Plus water system (Millipore Corporation, Bedford, MA, USA) before use. Tilmicosin enteric granules containing 20% tilmicosin (batch number 121 001) were supplied by Hubei Longxiang Pharmaceutical Co. Ltd. Tilmicosin premix containing 20% tilmicosin (batch number 20 140 404) was supplied by Ringpu Bio-Pharmacy Co. Ltd (Tianjin, China).

2.2 Equipment

The LC-MS/MS equipment (1 200-6 410A) consisted of a LC system with a binary pump (Model SL) and a triple quadrupole mass spectrometer with electrospray ionization (ESI) (Agilent Technologies Inc., Santa Clara, CA, USA). The system was controlled using MassHunter software (version B.01.04, Agilent Technologies Inc.).

2.3 Chromatographic and mass spectrometer conditions

The analysis was carried out using an Agilent Zorbax C_{18} column (2.1 mm × 30 mm, 3.5 μm); the mobile phase used for the analysis consisted of acetonitrile-water (90 : 10, v/v; water included 0.1% formic acid). The mobile phase was filtered before use to prevent entry of bubbles or impurities in the system. The mobile phase was delivered at a flow rate of 0.2 mL/min. The sample was injected at a volume of 2 μL at 25 ℃.

The mass spectrometer was operated in positive ion mode with an ESI interface. Quantitation was performed in multiple reaction monitoring mode. In the positive mode, the MS/MS parameters were as follows: capillary voltage, 4kV; cone voltage, 40V; source temperature, 100℃; desolvation temperature, 250℃; desolvation nitrogen gas flow, 10L/min; cone gas flow, 9 L/min. The optimized fragmentation voltages for tilmicosin and the IS were 100 and 80V, respectively. The delta electron multiplier voltage was 200V. Data were collected in multiple reaction monitoring mode using [M+H]$^+$ ions for tilmicosin and the IS, with collision energy of 62 and 54eV, respectively.

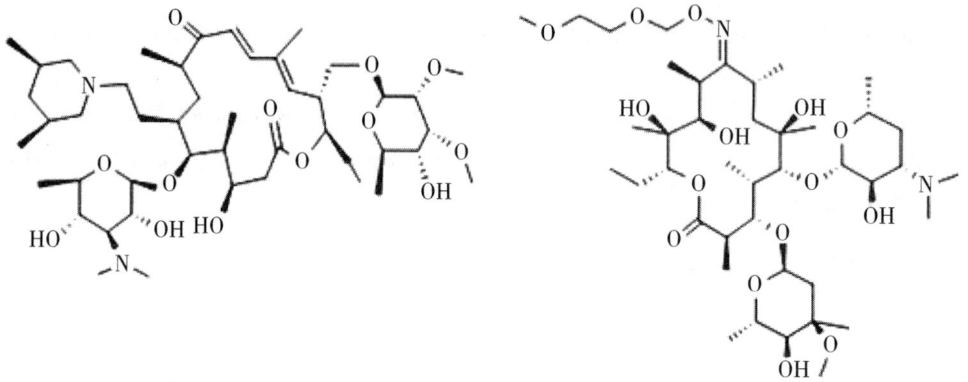

Tilmicosin (molecular weight: 869)   Roxithromycin (IS, molecular weight: 837)

Fig. 1 Chemical structures of tilmicosin and roxithromycin.

2.4 Preparation of standard solutions and quality control samples

For the standard solution of tilmicosin, 8 mg of tilmicosin was placed into a 100 mL brown volumetric flask, after which methyl alcohol was added to produce a stock solution of 80 μg/mL tilmicosin. A series of tilmicosin working standard solutions was prepared by diluting the stock solution with the mobile phase to obtain the following concentrations: 0.002, 0.01, 0.1, 4, 10, 20, 32 and 40 μg/mL.

For the IS solution, 5 mg of tylosin was placed into a 50mL brown volumetric flask, to which methyl alcohol was added to produce a stock solution of 100 μg/mL tylosin. Next, 8 mL of the stock tylosin solution was mixed with the mobile phase in a 50 mL volumetric flask to produce a solution of 16 μg/mL tylosin.

All solutions were stored at 4℃ and brought to room temperature before use. Plasma calibration standards of 0.5–2 000 ng/mL (0.5, 5, 200, 500, 1 000, 1 600 and 2 000 ng/mL) were prepared by spiking 200 μL aliquots of blank plasma with 10 μL of each of the standard solutions. Quality control (QC) samples were prepared in the same way at four concentrations: 0.5 ng/mL (QC-LLOQ), 5 ng/mL (QC-low), 1 000 ng/mL (QC-med), and 2 000 ng/mL (QC-high). The calibration standards and QC samples were applied for method validation and in the pharmacokinetic study.

2.5 Sample preparation

Plasma aliquots (200 μL) were spiked with 10 μL of methanol and tylosin (10 μL of the 16 μg/mL solution as an IS) in centrifuge tubes (when preparing the calibration and QC samples, the standard solution was added instead of methanol) and mixed. The centrifuge tubes were initially primed with 300 μL of methanol, followed by vortex mixing for 1min and centrifugation for 10 min at 5 000 rpm. Next, 10 μL of a solution of 0.1 M NaOH and 0.4 mg NaCl was added to the supernatant, followed by vortex mixing for 20 s. Next, 1mL of chloroform was added, followed by mixing for 1 min and centrifugation for 10 min at 3 500 rpm. The addition of chloroform and centrifugation were repeated, after which the solution was combined with the organic phase for evaporation. The organic phase was evaporated by use of a stream of nitrogen at 40 ℃. The residue was reconstituted with 100 μL of the mobile phase and immediately subjected to vortex mixing for 20 s, after which it was filtered through a 0.22 μm Millipore filter and injected into the LC-MS/MS system.

2.6 Validation

The LC-MS/MS method was validated in terms of linearity, specificity, LLOQ (lower limit of quantification), recovery, intra- and inter-day variation, accuracy and precision. The stability of the analyte during sample storage and processing procedures was also determined.

2.6.1 Selectivity

Selectivity was evaluated by comparing the chromatograms of six different batches of the blank plasma with corresponding standard plasma samples spiked with tilmicosin and theinternal standard (Fan, Li, Gu, Si, & Liu, 2012).

2.6.2 Linearity and LLOQ

A calibration curve was constructed from plasma standards at seven concentrations of tilmicosin ranging from 0.5 to 2 000 ng/mL. A calibration curve was constructed by plotting the peak area ratio of tilmicosin/IS vs the nominal concentration of tilmicosin. The correlation coefficient and linear regression equation were used to determine the analyte concentration in each sample. Weighted ($1/x^2$) linear leastsquares regression was used as the mathematical model. The LLOQ was determined as the lowest concentration that produced a signal/noise ratio of 5 (Pabbisetty et al., 2012). The limit of detection (LOD) was determined as the lowest concentration that produced a signal/noise ratio of 3 (Fan et al., 2012).

2.6.3 Accuracy and precision

xIntra-day accuracy and precision were evaluated by analyzing four QC concentrations (0.5, 5, 1 000 and 2 000 ng/mL; Table 1) with six determinations per concentration on the same day. Inter-day accuracy and precision were evaluated by analyzing four QC concentrations (0.5, 5, 1 000 and 2 000 ng/mL; Table 1) with six determinations per concentration over 3days. Precision and accuracy were based on the criterion that the relative standard deviation (RSD) for each concentration should be ⩽15%, except for that of the LLOQ (⩽20%; Shah et al., 2000).

Table 1 Intra- and inter-day precision and accuracy of tilmicosin (n=6) in pig plasma.

Concentration (ng/mL)		Intra-day precision and accuracy (n = 6)		Inter-day precision and accuracy (n = 6)	
		Accuracy(%) ± SD	RSD(%)	Accuracy(%) ± SD	RSD(%)
Tilmicosin	0.5	92.3 ± 2.0	2.2	93.0 ± 2.0	2.1
	5	93.4 ± 2.5	2.7	94.6 ± 3.0	3.2
	1 000	99.8 ± 1.9	1.9	98.5 ± 2.9	3.0
	2 000	97.3 ± 2.0	2.1	101.3 ± 2.6	2.6

2.6.4 Recovery and matrix effect

Recovery was determined in quadruplicate by comparing processed QC samples at three concentrations (low, medium and high) with reference solutions in blank plasma at the same concentrations. Matrix effects were determined by comparing the peak areas of each post-extraction spiked sample with those of the standards containing equal amounts of tilmicosin in the mobile phase. The experiments were performed at three concentrations in six different batches.

2.6.5 Stability

The stability of tilmicosin in pig plasma was assessed by analyzing replicates (n=6) of QC samples at concentrations of 5, 1 000 and 2 000ng/mL during the sample storage and processing procedure (Chien et al., 2013). For all stability studies, freshly prepared stability-QC samples were analyzed using a freshly prepared standard curve. The stability of stock solutions of tilmicosin was analyzed after storage at room temperature for 24 h and at 4℃ for 1month. Short-term stability was assessed after exposure of the plasma samples to ambient temperature for 24h. Long-term stability was assessed after storage of the plasma samples at −20℃ for 60days. Freeze−thaw stability was determined after three freeze−thaw cycles (room temperature to −20℃). Sample stability in the autosampler tray was evaluated at 4℃ for 24 h; this sample stability evaluation imitates the residence time of the samples in the autosampler for each analytical run.

2.7 Animal studies

The assay method described above was used to study the pharmacokinetics of tilmicosin enteric granules in pig plasma after oral administration. All experimental procedures were approved and performed in accordance with the Guidelines for the Care and Use of Laboratory Animals of the Lanzhou Institute of Animal Science and Veterinary Pharmaceutics. Healthy Duroc pigs (60 ± 1.27kg) were obtained from the Duroc Breeding Base (Zhangye, Gansu, China) and housed in a standard environmentally controlled animal room (temperature, 25 ± 2℃; humidity, 50% ± 20%) with a natural light−dark cycle for 1week before the experiment. The pigs werefastedfor 12 h beforedosing, but allowed free movement and access to water during the entire experiment.

A dual-period crossover trial design was utilized. All pigs (n=18) were randomly divided into two groups (n=9), one comprising four females and five males, and the other comprising five females and four males. The concentration of granules in the diet of each pig was adjusted so that the total daily intake of the test drug was 100mg/kg.

Tilmicosin premix was used as the control drug. For the first administration period, one group received tilmicosin enteric granules, while the other group received tilmicosin premix. After the first administration period, neither group received a drug for 15days. During the second administration period, the treatments of the groups were reversed, so that the group that received the tilmicosin granules received the tilmicosin premix and vice versa.

After a single dose of tilmicosin was administered, blood samples (3 mL) were collected in heparinized tubes via the jugular vein at 0.25, 0.5, 0.75, 1, 1.25, 1.5, 2, 3, 4, 5, 6, 7, 8, 10, 12, 24, 48, and 72 h post-administration. After all blood samples were centrifuged at 12 000 rpm for 10 min, plasma samples were collected and immediately stored in a −20℃ freezer until analysis by LC-MS/MS.

Pharmacokinetic parameters were calculated using WinNonlin Professional software version 5.2 (Pharsight, Mountain View, CA, USA). A compartmental model was utilized for data fitting and parameter estimation. The best pharmacokinetic model was confirmed by Akaike Information Criterion (AIC) (Klinkenberg, Tobias, Bouma, van Leengoed & Stegeman, 2014). Plasma AUC, plasma clearance ($C_{L/F}$), peak plasma concentration (C_{max}), elimination rate constant (K) and apparent volume of distribution (V) were calculated from the data. Half-life ($t_{1/2}$) was calculated directly according to the pharmacokinetic parameters.

3 RESULTS AND DISCUSSION

3.1 Mass spectrometric detection

In order to optimize positive ESI mode conditions, tilmicosin and the IS were dissolved in methanol and injected into the mass spectrometer for scanning in positive ion mode. When tilmicosin and the IS were injected directly into the mass spectrometer with the mobile phase, the analytes yielded predominantly $[M+H]^+$ ions at m/z 869.6 for tilmicosin and m/z 837.6 for the IS (Fig. 2). Each of the precursor ions was subjected to collision-induced dissociation to allow determination of the resulting product ions from the product ion mass spectra. The most abundant and stable fragment ions were generated at m/z 173.9 for tilmicosin and m/z 158.1 for the IS. Thus, the mass transitions chosen for quantitation were m/z 869.6 to 173.9 for tilmicosin and m/z 837.6 to 158.1 for the IS.

3.2 Chromatographic separation

High-performance LC-MS/MS separation was performed using a column packed with a small amount of the mobile phase and a shorter analysis time. A 30 mm column subjected to isocratic elution of the mobile phase for 3.5min at a flow rate of 0.2 mL/min was used for the chromatographic separation. A mobile phase consisting of a mixture of acetonitrile-water (water included 0.1% formic acid) was found to be suitable for separation and ionization of tilmicosin and the IS. Formic acid was found to increase ionization of all three tested Fig. 2 (Continued) compounds. Under optimized LC and MS conditions, tilmicosin and the IS were separated with retention times of 1.5 and 2.7 min, respectively, and endogenous substances in plasma did not interfere with target detection (Fig. 3).

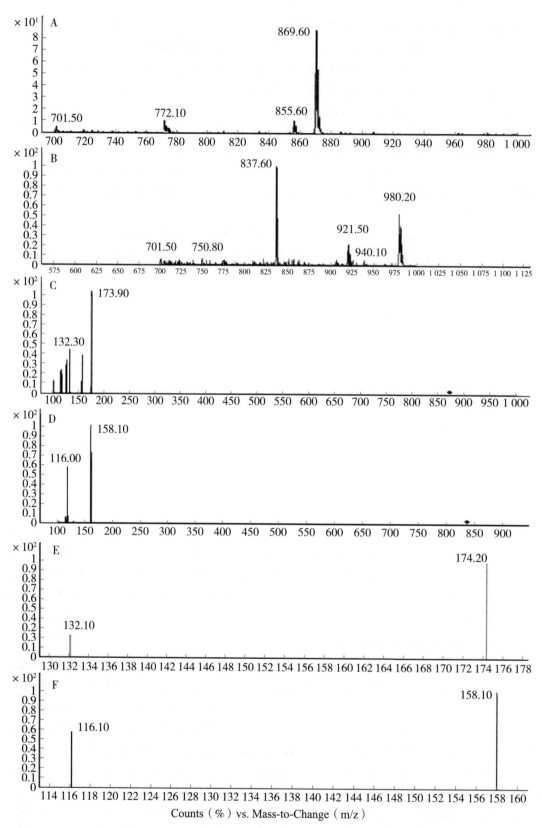

Fig. 2 Full mass spectra of tilmicosin (A) and the IS (B); product ions for tilmicosin (C) and the IS (D); multiple reaction monitoring for tilmicosin (E) and the IS (F).

3.3 Sample preparation

Recoveries of tilmicosin using acetonitrile (11%), ether (77%) and ethyl acetate (75%) were found to be less than that from chloroform (95%). Therefore, chloroform was selected as the extraction solvent for plasma because of its high recovery and sensitivity.

3.4 Method validation

3.4.1 Selectivity

The specificity of the method was evaluated by analyzing individual blank plasma samples from six different sources. All samples were found to have no interference from endogenous substances affecting the retention time of tilmicosin or the IS. There was good baseline separation of tilmicosin and the IS extracted from pig plasma. Representative chromatograms of blank plasma, blank plasma spiked with tilmicosin and the IS are shown in Fig. 3.

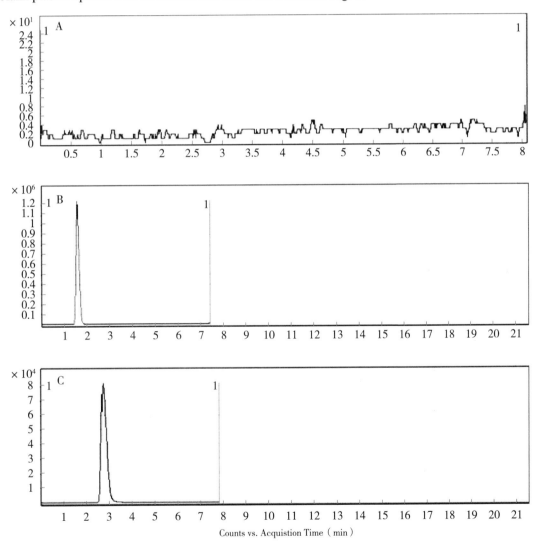

Fig. 3 Chromatograms. (A) Blank plasma; (B) tilmicosin in plasma; and (C) the IS in plasma.

3.4.2 Linearity and LLOQ

A calibration curve was constructed from plasma standards at six concentrations of tilmicosin ranging from 0.5 to 2 000 ng/mL. The ratio of the peak area of tilmicosin to that of the IS was used for quantification. The calibration model was selected based on analysis of the data by linear regression with intercepts and a $1/x^2$ weighting factor. A typical equation of the calibration curve for tilmicosin was $y=0.003\ 4\times +0.012\ 6$ ($r^2=0.999\ 8$), where y is the peak-area ratio of tilmicosin to the IS and x is the plasma concentration of tilmicosin. The calibration curve is shown in Figure 4. The LLOQ of tilmicosin was 0.5 ng/mL with an accuracy of 98.5%. The LOD of tilmicosin was 0.1 ng/mL.

3.4.3 Accuracy and precision

The intra- and inter-day precision and accuracy of the QC samples (0.5, 5, 1 000 and 2 000 ng/mL) are summarized in Table 1. These data demonstrate that the method reported herein has satisfactory accuracy, precision and reproducibility for the quantification of tilmicosin in pig plasma.

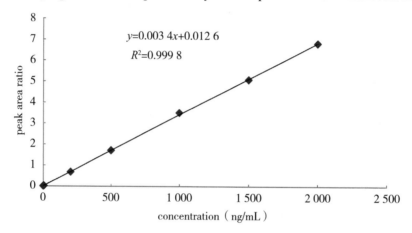

Figure 4 Calibration curve.

3.4.4 Recovery and matrix effect

The mean extraction recoveries of tilmicosin at concentrations of 5, 1 000 and 2 000 ng/mL were 95.0% ± 2.3%, 98.5% ± 2.1% and 96.7% ± 2.3%, respectively. The mean extraction recovery of the IS was 97.1% ± 1.9% at a concentration of 800 ng/mL. These results suggest that the recovery of tilmicosin was consistent and not concentration dependent. Recovery values are listed in Table 2.

The matrix effects for tilmicosin at 5, 1 000 and 2 000 ng/mL ranged from (4.1 ± 2.2) % to (95.8 ± 2.0) %, while the matrix effect of the IS was (93.5 ± 1.8) %. These results indicate that limited matrix effects were observed.

Table 2 Recovery of tilmicosin (n=6) from pig plasma.

Spiked concentration (ng/mL)		Recovery (%, n = 6)[a]	
Mean ± SD		RSD (%)[b]	
Tilmicosin	5	95.0 ± 2.2	2.3

(continued)

Spiked concentration (ng/mL) Mean ± SD		Recovery (%, n = 6) [a]	
		RSD (%) [b]	
	1 000	98.5 ± 2.1	2.1
	2 000	96.7 ± 2.3	2.4

[a] Recovery = ratio of the response of the spiked standard before extraction to that after extraction.
[b] RSD, relative standard deviation.

3.4.5 Stability

Stability tests performed using tilmicosin included freeze-thaw, shortterm, autosampler, and long-term stability tests (Table 3). Good stability was observed in all tests. Tilmicosin was stable in pig plasma following three freeze-thaw cycles (-20℃ to room temperature), repeated exposure to room temperature for 24h, storage in an autosampler tray at 4℃ for 24h, and storage at -20℃ for 60days. Tilmicosin stock solutions were stable at room temperature for 24h and at 4℃ for 1month. These results show that tilmicosin can be stored and processed under routine laboratory conditions without special attention.

Table 3 Stability of tilmicosin in pig plasma samples under various conditions (n=6).

Storage conditions	Concentration (ng/mL)	Accuracy (%) ± SD	RSD (%)
Three freeze-thaw cycles	5	90.4 ± 2.7	3.0
	1 000	93.7 ± 3.0	3.2
	2 000	96.1 ± 1.9	2.0
At room temperature for 24 h	5	92.6 ± 2.2	2.4
	1 000	98.3 ± 1.2	1.2
	2 000	97.5 ± 2.5	2.6
At 20℃ for 60 days	5	90.1 ± 2.1	2.3
	1 000	92.2 ± 2.0	2.2
	2 000	93.5 ± 2.3	2.5
At 4℃ in the autosampler for 24 h	5	97.3 ± 1.7	1.7
	1 000	101.2 ± 2.2	2.2
	2 000	95.6 ± 1.9	2.0
At 4℃ for 1 month	5	95.5 ± 1.8	2.0
	1 000	95.1 ± 2.4	2.5
	2 000	93.6 ± 2.8	3.0

3.5 Pharmacokinetic studies

The method described above was successfully validated and applied to quantitate tilmicosin in plasma samples after oral administration of tilmicosin enteric granules and tilmicosin premix to Duroc pigs at a dose of 100 mg/kg. The mean plasma concentration vs. time profiles for tilmicosin in pigs after oral administration of tilmicosin enteric granules And the control drug are shown in Figure 5. Based on the AIC (fitted value=62) , the plasma concentration-time curves for tilmicosin were adequately fitted by a one-compartment model. Major pharmacokinetic parameters were calculated using a one-compartment model (Table 4) .

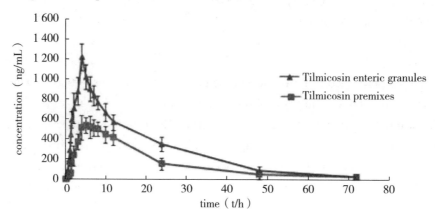

Figure 5　Mean plasma concentration-time profile after oral administration of tilmicosin enteric granules and tilmicosin premix to pigs (n=18) at a dose of 100 mg/kg.

Table 4　Pharmacokinetic parameters of tilmicosin after oral administration of tilmicosin enteric granules and tilmicosin premix to pigs (n=18) at a dose of 100 mg/kg.

Parameters	Mean ± SD	
	Tilmicosin enteric granules	Tilmicosin premix
AUC_{0-t} (ng/mL h)	19 371.24 ± 1 057.39	11 272.99 ± 1 222.31
$AUC_{0-\infty}$ (ng/mL h)	20 349.22 ± 1 102.03	12 046.43 ± 977.24
$C_{L/F}$ (mL/h/kg)	3 321.50 ± 412.62	5 970.32 ± 231.25
T_{max} (h)	4.28 ± 0.44	5.04 ± 0.69
C_{max} (ng/mL)	1 183.17 ± 44.36	516.65 ± 29.88
V_F (mL/kg)	4 970.32 ± 231.25	1 140.46 ± 228.57
K_a (h)	0.90 ± 0.14	0.71 ± 0.16
K (h)	0.05 ± 0.01	0.06 ± 0.01
$t_{1/2ka}$ (h)	0.77 ± 0.12	0.98 ± 0.28
$t_{1/2}$ (h)	13.08 ± 1.27	11.00 ± 1.06
T_{Lag}	0.50 ± 0.11	1.38 ± 0.38

　　AUC, area under the concentration-time curve. K, elimination rate constant. C_{max}, Peak plasma concentration. $C_{L/F}$, Plasma clearance. $t_{1/2}$, Elimination halflife. V_{ss}, Apparent volume of distribution.

In comparison with the control treatment, tilmicosin enteric granules produced faster tilmicosin absorption and a larger volume of distribution. The longer half-life ($t_{1/2}$) of tilmicosin following administration of tilmicosin enteric granules indicates that the compound is removed slowly from the blood and has a longer period of efficacy, in comparison with the control treatment.

No significant differences in pharmacokinetic parameters were found between males and females administered the tilmicosin enteric granules or the tilmicosin premix ($P > 0.05$, data not shown). The group that received a single oral dose of the enteric granules and the group that received the premix showed significant differences in most pharmacokinetic parameters of tilmicosin ($P < 0.05$). The AUC ($_{0-t}$), $t_{1/2}$ and C_{max} values for tilmicosin from enteric granules were 1.72, 1.19 and 2.29 times greater, respectively, than those of tilmicosin from the premix. These results show that tilmicosin from enteric granules was absorbed quickly, but eliminated slowly, indicating that oral administration of tilmicosin enteric granules can produce a more persistent therapeutic effect, in comparison with that of tilmicosin premix.

This study provides a basis for further research on tilmicosin enteric granules. Further development of the preparation method and evaluation of the pharmacokinetic properties of this formulation will facilitate the preparation of new formulations of similar drugs with improved pharmacokinetic profiles.

4 DISCUSSION

Methods for the quantification of tilmicosin primarily utilize residue determinations in various tissues; only a few studies have reported quantitation of tilmicosin in plasma (Abu-Basha, Idkaidek, & Al-Shunnaq, 2007; Shen et al., 2013) or serum (Herrera, Ding, McClanahan, Owens, & Hunter, 2007). Liu et al. (2011) and Shen et al. (2013) described an HPLC method with UV, which achieved an LLOQ for tilmicosin of 25ng/mL with a long cycle time of > 12min and a complex mobile phase of acetonitrile, tetrahydrofuran, dibutylamine, phosphate buffer and water. An LC-MS method for quantification of tilmicosin in canine serum achieved an LOQ of 50ng/mL using 0.2mL of serum (Herrera et al., 2007).

The primary advantage of the method developed in the present study over previously reported methods is the sample liquid-liquid extraction procedure (using only a single extraction with chloroform), which is simple and less laborious than other methods. Furthermore, the LC-MS/MS method described in this study achieved an LLOQ of 0.5 ng/mL for tilmicosin using a 200 μL plasma sample and a short run time of 4.5 min. Moreover, the LC-MS/MS method reported herein has satisfactory accuracy, precision and reproducibility for the quantification of tilmicosin in pig plasma. The LC-MS/MS method reported herein is simple, fast, less laborious than other methods, and able to achieve high sensitivity using a small sample volume, which is very advantageous in pharmacokinetic studies.

The pharmacokinetic properties obtained for tilmicosin in this study are in agreement with previously published data obtained following a single oral dose to healthy animals (Shen et al., 2005, 2013). The $t_{1/2}$ (13.08h) of tilmicosin reported herein is > 1.19 times that of the control drug, whereas the AUC $_{(0-t)}$ of tilmicosin is > 1.72 times that of the control drug, and the C_{max}

of tilmicosin is > 2.29 times that of the control drug. These results show that, in comparison with tilmicosin premix, tilmicosin enteric granules slowed the elimination rate of tilmicosin, prolonged its period of action and significantly improved its bioavailability.

5 CONCLUSIONS

The pharmacokinetic analysis of tilmicosin enteric granules reported herein relied on a highly sensitive assay to determine tilmicosin in pig plasma after oral administration. In this study, a rapid and sensitive LC–MS/MS method was developed, validated and successfully applied to evaluate the pharmacokinetic parameters of tilmicosin after oral administration of tilmicosin enteric granules to pigs. The method reported herein used a relatively small volume of plasma and was not susceptible to interference from the biological matrix. Sample preparation for the assay is simple and relatively quick. The analysis requires only 200 μL of plasma, has a short run time of 4.5 min and has excellent sensitivity, linearity, precision and accuracy. Currently, premix is the only preparation of tilmicosin available for clinical use. The LC–MS/MS method reported herein can be used to evaluate the pharmacokinetic properties and therapeutic potential of tilmicosin enteric granules.

ACKNOWLEDGMENTS

This work was supported by the earmarked fund for the China Agriculture Research System (cars-38) and the Special Fund for Agro-scientific Research in the Public Interest (no. 201303038-4).

（发表于《Biomedical chromatography》, SCI, IF: 1.729. DOI:1 002/bmc.3 825.）

Evaluation on Antithrombotic Effect of Aspirin Eugenol Ester from the View of Platelet Aggregation, Hemorheology, TX-B$_2$/6-keto-PGF$_1$α and Blood Biochemistry in Rat Model

MA Ning, LIU Xiwang, YANG Yajun, SHEN Dongshuai, ZHAO Xiaole, MOHAMED Isam, KONG Xiaojun, LI Jianyong[*]

(Key Lab of New Animal Drug Project of Gansu Province, Key Lab of Veterinary Pharmaceutical Development, Ministry of Agriculture, Lanzhou Institute of Husbandry and Pharmaceutical Science of CAAS, Lanzhou 730050, China)

Abstract: Background: Based on the prodrug principle, aspirin and eugenol, as starting precursors, were esterified to synthesize aspirin eugenol ester (AEE). The aim of the present study was to evaluate the antithrombotic effect of AEE in an animal disease model. In order to compare the therapeutic effects of AEE and its precursors, aspirin, eugenol and a combination of aspirin and eugenol were designed at the same molar quantities as the AEE medium dose in the control group.

After oral administration of AEE (dosed at 18, 36 and 72 mg/kg) for seven days, rats were treated with k-carrageenan to induce tail thrombosis. Following the same method, aspirin (20 mg/kg), eugenol (18 mg/kg) and 0.5% CMC-Na (30 mg/kg) were administered as control drug. Different drug effects on platelet aggregation, hemorheology, TXB$_2$/6-keto-PGF$_1$α ratio and blood biochemistry were studied.

AEE significantly inhibited ADP and AA-induced platelet aggregation in vivo. AEE also significantly reduced blood and plasma viscosity. Moreover, AEE down-regulated TXB$_2$ and up-regulated 6-keto-PGF$_1$α, normalizing the TXB2/6-keto-PGF$_1$α ratio and blood biochemical profile. In comparison with aspirin and eugenol, AEE produced more positive therapeutic effects than its precursors under the same molar quantity.

It may be concluded that AEE was a good candidate for new antithrombotic and antiplatelet medicine. Additionally, this study may help to understand how AEE works on antithrombosis in different ways.

Corresponding, E-mail: lijy1971@163.com.

© 2016 The Author(s). Open Access This article is distributed under the terms of the Creative Commons Attribution 4.0 International License (http://creativecommons.org/licenses/by/4.0/), which permits unrestricted use, distribution, and reproduction in any medium, provided you give appropriate credit to the original author(s) and the source, provide a link to the Creative Commons license, and indicate if changes were made. The Creative Commons Public Domain Dedication waiver (http://creativecommons.org/publicdomain/zero/1.0/) applies to the data made available in this article, unless otherwise stated.

Key words: Aspirin eugenol ester (AEE); Thrombosis; k-carrageenan; Rat; Platelet aggregation; Blood viscosity

1 BACKGROUND

In the last century, remarkable increases of life spans in dogs and cats have promoted the development of geriatric veterinary medicine [1, 2]. The findings have proved that age-related thrombin was a possible risk factor in dogs and cats [3, 4]. Thus, there is a need for anti-thrombotic drugs for animals. The physiological process of thrombosis is complex and influenced by many factors such as platelet aggregation, ratio of thromboxane B_2 (TXB_2) to 6-keto prostaglandin $F_1\alpha$ (6-keto-$PGF_1\alpha$), biochemical and hemorheological parameters [5-8]. Therefore, the impacts of drugs on platelet aggregation, hemorheology, TXB_2/6-keto-$PGF_1\alpha$ ratio and blood biochemistry can be employed for examining the effect of compound in thrombosis prevention [9].

Aspirin is extensively used in human and veterinary medicine to reduce inflammation, pain and fever. In cats, aspirin is not well tolerated because of the deficient in glucuronate and has a prolonged elimination half-time. The pharmacodynamics of aspirin are similar in human and dogs. Related researches have proved that prophylactic use of aspirin can reduce the risk of heart attack, stroke and some cancers [10, 11]. A number of reports have showed that eugenol has antiseptic, analgesic, antibacterial, antiplatelet aggregation and anticancer properties [12-14]. Notably, aspirin and eugenol produce similar pharmacological effects especially on inflammation and platelet aggregation.

Gastrointestinal damage caused by aspirin and vulnerability of eugenol limit their application. These side effects and structural instability are related to chemically carboxyl group of aspirin and hydroxyl group of eugenol. In order to increase therapeutic effect and stabilization, aspirin and eugenol as starting precursors were esterified to synthesize aspirin eugenol ester (AEE). AEE, a pale yellow, odorless crystal, overcomes the disadvantages of its precursors. Anti-ulcerogenic activity of eugenol has been confirmed in a previous study, in which eugenol (100 mg/kg, p.o.) significantly and dose-dependently reduced gastric ulcers induced by platelet activating factor and ethanol in Wistar rats [15]. So a combination of eugenol and aspirin could play complementary roles to reduce gastrointestinal damage and enhance the therapeutic effects. A 15-day sub-chronic toxicity study identified that no histopathological changes was observed in any organ from the rats following low-dose AEE (50 mg/kg), whereas mucosa height of stomach, duodenum and ileum were increased in rats following high-dose AEE exposure (2 000 mg/kg) [16].

Carrageenan-induced rat tail thrombosis model has been widely used to evaluate whether substances have antithrombotic effects during the drug discovery stage [17-20]. The length of thrombosis in rat tail directly reflects the effects of drug on thrombosis formation [21, 22]. The preventive effects of AEE on thrombosis formation have been confirmed in k-carrageenan-induced rat tail thrombosis model, in which AEE significantly inhibited thrombus formation and reduced the tail thrombosis length [23]. However, investigations of the preventive effects of AEE against thrombosis are not sufficient. In order to understand and illustrate the possible underlying mechanism of AEE, a follow-up study was carried out using the same animal disease model and drug dosage. The aim of the present study was to evaluate the effects of AEE on platelet aggregation, hemorheology, blood biochemistry and TXB_2/6-keto-$PGF_1\alpha$ ratio. To compare the effects of AEE and its precursors, equimolar doses of aspirin, eugenol and combination of aspirin and eugenol were also administered in the experiment.

2 METHODS

2.1 Chemicals and reagents

AEE (transparent crystal, 99.5% purity with RP-HPLC) was prepared in Key Lab of New Animal Drug Project of Gansu Province, Key Lab of Veterinary Pharmaceutical Development of AgriculturaL·ministry, Lanzhou Institute of Husbandry and Pharmaceutical Sciences of CAAS. CMC-Na was supplied by Tianjin Chemical Reagent Company (Tianjin, China). Aspirin and Tween-80 were obtained from Aladdin Industrial Corporation (Shanghai China). Eugenol was supplied by Sinopharm Chemical Reagent Co., Ltd. (Shanghai China). Adenosine diphosphate (ADP), arachidonic acid (AA) and k-carrageenan were purchased from Sigma (St. Louis, USA). Blood biochemical analysis reagents were obtained from Meikang Co., Ltd (Zhejiang, China). All chemical reagents were of analytical reagent grade. A Thermo Fisher Scientific Multiskan Go 1510 was used for spectrophotometry (USA).

2.2 Animals and treatment

Ninety male Wistar rats (approximately 160 g) were purchased from the animal breeding facilities of Lanzhou Army General Hospital (Lanzhou, China). Animals were housed in plastic cages (size: 50 cm × 35 cm × 20 cm, 10 rats per cage) with stainless steel wire cover and chopped bedding in a ventilated room. The light/dark regime was 12/12 h and living temperature was maintained at 22 ± 2℃ with relative humidity of (55 ± 10)%. The animals had free access to a standard diet and tap water (Standard compressed rat feed from Keao Xieli Co., Ltd., Beijing, China). Rats were allowed a 2-week acclimation period prior to the start of experiment.

2.3 Drug preparation

AEE and aspirin suspensions were prepared in 0.5% of CMC-Na. Eugenol and Tween-80 (mass ratio of 1∶2) were mixed with distilled water. Kappa-carrageenan (k-carrageenan) was dissolved in physiological saline.

2.4 Study design

Ninety rats were divided into 9 groups in the study. According to individual body weight (BW), rats were intra-gastrically (i.g.) administered with 18, 36 and 72 mg/kg AEE, 20 mg/kg aspirin or 18 mg/kg eugenol. For the comparability of the experimental results, medium-dose AEE, aspirin and eugenol were administered at the dose of 0.11 mmol. In order to compare AEE with its precursors, an equimolar combination of aspirin and eugenol was designed in the experiment. Rats in CMC-Na group were given 0.5% CMC-Na at the dose of 30 mg/kg BW, and the volume of CMC-Na was nearly equal in comparison with other drug suspensions. A detailed description of the study design is shown in Table 1. All drugs and CMC-Na were administered intragastrically for seven days, after which the rats were intraperitoneally injected with 20 mg/kg k-carrageenan to induce thrombosis.

Table 1 Design of the experiment.

Groups	Drug	Dosage (mg/kg)	Molar quantity (mmol)
Control	—	—	—
Model	—	—	—
CMC-Na	CMC-Na	30	—
Aspirin	aspirin	20	0.11
Eugenol	eugenol	18	0.11
AEE L	AEE	18	0.06
AEE M	AEE	36	0.11
AEE H	AEE	72	0.22
Combination	aspirin + eugenol	20 + 18	0.11

Ninety rats were divided into 9 groups (n = 10). CMC-Na was used as the vehicle. The volumes of CMC-Na and drug suspension were nearly equal. The molar quantity of medium-dose AEE, aspirin and eugenol are same at 0.11 mmol. An equimolar combination of aspirin and eugenol was used in the study. Rats in the model, CMC-Na, aspirin, eugenol, AEE and combination groups were intraperitoneally injected with 20 mg/kg k-carrageenan to induce thrombosis Carrageenan-induced rat tail thrombosis model Carrageenan-induced rat tail thrombosis was induced as described in the previous study [23].

2.5 Measurement of rat tail thrombosis length

After the last treatment with different drugs, each rat was intraperitoneally injected with k-carrageenan. Swelling and redness in rat tail were monitored. Thrombus lengths were measured and photographed at 24 and 48 h.

2.6 Blood sampling

After the last measurement of thrombosis length at 48 h, rats were anesthetized with 10% chloral hydrate and blood samples were taken from the heart. According to the requirement of the experiment, blood samples were collected in different volume with different anticoagulants. Serum for biochemical analysis and plasma for ELISA were stored at −80 ℃ until the day of analysis. All platelet aggregation and hemorheological tests were performed within 3 h of blood collection.

2.7 Hemorheological tests

Blood samples of approximately 2 mL were collected and anticoagulated with EDTA-K_2, after which 0.8 mL whole blood samples were prepared for whole blood viscosity examination. The remaining blood samples were centrifuged at 1 000 g for 10 min to obtain plasma, after which 0.5 mL of plasma was used for plasma viscosity examination. A kone plate viscometer LBY-N6A analytical instrument (Precil Company, Beijing, China) was used for the measurement of whole blood viscosity and plasma viscosity at 37.0 ± 0.5 ℃. For whole blood viscosity, measurements were made at three shear rates: 5, 100 and 200 s^{-1}. Plasma viscosity was measured at a shear rate of 100 s^{-1}.

2.8 Assay of platelet aggregation

In vivo assessment of the anti-platelet aggregation effect of AEE was performed using light transmission aggregometry in platelet-rich plasma (PRP). About 2 mL blood samples were collected and diluted by 3.8% sodium citrate on the proportion of 9 : 1 in vacuum tubes, which were prepared for platelet aggregation analysis. Platelet aggregation was assessed in PRP, which was obtained by centrifugation of citrated whole blood at room temperature for 10 min at 1 000 rpm. Then, PRP was placed into two cuvettes (0.25 mL PRP in each cuvette) and stirred with a rotor at 37 ℃ for 5 min, after which 5 μM ADP and 5 mM AA was added, respectively. The platelet-poor plasma (PPP) was obtained by centrifugation of PRP at room temperature for 10 min at 3 000 rpm, which was used to set zero. Aggregation was measured with a Chrono-log Platelet Aggregometer Model 700 (Chrono-log Corp., USA). After the addition of an aggregating agent, light transmission at maximal aggregation was recorded. The results are expressed as percentage of maximal aggregation.

2.9 Blood biochemical analysis

Blood samples of approximate 1 mL were collected into vacuum tubes without anticoagulant, and centrifuged at 4 000 rpm for 10 min to obtain serum. Serum was analyzed using an XL-640 automatic biochemistry analyzer (Erba, German). Levels of total bilirubin (T-BIL), total protein (TP), albumin (ALB), alanine transaminase (ALT), aspartate aminotransferase (AST), alkaline phosphatase (ALP), gamma glutamyl transpeptidase (GGT), lactate dehydrogenase (LDH), blood urea nitrogen (BUN), creatinine (CR), glucose (GLU), glutamic pyruvic transaminase (GPT), creatine kinase (CK), calcium (Ca), phosphorus (P), triglycerides (TG) and total cholesterol (TCH) were analyzed in the experiment.

2.10 Measurement of plasma TXB_2 and 6-keto-$PGF_1\alpha$

Samples anticoagulated with EDTA-K_2 were centrifuged at 1 000 g for 10 min to obtain 200 μL of plasma for TXB_2 and 6-keto-$PGF_1\alpha$ measurements. ELISA kit of TXB_2 was purchased from U.S.A TSZ Biological Trade Co., Ltd. (New Jersey, USA) and 6-keto-$PGF_1\alpha$ ELISA kit was from Abcam Co., Ltd (Cambridge, UK). The ELISA operation was followed the protocols included with each kit.

2.11 Statistical analysis

Statistical analyses were carried out using SPSS (Statistical Product and Service Solutions, IBM. Co., Armonk, NY, USA). Data obtained from experiment was expressed as mean ± standard deviation (SD). Statistical differences were evaluated by ANOVA with least significant difference (LSD) test. P-values less than 0.05 were considered to indicate statistical significance.

3 RESULTS

3.1 Effects on hemorheology

Hemorheological parameters were analyzed by routine laboratory assays. The results are shown in Table 2. In comparison with the control group, whole blood viscosity and plasma viscosity were significantly increased in model group ($P < 0.01$, or $P < 0.05$). There was no statistical difference between CMC-Na and model groups. In order to eliminate the influence of the vehicle, drug-treated groups were compared with CMC-Na group. Aspirin reduced whole blood viscosity at low shear rate ($P < 0.05$), but did not influence whole blood viscosity at medium and high shear rates. Whole blood viscosity at low and medium shear rates in groups supplemented with AEE were significantly lower than that of CMC-Na group ($P < 0.01$). Whole blood viscosity at high shear rate was also significantly reduced by low-dose AEE ($P < 0.05$). Aspirin and eugenol remarkably reduced plasma viscosity (aspirin: $P < 0.01$; eugenol: $P < 0.05$). Low and high doses of AEE showed strong effects on reducing the plasma viscosity values ($P < 0.01$). No change was observed on whole blood viscosity and plasma viscosity between CMC-Na and combination groups.

Table 2 Drug effects on hemorheological parameters in k-carrageenan-induced rat tail thrombosis model (n = 10).

Groups	WBV (mPa.s)			PV (mPa.s)
	Low shear rate ($5\ s^{-1}$)	Medium shear rate ($100\ s^{-1}$)	High shear rate ($200\ s^{-1}$)	
Control	18.56 ± 3.03##	4.63 ± 0.77##	3.55 ± 0.79##	1.55 ± 0.28#
Model	27.75 ± 3.27	674 ± 0.78	5.24 ± 0.63	2.04 ± 0.47
CMC-Na	26.42 ± 2.66	6.87 ± 0.62	5.22 ± 0.52	1.97 ± 0.25
Aspirin	22.75 ± 3.57*	6.21 ± 0.62	5.21 ± 035	1.39 ± 0.13**
Eugenol	28.69 ± 2.93	6.87 ± 0.72	5.52 ± 0.79	1.63 ± 0.25*
AEE L	20.03 ± 2.75**	5.38 ± 0.78**	4.58 ± 0.67*	1.32 ± 0.12**
AEE M	22.60 ± 2.28**	5.48 ± 0.86**	4.74 ± 0.52	1.68 ± 0.61
AEE H	22.25 ± 3.60**	5.14 ± 0.69**	4.63 ± 0.62	1.50 ± 0.14**
Combination	27.45 ± 2.90	6.58 ± 1.06	5.28 ± 0.63	2.19 ± 0.30

In order to find out the difference in results between AEE and its precursors, multiple comparisons were carried out. In the experiment, the molar quantities of aspirin, eugenol and medium-dose AEE were same. The results are shown in Fig. 1. AEE showed stronger effects on whole blood viscosity reduction at low shear rate than eugenol and combination groups ($P < 0.01$, Fig. 1a). In comparison with aspirin, eugenol and the combination groups, AEE significantly reduced whole blood viscosity at medium and high shear rates ($P < 0.01$, or $P < 0.05$ Fig. 1b and c).

Interestingly, the plasma viscosity values of each AEE group was significantly lower than that of combination group (P < 0.01, Fig. 1d). Moreover, the mean values of whole blood viscosity and plasma viscosity in AEE groups were lower than other drug-treated groups. Medium-dose AEE reduced whole blood viscosity more effectively than eugenol (P < 0.01, or P< 0.05 Fig. 1a, b and c). Moreover, medium-dose AEE showed better effects than aspirin on whole blood viscosity at medium shear rate (P < 0.01, Fig. 1b). There were significant differences on plasma viscosity and whole blood viscosity (at low and medium shear rates) between AEE M and combination groups (P < 0.01, Fig. 1a, b and d). To varying degrees, under the equimolar quantity of drugs, medium-dose AEE stronger reduced whole blood viscosity and plasma viscosity than aspirin, eugenol and combination of them from the different hemorheological parameters.

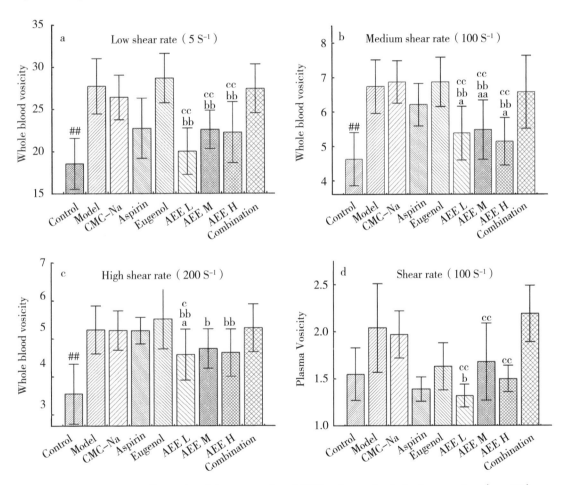

Fig. 1 Comparative effects of aspirin, eugenol and AEE on hemorheological parameters (n = 10).

Note: (a-c): whole blood viscosity at the shear rates of 5 s^{-1}, 100 s^{-1} and 200 s^{-1}. (d): plasma viscosity at the shear rate of 100 s^{-1}. $^{\#}P < 0.05$, $^{\#\#}P < 0.01$ compared with model group. $^{a}P < 0.05$, $^{aa}P < 0.01$ compared with aspirin group. $^{b}P < 0.05$, $^{bb}P < 0.01$ compared with eugenol group. $^{c}P < 0.05$, $^{cc}P < 0.01$ compared with combination group. Under the same molar quantity, medium dose of AEE showed better effects than its precursors on whole blood viscosity and plasma viscosity reduction in varying degrees. AEE L: AEE 18 mg/kg; AEE M: 36 mg/kg; AEE H: AEE 72 mg/kg. Combination: combination of aspirin and eugenol (molar ratio 1 : 1).

3.2 Effect of AEE on platelet aggregation in vivo

The results of platelet aggregation are shown in Table 3. Platelet aggregation induced by AA and ADP was significantly increased in model group ($P < 0.01$). No difference was observed between model and CMC-Na groups. AEE, eugenol and aspirin remarkably reduced AA-induced platelet aggregation than CMC-Na in varying degrees ($P < 0.05$, or $P < 0.01$). The mean values in aspirin and eugenol groups were 20.44 and 27.10, which indicated aspirin had stronger effects than eugenol on inhibiting AA-induced platelet aggregation. In regard to ADP-induced platelet aggregation, the values in aspirin, combination, medium and high-dose AEE groups were lower than CMC-Na group ($P < 0.01$).

The effects of different drug on platelet aggregation were evaluated in the study (seen in Fig. 2). Medium-and high-dose AEE showed stronger effects than aspirin on reducing AA-induced platelet aggregation values ($P < 0.01$). Meanwhile, three different doses of AEE displayed better effects than eugenol ($P < 0.01$). ADP-induced platelet aggregation values in low-dose AEE group were higher than aspirin and combination groups ($P < 0.01$). Medium- and After the last measurement of rat tail thrombosis length at 48 h, 0.8 mL of whole blood and 0.5 mL of plasma were collected for hemorheological parameter tests. Data are expressed as mean ± SD. $^{\#}P < 0.05$, $^{\#\#}P < 0.01$ compared with model group. $^{*}P < 0.05$, $^{**}P < 0.01$ compared with CMC-Na group. WBV whole blood viscosity, PV plasma viscosity. AEE L: AEE 18 mg/kg; AEE M: 36 mg/kg; AEE H: AEE 72 mg/kg; Combination: combination of aspirin and eugenol (molar ratio 1 : 1) high-dose AEE significantly inhibited ADP-induced platelet aggregation more effectively than eugenol ($P < 0.01$). The values of ADP-induced platelet aggregation in high-dose AEE group were lower than that in combination group ($P < 0.05$). Based on these results, under the same molar quantity, medium-dose AEE showed better effects than aspirin and eugenol on AA-induced platelet aggregation and possessed more effective effects of inhibiting ADP-induced platelet aggregation than eugenol.

Table 3 Effect of AEE on platelet aggregation in k-carrageenan-induced rat tail thrombosis model ($n = 10$).

Groups	AA-induced PAg	ADP-induced PAg
Control	$8.70 \pm 1.34^{\#\#}$	$31.90 \pm 2.77^{\#\#}$
Model	29.00 ± 1.49	50.10 ± 3.34
CMC-Na	30.10 ± 2.13	51.90 ± 2.28
Aspirin	$20.44 \pm 2.74^{**}$	$40.89 \pm 4.26^{**}$
Eugenol	$27.10 \pm 3.14^{*}$	50.10 ± 2.23
AEE L	$18.70 \pm 3.49^{**}$	49.30 ± 2.79
AEE M	$11.80 \pm 3.70^{**}$	$43.30 \pm 3.13^{**}$
AEE H	$15.70 \pm 3.43^{**}$	$42.20 \pm 3.26^{**}$
Combination	$11.50 \pm 2.50^{**}$	$45.50 \pm 3.17^{**}$

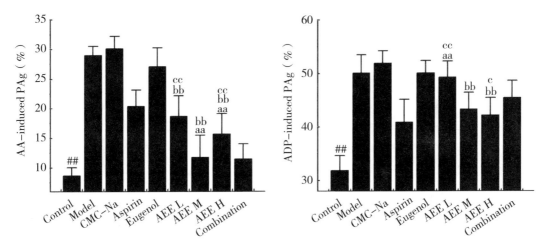

Fig. 2 Comparative effects of aspirin, eugenol and AEE on platelet aggregation in k-carrageenan-induced rat tail thrombosis model (n = 10).

Note: #P < 0.05, ##P < 0.01 compared with model group. aP < 0.05, aaP < 0.01 compared with aspirin group. bP< 0.05, bbP < 0.01 compared with eugenol group. cP < 0.05, ccP < 0.01 compared with combination group. No difference was observed between model and CMC-Na groups, which indicated that CMC-Na made no influence on PAg index. AEE showed strong effects than aspirin and eugenol at the same molar quantity. PAg: platelet aggregation. AEE L: AEE 18 mg/kg; AEE M: 36 mg/kg; AEE H: AEE 72 mg/kg; Combination: combination of aspirin and eugenol (molar ratio 1 : 1).

3.3 Blood biochemical results

The results of blood biochemistry are given in Table 4. In comparison with control group, TP, ALB, ALP, ALT, Ca and TG in model group were significantly reduced ($P < 0.01$), AST and TCH were remarkably elevated ($P < 0.05$ and $P < 0.01$). The results of ALB, ALT and LDH between model and CMC-Na groups showed some differences ($P < 0.05$, or $P < 0.01$). When compared with CMC-Na group, aspirin significantly reduced the levels of T-BIL, AST, LDH, TG and TCH, and elevated the levels of TP, ALB, BUN and Ca ($P < 0.05$, or $P < 0.01$). In addition, there was no significant difference between CMC-Na and eugenol groups.

In comparison with CMC-Na group, the levels of ALB and Ca were significantly increased in AEE groups ($P < 0.05$, or $P < 0.01$), whereas TCH was significantly reduced ($P < 0.05$, or $P < 0.01$). Low-and medium-dose AEE significantly increased the level of TP and reduced the level of T-BIL ($P < 0.01$). The TG and BUN levels in high-dose AEE group were significantly reduced ($P < 0.05$). In 5 μM ADP and 5 mM AA were added separately to platelet-rich plasma (PRP), which was used to assess platelet aggregation. Results are expressed as the percentage of maximal aggregation. Data are expressed as mean ± SD. #P < 0.05, ##P < 0.01 compared with the model group. *P < 0.05, **P < 0.01 compared with the CMC-Na group. PAg: platelet aggregation. AEE L: AEE 18 mg/kg; AEE M: 36 mg/kg; AEE H: AEE 72 mg/kg; Combination: combination of aspirin and eugenol (molar ratio 1 : 1) comparison with the CMC-Na group, the levels of ALB, AST, LDH and BUN in combination group were significantly elevated ($P < 0.01$). The differences in blood biochemical indexes among the groups treated with three doses of AEE were dose-independent.

Table 4 Biochemical parameters in rats intragastrically administered different drugs (n = 10) .

Variables	Unit	Control	Model	CMC-Na	Aspirin	Eugenol	AEE L	AEE M	AEE H	Combination
T-BIL	umol/L	0.90 ± 0.13	0.97 ± 0.19	1.00 ± 0.17	0.74 ± 0.12**	1.11 ± 0.17	0.72 ± 0.15**	0.65 ± 0.10**	0.86 ± 0.12	1.17 ± 0.17*
TP	g/L	58 ± 2.9##	53 ± 4.3	50 ± 3.4	60 ± 2.2**	50 ± 4.7	59 ± 2.4**	60 ± 3.3**	58 ± 4.1**	61 ± 3.9**
ALB	g/L	23 ± 1.0##	20 ± 1.4	18 ± 1.1##	20 ± 0.8*	19 ± 1.4	21 ± 2.3**	21 ± 0.7**	20 ± 1.1*	22 ± 1.2**
ALT	U/L	69 ± 10##	57 ± 9	67 ± 12#	61 ± 5	68 ± 13	65 ± 6	65 ± 6	58 ± 14	67 ± 13
AST	U/L	79 ± 11#	93 ± 10	85 ± 12	65 ± 8**	96 ± 14	76 ± 10	75 ± 16	74 ± 12	120 ± 14**
ALP	U/L	240 ± 19##	145 ± 22	156 ± 22	178 ± 27	137 ± 23	180 ± 25	181 ± 21	177 ± 11	172 ± 54
GGT	U/L	3.16 ± 0.49	3.34 ± 0.44	3.20 ± 0.25	3.44 ± 0.45	3.15 ± 0.67	3.55 ± 0.57	3.72 ± 0.64	3.61 ± 0.51	3.62 ± 0.60
LDH	U/L	141 ± 28	129 ± 29	161 ± 18#	123 ± 31*	190 ± 27	128 ± 20	126 ± 22	140 ± 38	227 ± 36**
BUN	mmol/L	8.0 ± 0.6	8.9 ± 1.7	9.9 ± 1.7	11.5 ± 1.4*	10.3 ± 1.9	9.4 ± 0.7	9.7 ± 1.5	8.3 ± 0.9*	12.5 ± 1.3**
CR	umol/L	27.3 ± 4.6	29.3 ± 4.9	29.0 ± 4.8	37.4 ± 3.6*	31.6 ± 4.1	24.9 ± 3.8	32.1 ± 4.6	26.2 ± 3.2	36.4 ± 4.7**
Glu	mmol/L	8.01 ± 0.68	8.11 ± 0.68	7.98 ± 0.53	7.78 ± 0.38	7.52 ± 0.30	7.50 ± 0.81	7.40 ± 0.25	7.57 ± 0.62	7.90 ± 0.81
Ca	mmol/L	2.42 ± 0.06##	2.08 ± 0.10	2.12 ± 0.15	2.30 ± 0.05**	2.12 ± 0.12	2.47 ± 0.12**	2.35 ± 0.05**	2.24 ± 0.14*	2.19 ± 0.06
P	mmol/L	3.31 ± 0.19	3.47 ± 0.25	3.50 ± 0.60	3.36 ± 0.40	3.64 ± 0.43	3.23 ± 0.45	3.36 ± 0.20	3.54 ± 0.15	3.53 ± 0.22
TG	mmol/L	1.19 ± 0.21##	0.66 ± 0.09	0.78 ± 0.06	0.56 ± 0.09*	0.69 ± 0.16	0.71 ± 0.19	0.59 ± 0.15	0.58 ± 0.27*	0.77 ± 0.21
TCH	mmol/L	1.37 ± 0.07##	1.98 ± 0.08	1.81 ± 0.23	1.53 ± 0.17**	1.93 ± 0.18	1.62 ± 0.12*	1.56 ± 0.18*	1.55 ± 0.26**	1.96 ± 0.29

3.4 Effects of AEE on TXB$_2$ and 6-keto-PGF$_1$α

The results of TXB$_2$ and 6-keto-PGF$_{1a}$ are shown in Table 5. TXB$_2$ and TXB2/6-keto-PGF$_1$α ratio were elevated in model group in comparison with those in control group, but 6-keto-PGF$_1$α was declined ($P < 0.01$). There was no difference in TXB$_2$ between CMC-Na and model groups. However, a significant decrease of 6-keto-PGF$_1$α and an increase of TXB$_2$/6-keto-PGF$_1$α ratio were observed ($P < 0.01$). Eugenol and aspirin significantly increased 6-keto-PGF$_1$α and lowered TXB$_2$/6-keto-PGF$_1$α ratio than CMC-Na ($P < 0.01$). In comparison with CMC-Na group, low-, medium- and high-dose AEE had significantly increased 6-keto-PGF$_1$α and reduced TXB$_2$ with a drop of TXB$_2$/6-keto-PGF$_1$α ratio ($P < 0.01$). No dose response relationship was observed in AEE groups on TXB$_2$ and 6-keto-PGF$_1$α.

Fig. 3 shows the effects different drugs on TXB$_2$ and 6-keto-PGF$_1$α. In comparison with aspirin and eugenol, AEE reduced TXB$_2$ and 6-keto-PGF$_1$α to varying degrees with the exception of 6-keto-PGF$_1$α in medium-dose AEE group. TXB$_2$ in high-dose AEE group were higher than combination group ($P < 0.05$). Three different doses of AEE remarkably lowered 6-keto-PGF$_1$α in comparison with combination group ($P < 0.05$ or $P < 0.01$). Under the same molar quantity, medium-dose AEE significantly reduced TXB$_2$ than aspirin and eugenol ($P < 0.05$ or $P < 0.01$). Results of 6-keto-PGF$_1$α in medium-dose AEE group were higher than eugenol and combination groups. Meanwhile, the TXB$_2$/6-keto-PGF$_1$α ratio in medium-dose AEE group was significantly lower than that in eugenol ($P < 0.01$). Therefore, equimolar AEE possessed better effects on reducing TXB$_2$, increasing 6-keto-PGF$_1$α and regulation of TXB$_2$/6-keto-PGF$_1$α ratio than its precursors.

Table 5 Effects of each treatment on TXB$_2$, 6-keto-PGF$_1$α and the TXB$_2$/6-keto-PGF$_{1a}$ ratio in k-carrageenan-induced rat tail thrombosis model ($n = 10$).

Groups	TXB$_2$	6-keto-PGF$_1$α	TXB$_2$/6-keto-PGF$_1$α
Control	498 ± 32$^{##}$	889 ± 28$^{##}$	0.56 ± 0.04$^{##}$
Model	696 ± 39	586 ± 24	1.19 ± 0.08
CMC-Na	703 ± 37	550 ± 22$^{##}$	1.28 ± 0.08$^{##}$
Aspirin	658 ± 44*	741 ± 23**	0.89 ± 0.06**
Eugenol	678 ± 58	699 ± 30**	0.97 ± 0.09**
AEE L	605 ± 22**	667 ± 17**	0.91 ± 0.04**
AEE M	612 ± 45**	737 ± 13**	0.83 ± 0.07**
AEE H	644 ± 23**	599 ± 26**	1.08 ± 0.07**
Combination	598 ± 51**	701 ± 31**	0.86 ± 0.10**

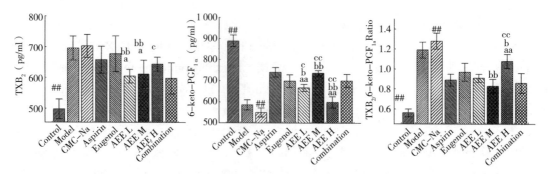

Fig. 3 Comparative effects of aspirin, eugenol and AEE on TXB_2, 6-keto-$PGF_1\alpha$ and TXB_2/6-keto-PGF_{1a} ratio in different groups (n = 10).

Note: #$P < 0.05$, ##$P < 0.01$ compared with model group. a$P < 0.05$, aa$P < 0.01$ compared with aspirin group. b$P < 0.05$, bb$P < 0.01$ compared with eugenol group. c$P < 0.05$, cc$P < 0.01$ compared with combination group. AEE L: AEE 18 mg/kg; AEE M: 36 mg/kg; AEE H: AEE 72 mg/kg; Combination: combination of aspirin and eugenol (molar ratio 1 : 1).

4 DISCUSSION

Several chemical drugs are used to prevent or treat thrombosis such as clopidogrel, ticlopidine, dabigatran and apixaban. However, available antithrombosis agents still have some limitations such as gastrointestinal damage and hemorrhage. Therefore, there is an obvious need for more efficacious and alternative treatment options for thrombosis. A previous study showed that AEE could reduce platelet aggregation in vitro, fibrinogen concentration and regulate coagulation parameters, indicating that AEE was a promising drug candidate for After the last measurement of thrombosis length at 48 h, blood was drawn from the heart. Serum samples were prepared by centrifugation and analyzed using an automatic biochemistry analyzer. #$P < 0.05$, ##$P < 0.01$ compared with the model group. *$P < 0.05$, **$P < 0.01$ compared with the CMC-Na group. AEE L: AEE 18 mg/kg; AEE M: 36 mg/kg; AEE H: AEE 72 mg/kg; Combination: combination of aspirin and eugenol (molar ratio 1 : 1) preventing cardiovascular diseases [23]. It was important to characterize its possible mechanism from different views.

Blood hemorheology is an integrated branch of physics and medicine. Increased blood viscosity is a risk factor for thrombosis, atherosclerosis and other cardiovascular events [24]. The increased blood and plasma viscosity in model group can be related to the acute inflammation caused by k-carrageenan injection [23, 25]. AEE decreased whole blood and plasma viscosity, which contributed to ameliorating blood circulation. It also proved Blood samples anticoagulated with EDTA-K_2 were centrifuged at 1 000 g for 10 min to obtain plasma, which was analyzed using ELISA kits. Data are expressed as mean ± SD. ##$P < 0.01$ compared with the model group. *$P < 0.05$, **$P < 0.01$ compared with the CMC-Na group. AEE L: AEE 18 mg/kg; AEE M: 36 mg/kg; AEE H: AEE 72 mg/kg; Combination: combination of aspirin and eugenol (molar ratio 1 : 1) that AEE could reduce fibrinogen concentration and normalize blood components [23]. Changes of blood components, especially the reduction of hematocrit, may be responsible for the decrease of viscosity.

Additionally, blood viscosity is also determined by shear rate, red blood cell deformability and aggregation. The effects of AEE and its dosage influence on hemorheological parameters should be assessed in the future studies. Some differences were observed between aspirin and combination groups in hemorheological parameters. The mean values of plasma component such as TP, ALB, TG and TCH in aspirin group were lower than those in combination group, which may be the reason for the decrease in plasma viscosity. It was speculated that the differences in RBC deformability and aggregation in aspirin and combination groups were responsible for the different results in whole blood viscosity.

Salicylic acid is the major metabolite of AEE confirmed by in vivo and in vitro experiments [26]. As a classic drug, aspirin is used therapeutically in the prevention of cardiovascular disease by blocking TXA2 synthesis. Salicylic acid, also the main metabolite of aspirin, is the primary substance responsible for pharmacological function of aspirin on platelet aggregation. AEE and aspirin share the same metabolite as salicylic acid; therefore, AEE and aspirin likely inhibit platelet aggregation via similar mechanism. The effects of eugenol on human platelet aggregation had been investigated and proved that eugenol was a strong platelet aggregation inhibitor [12, 13]. Vascular endothelial cell damage caused by k-carrageenan may be the reason for the increased platelet aggregation in model group [18]. The vehicle, CMC-Na, had no effect on platelet aggregation. In the case of equal molar quantity, AEE inhibited the platelet aggregation and showed stronger effects than eugenol on AA and ADP-induced platelet aggregation, and better effects than aspirin on AA-induced platelet aggregation. Based on the results, the effect of AEE on platelet aggregation is likely produced by both salicylic acid and eugenol. AEE is decomposed into salicylic acid and eugenol by the enzyme after absorption, after which salicylic acid and eugenol as the major metabolites showed their original activities and acted synergistically to increase the inhibitory effect on platelet aggregation. Previous studies have confirmed that platelet aggregation is an important influence in blood viscosity [27]. Thus, the reduced platelet aggregation caused by AEE was another factor contributing to the decreases of blood and plasma viscosity.

In order to explore the further mechanisms of inhibiting platelet aggregation, the levels of TXB_2 and 6-keto-$PGF_1\alpha$ in plasma were measured. TXB_2 and 6-keto-$PGF_1\alpha$ are the stable hydrolysis products of thromboxane A2 (TXA_2) and prostacyclin I2 (PGI_2), respectively. TXA_2 is a potent inducer of platelet aggregation and vasoconstriction [6]. PGI_2 is a powerful vasodilator that inhibits platelet aggregation [28]. Arachidonate is a precursor of both PGI_2 and TXA_2. Cyclooxygenase (COX) plays an important role in arachidonate metabolic pathways to generate PGI_2 and TXA_2 [29]. Some evidences have demonstrated that both aspirin and eugenol inhibit COX activity [30-32]. Aspirin, as an inhibitor of COX, could reduce both the level of TXB_2 and 6-keto-$PGF_1\alpha$. However, the level of 6-keto-$PGF_1\alpha$ in aspirin group was higher than that of CMC-Na group. The previous study showed that aspirin significantly decreased the length of rat tail thrombosis [23]. Thrombosis formation may produce the reductive effects on 6-keto-$PGF_1\alpha$. Aspirin inhibits the thrombosis formation in the rat tail, and then weakens the decreasing effects caused by k-carrageenan on 6-keto-$PGF_1\alpha$. Suppression of inflammation by aspirin may be the reason of reducing rat tail thrombosis length [33]. Platelets primarily process PGH_2 to TXA_2 and vascular endothelial cells primarily process PGH$_2$ to PGI$_2$. Aspirin selectively inhibits the production of TXA$_2$ in platelet, while sparing endothelial PGI$_2$ synthesis [34]. New cyclooxygenase could be produced in endothelial cells, but cannot in anucleate platelets. It was speculated that the different effects of aspirin on platelets and endothelial cells and the newly produced cyclooxygenase from endothelial cells may be the reasons for the raised 6-keto-PGF$_1$α in aspirin group. Under equal molar quantity, AEE produced better effects than single using of eugenol or aspirin on TXB$_2$ and 6-keto-PGF$_1$α. According to these results, AEE might significantly down-regulate TXB$_2$ and up-regulate 6-keto-PGF1α with the TXB$_2$/6-keto-PGF$_1$α decreased, indicating that the antithrombotic action of AEE was associated with the regulation of TXB$_2$ and 6-keto-PGF$_1$α.

Analysis of blood biochemical parameters could help veterinarians and breeders assess the general health status of animals [35, 36]. Injection of k-carrageenan disturbed the blood biochemical profile through the increase of AST and TCH and the reduction of TP, ALB, ALP, ALT, Ca and TG in model group. These alterations may be related to the stress caused by acute inflammation. Previous study showed that leukocyte and monocyte counts were significantly increased after k-carrageenan injection, which indicated that the rats suffered from acute and systemic inflammation [23]. Intense inflammation could cause the disorder of liver function, and then result in the changes of biochemical profile [37, 38]. In the present study, the reduction of TP, ALB, ALP and the increasing of AST in model group indicated abnormal liver function. Ca, a key factor in thrombosis formation, may be depleted in thrombosis formation process, which may be the reason for the reduction of Ca in model group. The increase of TCH and the reduction of TG may be related to rat body response [39]. It has been proved that AEE could regulate blood lipids in hyperlipidemic rats [40]. The results of TG and TCH indexes in this experiment were similar as found in previous studies [16, 40]. AEE showed positive effects on changing biochemical characters and possessed the ability to normalize the biochemical profile following inflammation, which may be supported by the prolonged time of drug action [41]. In the present study, the blood biochemical results in eugenol group showed no differences in comparison with CMC-Na group. Several reasons may be for these results. Blood biochemical parameters were drastically changed by k-carrageenan injection, which may mask the changes caused by eugenol. Moreover, the dosage of eugenol used in the experiment or eugenol itself was unable to affect biochemical parameters. Previous studies indicated that administration of 150 mg eugenol for 7 days had no significant effects on clinical biochemical parameters in humans [42].

CMC-Na is a reliable drug carrier that is used in a wide range of applications in the pharmaceutical industry [43, 44]. In this study, CMC-Na as vehicle had influence on 6-keto-PGF$_1$α, TXB$_2$/6-keto-PGF$_{1a}$ ratio, ALB, ALT and LDH. In order to demonstrate the activity of AEE, the comparisons were carried out between CMC-Na and treated groups. Therefore, the effect of CMC-Na was eliminated. Tween 80 is widely applied in emulsifying and dispersing substances in medicinal products. In the present study, Tween 80 was used to prepare eugenol. It had been confirmed that the body had a great tolerance to Tween 80 [45]. The effects of Tween 80 and combination of CMC-Na and Tween 80 were not investigated in this experiment.

AEE on the design of prodrug principle contains ester bond structure that is decomposed easily. Pharmacokinetics studies showed that the plasma concentration of AEE itself was extremely low; indeed it was not detectable in plasma. Therefore, it is speculated that the degradation products of AEE such as salicylic acid and eugenol are responsible for its effects. Salicylic acid and eugenol may play an efficient role and interact in a synergistic manner to prevent thrombosis formation. More studies should be conducted to investigate the action mechanism of AEE on thrombosis prevention. In addition, as a promising chemical compound, AEE should be studied in preclinical experiments to assess its therapeutic effects in various species.

5 CONCLUSION

The results obtained in our study showed that AEE had potent antithrombotic effects in rats with k-carrageenan-induced tail thrombosis. Moreover, AEE showed better antithrombotic effect than its precursors under same molar quantity. From the findings, the preventive effect of AEE may come from antiplatelet aggregation, reducing blood and plasma viscosity, balancing $TXB_2/6$-keto-PGF_{1a} ratio and normalizing blood biochemical parameters. These therapeutic effects of AEE may be related to the synergetic actions of aspirin and eugenol. It may be concluded that AEE was a good candidate for anti-thrombotic agent. These findings provide new insight into the action mechanism of AEE on preventing thrombosis. More studies are necessary to investigate its mechanism of action such as the protective effects of vascular endothelial cells, the way on anti-platelet aggregation and the influences on metabolic profile.

ABBREVIATIONS

6-keto-$PGF_1\alpha$, 6-keto prostaglandin $F_1\alpha$; AA, arachidonic acid; ADP, adenosine diphosphate; AEE, aspirin eugenol ester; ALB, albumin; ALP, alkaline phosphatase; ALT, alanine transaminase; AST, aspartate aminotransferase; BUN, blood urea nitrogen; Ca, calcium; CK, creatine kinase; CMC-Na, sodium carboxymethyl cellulose; CR, creatinine; GGT, gamma glutamyl transpeptidase; GLU, glucose; GPT, glutamic pyruvic transaminase; LDH, lactate dehydrogenase; P, phosphorus; PGI_2, prostacyclin I_2; PPP, platelet-poor plasma; PRP, platelet-rich plasma; T-BIL, total bilirubin; TCH, total cholesterol; TG, triglycerides; TP, total protein; TXA_2, thromboxane A_2; TXB_2, thromboxane B_2.

ACKNOWLEDGMENTS

The work was supported by the National Natural Science Foundation of China (No. 31572573).

FUNDING

The funders had no role in the design of the study and collection, analysis, and interpretation of data and in writing the manuscript.

AVAILABILITY OF DATA AND MATERIALS

The data supporting the findings is contained within the manuscript.

AUTHORS' CONTRIBUTIONS

JYL, YJY and NM designed and supervised the experiments. NM performed platelet aggregation test and ELISA. NM, XWL, IM and XLZ carried out the hemorheological test and blood biochemical analysis. NM, JYL, YJY, DSS and XJK performed data analysis and materials. NM, JYL, and YJY prepared the manuscript and reviewed the literature. All authors read and approved the final manuscript.

COMPETING INTERESTS

The authors declare that they have no competing interests.

CONSENT FOR PUBLICATION

Not applicable.

Ethics approval and consent to participate.

The experimental procedures were performed in compliance with the Guidelines for the care and use of laboratory animals as described in the US National Institutes of Health and approved by the Institutional Animal Care and Use Committee of Lanzhou Institute of Husbandry and Pharmaceutical Science of CAAS.

（发表于《BMC Veterinary Research》，院选SCI，IF：1.643.108.）

Multi-Residue Method for the Screening of Benzimidazole and Metabolite Residues in the Muscle and Liver of Sheep and Cattle Using HPLC/PDAD with DVB-NVP-SO₃Na for Sample Treatment

XIONG Lin[1, 2*], HUANG Lele[3], SHIMO Shimo-Peter[1, 2], LI Weihong[1, 2], YANG Xiaolin[1, 2], YAN Ping[1, 2]

(1. Lanzhou Institute of Husbandry and Pharmaceutical Sciences, Chinese Academy of Agricultural Sciences, 335 Jiangouyan St, Lanzhou, Gansu, China; 2. Laboratory of Quality and Safety Risk Assessment for Livestock Product (Lanzhou), Ministry of Agriculture, 335 Jiangouyan St, Lanzhou, Gansu, China; 3. Department of Unclear Medicine, Lanzhou University Second Hospital, 82 Cuiying St, Lanzhou, Gansu, 730050China)

Abstract: A high-performance liquid chromatography (HPLC) screening method with a photodiode array detector (PDAD) was established for the simultaneous determination of residues of 13 benzimidazoles (BZDs) and metabolites in sheep and cattle muscle and liver. Samples were extracted by ultrasonication in ethyl acetate and purified over DVB-NVP-SO₃Na sorbent. Under the optimized conditions, good linearities were obtained for BZDs and metabolites with correlation coefficients (R^2) greater than 0.991 1. The recoveries of the 13 BZDs and metabolites from spiked samples were 72.0%–119.3%, with intraday and interday relative standard deviations (RSDs) below 22.8%. The limits of detection (LODs) and quantitation (LOQs) were 0.8–4.9 and 2.6–18.2 $\mu g \cdot kg^{-1}$, respectively. The results clearly demonstrated that the developed approach enables reliable screening of 12 BZDs and metabolites except flubendazole (FLU) and could be used as a regulatory tool for the screening of BZD and metabolite residues in the muscle and liver of sheep and cattle.

Key words: Benzimidazole; Metabolites; Screening method; HPLC/PDAD; DVB-NVP-SO₃Na

1 INTRODUCTION

Benzimidazoles (BZDs) are broad-spectrum anthelmintics against nematodes, cestodes and trematodes and have been used for approximately 50 years [1]. Most BZD drugs used as anthelmintics are licensed for use in food-producing animals [2], but the residues of BZDs in food-producing

Corresponding author, E-mail: xionglin807@sina.com.
© Springer-Verlag Berlin Heidelberg 2016.

animals are of great concern for consumer safety [3]. Chronic exposure to BZDs can cause toxic effects, including teratogenic, embryotoxic, goitrogenic and mutagenic effects [4]. Maximum residue limits (MRLs) have been established in the EU for BZDs in edible tissues under Commission Regulation 2010/37/EC [5] in which residual definitions for albendazole (ABZ), fenbendazole (FBZ), oxfendazole (OBZ), Flubendazole (FLB) and thiabendazole (TBZ) were adopted. The inclusions of metabolites are important because they can be more toxic and persistent than the parent drug [4]. In cattle, albendazole-sulfoxide (ABZ-SO), albendazole sulphone (ABZ-SO$_2$), mebendazole (MBZ) and albendazole-2-aminosulfone (ABZ-2-NH2-SO$_2$) were found to be the major residues detected [6, 7]. Fenbendazole sulphone (FBZ-SO$_2$), thiabendazole-5-hydroxy (TBZ-5-OH), hydroxythiabendazole (TBZ-OH) and amino-flubendazole (NH$_2$FLU) were the important residues of metabolites [4].

The methods used to monitor BZD residues include high-performance liquid chromatography (HPLC) [8, 9], biosensors [4], capillary electrophoresis [10, 11] and liquid chromatography−mass spectrometry (LC−MS) [12, 13]. LC−MS is expensive and unavailable in many laboratories [14]. Biosensors and capillary electrophoresis have only been used in laboratories and not in-depth. In contrast, HPLC is inexpensive and highly reliable and is widely used in laboratories and testing agencies. To the best of our knowledge, there has been no report on the use of DVB-NVP-SO$_3$Na as a cleansing agent in the determination of BZD and metabolite residues in meat. The purpose of this study was to develop a multi-residue method for the determination of BZD and metabolite residues in the muscle and liver of sheep and cattle. The samples were prepared with DVB-NVP-SO$_3$Na and analyzed by HPLC/ PDAD. The developed method can be applied in routine laboratory analysis of large numbers of samples containing BZDs and metabolites.

2 EXPERIMENTAL

2.1 Chemicals, reagents and solutions

FBZ-SO$_2$ (99.9% ± 0.1%), FLBZ (99.9% ± 0.1%) and 2-NH$_2$-FLBZ (>99.9%) were provided by WITEGA Laboratorien Berlin-Adlershof GmbH (Berlin, Germany). ABZ-SO (98.5%), TBZ (99.0%), OBZ (98.0%), FBZ (98.5%), ABZ-SO$_2$ (98.3%), ABZ-2-NH$_2$-SO$_2$ (99.0%), OBZ (99.0%), TBZ-5-OH (99.1%), MBZ (99.5%) and ABZ (98.7%) were provided by Dr. Ehrenstorfer GmbH (Augsburg, Germany). Mixed working standard solutions (1 μg·mL^{-1}) were prepared by diluting single (stock) standard solutions in the mobile phase and stored at 20 ℃ for 6 months. HPLC-grade methanol and ace-tonitrile were obtained from Tedia Company Inc (Fairfield, USA). Ammonium hydroxide, ethyl acetate, 2,6-di-tertbutylphenol, hydrochloric acid and ammonium acetate were analytical grade and were obtained from Sigma-Aldrich (Santa Clara, USA). All other chemicals and solvents were of analytical grade from Sinopharm (Shanghai, China) and, unless stated otherwise, were used without further purification.

2.2 Instrumentation and apparatus

A Waters e2695 separations module with a photodiode array detector (Milford, USA) was used, and separation was performed on a Waters Xcharge C_{18} column (250 mm × 4.6 mm, 5 μm, 100 Å) (Milford, USA). An electronic analytical balance BSA224S-CW (Goetingen, Germany), a BÜCHI Mixer B-400 homogenizer (Flawil, Switzerland), a Labinco L24 vortex mixer (Breda, The Netherlands), a Yarong RE52A rotary evaporators (Zhengzhou, China), a Scientz Biotechnology ultrasonic apparatus (Ningbo, China), a ZP-200 shaker (Taichang, China), 0.22 μm Millipore Millex-GV membrane filters (MA, USA), a Millipore Milli-Q ultrapure water system (Molsheim, France), Omnifuge 2.ORS centrifuge (Osterode, Germany) and Peak Scientific N100DR nitrogen evaporators (Glasgow, UK) were used.

2.3 Preparation of DVB-NVP-SO$_3$Na and the cleansing principle

DVB-NVP-SO$_3$Na was prepared according to [15] in three steps. First, the intermediate DVB-NVP was prepared by polymerization of divinylbenzene and N-vinyl pyrrolidone. In the second step, DVB-NVP-SO$_3$H was prepared from DVB-NVP and concentrated sulfuric acid by sulfonation. Finally, the target compound DVB-NVP-SO$_3$Na was prepared via a neutralization reaction between DVB-NVP-SO$_3$H and sodium hydroxide solution. The BZD and metabolite molecules are quaternary ammonium salts under acidic conditions, and DVB-NVP-SO$_3$Na can absorb quaternary ammonium salts of BZD and metabolite molecules by chemical bonding. Other impurities do not bond with DVB-NVP-SO$_3$Na, and thus the BZD and metabolite residues were separated from the impurities by centrifugation. The BZD and metabolites molecules were then eluted from the DVB-NVP-SO$_3$Na using an alkaline solution and separated from the matrix. The benefit of using the developed sorbent instead of commercial SPEs was mainly embodied in reducing the cost. As far as we know, the commercial SPE columns are all very expensive. In contrast, the prepared DVB-NVP-SO$_3$Na is very inexpensive. The cost of DVB-NVP-SO$_3$Na is at most one tenth of that of using SPE. The developed method prepares the sample using only centrifugation, shaking and vortexing, avoiding the use of an expensive SPE column. Thus, the cost is decreased.

2.4 Sample preparation

Samples including cattle muscle, cattle liver, goat muscle and goat liver were frozen (20℃) until analysis. Samples were homogenized and weighed (5 ± 0.02 g) into 50 mL centrifuge tubes. Then, 20 mL ethyl acetate, 0.15 mL of 50% potassium hydroxide solution and 1 mL of 1% 2, 6-di-tert-butylphenol which was used to prevent oxidation of some BZD residues [16], were added. Next, the sample was extracted using an ultrasonic apparatus for 10 min, followed by centrifugation at 10 000 rpm for 5 min, and the supernatant was transferred to a bottle. The extraction was repeated twice. The combined supernatants were evaporated to near dryness. Then, 1.5 mL of acetonitrile solution was added, and the bottle was vortexed for 1 min, followed by the addition of 4.5 mL of 0.1 mol·L^{-1} hydrochloric acid solution and shaken for 10 min. The solution in the bottle was transferred to a centrifuge tube, and DVB-NVP-SO$_3$Na was added. The centrifuge tube was shaken for 10 min and centrifuged for 5 min at 5 000 rpm, and the supernatant was discarded. Then, 4.5 mL of methanol was added, followed by vortexing for 10 min and centrifugation at 5 000 rpm for 10 min. The supernatant was discarded, and 10% ammoniac acetonitrile was added as an eluent. The tube was capped immediately and shaken briefly by hand, followed by centrifugation at 5 000 rpm for 10 min. The supernatant was then transferred to a new 15 mL tube and evaporated under a nitrogen stream at 50 ℃. The residue was dissolved in 0.5 mL of acetonitrile by ultrasonication for 5 min, and 1.5 mL of 0.025 mol·L^{-1} ammonium acetate solution was added. The mixture was shaken and filtered into LC vials.

Table 1　Gradient elution produce.

Time/min	A%(v)	B%(v)
	20.0	80.0
15.00	25.0	75.0
28.00	30.0	70.0
38.00	60.0	40.0
38.10	20.0	80.0
45.00	20.0	80.0

2.5　LC operating conditions

The mobile phase was consisted of acetonitrile (A) and 0.025 mol·L^{-1} ammonium acetate solution (B). The gradient elution profile is presented in Table 1, and the eluent flow rate was 1.0 mL·min^{-1}. The wavelength was 292 nm. The injection volume was 100 μL, and the column and ambient temperature were maintained at (35 ± 5) ℃.

3　RESULTS AND DISCUSSION

3.1　Precision amongst different batches of DVB-NVP-SO$_3$Na

DVB-NVP-SO$_3$Na from three consecutively manufactured batches was used to measure the precision amongst different batches. For the experiments, six spiked samples of cattle muscle were prepared with each batch of cleaning agent and the recoveries and RSDs were determined. The average recoveries were 81.7%-90.5%, 82.5%-92.1% and 80.1%-83.9%. The RSDs were 3.5%-8.5%, 2.1%-9.4% and 2.8%-7.5%. Based on the above results, one can initially draw a conclusion that different batches of DVB-NVP-SO$_3$Na are all capable of high precision (2.1%-8.5%).

3.2　Optimization of the HPLC parameters

The following parameters were assessed: (A) A : B (50 : 50, v : v) with isocratic elution; (B) A : B (80 : 20, v : v) with isocratic elution and (C) A : B (80 : 20, v : v) with gradient elution. Chromatograms of cattle muscle samples spiked with 13 BZDs and metabolites (160 μg·kg^{-1}) are shown in Fig. 1. The C conditions exhibited a high degree of separation and high selectivity in the analysis of the 13 BZDs and metabolites, and all compounds were separated from the baseline and exhibited fine peak shapes. However, poor separation of the 13 BZDs and metabolites was observed for the other two sets of parameters.

3.3 Optimization of sample preparation

The dosage of the cleansing agent and eluent were optimized using single-factor experiments. The parallel cattle muscle samples (5 ± 0.02 g) were placed in a centrifuge tube in each group experiment, and 1 mL of a 100 μg·mL^{-1} mixed solution of the 13 standards was added.

3.4 Optimization of the dosage of DVB-NVP-SO$_3$Na

All parameters were identical, except that the dosage of DVB-NVP-SO$_3$Na in the 5 g samples was varied: 0.1, 0.2, 0.3, 0.4, 0.5 and 0.6 g. Analysis of the six prepared samples revealed recoveries of 86%–112% for the 13 BZDs when 0.3 g of DVB-NVP-SO$_3$Na was used. When the dosage of DVB-NVP-SO3Na was more than 0.3 g, the recoveries of the 13 BZDs and metabolites were 86%–112% and did not exhibit any significant variation. When the dosage of DVB-NVP-SO$_3$Na was less than 0.3 g, the recoveries were < 84%. Consequently, to minimize cost, 0.3 g of DVB-NVP-SO$_3$Na was used in the final optimized method.

3.5 Optimization of the dosage of eluent

The dosage of eluent was set at six levels: 5, 8, 11, 14, 17 and 20 mL. Recoveries of 88%–100% were obtained for the 13 BZDs and metabolites when 11 mL of eluent was used. When the dosage of eluent was greater than 11 mL, the recoveries of the 13 BZDs and metabolites were 90%–106% and did not differ significantly. When the dosage of eluent was less than 11 mL, the recoveries were < 80% because the volume of eluent was insufficient to wash all of the spiked BZDs and metabolites bonded to DVB-NVP-SO$_3$Na. To minimize cost, 11 mL was chosen as the optimal parameter.

3.6 Effect of pH value on cleanup efficient

The concentration of hydrochloric acid solution decides the pH value of cleanup solution in the same volume. Therefore, a study of concentration of the hydrochloric acid solution on cleanup efficiency was carried out. The concentration of hydrochloric acid solution was set at five levels: 0.05 (pH 1.3), 0.1 (pH 1.0), 0.2 (pH 0.7), 0.4 (pH 0.4) and 1.0 mol·L^{-1} (pH 0.0). The recoveries were 72%–95%, 85%–93%, 83%–92%, 85%–95% and 75%–85%. The impurity peaks were prominent in the chromatograms when 1.0 mol·L^{-1} (pH 0) and 0.05 mol·L^{-1} (pH 1.3) hydrochloric acid solution were used. In the other comparison tests, their chromatograms were very similar, smooth and concise with no interfering peaks. The validation data were based on the condition without the addition of hydrochloric acid, and the pH values can also be adjusted using other strong acids. Overall, good cleanup efficiencies can be obtained when the pH 0.4–1.0 strong acid used.

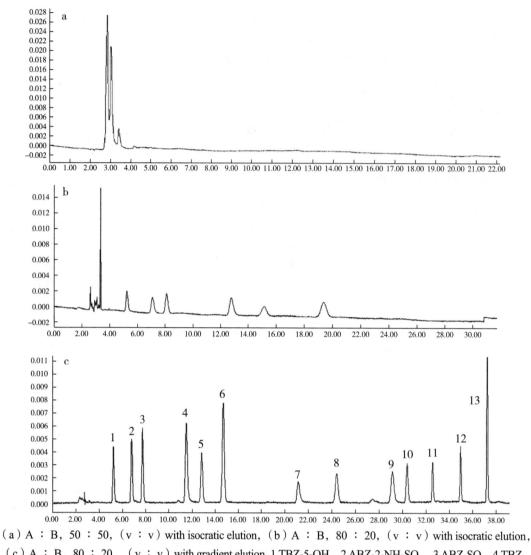

(a) A : B, 50 : 50, (v : v) with isocratic elution, (b) A : B, 80 : 20, (v : v) with isocratic elution, (c) A : B, 80 : 20, (v : v) with gradient elution. 1 TBZ-5-OH, 2 ABZ-2-NH-SO_2, 3 ABZ-SO, 4 TBZ, 5 ABZ-SO_2, 6 OBZ, 7 2-NH_2-FLBZ, 8 FBZ-SO_2, 9 OFZ, 10 MBZ, 11 FLBZ, 12 ABZ and 13 FBZ.

Fig. 1 Effect of the mobile phase on the chromatogram of the BZDs.

3.7 Purifying effect of DVB-NVP-SO_3Na

Cattle muscle was selected as the matrix to assess the removal of impurities by DVB-NVP-SO_3Na (Fig. 2). Spiked samples (50 μg·kg^{-1}) cleaned with DVB-NVP-SO_3Na were compared with spiked samples (50 μg·kg^{-1}) that were not cleaned with DVB-NVP-SO_3Na, and blank samples cleaned by DVB-NVP-SO_3Na were compared with blank samples that were not cleaned with DVB-NVP-SO_3Na. Both the blank samples and the spiked samples cleaned with DVB-NVP-SO_3Na contained few impurities, and their chromatograms were smooth and concise with no interfering peaks at the retention times of all 13 BZDs and metabolites. Therefore, using DVB-NVP-SO3Na for purification of the BZD and metabolite residues in the samples was successful.

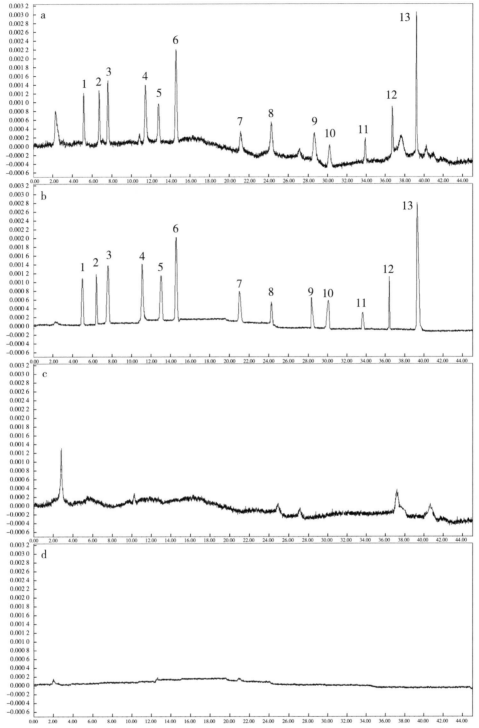

(a) sample spiked with 50 μg·kg^{-1} but not cleaned, (b) cleaned sample spiked with 50 μg·kg^{-1}, (c) blank sample that was not cleaned, (d) cleaned blank sample. 1 TBZ-5-OH, 2 ABZ-2-NH-SO$_2$, 3 ABZ-SO, 4 TBZ, 5 ABZ-SO$_2$, 6 OBZ, 7 2-NH$_2$-FLBZ, 8 FBZ-SO$_2$, 9 OFZ, 10 MBZ, 11 FLBZ, 12 ABZ and 13 FBZ

Fig. 2 Purifying effect of DVB-NVP-SO$_3$Na.

3.8 Linearity

The linearity of the PDAD chromatographic response was tested in the concentration range of 10–160 μg·kg^{-1}. The linear regression coefficients of the calibration curves (Table 2) indicated that good results were achieved, with R^2 between 0.991 1 and 0.999 6.

3.9 Recovery

Blank samples were spiked with BZDs and metabolites at three concentrations (10, 50 and 100 μg·kg^{-1}) for intraday (n 18) and interday experiments (n 6). The recoveries were 74.7%–119.3% for cattle muscle, 76.7%–118.0% for goat muscle, 72.0%–113.3% for cattle liver and 74.7%–95.4% for goat liver. Overall, the recoveries of the 13 BZDs and metabolites in the cattle and goat samples were in the ranges of 72.0%–119.3 and 74.7%–118.0%, respectively.

Table 2 Linear ranges, regression coefficients (R^2), LODs and LOQs for BZDs.

Compound	Sample	R^2	Linear range (μg·kg^{-1})	LODs (μg·kg^{-1})	LOQs (μg·kg^{-1})
TBZ-5-OH	C.M	0.999 5	10-160	2.1	7.1
	C.L	0.999 0	10-160	2.5	8.4
	S.M	0.999 6	10-160	2.3	7.8
	S.L	0.991 2	10-160	2.4	7.9
ABZ-2-NH$_2$-SO$_2$	C.M	0.998 4	10-160	2.1	7.2
	C.L	0.997 2	10-160	2.3	7.6
	S.M	0.999 1	10-160	2.2	7.3
	S.L	0.999 0	10-160	2.3	7.5
ABZ-SO	C.M	0.998 0	10-160	1.9	6.3
	C.L	0.996 5	10-160	2.1	6.9
	S.M	0.997 9	10-160	2.1	7.1
	S.L	0.992 3	10-160	2.2	7.3
TBZ	C.M	0.999 2	10-160	2.0	6.7
	C.L	0.998 7	10-160	2.0	6.8
	S.M	0.997 8	10-160	2.2	7.4
	S.L	0.999 1	10-160	2.2	7.2
ABZ-SO$_2$	C.M	0.998 8	10-160	2.4	7.9
	C.L	0.992 0	10-160	2.6	8.7
	S.M	0.998 6	10-160	2.3	7.7
	S.L	0.996 5	10-160	2.1	7.1

(continued)

Compound	Sample	R^2	Linear range ($\mu g \cdot kg^{-1}$)	LODs ($\mu g \cdot kg^{-1}$)	LOQs ($\mu g \cdot kg^{-1}$)
OBZ	C.M	0.998 7	10-160	1.4	4.8
	C.L	0.996 5	10-160	1.4	4.6
	S.M	0.999 2	10-160	1.3	4.3
	S.L	0.993 3	10-160	1.5	5.0
2-NH_2-FLBZ	C.M	0.998 1	10-160	5.5	18.2
	C.L	0.992 5	10-160	4.9	16.3
	S.M	0.999 3	10-160	3.9	13.0
	S.L	0.991 3	10-160	4.2	14.0
FBZ-SO_2	C.M	0.998 7	10-160	4.1	13.5
	C.L	0.992 5	10-160	4.3	14.3
	S.M	0.998 6	10-160	3.8	12.5
	S.L	0.994 4	10-160	3.8	12.6
OFZ	C.M	0.998 7	10-160	3.9	13.1
	C.L	0.991 8	10-160	3.8	12.5
	S.M	0.999 2	10-160	3.8	12.8
	S.L	0.996 5	10-160	3.5	11.6
MBZ	C.M	0.998 8	10-160	3.9	13.1
	C.L	0.994 7	10-160	3.6	12.0
	S.M	0.998 7	10-160	3.4	11.2
	S.L	0.991 1	10-160	3.3	11.1
H-BZ	C.M	0.998 7	10-160	4.0	13.4
	C.L	0.998 4	10-160	4.3	14.2
	S.M	0.997 5	10-160	4.3	14.4
	S.L	0.994 4	10-160	3.7	12.4
ABZ	C.M	0.996 2	10-160	2.6	8.8
	C.L	0.998 2	10-160	2.7	9.1
	S.M	0.998 5	10-160	3.7	12.4
	S.L	0.999 1	10-160	2.2	7.3

Table 3 continued.

Compound	Sample	R^2	Linear range ($\mu g \cdot kg^{-1}$)	LODs ($\mu g \cdot kg^{-1}$)	LOQs ($\mu g \cdot kg^{-1}$)
FBZ	C.M	0.999 2	10-160	0.8	2.6
	C.L	0.994 8	10-160	0.8	2.7
	S.M	0.999 4	10-160	0.9	3.1
	S.L	0.995 1	10-160	0.8	2.6

(1) TBZ-5-OH, (2) ABZ-2-NH-SO$_2$, (3) ABZ-SO, (4) TBZ, (5) ABZ-SO$_2$, (6) OBZ, (7) 2-NH2-FLBZ, (8) FBZ-SO$_2$, (9) OFZ, (10) MBZ, (11) FLBZ, (12) ABZ and (13) FBZ

C.M cattle muscle, C.L cattle liver, S.M sheep muscle, S.L sheep liver

3.10 Precision

The interday and intraday RSDs ranged from 1.0% to 22.8% and 4.5% to 17.5%, respectively, for the cattle muscle samples; from 1.0% to 14.8% and 5.3% to 9.5%, respectively, for the cattle liver samples; from 2.9% to 20.4% and 2.76% to 14.3%, respectively, for the goat muscle samples; from 0.7% to 14.8% and 5.3% to 11.7%, respectively, for the goat liver samples. All RSDs were lower than 22.8%.

3.11 LODs and LOQs

LODs and LOQs are also shown in Table 2. The LODs in the cattle and goat samples were 0.8-4.9 and 0.8-4.2 $\mu g \cdot kg^{-1}$, respectively. The LOQs in the cattle and goat samples were 2.6-18.2 and 2.6-14.4 $\mu g \cdot kg^{-1}$, respectively. The recommended MRLs for BZDs established by the EU and China range from 50 to 400 $\mu g \cdot kg^{-1}$ [5] and 60 to 5 000 $\mu g \cdot kg^{-1}$ [17]. FLB does not yet have an MRL in cattle/sheep muscle/liver in China and EU; if the limit of 10 $\mu g \cdot kg^{-1}$ is adopted, the developed method is not suitable for monitoring FLB in these matrices as the LOQ (12.4-14.4 $\mu g \cdot kg^{-1}$) is not good enough. Therefore, the developed method was adequate for routine residue monitoring of 12 BZDs in cattle and goat samples, with the exception of FLB.

3.12 Application of the method to real samples

The effectiveness of the presented method was demonstrated by analysing 20 goat samples and 20 cattle samples obtained from local dairy farmers. All samples were processed according to the method described, and no residues of the target analytes were detected in any of the 40 samples.

4 CONCLUSIONS

This paper reports the development and validation of an LC method for the simultaneous determination of trace amounts of 12 BZD and metabolite residues in the muscle and liver of sheep and cattle; FLB could not be determined with sufficiently high precision. We prepared DVB-NVP-SO$_3$Na as the cleansing agent for the BZDs and metabolites. Compared with traditional analytical procedures for BZD and metabolite residue analysis, the manual operation and sample preparation time were reduced. Good results, including for the recoveries, precisions, linearities, LODs and LOQs, were achieved. In addition, compared with traditional strong cation exchange SPE, the developed method prepares the sample avoiding the use of an expensive SPE column. The proposed method is suitable for screening in routine analysis.

ACKNOWLEDGMENTS

This study was funded by research Grants from the Chinese Academy of Agricultural Sciences (No. 1610322014014).

Compliance with Ethical Standards

Conflict of interest The authors have no conflicts of interest.

（发表于《Chromatographia》，SCI，IF：1.332. 1 373–1 380.）

Treatment of the Retained Placenta in Dairy Cows: Comparison of a Systematic Antibiosis with an Oral Administered Herbal Powder Based on Traditional Chinese Veterinary Medicine

CUI Dongan[1, 2], WANG Shengyi[1, 2], WANG Lei[1, 2], WANG Hui[1, 2], LI Jianxi[1, 2], TUO Xin[1, 2], HUANG Xueli[1, 3], LIU Yongming[1, 2, *]

(1. Lanzhou Institute of Husbandry and Pharmaceutical Sciences of Chinese Academy of Agricultural Sciences, Lanzhou 730050, China; 2. Engineering & Technology Research Center of Traditional Chinese Veterinary Medicine of Gansu Province, Lanzhou 730050, China; 3. College of Veterinary Medicine, Northwest A & F University, Yangling, Shaanxi 712100, China)

Abstract: Cows affected with retained placenta are at a higher risk of developing puerperal metritis. Herbal remedies bear a high potential to treat postpartum uterine diseases in cows. The aim of this randomized clinical trial was to compare an herbal powder and ceftiofur hydrochloride in the treatment of cows affected with retained placenta and for puerperal metritis prevention. The herbal powder was prepared from a combination of Leonurus artemisia (Laur.) S.Y. Hu F, Angelica sinensis (OLIV.) DIELS (radix), Ligusticum chuanxiong HORT (radix), Sparganiumstoloniferum (Graebn.) Buch.-Ham.exJuz (radix), Curcuma zedoaria (Christm.) ROSC (radix), Cyperu srotundus Linn. (radix), and Glycyrrhiza uralensis FISCH (radix). A total of 157 cows diagnosed with retained placenta were randomly divided into 2 treatment groups. Cows in the herbal group (n=85) were treated with an oral dose of 0.5 g crude herb/kg bw once daily for 1-3 day (s), and cows in the control group (n=72) were treated with ceftiofur hydrochloride (2.2 mg/kg bw, i. m.) twice daily for 3 consecutive days. Seventy-three cows had total expulsion of the placenta within 72 h following initial herbal treatment, yet no cows in the control group expelled the placenta during the same time period, and 50 out of 73 cows achieved total expulsion of the placenta following only one herbal treatment. The median time of retained placenta shedding (20.0 vs. 101.5 h; $P < 0.01$) was shorter in the herbal group than in the control group. The logistic regression analysis indicated that the oral administration of the herbal powder tended to have superior clinical efficacy in metritis prevention compared to the systemic administration of ceftiofur hydrochloride in cows affected with retained placenta (8.2% vs. 11.1%, P=0.057, OR 5.771) within 21 days after parturition. Additionally, fewer cows in the herbal group required ad-

---

Corresponding author, E-mail: Yongmliu@126.com.

ditional therapeutic antibiotics compared to the controls (8.2% vs. 26.4%, P=0.003). Evidence from this randomized controlled clinical trial suggested that the herbal powder is a clinically effective treatment for retained placenta and the prevention of puerperal metritis and, thus, might have great potential for the medical management of retained placenta in dairy cows.

Key words: Retained placenta; Herbal remedy; Herbal powder; Oral administration; Dairy cow

1 BACKGROUND

Retained placenta is a frequently diagnosed uterine disease in early-postpartum cattle (Beagley et al., 2010). Cows with retained placenta are at a higher risk of developing puerperal metritis (Han and Kim, 2005; Sandals et al., 1979). Puerperal metritis has been identified as one of the main reasons for reduced fertility in cows with retained placenta. Consequences of cases of retained placenta on farm profitability are partly dependent on when metritis occurs (Laven and Peters, 1996). Thus, an ideal medical therapy for retained placenta should prevent puerperal metritis in cows.

Generally, the dominant approach to retained placenta in cattle in the field is to locally or systemically administer antibiotics. Goshen and Shpigel (2006) demonstrated that intrauterine antibiotic treatment is beneficial for metritis prevention in cows affected with retained placenta. As a third generation cephalosporin, ceftiofur does not result in antibiotic residues in the milk when used according to the label directions, and it has been widely applied to treat cows affected with retained placenta and metritis prevention in clinical practice (Liu et al., 2011; Risco and Hernandez, 2003). However, current reports indicate that antibiotic therapy, including intrauterine antibiotics and systemic antibiotics, generally has low efficacy in hastening the separation and expulsion of the retrained placenta (Drillich et al., 2006, 2007; Haimerl and Heuwieser, 2014; Risco and Hernandez, 2003; Stevens et al., 1995). Therefore, the challenge is to develop alternative strategies to manage these two conditions in cows. Ethnoveterinary medicines play a pivotal role in animal health care worldwide (Ayrle et al., 2016; Lans et al., 2007; McGaw and Eloff, 2008; Mayer et al., 2014; Pieroni et al., 2006; Van der Merwe et al., 2001), and medicinal plants are often used to treatment livestock affected with retained placenta (Cui et al., 2014b; Fall and Emanuelson, 2009; Mohan and Bhagwat, 2007; Moreki et al., 2012). Traditional Chinese veterinary medicine (TCVM) has always focused more on the body's response to pathogenetic factors than on pathological mechanisms (Editorial Committee of Encyclopedia of China's Agriculture, 1991) and, thus, offers new options for the treatment of postpartum uterine diseases in cows (Song, 1988; Yang et al., 2006).

In TCVM theory, blood stasis is an important underlying pathology of certain postpartum diseases (Bensky and Gamble, 1993; Luo, 1986; Yang et al., 2006), and retained placenta and/or puerperal metritis both fall within the blood stasis syndrome category (Editorial Committee of Encyclopedia of China"s Agriculture, 1991). Evidence from clinical trials suggests that there are beneficial effects of herbal remedies for postpartum uterine diseases in dairy cows (Cui et al., 2014a; Cui et al., 2014b; Cui et al., 2015; Mohan and Bhagwat, 2007). The herbal powder used in the present study was designed according to the therapeutic principle of "promoting blood circulation and removing blood stasis, regulating qi flow and relieving pain" for the treatment of postpartum uterine diseases, including retained placenta and puerperal metritis in cattle, and consists of Leonurus artemisia (Laur.) S.Y. Hu F, Angelica sinensis (OLIV.) DIELS (radix), Ligusticum chuanxiong HORT (radix), Sparganiumstoloniferum (Graebn.) Buch.-Ham.exJuz (radix), Curcuma zedoaria (Christm.) ROSC (radix), Cyperu srotundus Linn. (radix), and Glycyrrhiza uralensis FISCH (radix) (Appendix Table A1). According to traditional Chinese veterinary medicine theory, these herbs, mixed as presented, could resolve the blood stasis syndrome caused by retained placenta and, thus, can be used for the treatment of retained placenta and to lower puerperal metritis risks in postpartum cows.

The hypothesis of the present study is that cows with retained placentas treated with an oral administered herbal powder will experience shorter time to recovery, a reduced incidence of metritis subsequent to retained placenta, and will require fewer antibiotics compared to cows treated with ceftiofur hydrochloride. The aim of this randomized trial was to compare an oral administered herbal powder and ceftiofur hydrochloride in the treatment of cows with retained placenta and for the prevention of puerperal metritis under field conditions.

2 MATERIALS AND METHODS

2.1 Herbal powder preparation

The herbal powder used in the present study was composed of seven dried herbs (Appendix Table A1) that were obtained from Hui Ren Tang Chinese Medicine Co., Ltd. (Lanzhou, China), and the herb quality criteria were congruent with Chinese pharmacopoeia (The Pharmacopoeia Commission of PRC, 2010). After pre-processing by washing and drying, all herbs were ground to a homogenous powder and mixed to obtain the herbal powder with 80–100 mesh particle size (150–180 μm).

2.2 Animals and herds

The present study was approved by the Institutional Animal Care and Use Committee of Lanzhou Institute of Husbandry and Pharmaceutical Sciences of the Chinese Academy of Agricultural Sciences (SCXK20008-0003). This trial was conducted in two large commercial dairy farms with similar management practices in Lanzhou city, China. The animals were housed in free-stall facilities with cubicles, rubber mats and slotted floors and moved into loose housing systems in large, completely covered open sheds with straw bedding at least 1 week before the expected calving. The cows were fed hay ad libitum and a total mixed ration (TMR) consisting of corn silage, corn meal, barley and mineral supplements. Cows were milked three times daily in computer controlled milking parlors, and the herd-average milk yield varied between 7 200 and 9 100 kg per lactation in both herds. All cows were identified by ear tags and freeze marking.

2.3 Enrollment criteria

The treatment groups included cows with retained placenta after calving between August 2014 and December 2015. In this trial, retained placenta was diagnosed as retention of fetal membranes at 12-24 h after delivery of the calf, and the clinical diagnosis was established by Doctor Li or Tuo in the two farms respectively, who also diagnosed, treated and recorded all periparturient disease conditions. The enrolled cows were 3-7 years old (2-5 lactations) and had a BCS (Body Condition Score) between 2.5 and 3.5 during the peripartal stage (Edmonson et al., 1989). In addition, all cows involved in this trial were studied for only 1 lactation session and all newborns were delivered in good condition.

2.4 Exclusion criteria

Any cows suffering from conditions such as caesarean section during a previous or current calving, displaced abomasum, laminitis, postpartum acute diarrhea or ketosis were excluded from the study due to the possible influence of these conditions on the results. Additionally, cows with incomplete treatments or other deviations from the treatment protocol were retrospectively removed from the trial.

2.5 Study design and clinical examination schedules

During the study periods, a total of 2 125 cows calved and retention of the fetal membranes occurred in 221 cows. After completing the baseline evaluation, 141 cows having spontaneous delivery, 9 cows having twins, and 22 cows given with dystocia were enrolled in this trial. The 141 cows allocated to one of two groups based on their ear-tag numbers (the odd-numbered cows in the herbal group, and even-numbered cows in the control group), and the 9 cows and 22 cows were randomly assigned to 1 of the two groups by using random-number tables with a block size of 4 (SPSS software, version 17.0), 91 cows in herbal group, 81 cows in the control group. After treatment, fifteen cows were withdrawn from the final analyses. In herbal group, two cows were excluded for displaced abomasum, three cow for acute mastitis and one cows for serious lameness. In the control group, three cows were excluded for acute mastitis, four cow for displaced abomasum and two cows with interrupted treatment for application problems. Finally, two groups of animals (herbal group, n=85; control group, n=72) were created for further statistical analysis. The Fig. 1 shows the disposition of the enrolled cows.

The day of the retained placenta diagnosis was considered day 1 and was also the first day of treatment. Animals treated with herbal powder or ceftiofur hydrochloride were submitted for veterinary examination on the next routine 2-h interval visit. Obstetrical examinations were performed 72 h after initial treatment by palpation of the uterus per the rectum and/or vaginoscopy to evaluate uterine discharge and uterus status after thoroughly cleaning and disinfecting the perineal area with antiseptic solution. No attempt was made, at any stage, to manually remove the retained placenta.

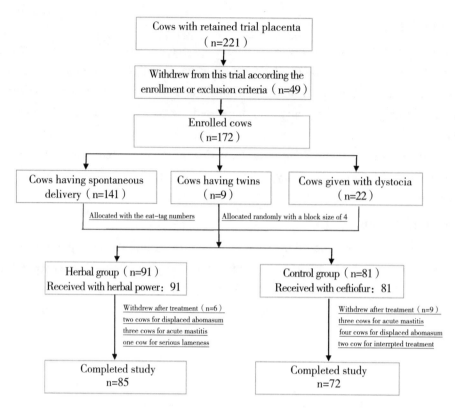

Fig. 1 Flow diagram showing the cows affected with retained placenta, randomization of cows with retained placenta, reasons for excluding cows from analyses and cows completed study for statistical analysis.

After diagnosis of retained placenta, the cows have been exami-nated every 48 (±6) h within 21 days post partum by Doctor Wang (one of the authors) who did not know the treatment assignment of the cows, and the cows with unfavorable evolution were submitted to a final clinical examination for possible puerperal disorders. The diagnosis of puerperal metritis was based on the combined characteristics of vaginal discharge, rectal temperature, uterine discharge and uterus status by transrectal palpation and vaginoscopy according to the protocol described by Sheldon et al. (2006). We defined as puerperal metritis as cows with a flaccid, nonretractable uterus that was located in the abdomen, a rectal temperature > 39.5 ℃, and a watery or purulent, fetid vaginal discharge within 21 days post partum. In both groups, if developed metritis, the cows were treated systemically with ceftiofur hydrochloride, nonsteroidal anti-inflammatory medication, and energy-calcium supplements by farm health technicians according to the established farm health protocols for postpartum cows. Data on retained placenta cases, the clinical scores, time to recovery from retained placenta conditions (from the initial treatment to recovery), and puerperal metritis cases were recorded for further statistical analysis.

2.6 Treatment protocol

After enrollment, the cows received either an oral dose of 300 g crude herb/600 kg BW (equal to 0.5 g crude herb/kg bw) once daily for 1-3 day (s) in the herbal group or systemic antibiotic treatment of 2.2 mg of ceftiofur/kg of BW (Qilu Pharmaceutical Co., Ltd. Jinan,

China) twice daily for 3 consecutive days in the control group. Herbal powder was administered via esophageal tubing inserted directly into the rumen through a stainless steel speculum after dispersing the herbal powder in lukewarm water (36 ± 2℃) for a total volume of 5.0 L and was vigorously stirred immediately. To ensure that the animal received the complete dose, an additional 0.5 L of water was used to rinse the container and tubing after administration of the 5.0 L. In addition, the herbal treatment was suspended until the placenta was expelled. No cows affected with retained placenta were used in the untreated control group for ethical reasons.

2.7 Efficacy measurements

The primary efficacy end point was time to the expulsion of the retained placenta. Clinical cure is defined as the expulsion of the placenta within 72 h after initial treatment according to the protocol described by Cui et al. (2014b). The secondary outcome was the proportion of cows that were free of puerperal metritis consequent to retained placenta within 21 days after parturition.

2.8 Statistical analysis

Statistical analyses were performed using SPSS for Windows (version 17.0, IBM SPSS Statistics; USA). The baseline characteristics, including the duration of retained placenta prior to randomization, BCS at calving, parity, and age were compared between the two groups by a one-way analysis of variance or a Kruskal-Wallis rank sum test. The proportion of twins, proportion of assistance at parturition, proportion of abortion and proportion of cows requiring extra therapeutic antibiotics were analyzed by the Chi-squared test. The risk of puerperal metritis between the groups was analyzed using logistic regression including treatment group, the expulsion of the placenta or not after treatment, twins, BCS at calving and assistance at parturition as covariates, and the odds ratios are reported. Box plots of the median time of retained placenta shedding following the initial herbal powder or ceftiofur treatment are displayed. All data were analyzed on an intention-to-treat basis and a p value < 0.05 was considered statistically significant.

3 RESULTS AND DISCUSSION

3.1 Herbal powder facilitated the expulsion of the retained placenta

In the herbal group, 73 cows out of 85 cows had total expulsion of the placenta within 72 h of initial treatment, whereas in the control group, none of the cows had total expulsion of the placenta within this time period. In the herbal powder, a large dose of Leonurus artemisia (Laur.) S.Y. Hu F was believed to modulate uterine contractions, and improve uterine hemorheology and microcirculation (Wojtyniak et al., 2013). As classical herb pair used in traditional Chinese medicine, Sparganiumstoloniferum (Graebn.) Buch.-Ham.exJuz (radix) -Curcuma zedoaria (Christm.) ROSC (radix) and Angelica sinensis (OLIV.) DIELS (radix) -Ligusticum chuanxiong HORT (radix) are both traditionally used for activating blood circulation to remove blood stasis, which might prevent the development of retained placenta. Additionally, it can be suggested that Glycyrrhiza uralensis FISCH (radix), a tonifying herbal medicine extensively used in TCVM, modulates host immunity through anti-inflammatory, immunomodulatory and anti-oxidative properties (Cheng et al., 2008; Gao et al., 2009; Zhang and Ye, 2009). Based on the above potential mechanisms of single herbs, the herbal powder seemed not only to regulate uterine contraction and improve the uterine condition but also improved the overall physical health of cows with retained placenta, thus providing a more favorable environment for earlier placenta detachment and the spontaneous release of the retained placenta. This trial also found that 50 out of 73 cows had total expulsion of the placenta following only one herbal treatment. Additionally, the herbal treatment produced a significantly higher reduction in the time to recovery from the retained placenta condition compared to the control group, as presented in the Fig. 2. Although the use of ceftiofur hydrochloride for the treatment of cows affected with retained placenta has been advocated by veterinar-ians (Beagley et al., 2010), this therapy is unlikely to result in an earlier release of fetal membranes, and the present study confirms this result. Thus, the oral administration of herbal powder is clinically effective for the management of a retained placenta.

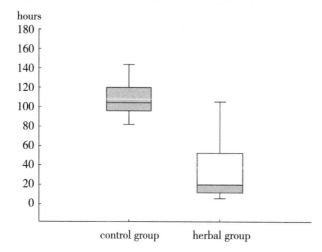

Fig. 2　Time of retained placenta shedding from initial treatment of box plots indicate the median shedding time of retained placenta with the 25th and 75th percentile values at the bottom and top of each box in the herbal and control groups (outliers and extremes not shown).

3.2　Herbal powder lowered the puerperal metritis risks subsequent to retained placenta

Systemic administration of ceftiofur hydrochloride is believed to reduce putrefaction of the placental tissue due to its broad spectrum activity against Gram-positive organisms, beta lactamase producing bacteria, anaerobic bacteria and Gram-negative aerobic bacteria. Risco and Hernandez (2003) demonstrated that the systemic administration of ceftiofur hydrochloride is clinically effective for the prevention of metritis subsequent to retained placenta in cows (Risco and Hernandez, 2003). In the present study, the logistic regression analysis indicated that the oral administration of herbal powder tended to have superior clinical efficacy in the prevention of puerperal metritis compared to ceftiofur hydrochloride in cows affected with retained placenta (8.2% vs. 11.1%, P=0.057, OR 5.771) (Table 1). Typically, 25%–50% of cows with retained placenta develop metritis during the early postpartum stage (Goshen and Shpigel, 2006; LeBlanc, 2008; Sandals et al., 1979). In the herbal powder, Herba Leonuri and Radix Angelicae Sinensis could modulate uterine contraction (Liu et al., 2015; Su et al., 2009; Wojtyniak et al., 2013), which might have contributed to the elimination of bacterial contamination of the uterus in postpartum cows. The earlier placental detachment and timely elimination of bacterial contamination could be a significant contributor to an improved uterine environment, and thus lower postpartum metritis risks in cows affected with retained placenta. Additionally, fewer cows in the herbal group required additional therapeutic antibiotics compared to the controls (8.2% vs. 26.4%, P=0.003), which is of interest in a time of increased public concern about the use of antimicrobial drugs in food-producing animals. From a clinical perspective, the herbal powder might have high potential for the medical management of dairy cows with retained placenta.

Table 1 The outcomes of the herbal and control groups.

Indexes	Herbal group (n=85)	Control group (n=72)
Median time of retained placenta shedding [a] (interquartile range), hours	20.0[A] (12.0, 52.5)	101.5[B] (96.0, 120.0)
Incidence of puerperal metritis secondary to retained placenta, %	8.2	11.1
Need for add extra therapeutic antibiotics[b], %	8.2[A]	26.4[B]

Values within the same row marked with different letters in superscript differ significantly: a, b=$P < 0.05$; A, B=$P < 0.01$.

[a] Time of retained placenta shedding was defined as the time from initial treatment to placental release.

[b] Need for add extra therapeutic antibiotics represents that the cows received ceftiofur hydrochloride and/or nonsteroidal anti-inflammatory medication, and energy-calcium supplements after the treatment scheme of the present study due to puerperal metritis or rectal temperatures > 39.5 ℃.

4 CONCLUSION

In agreement with the working hypothesis of our study, the results of the present trial suggest the herbal powder is clinically effective for the management of a retained placenta, including higher clinical cure rates, shorter time to recovery, a tendency of lowered risk of occurrence of puerperal metritis within 21 days after parturition, and a reduced need for extra therapeutic antibiotics. Thus, the herbal powder might have great potential for the medical management of dairy cows with retained placenta. The present results provide valuable information regarding the use of herbal remedies for the treatment of retained placenta and the prevention of puerperal metritis in cows. The mechanism of the effect of herbal powder on placental separation needs to be further studied.

ACKNOWLEDGMENTS

This paper was financially supported by the Special Fund of the Chinese Central Government for Basic Scientific Research Operations in Commonwealth Research Institutes (1610322016004). The authors thank Mr. Jianping Zhu and Junbao Wang, who assisted in conducting the clinical trial.

APPENDIX

See Table 2.

Table 2 Components of the herbal powder used in this trial.

Herb name		Role in herbal power	Standard dose (g)
Chinese name	Latin name		
Yi Mu Cao	Leonurus artemisia (Laur.) S.Y. Hu F	Monarch drug[a]	120.0
San Leng	Sparganiumstoloniferum (Graebn.) Buch.-Ham.exJuz (radix)	Minister drug[b]	40.0
E Zhu	Curcuma zedoaria (Christm.) ROSC (radix)	Minister drug	40.0
Dang Gui	Angelica sinensis (OLIV.) DIELS (radix)	Minister drug	30.0
Chuan Xiong	Ligusticum chuanxiong HORT (radix)	Minister drug	25.0
Xiang Fu	Cyperu srotundus Linn, (radix)	Adjuvant drug[c]	25.0
Gan Cao	Glycyrrhiza uralensis FISCH (radix)	Guide drug[d]	20.0

A traditional herbal formula generally contains different quantities of several herbs with different roles (Monarch, Minister, Adjuvant and Guide). In our herbal powder, the large dose of Leonurus artemisia (Laur.) S.Y. Hu F can modulate uterine contractions and improve uterine hemorheology and microcirculation against the principal symptom of cows affected with retained placenta. The two herb pairs, Sparganiumstoloniferum (Graebn.) Buch.-Ham.exJuz (radix) -Curcuma zedoaria (Christm.) ROSC (radix) and Angelica sinensis (OLIV.) DIELS (radix) -Ligusticum chuanxiong HORT (radix) were used for activating blood circulation to remove blood stasis, which assists the Leonurus artemisia (Laur.) S.Y. Hu F in regulating uterine contraction and improving the uterine condition. Cyperu srotundus Linn. (radix) modulates qi circulation and resolves stagnation to alleviate pain. Glycyrrhiza uralensis FISCH (radix), a tonifying herbal medicine, harmonizes the herbs and minimizes any potential side effects. All of these herbs work together as a formula to vitalize the blood, transform stasis, contract the uterus, warm the channels and alleviate pain to resolve postpartum blood stasis syndrome. The seven herbs work together exhibiting a synergistic action of "promoting blood circulation and removing stasis, facilitating the flow of qi and relieving pain" to resolve blood stasis suffered from the retained placenta and to regulate uterine contraction and improve the uterine condition, thus creating a more favorable environment for an earlier placenta detachment and the spontaneous release of the retained placenta (Editorial Committee of Encyclopedia of China's Agricultural, 1991).

[a] The Monarch drug is the key component of an herbal formula, and it provides the principal therapeutic effects against a related disease.

[b] The Minister drug is the synergistic secondary element of an herbal formula. It potentiates the effects of the Monarch drug or treats the accompanying symptoms.

[c] The Adjuvant drug is generally intended to enhance the therapeutic effects and mitigate any adverse effects of the Monarch drug and/or Minister drug.

[d] The Guide drug is generally used to harmonize the other herbs; it may also be used to facilitate the uptake of other herbs by a specific organ or target tissue.

(发表于《Livestock Science》,院选SCI, IF: 1.293. http://dx.doi.org/10.1016/j.livsci.2016.12.008.)

The Complete Mitochondrial Genome of Hequ Tibetan Mastiff Canis Lupus Familiaris (Carnivora: Canidae)

GUO Xian[1*], PEI Jie[1], BAO Pengjia[1], YAN Ping[1], LU Dengxue[2*]

(1. Lanzhou Institute of Husbandry and Pharmaceutical Sciences, Chinese Academy of Agricultural Sciences, Lanzhou, China; 2. Institute of Biology, Gansu Academy of Sciences, Lanzhou, 730050 China)

Abstract: The Hequ Tibetan Mastiff Canis lupus familiaris (Carnivora: Canidae) is a primitive breed of large dogs native to the northeastern Qinghai-Tibetan Plateau of China. In this study, its complete mitochondrial genome sequence has been assembled and characterized using high-through-put Illumina sequencing technology. The circular genome is 16 730 bp in length, and possesses all genomic components as typically found in most other metazoan mitogenomes. The gene arrangement is identical to those of most other vertebrates. Except for ND4L with GTG as its start codon, all the other PCGs are initiated with an ATR (ATA/ATG) codon. Three distinct stop codons are employed, i.e. AGA for CYTB, TAA for ATP6, ATP8, COX1, COX2, ND1, ND4L, ND5 and ND6, and an incomplete stop codon T for COX3, ND2, ND3 and ND4. The nucleotide composition is asymmetric (31.6% A, 25.5% C, 14.2% G and 28.7% T) with an overall A + T content of 60.3%. These data would contribute to our better understanding its evolutionary history.

Key words: Canis lupus familiaris; Hequ Tibetan Mastiff; Illumina sequencing; Mitogenome

The Hequ Tibetan Mastiff (Canis lupus familiaris; Carnivora: Canidae) is a primitive breed of large dogs originating from the northeastern Qinghai-Tibetan Plateau of China. It has long been widely used as a guardian of herds and private property by local Tibetans. Here, we present its complete mitochondrial genome, which is publicly available from the GenBank under the accession no. KT591870.

Total genomic DNA was isolated from the blood sample of a single individual from Maqu County (Gansu Province, China; 33° 26′N, 102° 34′E) using the Genomic DNA Isolation Kit (Tiangen Biotech, Beijing, China), and used for the shotgun library construction according to the manufacturer's manual for the Illumina Hiseq 2 500 Sequencing System (Illumina, San Diego, CA). A voucher blood sample (HTM20150701) is held in Lanzhou Institute of Husbandry

---

* Corresponding author, E-mail: guoxian@caas.cn (X GUO); ludengxue@126.com (D XLU).

and Pharmaceutical Sciences, Chinese Academy of Agricultural Sciences, China. A total of 15.09 M 125-bp raw reads were generated, and were trimmed with default parameters using CLC Genomics Workbench v8.0 (CLC Bio, Aarhus, Denmark). The trimmed dataset was then used for mitogenome assembly with MITObim v1.8 (Hahn et al., 2013). Genome annotation was done by aligning with the mitogenomes of related taxa.

The circular genome of Hequ Tibetan Mastiff is 16 730 bp in length. The nucleotide composition is asymmetric (31.6% A, 25.5% C, 14.2% G, and 28.7% T) with an overall A + T content of 60.3%. The genome possesses all 37 genes as reported for most metazoan mitogenomes, including 22 transfer RNAs (tRNAs), 13 protein-coding genes (PCGs), two ribosomal RNAs (12S rRNA and 16S rRNA) and a non-coding control region. Most genes are encoded on the heavy strand except for one PCG (ND6) and eight tRNAs (tRNA-Ala, -Asn, -Cys, -Gln, -Glu, -Pro, -Ser and -Tyr).

Except for ND4L with GTG as its start codon, all the other PCGs are initiated with an ATR (ATA/ATG) codon. Three distinct stop codons are employed, i.e. AGA for CYTB, TAA for ATP6, ATP8, COX1, COX2, ND1, ND4L, ND5 and ND6, and an incomplete stop codon T-- for COX3, ND2, ND3, and ND4. The total length of PCGs is 11 336 bp, accounting for 67.8% of the total genomic size. The two rRNAs are separated by tRNA-Val, and are 954 bp (12S) and 1 580 bp (16S) in length, respectively. Twenty-two tRNAs are interspersed throughout the whole mito-chondrial genome, ranging in size from 60 to 75 bp. The control region is located between tRNA-Phe and tRNA-Pro with a length of 1 270 bp. A putative origin of the light strand replication is located between tRNA-Asn and tRNA-Cys (37 bp).

To confirm its taxonomic identity, a phylogenetic tree was reconstructed based on the maximum likelihood (ML) analysis of 12 protein-coding genes (10 864 bp) located on the heavy strand (i.e. ATP6, ATP8, COX1, COX2, COX3, CYTB, ND1, ND2, ND3, ND4, ND4L and ND5) for a group of 12 confamilial species with MEGA6 (Tamura et al., 2013). As shown in Figure 1, Hequ Tibetan Mastiff is closely related to the congeneric coyote Canis latrans. The intergeneric relationships revealed in this study are in agreement with those of Bardeleben et al. (2005).

Declaration of interest

The authors report no conflicts of interest. The authors alone are responsible for the content and writing of the paper. The work was supported by the National Natural Science Foundation of China (31301976), the Central Public-Interest Scientific Institution Basal Research Fund of China (1610322014010), and the Innovation Project of Chinese Academy of Agricultural Sciences (CAAS-ASTIP-2014-LIHPS-01).

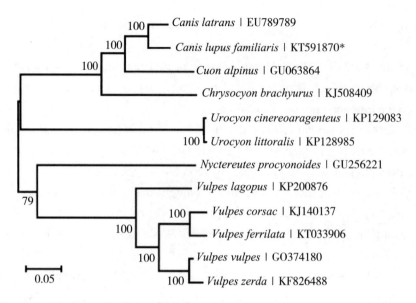

Figure 1 Maximum likelihood (ML) phylogeny of 12 species within the family Canidae.

Note: The GTR + G + I model was implemented as suggested by the 'Model Selection (ML)' function of MEGA6 (Tamura et al., 2013). The bootstrap values were based on 500 resamplings, and are indicated next to the branches. Codon positions included were 1st + 2nd + 3rd.

(发表于《Mitochondrial DNA》, SCI, IF: 1.209. 1-2.)

Molecular Characterization and Phylogenetic Analysis of Porcine Epidemic Diarrhea Virus Samples Obtained from Farms in Gansu, China

HUANG M.Z.[1,2,3], WANG H.[1,2,3], WANG S.Y.[1,2,3],
CUI D.A.[1,2,3], TUO X.[1,2,3], LIU Y.M.[1,2,3*]

(1. Engineering and Technology Research Center of Traditional Chinese Veterinary Medicine of Gansu Province, Lanzhou Institute of Husbandry and Pharmaceutical Sciences of Chinese Academy of Agricultural Sciences, Lanzhou, Gansu, China; 2. Key Lab of New Animal Drug Project of Gansu Province, Lanzhou Institute of Husbandry and Pharmaceutical Sciences of Chinese Academy of Agricultural Sciences, Lanzhou, Gansu, China; 3. Key Lab of Veterinary Pharmaceutical Development of Ministry of Agriculture, Lanzhou Institute of Husbandry and Pharmaceutical Sciences of Chinese Academy of Agricultural Sciences, Lanzhou, Gansu, 730050 China)

Abstract: Porcine epidemic diarrhea poses significant sanitation problems in the porcine industry, and has negatively affected the economy in recent years. In this study, 48 fecal specimens were collected from piglets from four intensive swine farms located in the Gansu Province of China. The molecular diversity and phylogenetic relationships between porcine epidemic diarrhea viruses (PEDV) prevalent in Gansu were probed, and the resultant proteins were characterized. Sequence analysis of the spike protein (S) genes showed that each specimen had unique characteristics, and that the PEDV1/S/4 strain could be differentiated from the others via a unique mutation of the S gene. The phylogeny of S glycoprotein showed that all strains were clustered into two major groups. The four Gansu PEDV field strains were characterized into different groups; this finding was consistent with the results of the protein characterization prediction. This analysis additionally revealed the unique characteristics of each specimen. The results of this study could be used to elucidate the prevalence of PEDV and contribute to the prevention of PEDV in Gansu.

Key words: Porcine epidemic diarrhea; Virus; Gene; Phylogenetic analysis

1 INTRODUCTION

Porcine epidemic diarrhea (PED) is a disease that is considered to be devastating for pig farmers with a rapid spread of infection (Wang et al., 2013; Alonso et al., 2014; Jung and Saif, 2015). The clinical symptoms of PED include severe enteritis, vomiting, and watery di-

---

* Corresponding author, E-mail: myslym@sina.com.
DOI: http://dx.doi.org/10.4238/gmr.15017696.

arrhea, with high infectivity and lethality in piglets. Since the 2010s, there have been continuous outbreaks of this disease in swine farms in nearly all provinces in China (Chen et al., 2010; Sun et al., 2012; Wang et al., 2013; Zhang et al., 2013; Zhao et al., 2013; Jung and Saif, 2015; Song et al., 2015a). Specifically, during the initial prevalence of the disease in farrowing barns, the mortality of newborn piglets has been found to be approximately 100%, which is responsible for considerable financial losses faced by the swine industry (Shibata et al., 2000; Li et al., 2012; Jung and Saif, 2015).

The causative agent of PED, the PED virus (PEDV), is an enveloped virus possessing an approximately 28-kb, positive-sense, single-strand RNA genome with a 5' cap and a 3' polyadenylated tail. The genome comprises a 5' untranslated region (UTR), a 3' UTR, and at least seven open reading frames that encode 4 structural proteins: spike protein (S), envelope protein, membrane protein, and the nucleocapsid protein (Bosch et al., 2003; Park et al., 2011b; Bi et al., 2012; Song and Park, 2012; Kim et al., 2015; Song et al., 2015a).

The PEDV S protein, which is a type I glycoprotein composed of 1383 amino acids (aa), plays a pivotal role in regulating the interactions with specific host cell receptor glycoproteins in order to mediate the viral entry and stimulate the induction of neutralizing antibodies in the natural host (Park et al., 2011a; Shirato et al., 2011; Cho et al., 2014). Therefore, the S protein must be thoroughly researched to understand the genetic relationships and diversity of PEDV isolates.

In addition, the S glycoprotein could be used as a primary target for the development of effective vaccines against PEDV. Although most large-scale pig farms were vaccinated according to the proper immune program, immunized swine herds continue to be infected by PEDV (Li et al., 2012; Zhao et al., 2013; Jung and Saif, 2015; Song et al., 2015b) (Li et al., 2012; Zhao et al., 2013; Jung and Saif, 2015; Song et al., 2015b). Therefore, the genetic relationships between different strains must be elucidated to effectively control and prevent PEDV infection. In this study, we investigated the molecular epidemiology and analyzed the protein characteristics of the S protein obtained from Gansu PEDV field samples. PEDV S protein plays a vital role in viral function and higher variation; therefore, the study mainly focused on the S gene.

2 MATERIAL AND METHODS

2.1 Sample collection

The 48 fecal specimens were individually obtained from piglets grown in four different intensive swine farms in the Gansu Province of China from August 2014 to May 2015. The piglets presented severe enteritis, vomiting, and watery diarrhea. All feces samples were homogenized with twice the amount of phosphate buffered saline. The suspensions were vortexed for 1 min and clarified by centrifugation for 10 min at 4 000 g. The supernatants were stored at −80℃ until further use.

2.2 RNA extraction

PEDV RNA was extracted from the supernatants of homogenized samples with MiniBEST Viral RNA/DNA Extraction Kit Ver.5.0 (Takara, Otsu, Japan) according to the manufacturer protocols. The RNA products were dissolved in 40 μL RNase free dH$_2$O and stored at -80℃.

2.3 Primers

One set of primers was designed and synthesized by the Beijing Genomic Institute (Beijing, China) to amplify the S genes based on the genome of PEDV CV777 (GenBank No.: AF353511.1): forward primer: 5'-ATGAGGTCTTTACTTCTGGTTG-3', reverse primer: 5'-TCACTGCACGTGGAC CTT-3'.

2.4 RT-PCR, DNA cloning, and sequence analysis

RT-PCR was conducted to amplify the complete S genes from the isolated RNA, as described in the manual of the Primescript TM One step RT-PCR kit v.2 (TaKaRa) under the following conditions: reverse transcription at 50℃ for 30 min, denaturation at 94℃ for 2 min, and 32 cycles of denaturation at 94℃ for 30 s, annealing at 56℃ for 30 s, and extension at 72℃ for 1 min.

RT-PCR products were identified by electrophoresis on a 1% agarose gel, and cloned using a PMD19/T vector (TaKaRa). The recombinant vector was identified by PCR and enzyme digestion. The positive clones were sent to Beijing Genomics Institute for sequencing. All sequencing reactions were performed in triplicate. The PCR product sizes (4 141-bp) were validated.

The nucleotide sequence and the deduced amino acid sequence of complete S genes of the four strains obtained from four different intensive swine farms (designated as PEDV1/S/4, PEDV3/S/1, PEDV4/S/3, and PEDV2/S/5), were aligned, and analyzed using the Megalin software (DNA Star). Phylogenetic trees were constructed using the Molecular Evolutionary Genetics Analysis (MEGA) software (version 6.0), by the neighbor-joining method based on the predicted amino acid sequence of the completed S proteins. The reference strains used for sequence alignment and phylogenetic analyses of the four PEDV strains are presented in Table 1.

Table 1 PEDV strains used for sequence alignment and phylogenetic analysis.

Isolate/strain	Accession No.	Origin
PEDV3/S/1	KT313037	In this study
PEDV4/S/3	KT313038	In this study
PEDV1/S/4	KT313039	In this study
PEDV2/S/5	KT313036	In this study
FJ/LY 2013	KJ646584.1	Southern China
HB/2012/2	JX435303.1	Southern China
BJ/2011/3	JX435298.1	Northern China

(continued)

Isolate/strain	Accession No.	Origin
CH17/GZ	JQ979289.1	Southern China
FJ/QZ 2013	KJ646605.1	Southern China
CH/QTC/2015	KR296670.1	Southern China
JY5C	KF177254.1	Western China
CV777	JN599150.1	Europe
CH/HGC/01/2015	KR296667.1	Southern China

2.5 Protein characterization prediction

The aa sequences were characterized based on their predicted protein isoelectric points (pl), coiled coil regions, transmembrane helices in proteins, the presence and location of signal peptide cleavage sites in amino acid sequences, structure domain, threonine and tyrosine phosphorylation sites, and asparagine (N) -linked glycosylation sites (N-Glyc); these were determined using DNA star and the ExPASy - SIB bioinformatics resource portal online tools.

3 RESULTS

3.1 Sequence analysis of the complete S gene

The nucleotide sequences of the complete S genes of PEDV1/S/4 (GenBank accession No. KT313039), PEDV2/S/5 (GenBank accession No. KT313036), PEDV3/S/1 (GenBank accession No. KT313037), and PEDV4/S/3 (GenBank accession No. KT313038) were 4 161, 4 161, 4 155, and 4 155 bp long, respectively. S proteins of PEDV3/S/1 and PEDV4/S/3 were 1384 aa in length, with predicted Mrs of 151.5 and 151.4 kDa, respectively. S proteins of PEDV2/S/5 and PEDV1/S/4 were 1386 aa in length with predicted Mrs of 151.6 and 151.5 kDa, respectively. The S gene sequences of the four Gansu PEDV field samples were compared to those of chosen PEDV reference strains [excluding CV777 (GenBank accession No. JN599150.1) which was only 99% similar]. CV777 had unique mutations that were different from those seen in all other stains. The alignment analysis of the deduced amino acid sequences showed that most of the mutations were observed between positions 230 and 380.

PEDV3/S/1 and PEDV4/S/3 displayed 7 common point mutations and deletion sites at the 380th position (Fig. 1); moreover, PEDV3/S/1 and PEDV4/S/3 had unique point mutations at positions 1358 (C→G) and 1359 (C®R), respectively. In addition, PEDV2/S/5 and PEDV1/S/4 also displayed special mutations. Results of sequence identity based on the complete S gene of all PEDV strains are summarized in Table 2. In brief, the S gene of 4 Gansu PEDV field samples and chosen PEDV reference strains (except CV777) had very high homology; however, they also had their own characteristics.

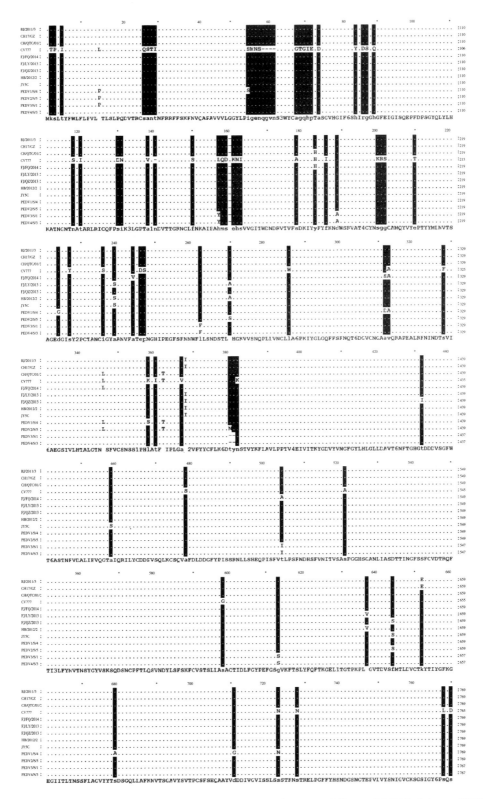

Fig. 1 Amino acid sequence alignment of the S glycoprotein genes of Gansu PEDV isolates and PEDV strains. The dashes represent deleted amino acids. The shadows indicate the unique substitutions present within each strain.

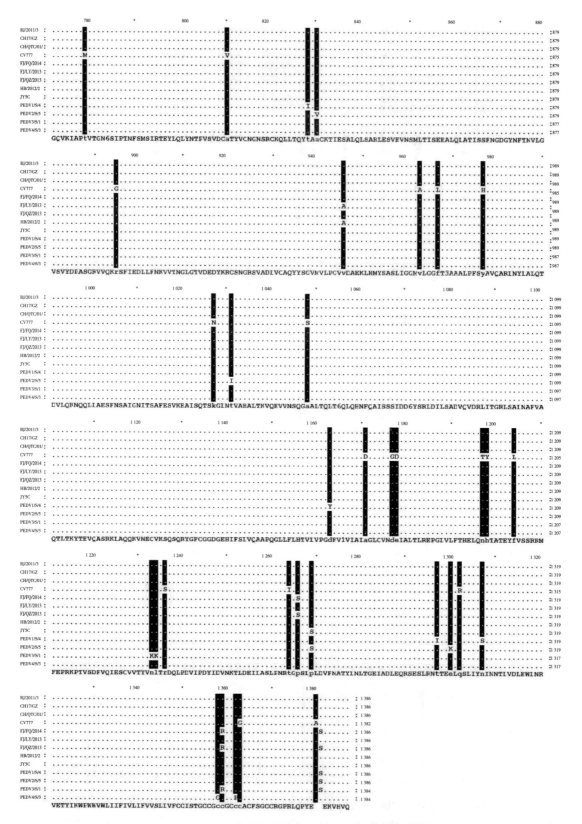

Fig. 1. Continued.

Table 2 Comparison of deduced amino acid sequences of S genes of Gansu PEDV strains and PEDV reference strains.

Strain/isolate	1	2	3	4	5	6	7	8	9	10	11	12	13
1 BJ/2011/3	—	99.7	99.4	92.8	99.1	99.6	99.4	99.6	99.4	98.4	98.9	98.8	99.1
2 CH17/GZ	0.3	—	99.2	93	98.9	99.4	99.3	99.4	99.3	98.3	98.8	98.8	98.9
3 CH/QTC/2015	0.7	0.8	—	93.1	99.1	99.2	99.1	99.2	99.1	98.6	99.1	98.8	99
4 CV777	7.4	7.2	7.1	—	92.9	92.8	92.5	92.8	92.7	92.5	92.6	92.2	92.3
5 FJ/FQ/2014	0.9	1.1	0.9	7.3	—	98.9	99.2	98.9	98.8	98.4	98.7	98.7	98.6
6 FJ/LY 2013	0.4	0.6	0.8	7.4	1.1	—	99.3	100	99.5	98.3	98.8	98.7	98.9
7 FJ/QZ 2013	0.6	0.7	0.9	7.7	0.8	0.7	—	99.3	99.3	98.3	98.9	98.8	98.8
8 HB/2012/2	0.4	0.6	0.8	7.4	1.1	0	0.7	—	99.5	98.3	98.8	98.7	98.9
9 JY5C	0.6	0.7	0.9	7.5	1.2	0.5	0.7	0.5	—	98.1	99	98.6	98.8
10 PEDV1/S/4	1.6	1.7	1.4	7.7	1.6	1.8	1.8	1.8	1.9	—	98.5	98.1	98
11 PEDV2/S/5	1.1	1.2	0.9	7.6	1.3	1.2	1.1	1.2	1	1.5	—	98.7	98.6
12 PEDV3/S/1	1.2	1.2	1.2	8	1.3	1.3	1.2	1.3	1.5	1.9	1.3	—	99.5
13 PEDV4/S/3	0.9	1.1	1	7.9	1.4	1.1	1.2	1.1	1.2	2	1.4	0.5	—

Percent similarity in upper triangle. Divergence in lower triangle.

3.2 Phylogenetic analysis

The phylogenetic relationships among the four Gansu samples, other PEDV strains isolated from various regions in China, and a CV777 strain were determined. A phylogenetic tree was constructed based on the deduced amino acid sequences of the complete S gene (Fig. 2).

The phylogenetic relationships based on the S glycoprotein showed that all strains fell into two major groups. Group I consisted of 11 strains, while group II only contained CV777 and PEDV1/S/4. PEDV3/S/1 and PEDV4/S/3 formed a subgroup, and PEDV2/S/5 was closely related to PEDV3/S/1 and PEDV4/S/3, which were clustered into group I. The results were consistent with the findings from sequence analysis.

3.3 Protein characterizations

The S protein characterization of the four Gansu PEDV field strains is shown in Table 3. The structure of the S protein was similar across the Gansu field PEDV strains, and consisted of a signal peptide (1 to 20 aa), 2 conserved structure domains corona-S1 (234 to 736 aa) and corona-S2 (744 to 1 385 aa), and a transmembrane region (1 326 to 1 348 aa), which corresponded

Continued on next page with the results of previous studies (Lee et al., 2010).

The prediction of coiled coil regions agreed with those of transmembrane regions; coiled coil regions existed between 1 326 and 1 348-aa in the 4 Gansu PEDV field strains.

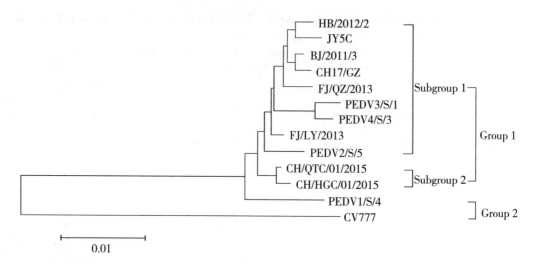

Fig. 2 Phylogenetic relationships between Gansu PEDV isolates and other reference strains, based on the S glycoprotein amino acid sequences. The GenBank accession Nos. of these genes are listed in Table 1.

Table 3 Predicted protein characteristics of the deduced protein sequence of the complete S gene.

Strains	pI value	Antigenic peptide		Transmembrane regions between positions	Coiled coil regions between positions	Signal peptide
		Number of hits	Position			
PEDV1/S/4	5.28	5	564-571 912-921 1 125-1 130 1 206-1 219 1 285-1 301	1 326-1 348	1 326-1 348	1-20
PEDV2/S/5	5.28	6	564-571 911-920 1 123-1 128 1 203-1 218 1 230-1 237 1 284-1 299	1 326-1 348	1 326-1 348	1-20
PEDV3/S/1	5.32	6	564-571 911-920 1 123-1 128 1 203-1 218 1 230-1 237 1 284-1 298	1 326-1 348	1 326-1 348	1-20
PEDV4/S/3	5.22	6	562-570 912-919 1 124-1 128 1 230-1 237 1 203-1 217 1 284-1 299	1 326-1 348	1 326-1 348	1-20

The pI of S protein varied from 5.22 to 5.32 among the four Gansu PEDV field strains; PEDV2/S/5 and PEDV1/S/4 had the same pI value (5.28). The four Gansu PEDV field strains shared a similar epitope, except PEDV1/S/4. PEDV1/S/4 was short of an antigenic peptide position (1 230 to 1 237 aa) compared to the other PEDV field strains. We observed 22 potential N-Glyc in PEDV4/S/3, and 21 such sites in the other three Gansu PEDV field strains (Table 4). The identified potential phosphorylated sites showed the presence of strain-specific sites (16 in PEDV4/S/3, 14 in PEDV2/S/5, and 15 in PEDV1/S/4 and PEDV3/S/1; Table 4). These variations were attributed to mutations in the gene sequences of the PEDV field strains, consistent with the results of previous sequence analyses.

Table 4 Prediction of the post-transcriptional modification sites in the deduced protein sequence of the complete S gene.

Strains	Protein modification sites			
	No. of N-Glyc sites	No. of phosphorylated sites		
		Serine	Threonine	Tyrosine
PEDV1/S/4	21	8	3	4
PEDV2/S/5	21	7	3	4
PEDV3/S/1	21	7	4	4
PEDV4/S/3	22	8	4	4

4 DISCUSSION

As seen in previous studies with coronavirus S proteins, the glycoprotein of PEDV is known to play a pivotal role in the interaction of cell with the cellular receptor, mediating viral entry and inducing neutralizing antibodies in the natural host. Moreover, the S gene of PEDV is considered to be the most useful in revealing the genetic diversity of PEDV (Chang et al., 2002; Bosch et al., 2003; Nam and Lee, 2010; Park et al., 2011a; Shirato et al., 2011; Li et al., 2012). Therefore, the diversity of PEDV strains may be studied based on the S gene. However, most studies have only assessed the partial or full S genes, with the full S protein characterization being conducted rarely (Lee et al., 2010; Li et al., 2012; Gao et al., 2013). In this study, the complete nucleotide and deduced peptide sequences of S protein genes of the four Gansu PEDV field strains were compared to those of published PEDV reference strains, and the S protein characteristics were predicted. We demonstrated that the S glycoprotein genes of Gansu PEDV field strains were diverse, leading to variations in the predicted protein characteristics, similar to the results seen in previous studies.

Despite the presence of many mutations in the S protein gene between the four Gansu PEDV field strains and published reference strains, we observed that the variations were focused primarily on the N-terminal region of conserved structure domain corona-S1, which agreed with the results of previous reports (Lee et al., 2010). The gene segment coding for the N-terminal conserved structure domain corona-S1 may therefore be an ideal molecular target for molecular epidemiology research.

Phylogenetic trees constructed based on deduced amino acid sequences indicated that PEDV1/S/4 was relatively close to CV777, but distantly related to group I. The other Gansu PEDV field strains were clustered into group I with PEDV3/S/1 and PEDV4/S/3 forming a subgroup. In all, the phylogenetic analysis suggested that there was high sequence variation among the Gansu PEDV field strains. The CV777 strain has been used to develop a PEDV vaccine in China; the results of this study indicate the need to determine if the range of PEDV1/S/4 infection is relatively smaller in Gansu.

The Gansu PEDV field strains presented similar functional domains, a signal peptide (1 to 20 aa) region, two conserved structural domains, and a transmembrane region; however, the protein characterization of these strains showed the presence of only one unique difference in the form of an amino acid substitution. S protein gene mutations may alter some functional sites and protein modification sites. The asparagine (N)-linked glycosylation sites and phosphorylated sites differed among the Gansu PEDV field strains, which was consistent with the results of a previous study (Lee et al., 2010; Chen et al., 2012). The prediction of epitopes, the pivotal functional sites closely associated with immunological recognition, should contribute to the design and development of epitope vaccines and antibodies (Chen et al., 2009; Pasick et al., 2014; Farrell and Gordon, 2015). PEDV1/S/4 had fewer epitopes than the other PEDV field strains, which further confirmed that PEDV1/S/4 was distantly related to the other three Gansu PEDV field strains.

In conclusion, the Gansu PEDV field strains underwent genetic diversity in their S glycoprotein genes and were clustered into different groups. PEDV2/S/5, PEDV3/S/1, and PEDV4/S/3 were closely related to the strains isolated from other Chinese provinces; however, they also displayed unique characteristics. PEDV1/S/4, on the other hand, was phylogenetically related to CV777. The results of this study indicated the importance of sequence analysis and protein characteristic prediction in understanding the genetic diversity among PEDV isolates, and its role in the development of an effective vaccine.

Conflicts of interest

The authors declare no conflict of interest.

ACKNOWLEDGMENTS

Research supported by the Special Fund for Agro-scientific Research in the Public Interest (#201303040-18) and the National Key Technology Research and Development of the Ministry of Science and Technology of China (#2012BAD12B03). The authors would like to thank the personnel of the Key Laboratory of Veterinary Pharmaceutical Development of Ministry of Agriculture, Key Laboratory of New Animal Drug Project of Gansu Province and Engineering and the Technological Research Center of Traditional Chinese Veterinary Medicine of Gansu Province for their contribution to this project.

（发表于《Genetics and Molecular Research》, SCI, IF: 0.775. gmr.15017696.）

Association of Genetic Variations in the ACLY Gene with Growth Traits in Chinese Beef Cattle

LI M.N., GUO X., BAO P.J., WU X.Y., DING X.Z., CHU M., LIANG C.N., YAN P.*

(Key Laboratory for Yak Breeding Engineering of Gansu Province, Lanzhou Institute of Husbandry and Pharmaceutical Sciences, Chinese Academy of Agricultural Sciences, Lanzhou, 730050 China)

Abstract: ATP citrate lyase (ACLY) is the primary enzyme responsible for the synthesis of cytosolic acetyl-CoA, which is a key precursor of both fatty acid and mevalonate synthesis pathways. Genetic variation of the ACLY gene may influence multiple traits associated with animal production. Here, we identified three non-synonymous mutations in ACLY exons in five beef cattle populations using DNA pool sequencing and high-resolution melting analysis. Results from association analyses revealed that the single nucleotide polymorphism (SNP) g.17127C>T is significantly associated with chest girth ($P < 0.01$) and body height ($P < 0.05$) in the Fleckvieh x Zhangye local crossbred cattle, and with body slanting length ($P < 0.05$) in the Simmental x Guyuan local crossbred cattle. SNP g.40427T>C is significantly associated with an increase in chest girth ($P < 0.05$) in the Simmental x Huzhu cattle population. These results provide preliminary evidence that polymorphisms in the bovine ACLY gene are associated with growth traits in beef cattle in northwest China. However, a larger sample set is needed to validate these findings.

Key words: ACLY gene; Beef cattle; High-resolution melting; Growth traits

1 INTRODUCTION

ATP citrate lyase (ACLY), encoded by the ACLY gene, is also called ATP citrate synthase, and was first identified in pigeon liver (Srere and Lipmann, 1953). ATP citrate lyase is a member of the acyl-CoA synthetase (ADP-forming) superfamily (Sánchez et al., 2000), and consists of ACLA and ACLB subunits (Fatland et al., 2002). In mammals, ACLY is a homotetramer, containing an ATP-grasp domain (N-terminal region), a CoA binding domain, a CoA-ligase, and citrate synthetase (C-terminal region) (Morita et al., 2014). Sun et al. (2010, 2011) successfully identified the citrate and ATP binding site of chymotrypsin-trun-

---

* Corresponding author, E-mail: pingyanlz@163.com
DOI http://dx.doi.org/10.4238/gmr.15028250

cated human ACLY using X-ray crystallography, revealing the partial residue conformation of the 1101-amino acid protein.

ACLY is an enzyme that catalyzes the conversion of citrate to oxaloacetate and acetyl-CoA coupled with the hydrolysis of ATP to ADP in many oleaginous species (Watson et al., 1969). Acetyl-CoA is produced in the mitochondria and is required for fatty acid and cholesterol biosynthesis pathways (Chypre et al., 2012). Acetyl-CoA is transformed to citrate via the activity of citrate synthase and then exported to the cytoplasm, where ACLY regenerates acetyl-CoA (Sun et al., 2011). ACLY is believed to play an important role in lipid metabolism (Khwairakpam et al., 2015).

High-starch diets markedly increase the expression of ACLY in the longissimus lumborum of cattle at 56 days of feeding (Graugnard et al., 2010), and gradual increases in glucose and insulin upregulate ACLY expression in subcutaneous adipose tissue of mid-lactation dairy cows (Carra et al., 2013). Genome-wide linkage analysis demonstrated that ACLY located on BTA19 was involved in the biosynthesis of milk fat (Bouwman et al., 2011). Several investigations have focused on ACLY polymorphisms and their association with growth and carcass quality traits in pig (Muñoz et al., 2013; Davoli et al., 2014). In addition, the expression of ACLY is associated with intramuscular fat percentage in sheep (Guo et al., 2014). However, studies investigating the association of polymorphisms with growth traits in beef cattle in northwest China are limited.

The aim of our study was to investigate genetic variation in the ACLY gene in five beef cattle hybrid populations of northwest China using high-resolution melting (HRM) analysis, and to investigate the association between mutations and growth traits. The results of this study provide some useful information on cattle genetic resources. Moreover, the results helped to select the candidate marker for breeding excellent beef cattle breeds of northwest China.

2 MATERIAL AND METHODS

2.1 DNA isolation and data collection

A total of 170 beef cattle of various hybrid populations from northwest China were investigated at random. These crossbred cattle included Angus x Guyuan local hybrid cattle (AG, N = 41, 12 ± 2 months old, Guyuan district of Ningxia Province), Simmental x Guyuan local crossbred cattle (SG, N = 38, 18 ± 2 months old, Guyuan district of Ningxia Province), Simmental x Huzhu local crossbred cattle (SH, N = 30, 18 ± 2 months old, Huzhu district of Qinhai Province), South Devon x Pingliang local crossbred cattle (SDP, N = 31, 6 ± 2 months old, Pingliang district of Gansu Province), and Fleckvieh x Zhangye local crossbred cattle (FZ, N = 30, 12 ± 3 months old, Zhangye district of Gansu Province). Blood samples were collected from the jugular vein and treated with acid-citrate-dextrose anti-coagulation. Genomic DNA was isolated using a genomic DNA isolation kit (Tiangene, Beijing, China) according to the manufacturer instructions, and all DNA samples with OD 260/280 ratios >1.8 were diluted to 20 ng/μL. All DNA samples were stored at −20℃ for subsequent analysis. The body dimensions of each animal were measured, including body height (BH), body slanting length (SL), chest girth (CG), and cannon circumference (CC). In order to minimize errors, one person was assigned to measure each trait.

2.2 PCR primers and amplification

In this study, DNA pool sequencing was performed to identify single nucleotide polymorphisms (SNPs) in the bovine ACLY gene. The basic DNA pool unit consisted of DNA samples from 70 individuals that were selected at random. Based on the sequence of bovine ACLY (GenBank accession No. NC_007317.5), 19 pairs of primers were designed using the Primer Premier 5.0 software to amplify the 28 ACLY exons including exon-intron boundaries (Table 1). Primers were synthesized by BGI (Beijing, China). The DNA pool described above was used as a template to amplify these regions.

Table 1 PCR primers used for sequencing and small amplicons in the high-resolution melting (HRM) analysis.

Primer name	Primer sequence (5' →3')	Annealing temperature (℃)	Amplicon type	Amplicon size (region)
P1	F: GCGGTCAGGATAGGGAATG R: GGAAGGCAGGCAACAACG	58.4	Sequencing	474 (exon 1)
P2	F: CGCACCCTTTGACCAGC R: GGATTCTATCAGCCATCTACAC	58.4	Sequencing	611 (exon 2)
P3	F: GATACCACATAGGGAGGG R: GCAATCGGACCAGTCAT	56.8	Sequencing	670 (exon 3)
P4	F: CCCATTGCCCTAGTTTCT R: GTGCTGGTTCCTTTCTGC	56.8	Sequencing	449 (exon 4)
P5	F: TTTGCGGCAGAAAGGAA R: CAACTGTGGCGGGTCAA	58.4	Sequencing	1 346 (exon 5, 6)
P7	F: AGAGTTGTGGATGGGTGAA R: CCAGGGAGGTTGATTGAGA	58.4	Sequencing	653 (exon 7)
P8	F: GGCGGTGACTGGCTGTA R: CGAGGGTTCCTGTTCTGT	56.8	Sequencing	439 (exon 8)
P9	F: GGGGCTTGAGTGTCTGA R: TTCCCTTCCCTGCTTCC	58.4	Sequencing	1071 (exon 9, 10)
P11	F: TGTTCTCACCTCAGCACCAT R: CCATCTTACCGACTTATCCC	52.5	Sequencing	475 (exon 11)
P12	F: GGTGTCCATAGAGCCATTT R: TCTCCGTGTCCCAITCC	58.4	Sequencing	1 064 (exon 12, 13)
P14	F: CATCCTCTTGCCTTCCT R: TCCCACCACTTTCACCTAA	58.8	Sequencing	830 (exon 14, 15)
P16	F: AGAGGCTGGTGGGAAGA R: GCAGGAGCAGTCAGAACA	56.8	Sequencing	381 (exon 16)

(continued)

Primer name	Primer sequence (5'→3')	Annealing temperature (℃)	Amplicon type	Amplicon size (region)
P17	F: GGGACCTTTCGTCTTGG R: ATGCCTGAGTTCCTTCG	52.5	Sequencing	833 (exon 17)
P18	F: GTTAGGTCCATACCATTTCT R: GACGCCATCTCAACTCAT	56.0	Sequencing	1 306 (exon 18, 19)
P20	F: AATCCGTTCTCCTTTGC R: CATCTGCTACCCAITGTTC	51.4	Sequencing	1 143 (exon 20, 21)
P22	F: CTGCCTGAGTCCACATTC R: GCCTTCTTTGCTAACCCT	56.0	Sequencing	742 (exon 22)
P23	F: GTGTTAGACCTGGATTGGG R: AAGAGTCACCTGGGAAGC	56.0	Sequencing	880 (cxon23, 24)
P25	F: TGCTTGACGCTGTAGGA R: TTAGGACCCAGATTTGAC	49.2	Sequencing	1 380 (exon 25, 26, 27)
P28	F: CTTCCGCTTTGTCCTTG R: GGCTTTAGTTGCTCCTC	51.4	Sequencing	1 420 (exon 28)
PH1	F: TCCTCACAGGGAAGACCAC R: AGAGCAATGCCCACGATG	57.0	IIRM	82
PH2	F: GTTCTCCTTTACATTTCAGGTCCA R: AACACTCCAGCCTCCTTCAA	57.8	IIRM	109
PH3	F: CAGCCAAGATGTTCAGCAAG R: GGTGACCAATGCCCATGATAA	58.9	HRM	102

P1-28 refer to the pairs of primers for scanning SNPs, PH1-3 refer to the pairs of primers for HRM analysis.

Amplifications were performed in 12.5 μL reactions containing 1 μL 20 ng genomic DNA, 0.4 μM each primer, and 6.25 μL Taq PCR MasterMix (Tiangene). PCR was performed using the following program: 3 min at 94℃, followed by 25 cycles of 30 s at 94℃, 30 s at the corresponding temperature (Table 1), and 40 s at 72℃, with a final extension of 5 min at 72℃. Amplifications were performed in a thermal cycler (Bio-Rad, USA).

PCR amplicons were purified with a Gel Extraction Mini Kit (Tiangene) and sequenced in both directions by BGI. The sequences were edited by the Chromas software, and then imported into MEGA 5 to identify SNPs.

2.3 HRM analysis of small amplicons

Based on the polymorphic sites detected, three pairs of primers were designed using the LightScanner Primer Design Software (Idaho Technology, USA) (Table 1). High and low calibrators were synthesized by Sangon (Bejing, China) according to the method described by Gundry et al. (2008).

HRM-PCR was performed in 10-μL reactions containing 1 μL 20 ng genomic DNA, 0.2 μM each primer, 5 μL Taq PCR MasterMix (Tiangene), and 1 μL 10X DNA dye LC Green (Idaho

Technology). PCR was performed in a thermal cycler (Bio-Rad) using the following program: 5 min at 95℃, followed by 35 cycles of 20 s at 94℃, 20 s at the corresponding temperature (Table 1), and 20 s at 72℃, with a final extension of 10 min at 72℃. Next, 0.1 μM each calibrator and 1 mM NaCl₂ was included in the amplification reactions. After this, a final denaturation and reannealing protocol was performed, which increased the temperature to 95℃ for 30 s followed by a 25℃ hold for 30 s. HRM analysis using small PCR amplicons was performed on a 96-well plate LightScanner (Idaho Technology). Data were analyzed with the LightScanner software (Idaho Technology).

2.4 Statistical analysis

Genotype and allelic frequencies were determined by counting and compared using the chi-square test among groups. Hardy-Weinberg equilibrium and haplotypes for each pair of segregating sites were determined using SHEsis. The Minitab software (version 16) was used to analyze the relationship between different genotypes of ACLY and the four growth trait parameters (BH, SL, CG, CC) recorded in various hybridized combinations. The following statistical model was used:

$$Y_{ij} = \mu + G_i + S_j + e_{ij} \qquad \text{(Equation 1)}$$

where Y_{ij} is the phenotypic value of the target trait, μ is the population mean, G_i is the ith genotypes, S_j is the effect of jth sex, and e_{ij} is the random error.

3 RESULTS

3.1 Analysis of sequence variation in the bovine ACLY gene

All of the coding exons of ACLY were successfully amplified using pooled DNA as a template, and three polymorphic sites were detected by DNA pool sequencing. SNP g.17127C>T was located in exon 12, SNP g.35520C>T was located in exon 20, and SNP g.40427T>C was located in exon 24. Amino acid sequence analysis at all SNPs revealed that the substitutions were synonymous and did not cause an amino acid change. Corresponding to human ACLY isoform 1 (GenBank accession No. NP_001087), SNP g.17127C>T and g.35520C>T were located at domain 5 and 2, respectively.

Based on the melting temperature, different melting curves representing different SNP genotypes were exhibited upon HRM analysis. The melting curves for T/C variants are shown in Figure 1. Genotype distribution, allelic frequencies, and Hardy-Weinberg equilibrium values are shown in Table 2. Among the five populations investigated, allele C was found to be common at the loci g.17127C>T, g.35520C>T, and g.40427T>C, and allele T for all mutations was rarer. CC and CT genotypes were detected in three variants of all five populations, and the TT genotype was not found. All three loci were in Hardy-Weinberg equilibrium ($P > 0.05$). Genotypic distributions of SNP g.17127C>T and SNP g.35520C>T were significantly different among the five cattle populations ($P < 0.05$). For the g.17127C>T locus, the CC genotype frequency in the FZ group was significantly higher than that in the AG and SG groups ($P < 0.01$), which was significantly lower than the frequency in the SH group ($P < 0.05$). A significantly higher frequency of the CC genotype of g.35520C>T was observed in the AG group than in the SDP group ($P < 0.05$). However, no significant differences were found at the g.40427T>C locus.

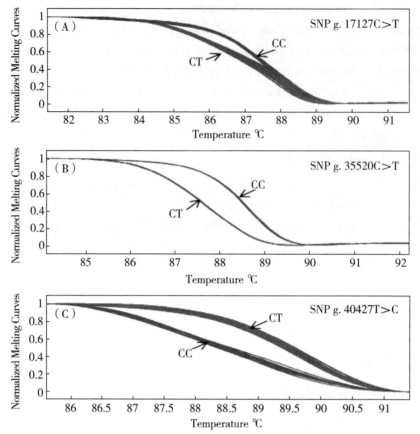

(A) SNP g.17127C>T; (B) SNP g.35520C>T; (C) SNP g.40427T>C. The curves are labeled by the SNP genotypes. The CT genotypes are presented as a red curve; the CC genotypes are presented as a gray curve.

Fig. 1 High-resolution melting curves (normalized) of the single nucleotide polymorphisms (SNPs) detected in this study.

Table 2 Genotype distribution and allelic frequencies at polymorphic sites of the ACLY gene in five populations.

Loci	Population	Genotype frequencies (%)		P value	x^2 (HWE)	Allele frequencies (%)		P value
		CC	CT			C	T	
	AG	70.7[a]	29.3		P>0.05	85.4	14.6	
	SG	71.1[a]	28.9		P>0.05	85.5	14.5	
g.17127C>T	SH	83.3[b]	16.7	P<0.01	P>0.05	91.7	8.3	P>0.05
	SDP	77.4[ab]	22.6		P>0.05	88.7	11.3	
	FZ	86.7[B]	13.3		P>0.05	93.3	6.7	

(continued)

Loci	Population	Genotype frequencies (%)		P value	x^2 (HWE)	Allele frequencies (%)		P value
		CC	CT			C	T	
		CC	CT			C	T	
g.35520C>T	AG	92.7[b]	7.3		P > 0.05	96.3	3.7	
	SG	89.5[ab]	10.5		P > 0.05	94.7	5.3	
	SH	86.7[ab]	13.3	P<0.05	P > 0.05	93.3	6.7	P > 0.05
	SDP	80.6[a]	19.4		P > 0.05	90.3	9.7	
	FZ	90.0[ab]	10.0		P > 0.05	95.0	5.0	
		CC	CT			C	T	
g.40427T>C	AG	68.3	31.7		P > 0.05	84.1	15.9	
	SG	76.3	23.7		P > 0.05	88.2	11.8	
	SH	70.0	30.0	P>0.05	P > 0.05	85.0	15.0	P > 0.05
	SDP	67.7	32.3		P > 0.05	83.9	16.1	
	FZ	76.7	23.3		P > 0.05	88.3	11.7	

Lowercase letters "a" and "b" denote significant different values at P < 0.01; uppercase "B" denotes significantly different values at P < 0.05. AG = Angus x Guyuan local hybrid cattle, SG = Simmental x Guyuan local crossbred cattle, SH = Simmental x Huzhu local crossbred cattle, SDP = South Devon x Pingliang local crossbred cattle, FZ = Fleckvieh x Zhangye local crossbred cattle.

3.2 Haplotype and linkage disequilibrium analysis of ACLY in five cattle populations

Six haplotypes were identified for the ACLY gene in five cattle populations (Table 3). Haplotype CCC was the dominant haplotype among the five populations, occurring at a frequency of 0.662-0.796. Haplotype CTC occurred at a significantly higher frequency in the SDP cattle than in the AG cattle. The frequency of the haplotype TCC was significantly higher in the AG and SG cattle than in the SH cattle. Linkage disequilibrium between the SNPs in all cattle populations was estimated (Table 4), and the results indicated that SNPs g.17127C>T and g.40427T>C were in moderate linkage disequilibrium in the SH population ($r^2 = 0.280$), and little linkage disequilibrium was found among the three sites in other populations.

Table 3 Haplotype distribution of three SNPs in the five populations.

Breed	Haplotype					
	CCC	CCT	CTC	CTT	TCC	TCT
AG	0.713	0.104	0.014[a]	0.022	0.114[b]	0.032
SG	0.707	0.095	0.053[ab]	0.000	0.122[b]	0.023
SH	0.776	0.074	0.055[ab]	0.011	0.018[a]	0.065

(continued)

Breed	Haplotype					
	CCC	CCT	CTC	CTT	TCC	TCT
SDP	0.662	0.128	0.097[b]	0.000	0.080[ab]	0.033
FZ	0.796	0.087	0.050[ab]	0.000	0.037[ab]	0.029

Lowercase letters "a" and "b" denote significantly different values at P < 0.05. AG = Angus x Guyuan local hybrid cattle, SG = Simmental x Guyuan local crossbred cattle, SH = Simmental x Huzhu local crossbred cattle, SDP = South Devon x Pingliang local crossbred cattle, FZ = Fleckvieh x Zhangye local crossbred cattle.

Table 4 Linkage disequilibrium among three loci within the ACLY gene in five cattle populations.

Population	Loci	D'		r^2	
		g.35520C>T	g.40427T>C	g.35520C>T	g.40427T>C
AG	g.17127C>T	1.000	0.031	0.007	0.001
	g.35520C>T	-	0.525	-	0.056
SG	g.17127C>T	1.000	0.074	0.009	0.004
	g.35520C>T	-	1.000	-	0.007
SH	g.17127C>T	1.000	0.738	0.006	0.280
	g.35520C>T	-	0.442	-	0.002
SDP	g.17127C>T	1.000	0.184	0.014	0.022
	g.35520C>T	-	1.000	-	0.021
FZ	g.17127C>T	1.000	0.369	0.004	0.074
	g.35520C>T	-	1.000	-	0.007

AG = Angus x Guyuan local hybrid cattle, SG = Simmental x Guyuan local crossbred cattle, SH = Simmental x Huzhu local crossbred cattle, SDP = South Devon x Pingliang local crossbred cattle, FZ = Fleckvieh x Zhangye local crossbred cattle.

3.3 Association analysis of single markers and combined genotypes

Association analyses between ACLY polymorphisms and growth traits (BH, SL, CG, and CC) are shown in Table 5. For the g.17127C>T locus, individuals with the genotype CT had greater CG (P < 0.01) than those with genotype CC in the FZ population, and the animals with genotype CT had significantly greater BH (P < 0.05) than those with genotype CC. Individuals with genotype CT had greater SL (P < 0.05) than those with genotype CC in the SG population, demonstrating that the g.17127C>T polymorphism is significantly associated with growth traits in SG and FZ beef cattle. Furthermore, the SNP g.40427T>C was significantly associated with an increase in CG (P < 0.05) in the SH cattle population. However, no significant association was detected between the SNP markers and measured traits in the AG and SDP populations. These results reveal different behaviors among the five analyzed cattle populations, and the heterozygous individuals had higher body dimensions than individuals with homozygous genotypes.

Table 5 Associations between single SNPs and phenotypic traits in five populations of cattle.

Loci	Population	Genotypes	BH(cm)	SL(cm)	CG(cm)	CC(cm)
g.17127C>T	AG	CC(29)	117.1 ± 2.194	132.1 ± 4.463	162.0 ± 5.168	16.1 ± 0.565
		CT(12)	114.5 ± 2.921	143.5 ± 5.898	165.7 ± 7.312	15.6 ± 0.936
	SG	CC(27)	120.9 ± 2.584	140.4 ± 3.548^b	174.2 ± 5.148	19.7 ± 0.657
		CT(11)	125.2 ± 3.499	148.6 ± 4.804^a	183.0 ± 7.011	20.9 ± 0.887
	SH	CC(25)	121.0 ± 2.493	143.1 ± 4.845	175.1 ± 5.413	16.6 ± 0.486
		CT(5)	121.2 ± 4.045	140.5 ± 7.862	166.2 ± 8.745	17.7 ± 0.856
	SDP	CC(24)	103.8 ± 3.087	114.6 ± 4.908	128.2 ± 5.470	15.2 ± 0.465
		CT(7)	104.6 ± 5.292	115.8 ± 8.414	128.1 ± 9.376	14.5 ± 0.797
	FZ	CC(26)	107.0 ± 3.419b	129.4 ± 5.029	146.5 ± 4.262^B	17.6 ± 0.735
		CT(4)	113.8 ± 2.057a	135.3 ± 3.027	160.8 ± 2.565^A	18.9 ± 0.442
g.35520C>T	AG	CC(38)	-	-	-	-
		CT(3)	-	-	-	-
	SG	CC(34)	123.7 ± 1.916	143.4 ± 2.630	178.4 ± 3.815	19.6 ± 0.457
		CT(4)	122.4 ± 4.446	145.6 ± 6.103	178.8 ± 8.896	21.0 ± 1.162
	SH	CC(26)	120.5 ± 1.846	142.3 ± 3.589	172.6 ± 3.962	16.9 ± 0.442
		CT(4)	121.6 ± 4.305	141.2 ± 8.367	168.6 ± 9.222	17.4 ± 0.813
	SDP	CC(25)	108.3 ± 2.585	119.6 ± 4.110	132.1 ± 4.580	15.4 ± 0.389
		CT(6)	100.1 ± 5.866	110.9 ± 9.326	124.2 ± 10.393	14.4 ± 0.884
	FZ	CC(27)	-	-	-	-
		CT(3)	-	-	-	-
g.40427T>C	AG	CC(28)	115.7 ± 2.121	141.3 ± 4.049	160.5 ± 5.064	15.7 ± 0.676
		CT(13)	115.8 ± 2.938	134.3 ± 6.180	167.1 ± 7.145	15.9 ± 0.720
	SG	CC(29)	122.7 ± 2.714	143.0 ± 3.726	175.9 ± 5.407	19.9 ± 0.695
		CT(9)	123.4 ± 3.466	146.1 ± 4.759	181.4 ± 6.939	20.7 ± 0.868
	SH	CC(21)	120.4 ± 3.074	138.2 ± 5.975	160.9 ± 6.758^b	17.3 ± 0.568
		CT(9)	121.8 ± 3.094	145.4 ± 6.013	180.4 ± 6.712^a	17.1 ± 0.716
	SDP	CC(21)	102.7 ± 3.350	109.7 ± 5.326	122.1 ± 5.935	14.4 ± 0.505
		CT(10)	105.6 ± 4.935	120.7 ± 7.847	134.2 ± 8.744	15.3 ± 0.743
	FZ	CC(23)	110.2 ± 2.321	132.6 ± 3.414	152.8 ± 2.893	18.4 ± 0.499
		CT(7)	110.5 ± 2.966	132.2 ± 4.364	154.5 ± 3.698	18.1 ± 0.638

Results reported as means ± standard error. The genotypes (N < 3) were neglected in this analysis. Lowercase letters "a" and "b" denote significantly different values at P < 0.05; uppercase letters "A" and "B" denote values significantly different at P < 0.01. AG = Angus x Guyuan local hybrid cattle, SG = Simmental x Guyuan local crossbred cattle, SH = Simmental x Huzhu local crossbred cattle, SDP = South Devon x Pingliang local crossbred cattle, FZ = Fleckvieh x Zhangye local crossbred cattle, BH = body height, SL = body slanting length, CG = chest girth, CC = cannon circumference.

4 DISCUSSION

ACLY acts as a cross-link between pathways involved in carbohydrate metabolism and the production of fatty acids (Chypre et al., 2012). As a key precursor for fatty acid and mevalonate synthesis pathways (Zaidi et al., 2012), acetyl-CoA, which is produced in mitochondria, is transformed to citrate by citrate synthase, then citrate is exported to the cytoplasm where ACLY catalyzes the conversion of citrate into oxaloacetate and acetyl-CoA (Sun et al., 2011). The ACLY gene, which encodes ATP citrate lyase and is associated with the tricarboxylic acid cycle, might help to sustain the lipogenic process (Graugnard et al., 2010). In addition, the expression of ACLY has been strongly correlated with intramuscular fat in sheep and cattle (Guo et al., 2014). Davoli et al. (2014) investigated two SNPs in the porcine ACLY gene and found significant associations between polymorphisms and the productive traits average daily gain, ham weight, and back fat thickness. Therefore, ACLY might be an important biomarker for growth traits in animals.

Bovine ACLY was mapped to chromosome 19, and contains 28 exons and 27 introns. To better understand the distribution of genetic variation in bovine ACLY, diversity was explored in five beef cattle hybrid populations from northwest China including Angus, Simmental, South Devon, and Fleckvieh. As a result, limited polymorphisms were revealed within the ACLY coding region, with all three variants identified being synonymous. Those results showed that bovine ACLY has low polymorphism, and that the coding region is relative conservative (Li et al., 2012). Synonymous variants do not result in an amino acid change in the protein, although they may affect the level of protein expression, protein conformation, or function (Sauna and Kimchi-Sarfaty, 2011). More SNP markers may be identified as more samples are collected from more breeds.

ACLY is a member of the acyl-CoA synthetase superfamily (ADP-forming), which consists of separate α and β subunits, or a fusion of α-β protein (Sánchez et al., 2000). All five domains of human ACLY are common constitutes of this superfamily. Corresponding to human ACLY isoform 1 (GenBank accession No. NP_001087), residues 2-425 form domains 3, 4, and 5, and all three of these domains are homologous to the β subunit of succinyl-CoA synthetase. Residues 487-820 form domains 1 and 2, which are homologous to the α subunit of succinyl-CoA synthetase (Sun et al., 2010). In the present study, two SNPs g.17127C>T and g.35520C>T identified in bovine ACLY were found to be located in domains 5 and 2 of human ACLY isoform 1, respectively. Domain 5 forms the citrate binding site, and domain 2 possesses a phosphohistidine loop, which binds in the C termini of β strands corresponding to the structure of the human ACLY protein (MMDB ID: 94227). Of note, the SNPs in bovine ACLY may be relevant to the binding site of ATP and citrate in bovine, and may have functional relevance for the role of the bovine ACLY protein.

Two SNPs in porcine ACLY were found to be associated with important productive traits in Italian Large White and Duroc pigs (Davoli et al., 2014). We further evaluated whether three mutations in ACLY could account for differences in the measured traits of the five hybrid populations. In this study, gene-specific SNP marker analysis showed that SNP g.17127C>T is significantly associated with CG ($P < 0.01$) and BH ($P < 0.05$) in the FZ population, and is also significantly associated with SL ($P < 0.05$) in the SG population, and the SNP g.40427T>C is significantly associated with an increase in CG ($P < 0.05$) in the SH cattle population. However, no significant association was detected between the markers and measured traits in AG and SDP populations. It is possible that SDP beef cattle (6 ± 2 months old) were younger than the other cattle, and that the expression of ACLY differs during animal development (Chypre et al., 2012). Another explanation for this might be the different breeding history among the five populations. All of the five populations presented in this study were different cross-combinations. No significant association was detected between SNP g.35520C>T and measured traits in five populations; however, the number of samples used in this study was limited. To our knowledge, this is the first report describing SNPs in the whole exon region of the bovine ACLY gene and their effects on growth traits. These results provide preliminary evidence that bovine ACLY gene polymorphisms are associated with growth traits in beef cattle of northwest China; however, a larger set of samples is needed to validate the results.

Conflicts of interest

The authors declare no conflict of interest.

ACKNOWLEDGMENTS

Research supported by the Innovation Project of Chinese Academy of Agricultural Sciences (#CAAS-ASTIP-2014-LIHPS-01) and the Program of National Beef Cattle and Yak Industrial Technology System (#CARS-38).

（发表于《Genetics and molecular research》，SCI，IF：0.764. gmr.15028250.）

PPARα Signal Pathway Gene Expression is Associated with Fatty Acid Content in Yak and Cattle Longissimus Dorsi Muscle

QIN W.[*], LIANG C.N., GUO X., CHU M., PEI J.,
BAO P.J., WU X.Y., LI T.K., YAN P.[*]

(Key Laboratory of Yak Breeding Engineering Gansu Province, Lanzhou Institute of Husbandry and Pharmaceutical Sciences, Chinese Academy of Agricultural Science, Lanzhou, 730050 China)

Abstract: Intramuscular fatty acid (FA) is related to meat qualities such as juiciness, tenderness, palatability, and shear force. PPARα plays an important role in lipid metabolism in the liver and skeletal muscle. This study investigated FA composition in yaks and cattle, in order to ascertain whether a correlation between PPARα signal pathway genes as candidate genes and meat FA composition in yaks and cattle exists. Statistical analyses revealed that levels of monounsaturated fatty acid (MUFA) and polyunsaturated fatty acid (PUFA) in yaks were significantly higher than those in cattle ($P < 0.01$), whereas saturated fatty acid (SFA) levels were significantly lower than those in cattle ($P < 0.05$). The mRNA expression levels of FABP4 ($P < 0.05$), SCP2 ($P < 0.05$), and APOA1 ($P < 0.01$) in yaks were significantly lower than those in cattle. However, LPL expression in yaks was significantly higher than that in cattle ($P < 0.05$). In yaks, the expression levels of FABP3 ($P < 0.05$) and LPL ($P < 0.01$) were negatively correlated with MUFA, and those of FABP4 and SCD were positively correlated with PUFA ($P < 0.01$). In cattle, the mRNA level of PLTP was positively correlated with SFA ($P < 0.05$), and LPL was positively correlated with MUFA ($P < 0.05$). These results suggest that these genes may participate in the regulation and control of intramuscular FA metabolism in yaks, so they could be used as candidate markers to improve yak meat quality.

Key words: PPARα signal pathway gene; mRNA expression; Yak; Cattle; Fatty acid composition

1 INTRODUCTION

In recent years, meat quality has become an important component of consumer demand, particularly yak meat, which is low in fat and cholesterol, and high in protein, vitamins, and essentiaL·minerals such as copper, zinc, iron, and potassium (Wan et al., 2012). Yak meat

---

Corresponding authors, E-mail: pingyanlz@163.com (P. Yan,) / qinwen_614@163.com (W. Qin)
DOI http://dx.doi.org/10.4238/2015.November.18.9

is a staple source of animal protein, and the most important component of economic income for Tibetans living between 2 500 and 5 500 m above sea level. Despite the fact that yak meat is not as tender as cattle meat, the economic value of yak meat is high (Niu et al., 2009).

Intramuscular fatty acid (FA) is an important factor that affects meat quality variables such as juiciness, tenderness, palatability, shear force, and muscle pH value (Goodson et al., 2002; Hausman et al., 2009). Intramuscular FA is composed of 60%-70% phospholipids. Several FA groups and single FAs have different physiological effects, and can function as physiological regulators (Cao et al., 2008). The content and composition of intramuscular FA are influenced by several factors, such as breed (Wang et al., 2011), gender, age (Bednárová et al., 2013), and nutrition (Yang et al., 2006).

The content and composition of intramuscular FA are mainly determined by lipid metabolism. Peroxisome proliferator-activated receptors (PPARs), which have three subtypes (PPARα, PPARβ/δ, and PPARγ), are nuclear hormone receptors that are activated by FAs and their derivatives (Takahashi, 2005). PPARα plays a role in the clearance of circulating or cellular lipids by the regulation of gene expression involved in lipid metabolism in the liver and skeletal muscle. The expression of many PPARα signaling pathway genes in the longissimus muscles, including those involved in lipid transport [phospholipid transfer protein (PLTP) and apolipoprotein A-I (APOA1)], lipogenesis [malic enzyme 1 (ME1) and stearoyl-CoA desaturase (SCD)], FA transport [diazepam binding inhibitor (DBI), FA binding protein 3 (FABP3), thrombospondin receptor (CD36), and lipoprotein lipase (LPL)], FA oxidation sterol carrier protein 2 (SCP2), and adipocyte differentiation [FA binding protein 4 (FABP4)] may contribute to FA deposition. However, whether the expression of lipid metabolic genes is associated with FA deposition within steers is unknown.

This study was conducted in order to identify PPARα signaling pathway genes that are associated with FA content in the longissimus dorsi muscle of yaks and cattle. The mRNA expression levels of 11 PPARα signaling pathway genes, including those involved in lipid transport, lipogenesis, FA transport, FA oxidation, and adipocyte differentiation were measured. Correlations between gene expression levels and FA content were analyzed, and differences in gene expression levels and FA content between yaks and cattle were investigated.

2 MATERIAL AND METHODS

2.1 Animals

The eight yaks and eight cattle used were 3 years old and from pastoral areas in the Tibetan Autonomous Prefecture of Gannan, Gansu Province, China. Muscle tissue samples were taken from the right side of the longissimus dorsi muscle at the 12th/13th rib. Samples for RNA extraction were collected after slaughter, immediately frozen in liquid nitrogen, and stored at −80℃. Samples for FA composition analysis were collected 24 h after carcass cooling and maintained at 4℃. RNA extraction and real-time polymerase chain reaction (PCR) Total RNA was extracted from the tissues using an RNAprep Pure Kit (For Tissue) (Tiangen), according to the manufacturer instructions. Total RNA was quantified by absorbance at 260 nm, and its integrity was checked by agarose gel electrophoresis and ethidium bromide staining of the 28 and 18S bands. Total RNA was reverse-transcribed into cDNA using a PrimeScript™ RT reagent kit (Takara), according to manufacturer instructions. A real-time PCR was conducted with 12.5 μL SYBR® Premix Ex Taq™ Ⅱ (Takara), 0.2 μL 10 μM primers, 9.5 μL ddH$_2$O, and a 1-μL total reaction volume that contained 100 ng cDNA. The thermal cycling parameters were as follows: 95℃ for 5 s, followed by 40 cycles at 95℃ for 30 s and Tm for 30 s. All of the primers were designed using integrated mRNA sequences based on sequences published by the National Center for Biotechnology Information (NCBI) (www.ncbi.nlm.nih.gov; Table 1). The $2^{-\Delta\Delta Ct}$ method was used to determine the relative fold-changes (Schmittgen and Livak, 2008), and all of the data were normalized with the housekeeping glyceraldehyde-3-phosphate dehydrogenase gene.

Table 1 Primer sequences used in the real-time polymerase chain reaction.

Gene name and symbol	GenBank ID	5'→3'	Sequence	Amplicon size (bp)
Glyceraldehyde-3-phosphate dehydrogenase, GAPDH	NM_001034034	Forward	ccacgagaagtataacaacacc	120
		Reverse	gtcataagtccctccacgat	
Phospholipid transfer protein, PLTP	NM_001035027	Forward	tccatttccagccagacca	102
		Reverse	ccccatcatagaagaaccagtagag	
Fatty acid binding protein 3, FABP3	NM_174313	Forward	tgtgcgggagatggttga	146
		Reverse	tgccgagtccaggagtagcc	
Fatty acid binding protein 4, FABP4	NM_74314	Forward	caaattgggccaggaatttga	197
		Reverse	tctcataaactctggtggcagtgac	
Lipoprotein lipase, LPL	NM_001075120	Forward	acttgccacctcattcctg	119
		Reverse	acccaactctcatacattcctg	
Thrombospondin receptor, CD36	NM_174010	Forward	ggtccttacacatacagagttcg	115
		Reverse	atagcgagggttcaaagatgg	
Malic enzyme 1, ME1	NM_001144853	Forward	tgctgcgattggtggtgc	191
		Reverse	tcggaagggtaacgggat	
Stearoyl-CoA desaturase, SCD	NM_173959	Forward	actgcggtccaagtcgtt	164
		Reverse	cagccttgtctggagtcatc	
Sterol carrier protein 2, SCP2	NM_001033990	Forward	tgaactcccttttgcctcctttt	171
		Reverse	caggttctattcacccagcactt	
Apolipoprotein A-I, APOA1	NM_174242	Forward	accgtgtatgtggaagcaatcaag	107
		Reverse	tcccagttgtccaggagtttcag	

(continued)

Gene name and symbol	GenBank ID	5'→3'	Sequence	Amplicon size (bp)
Peroxisome proliferator-activated receptor alpha, PPARa	NM_001034036	Forward	gaatcggaataagtgcca	156
		Reverse	gtttcggaatcttctaggtc	
Diazepam binding inhibitor, DBI	NM_001113321	Forward	gcatcttaagaccaagccagcag	117
		Reverse	ttgcctttgaagtccaacattcc	

2.2 FA composition of intramuscular fat

Total lipids were extracted from approximately 10 g longissimus dorsi muscle samples with a chloroform-methanol (1 : 1) solvent, and the samples were homogenized and then extracted for 24 h (Yan et al., 2005). Lipid fractions were hydrolyzed in 5 mL 2 M KOH and CH_3OH (1 : 1) after being dissolved in 2 mL chloroform and petroleum ether-benzene (1 : 1). After stratification of the petroleum ether methyl ester solution by distilled water, the supernatant was used for gas chromatography-mass spectrometry. The chromatographic conditions were as follows: 160℃ for 2 min, increasing at 5℃/min to 220℃ for 1 min, then increasing at 8℃/min to 230℃ for 1 min; the injection port was at 250℃, the diversion ratio was 60 : 1, the carrier gas was He, and the flow rate was 1.2 mL/min. The mass spectrometry conditions were as follows: electron ionization mode was set; the ion source temperature was 250℃, the electron energy was 70 eV, the solvent latency was 1.8 min, and the electron multiplier was 1 x 105 kV; full-scan mode was set; the scanning range was 30-450 aum, and the scanning speed was 563/s (Yang et al., 2008).

2.3 Statistical analyses

Data are reported as means and standard deviations. Analyses of variance, independent-sample Student t-tests, and correlation analyses were performed in SPSS 18.0. Significance was set at the 0.05 level, and P values lower than 0.01 were considered to be extremely significant.

3 RESULTS

3.1 FA content and composition

FA content and composition in the longissimus dorsi muscles of yaks were significantly different from those of cattle (Table 2). Yaks contained more C15 : 0, C16 : 1, C17 : 0, C20 : 0, and C24 : 0 than did cattle, but had significantly lower levels of saturated fatty acid (SFA) ($P < 0.05$). Yaks had significantly higher levels of C18 : 1, C18 : 2n6t, C18 : 3, C20 : 4, conjugated linoleic acid (CLA), monounsaturated fatty acid (MUFA), and polyunsaturated fatty acid (PUFA) than cattle, whereas C16 : 0 levels were lower than in cattle ($P < 0.01$). Eicosapentaenoic acid (EPA) and docosahexaenoic acid (DHA) were not detected.

Table 2 Fatty acid composition in longissimus dorsi of yaks and cattle (means ± SD).

Fatty acid (%)	Yaks	Cattle
SFA	49.30 ± 4.13*	56.11 ± 5.09
C10:0	0.07 ± 0.03	0.05 ± 0.02
C12:0	0.06 ± 0.01	0.06 ± 0.02
C14:0	2.49 ± 0.39	2.32 ± 0.21
C15:0	0.42 ± 0.18*	0.16 ± 0.04
C16:0	21.09 ± 3.10**	30.82 ± 3.21
C17:0	1.12 ± 0.32*	0.55 ± 0.14
C18:0	22.72 ± 2.33	21.16 ± 2.25
C20:0	0.57 ± 0.22*	0.41 ± 0.03
C22:0	0.17 ± 0.04	0.20 ± 0.02
C24:0	0.58 ± 0.15*	0.38 ± 0.02
MUFA	39.91 ± 2.32**	32.03 ± 1.30
C16:1	4.69 ± 1.10*	3.24 ± 0.18
C17:1	0.87 ± 0.24	0.76 ± 0.12
C18:1	33.95 ± 1.48**	27.75 ± 1.18
PUFA	6.81 ± 0.72**	3.28 ± 0.80
C18:2n6c	1.58 ± 0.46	1.30 ± 0.61
C18:2n6t	0.90 ± 0.27**	0.32 ± 0.09
C18:3	0.98 ± 0.14**	0.57 ± 0.24
C20:4	1.56 ± 0.05**	0.38 ± 0.02
CLAC18:2	1.09 ± 0.28*	0.71 ± 0.03
EPAC20:5	0.41 ± 0.05**	ND
DHA C22:6	0.29 ± 0.05**	ND

SFA, saturated fatty acid; MUFA, monounsaturated fatty acid; PUFA, polyunsaturated fatty acid; CLA, conjugated linoleic acid; EPA, eicosapentaenoic acid; DHA, docosahexaenoic acid. *Significantly different ($P < 0.05$); **extremely significantly different ($P < 0.01$). ND = not detected.

3.2 PPARα signaling pathway gene expression levels

The expression levels of FABP4 ($P < 0.05$), SCP2 ($P < 0.05$), and APOA1 ($P < 0.01$) in yaks were significantly lower than those in cattle. However, LPL expression in yaks was significantly higher than in cattle ($P < 0.0$; Figure 1).

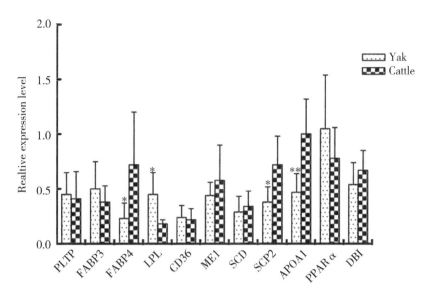

Fig. 1 Expression of PPARα signaling pathway genes mRNA in longissimus dorsi of yak and cattle. For abbreviations, see Table 1.

3.3 Correlation between gene expression levels and FA content and composition

Gene expression levels were correlated with FA content. Pearson's correlation coefficients between yak FA content and the expression of each gene are shown in Table 3. Overall, MUFA levels were negatively correlated with the expression levels of FABP3 ($P < 0.05$) and LPL ($P < 0.01$), and PUFA levels were positively correlated with those of FABP4 and SCD ($P < 0.01$). CD36 and APOA1 were both positively correlated with C12:0 and C15:0 ($P < 0.05$), and SCP2 was positively correlated with C10:0, C12:0, and C14:0 ($P < 0.05$). C22:0 was negatively correlated with FABP4 and SCD ($P < 0.05$). PLTP mRNA expression levels were positively correlated with C18:2n6t ($P < 0.05$), whereas DBI mRNA expression levels were negatively correlated with C20:0 ($P < 0.05$). No correlations between PPARα expression levels and FAs were found.

Pearson's correlation coefficients between cattle FA content and the expression of each gene are shown in Table 4. The mRNA abundance of PLTP was positively correlated with SFA and C16:0 ($P < 0.05$), and LPL was positively correlated with MUFA and C20:0 but negatively correlated with C10:0 ($P < 0.05$). APOA1 expression levels were negatively correlated with SFA, C16:0, and C18:0 ($P < 0.05$).

Table 3 Correlation coefficients between gene expression level and fatty acid composition in longissimus dorsi of yaks.

Fatty acid (%)	PLTP	FABP3	FABP4	LPL	CD36	ME1	SCD	SCP2	APOA1	PPARa	DBI
SFA	0.400	−0.387	−0.119	−0.212	−0.398	0.072	0.044	0.150	−0.253	−0.448	0.063
C10:0	0.482	0.362	0.320	0.264	0.556	0.489	0.463	0.820*	0.497	0.391	0.740

(continued)

Fatty acid (%)	PLTP	FABP3	FABP4	LPL	CD36	ME1	SCD	SCP2	APOA1	PPARa	DBI
C12:0	0.073	0.557	0.280	0.363	0.897**	0.328	0.273	0.842*	0.858*	0.383	0.368
C14:0	0.533	0.564	0.183	0.709	0.463	0.268	0.368	0.794*	0.495	0.614	0.721
C15:0	-0.062	0.635	0.146	0.651	0.806*	0.255	0.432	0.599	0.768*	0.346	0.214
C16:0	0.581	-0.190	-0.544	-0.010	-0.542	0.441	-0.422	-0.296	-0.501	-0.205	-0.035
C17:0	-0.239	0.358	0.528	0.277	0.838*	0.159	0.065	0.630	0.716	0.238	0.342
C18:0	-0.259	-0.572	0.520	-0.556	-0.143	-0.620	0.542	0.362	0.025	-0.637	0.002
C20:0	-0.122	-0.094	-0.729	0.063	-0.305	-0.173	-0.793*	-0.315	-0.060	-0.559	-0.882*
C22:0	0.092	0.026	-0.779*	0.228	-0.177	0.398	-0.813*	-0.420	-0.126	-0.328	-0.648
C24:0	0.705	0.261	-0.342	0.177	-0.039	0.801*	-0.345	-0.033	-0.174	0.444	0.404
MUFA	-0.268	-0.797*	0.618	-0.985**	-0.486	-0.709	0.351	-0.070	-0.436	-0.438	0.109
C16:1	-0.316	-0.772	0.401	-0.841*	-0.578	-0.491	0.025	-0.540	-0.667	-0.235	-0.022
C17:1	-0.516	-0.057	0.438	-0.251	0.568	-0.407	0.215	0.608	0.679	-0.340	-0.213
C18:1	-0.096	-0.672	0.588	-0.884*	-0.399	-0.666	0.514	0.215	-0.269	-0.449	0.224
PUFA	-0.498	-0.439	0.988**	-0.539	0.092	-0.546	0.907**	0.154	-0.036	-0.072	0.403
CLA18:2	-0.186	-0.577	0.673	-0.572	-0.169	-0.544	0.733	0.319	-0.091	-0.458	0.256
C18:2n6c	0.076	-0.437	0.665	-0.571	-0.316	-0.094	0.530	-0.171	-0.530	0.222	0.664
C18:2n6t	0.845*	-0.182	0.527	-0.298	0.411	-0.666	0.293	0.221	0.459	-0.351	0.368
C18:3	-0.122	0.463	0.447	0.401	0.688	-0.052	0.591	0.332	0.558	0.450	0.260
C20:4	-0.468	0.349	0.104	0.487	0.565	-0.028	0.286	0.130	0.512	0.054	-0.207
EPA20:5	-0.044	-0.061	0.332	0.036	-0.050	0.240	0.497	-0.352	-0.339	0.312	0.420
DHA22:6	-0.662	-0.014	0.086	0.124	0.165	-0.542	0.164	-0.065	0.260	-0.313	-0.514

SFA, saturated fatty acid; MUFA, monounsaturated fatty acid; PUFA, polyunsaturated fatty acid; CLA, conjugated linoleic acid; EPA, eicosapentaenoic acid; DHA, docosahexaenoic acid. *Significantly different ($P < 0.05$); **extremely significantly different ($P < 0.01$).

Table 4 Correlation coefficients between gene expression level and fatty acid composition in longissimus dorsi of cattle.

Fatty acid (%)	PLTP	FABP3	FABP4	LPL	CD36	ME1	SCD	SCP2	APOA1	PPARa	DBI
SFA	0.918*	0.056	0.446	0.414	0.432	0.067	-0.034	0.356	-0.881*	-0.410	0.273
C10:0	-0.673	-0.501	-0.101	-0.858*	-0.166	-0.107	-0.132	-0.401	0.740	0.148	-0.784
C12:0	-0.837	-0.386	-0.249	-0.787	-0.291	-0.104	-0.083	-0.425	0.870	0.269	-0.678
C14:0	0.201	0.564	-0.177	0.754	-0.089	0.083	0.182	0.255	-0.311	0.096	0.759
C15:0	-0.360	0.443	-0.390	0.409	-0.309	0.034	0.171	0.011	0.245	0.303	0.490
C16:0	0.920*	0.224	0.372	0.623	0.385	0.089	0.022	0.407	-0.917*	-0.362	0.491

(continued)

Fatty acid(%)	PLTP	FABP3	FABP4	LPL	CD36	ME1	SCD	SCP2	APOA1	PPARa	DBI
C17:0	0.483	-0.391	0.424	-0.301	0.349	-0.020	-0.160	0.052	-0.374	-0.340	-0.397
C18:0	0.739	-0.224	0.475	-0.003	0.421	0.018	-0.118	0.202	-0.653	-0.405	-0.129
C20:0	0.432	0.561	-0.058	0.835*	0.023	0.098	0.167	0.335	-0.526	-0.011	0.805
C22:0	-0.265	-0.567	0.147	-0.780	0.060	-0.088	-0.179	-0.278	0.371	-0.069	-0.776
C24:0	0.689	-0.266	0.468	-0.074	0.408	0.009	-0.129	0.169	-0.596	-0.394	-0.194
MUFA	0.798	0.425	0.208	0.817	0.258	0.106	0.098	0.422	-0.841	-0.236	0.717
C16:1	-0.775	0.190	-0.478	-0.053	-0.428	-0.025	0.109	-0.226	0.694	0.411	0.077
C17:1	-0.036	-0.547	0.249	-0.673	0.161	-0.071	-0.185	-0.190	0.152	-0.164	-0.702
C18:1	0.857	0.362	0.272	0.765	0.309	0.102	0.073	0.425	-0.884*	-0.287	0.651
PUFA	0.226	-0.488	0.347	-0.511	0.262	-0.048	-0.179	-0.076	-0.108	-0.259	-0.574
C18:2 CLA	-0.155	-0.561	0.198	-0.733	0.110	-0.080	-0.183	-0.238	0.267	-0.116	-0.745
C18:2n6c	0.178	-0.502	0.330	-0.544	0.244	-0.052	-0.181	-0.098	-0.059	-0.242	-0.601
C18:2n6t	-0.251	0.481	-0.355	0.493	-0.271	0.045	0.178	0.064	0.133	0.267	0.560
C18:3	0.352	-0.446	0.387	-0.416	0.306	-0.035	-0.172	-0.015	-0.237	-0.300	-0.495
C20:4	0.689	-0.266	0.468	-0.074	0.408	0.009	-0.129	0.169	-0.596	-0.394	-0.194

SFA, saturated fatty acid; MUFA, monounsaturated fatty acid; PUFA, polyunsaturated fatty acid; CLA, conjugated linoleic acid; EPA, eicosapentaenoic acid; DHA, docosahexaenoic acid. *Significantly different ($P < 0.05$); **extremely significantly different ($P < 0.01$).

4 DISCUSSION

FA composition and content in yaks and cattle differed significantly, with higher levels of MUFA and PUFA and lower levels of SFA in yaks. These results agree with those of previous studies, which reported that yaks had more unsaturated fatty acids in their longissimus dorsi muscle than did cattle (Li et al., 2008; Wan et al., 2012). MUFA has a positive effect on the absorption of other fatty acids, so that it can increase the mobility and metabolism of fat globules (Schmid et al., 1998). Yak meat contained a certain amount of EPA and DHA, which were not detected in beef. The main physiological functions of EPA are lowering blood fat and cholesterol, anticarcinogenic activity, and improving brain function; DHA promoting intelligence and healthy growth and development (Dai et al., 1998). CLA is present in ruminant meat and milk, and enhances immunity, is anticarcinogenic, lowers cholesterol, and improves meat quality (Wei and Wang, 2002); the CLA content in yaks was higher than in cattle. PUFA plays a variety of physiological roles, and is extremely important in biological systems (Yu, 1998); yak meat is more nutritious and healthier than beef because of its high PUFA levels.

The mRNA expression levels of the PLTP gene were positively correlated with C18:2n6t levels in yaks, and with SFA levels and C16:0 levels in cattle. PLTP mediates the exchange of phospholipids between lipoproteins, and plays an important role in regulating the metabolism of high-density lipoproteins and very-low-density lipoproteins (Albers and Cheung, 2004). A recent study reported that PLTP affects the n-6/n-3 ratio (Dunner et al., 2013). APOA1 (another lipid transport gene) mRNA expression was positively correlated with C12 : 0 and C15 : 0 in yaks, but was negatively correlated with SFA, C16 : 0, and C18 : 1 in cattle. APOA1 is an important component of high-density lipoprotein, which is a key factor in reverse cholesterol transport, and dissociative APOA1 is a cholesterol and phospholipid receptor (Wang et al., 2008). Our findings suggest that PLTP and APOA1 may play pivotal roles in lipid transport in the longissimus dorsi muscle of yaks and cattle. We also found that APOA1 expression levels in yaks were significantly lower than those in cattle, suggesting that APOA1 may be a genetic marker that is predictive of FA deposition; further research is required to identify APOA1 gene markers, such as SNPs, that are associated with FA content.

The ME1 protein is a part of the tricarboxylic acid shuttle; the nicotinamide adenine dinucleotide phosphate and Coenzyme A (CoA) produced by the process of releasing CoA by ME1 from mitochondria into the cytoplasm are used in the biosynthesis of FAs, and in many other metabolic processes (Vidal et al., 2006). ME1 gene expression level is significantly positively related to FA synthesis in rat adipose tissue (Stelmanska et al., 2004). In the present study, ME1 expression was positively associated with C24 : 0 in yaks. SCD is a rate-limiting enzyme that is responsible for the conversion of SFA into MUFA by inserting a double bond between carbons 9 and 10 of the fatty acyl chain to affect the FA composition of membrane phospholipids, triglycerides, and cholesterol esters (Ntambi and Miyazaki, 2004). Previous studies have found significant differences between different genotypes of the SCD gene and FA content and composition in cattle milk and meat (Orrù et al., 2011; Li et al., 2012). In the present study, SCD expression was found to be significantly negatively associated with C20 : 0 and C22 : 0 in the yak; however, it was even more significantly positively associated with PUFA in the yak. Neither ME1 nor SCD were correlated with FA composition in cattle, so ME1 and SCD may have much more influence on FA content and composition in yaks than in cattle.

FABPs are widely distributed in a variety of animal tissues, and have high affinity with long-chain FAs and play an important role in the oxidation, esterification, and metabolism of FAs (Ockner et al., 1972; Zimmerman and Veerkamp, 1998). Fifteen types of FABP have been found, including FABP3 and FABP4, which are heart-type FA binding proteins (H-FABPs) and fat-type FA binding proteins (A-FABPs), respectively (Chmurzyńska, 2006). There is a significant association between FABP4 polymorphisms and milk FA composition in bovine milk (Nafikov et al., 2013). In yaks, we found that FABP3 mRNA expression levels were significantly negatively correlated with MUFA levels, while FABP4 mRNA expression levels were significantly negatively correlated with C22 : 0 and significantly positively associated with PUFA. We did not find a correlation between FABP3 or FABP4 and FA composition in cattle. Overall, FABP3 and FABP4, particularly FABP4, may have a very important effect on FA composition in yaks, because the mRNA expression level of FABP4 in yaks was lower than in cattle.

LPL is a triglyceride-acyl hydrolase protein of the hydrolase family (Emmerich et al.,

1992) that is a rate-limiting enzyme that decomposes chylomicrons in circulating lipoproteins and triglycerides of very-low-density lipoproteins, and releases FAs and glycerol. LPL was significantly negatively correlated with MUFA, C16:1, and C18:1 in yaks, while it was significantly negatively correlated with C10:0 and significantly positively correlated with C20:0 in cattle. In accordance with our data, Zhu et al. (2013) found that LPL activity is correlated with FA composition in meat. Correlation analyses revealed positive correlations between CD36 gene expression levels and C12:0, C15:0, and C17:0 in yaks, while the DBI gene was negatively correlated with C20:0. However, there were no significant correlations between CD36 or DBI expression levels and FA content in cattle. These results provide a theoretical basis for the further study of the molecular mechanisms that underlie FA metabolism in meat (Guidotti et al., 1983; Rosendal et al., 1993; Silverstein and Febbraio, 2009).

SCP2, known as the nonspecific lipid transfer protein, enhances transport between plasma membranes and plays an important role in lipid metabolism (Starodub et al., 2000; Stolowich et al., 2002). McLean et al. (1995) suggested that SCP2 expression levels are altered in a number of diseases in which lipid metabolism is abnormal. In this study, a significant association was observed between SCP2 gene expression level and C10:0, C12:0, and C14:0 in yaks; there was no significant association between SCP2 expression level and FA content in cattle. After comparing SCP2 mRNA expression levels in yaks and cattle, we found that the SCP2 expression level in yaks was significantly lower than in cattle. These results suggest that SCP2 is important for lipid metabolism in the yak.

In conclusion, we found differences in FA composition of the longissimus dorsi muscle of yaks and cattle. PUFA is beneficial to human health, and we found higher levels of PUFA in yak meat than in beef. Of the 11 PPARα genes that we examined, the mRNA expression levels of the lipid transport gene APOA1 may have the most important influence on the FA composition of yak and cattle meat. The mRNA expression levels of FABP4, LPL, and SCP2 were also important predictors of FA composition. Our results indicate that these genes may participate in the regulation and control of intramuscular FA metabolism in yaks, and they can be used as candidate genes for FA selection in yaks.

Conflicts of interest

The authors declare no conflict of interest.

ACKNOWLEDGMENTS

Research supported by the National Beef Cattle Industry Technology System (#nycytx-38) 14477 PPARα signal pathway genes and FA in yaks and cattle and the "Five-Twelfth" National Science and Technology Support Program (#2012BAD13B05).

（发表于《Genetics and molecular research》, SCI, IF: 0.764. 14 469–14 478.）

Genetic Characterization of Antimicrobial Resistance in *Staphylococcus aureus* Isolated from Bovine Mastitis Cases in Northwest China

YANG Feng[1,**], WANG Qi[2,**], WANG Xurong[1], WANG Ling[1], LI Xinpu[1], LUO Jinyin[1], ZHANG Shidong[1], LI Hongsheng[1]

(1. Key Laboratory of Veterinary Pharmaceutical Development, Ministry of Agriculture/Engineering & Technology Research Center of Traditional Chinese Veterinary Medicine of Gansu Province/Lanzhou Institute of Husbandry and Pharmaceutical Sciences, Chinese Academy of Agricultural Sciences, Lanzhou 730050, China; 2. State Key Laboratory of Applied Organic Chemistry, Lanzhou University, Lanzhou 730000, China)

Abstract: *Staphylococcus aureus* is the most common etiological pathogen of bovine mastitis. The resistant strains make the disease difficult to cure. The aim of this study was to characterize the genetic nature of the antimicrobial resistance in *S. aureus* cultured from bovine mastitis in northwest China in 2014. A total of 44 S. aureus were isolated for antimicrobial resistance and resistance-related genes. Antimicrobial resistance was determined by disc diffusion and the corresponding resistance genes were detected by PCR. Phenotype indicated that S. aureus isolates were resistant to penicillin (84.09%), erythromycin (20.45%), tetracycline (15.91%), gentamicin (9.09%), tobramycin (6.82%), kanamycin (6.82%) and methicillin (2.27%). 9.09% of the S. aureus isolates were classified as multidrug resistant. In addition, genotypes showed that the isolates were resistant to rifampicin (100%, rpoB), penicillin (95.45%, blaZ), tetracycline (22.73%, tetK, tetM, alone or in combination), erythromycin (22.73%, ermB or ermC), gentamicin/tobramycin/kanamycin (2.27%, aacA-aphD), methicillin (2.27%, mecA) and vancomycin (2.27%, vanA). Resistance to tetracycline was attributed to the genes tetK and tetM ($r=0.558$, $P<0.001$). This study noted high-level geno- and phenotypic antimicrobial resistance in S. aureus isolates from bovine mastitis cases in northwest China.

Key words: Bovine mastitis; *Staphylococcus aureus*; Antimicrobial resistance; Phenotype; Genotype

---

Corresponding authors, E-mail: lihsheng@sina.com
* These authors contributed equally to this study.
© 2016, CAAS. All rights reserved. Published by Elsevier Ltd.
doi: 10.1016/S2095-3119(16)61368-0

1 INTRODUCTION

Bovine mastitis is the most costly disease for dairy industry. *Staphylococcus aureus* is frequently recognized as a major contagious pathogen in bovine mastitis (Feßler et al. 2010). This bacterium can be transmitted among cows through contact with contaminated milk and results in subclinical mastitis that is often difficult to detect at the earliest stage. This results in serious economic losses being incurred in the dairy industry (Sharif et al. 2009).

Antimicrobial chemotherapy is the primary approach to treat staphylococcal bovine mastitis (McDougall et al. 2014) and susceptibility tests can guide the veterinarian to select the most appropriate antimicrobial agent (Moroni et al. 2006). However, due to the widespread use of antimicrobial compounds, the occurrence of antimicrobial-resistant *S. aureus* cultured from cows with mastitis has increased (El-Jakee et al. 2011), thereby compromising this approach to treatment (Unakal and Kaliwal 2010).

S. aureus has evolved four general resistance mechanisms, including trapping of the drug, modification of the drug target, enzymatic inactivation of the drug and the activation of transmembrane efflux pumps, to fend off attack from antimicrobials (Pantosti et al. 2007). Various genetic determinants such as mecA and blaZ (penicillins), aacA-aphD (aminoglycosides), ermA/B/C (macrolides), tetK/M (tetracyclines), vanA (vancomycin), fusB (fusidic acid), ileS (mupirocin) and rpoB (rifampicin) are reported to be responsible for the corresponding antimicrobial resistance mechanisms in *S. aureus* (Jensen and Lyon 2009). These genetic determinants enable *S. aureus* to reside for a long time inside the host or herd environment and avoid antimicrobial therapy (Kumar et al. 2011). Few antimicrobial resistance genes detected in S. aureus had been reported in China, and little is known about the genetic background of S. aureus resistance in northwest China (Memon et al. 2013; Wang et al. 2015). The focus of this short study was to characterize the genetic nature of antimicrobial resistance among S. aureus cultured from bovine mastitis in northwest China.

2 MATERIALS AND METHODS

2.1 Sample collection and bacterial strains

A total of 44 *S. aureus* strains were obtained from cows presenting mastitis in northwest China during 2014. These included 32 strains isolated from clinical cases and 12 strains isolated from subclinical cases. The regions and number of isolates included Gansu (23), Qinghai (11) and Ningxia (10). Mastitis cases were confirmed by California mastitis test (CMT). Milk samples were collected aseptically for bacteriological assay as described by Pitkälä et al. (2004). Before sampling, the first streams of milk were discarded, and teat ends were disinfected with cotton swabs soaked in 70% (v/v) alcohol and allowed to dry. Then a 5-mL secretion was collected into a sterile 10-mL tube. Any samples that were not processed immediately were kept at 4°C for microbiological examination within 18 h of collection.

Bacterial identification was performed on sheep blood agar plates, Chapman's agar plate and Baird Parker agar, incubated at 37°C for 24–48 h. After overnight growth, presumptive staphylococci positive for Gram staining and catalase were tested for coagulase, hemolysis, DNAse, acetoin and anaerobic fermentation of mannitol. The confirmed S. aureus strains were preserved at −79 ℃ for subsequent study.

2.2 Antimicrobial susceptibility test

Antimicrobial susceptibility for S. aureus was determined using disc diffusion method on Mueller-Hinton agar (Oxoid, United Kingdom) with commercial available discs (Oxoid, United Kingdom) according to Clinical and Laboratory Standards Institute (CLSI 2010). The antimicrobial agents included penicillin, cefoxitin, erythromycin, gentamicin, tobramycin, kanamycin, tetracycline, vancomycin, fusidic acid, mupirocin and rifampicin. Cefoxitin was used to investigate methicillin susceptibility of S. aureus. The minimal inhibitory concentration (MIC) for vancomycin, if needed, was determined by E-test method (Biomerieux, France). The criteria for the interpretation of zone diameter used in this study were described in Table 1.

Table 1 Zone of inhibition diameter interpretive criteria for Staphylococcus aureus.

Antibiotic family	Antibiotics	Disc content	R (mm)	I (mm)	S (mm)	References
Penicillins	Penicillin	10 IU	≤28	-	>29	CLSI (2010)
	Cefoxitin	30 μg	≤19	-	>20	Tiwari et al. (2009)
Macrolides	Erythromycin	15 μg	≤13	14-22	>23	CLSI (2010)
Aminoglycosides	Gentamicin	10 μg	≤12	13-14	>15	CLSI (2010)
	Tobramycin	10 μg	≤12	13-14	>15	CLSI (2010)
	Kanamycin	30 μg	≤13	14-17	>18	CLSI (2010)
Tetracyclines	Tetracycline	30 μg	≤14	15-18	>19	CLSI (2010)
Glycopeptides	Vancomyci	30 μg	-	-	>15	CLSI (2010)
Others	Fusidic acid	5 μg	≤17	18-19	>20	Skov et al. (2001)
Monoxycarbolic acid	Mupirocin	5 μg	≤13	-	>14	Finlay et al. (1997)
Ansamycins	Rifampicin	5 μg	≤16	17-19	>20	CLSI (2010)

R, resistant; I, intermediate; S, sensitive. The same as below.

2.3 PCR amplification of antimicrobial resistance genes

Preparation of bacterial genomic DNA was performed using the Bacterial DNA Kit (Omega Bio-Tek, USA) according to the manufacturer's recommendation. All isolates were tested by simplex PCR amplification of genes that confer resistance to penicillin (blaZ), methicillin (mecA), erythromycin (ermA, ermB, ermC), gentamicin/tobramycin/kanamycin (aacA-aphD), tetracycline (tetK, tetM), vancomycin (vanA), fusidic acid (fusB), mupirocin (ileS), rifampicin (rpoB). Primers for these resistance genes were used as Table 2. The PCR mixtures (25 μL) used to detect the antimicrobial resistance genes in all strains contained 12.5 μL reaction mixtures (Premix Ex Taq^{TM} ver. 2.0, TaKaRa, China), 0.5 μL primer 1, 0.5 μL primer 2, 2 μL genomic DNA and 9.5 μL ddH$_2$O. PCR products (5 μL) were analysed by electrophoresis on a 1.2% agarose gel and stained with ethidium bromide.

Table 2 Target genes and oligonucleotide primers used to amplify antimicrobial resistance genes of S. aureus isolates.

Target genes	Forward and reverse primers (5′→3′)	Amplimer size (bp)	Annealing temp (℃)	References
blaZ	TAAGAGATTTGCCTATGCTT TTAAAGTCTTACCGAAAGCAG	377	48	Olsen et al. (2006)
mecA	GTGAAGATATACCAAGTGATT ATGCGCTATAGATTGAAAGGAT	147	57	Choi et al. (2003)
ermA	TCTAAAAAGCATGTAAAAGAA CTTCGATAGTTTATTAATATTAGT	645	45	Sutcliffe etal. (1996)
ermB	GAAAAGGTACTCAACCAAATA AGTAACGGTACTTAAATTGTTTAC	639	45	Sutcliffe etal. (1996)
ermC	TCAAAACATAATATAGATAAA GCTAATATTGTTTAAATCGTCAAT	642	45	Sutcliffe etal. (1996)
aacA-aphD	GAAGTACGCAGAAGAGA ACATGGCAAGCTCTAGGA	491	45	Strommenger et al. (2003)
tetK	GTAGCGACAATAGGTAATAGT GTAGTGACAATAAACCTCCTA	360	48	Strommenger et al. (2003)
tetM	AGTGGAGCGATTACAGAA CATATGTCCTGGCGTGTCTA	158	48	Strommenger et al. (2003)
vanA	GGGAAAACGACAATTGC GTACAATGCGGCCGTTA	732	50	Dutka-Malen etal. (1995)
fusB	ATTCAATCGGAAACCTATATGATA TTATATATTTCCGATTTGATGCAAG	292	48	O'Neill etal. (2004)
ileS	TATATTATGCGATGGAAGGTTGG AATAAAATCAGCTGGAAAGTGTTG	458	50	Anthony et al. (1999)
rpoB	AGTCTATCACACCTCAACAA TAATAGCCGCACCAGAATCA	702	50	Aubry-Damon et al. (1998)

2.4 Statistical analysis

SPSS.17.0 (Bivariate correlations) was used to analyze the associations between phenotypic and genotypic resistance patterns. The intermediate isolates were considered to be resistant.

3 RESULTS

3.1 Antimicrobial susceptibility

The antimicrobial resistances of S. aureus cultured from bovine mastitis were shown in Table 3. Most were resistant to penicillin (84.09%). Less than half of these isolates expressed resistance to erythromycin (20.45%) and tetracycline (15.91%). Few were resistant to gentamicin

(9.09%), tobramycin (6.82%), kanamycin (6.82%) and methicillin (2.27%). None of these isolates were found to be resistant to fusidic acid, mupirocin, rifampicin and vancomycin. Five of the 44 isolates were susceptible to all antimicrobial compounds tested, while the remaining strains were resistant to various combinations.

Table 3 Antimicrobial susceptibility of S. aureus (n=44) from bovine mastitis.

Antimicrobials	Antimicrobial susceptibility					
	R		I		S	
	n	%	n	%	n	%
Penicillin	37	84.09			7	15.91
Methicillin	1	2.27	1	2.27	42	95.45
Erythromycin	9	20.45	2	4.55	33	75.00
Gentamicin	4	9.09			40	90.91
Tobramycin	3	6.82			41	93.18
Kanamycin	3	6.82			41	93.18
Tetracycline	7	15.91	1	2.27	36	81.82
Vancomycin					44	100
Fusidic acid			1	2.27	43	97.73
Mupirocin					44	100
Rifampicin					44	100

3.2 Antimicrobial resistance genes

As shown in Table 4, all S. aureus included in this study contained drug-specific resistance gene for rifampicin (rpoB). Most carried the penicillin-resistant gene (blaZ, 95.45%). 22.73% of the isolates in the collection possessed the erythromycin-resistant genes (ermB, 9.09%; ermC, 13.64%, alone), while ermA was not detected in any of the isolates. Also 22.73% strains contained tetracycline-resistant genes (tetK, 22.73%; tetM, 2.27%, alone or in combination), and 2.27% possessed the aminoglycoside-resistant gene (aa-cA-aphD), methicillin-resistant gene (mecA) mupirocin-resistant gene (ileS) and vancomycin-resistant gene (vanA), respectively. None of the tested strains were positive for fusidic acid-resistant gene (fusB).

3.3 The association between phenotypic and genotypic resistance

In this study, the correlation between phenotypic and genotypic resistance was moderate for penicillin ($r=0.558$, $P<0.001$). Whereas for the other 12 tested antimicrobials, no correlations were observed (Table 4).

Table 4 Comparison of phenotypic and genotypic testing for antimicrobial resistance in *S. aureus* from bovine mastitis.

Antimicrobials	Gene(s)	Characteristics of S. aureus isolates[1]					Association[2]	
		P⁺/G⁺	P⁻/G⁻	P⁺/G⁻	P⁻/G⁺	G+	r	P
		n	n	n	n	%		
Penicillin	Any	36	1	1	6	95.45	0.203	0.185
	blaZ	36			6	95.45		
Methicillin	Any	0	41	2	1	2.27	−0.041	0.790
	mecA	0			1	2.27		
Erythromycin	Any	3	26	8	7	22.73	0.063	0.686
	ermA	0			0			
	ermB	1			3	9.09		
	ermC	2			4	13.64		
Gentamicin	Any	0	39	4	1	2.27	−0.048	0.756
	aacA-aphD	0			1	2.27		
Tobramycin	Any	0	40	3	1	2.27	−0.041	0.790
	aacA-aphD	0			1	2.27		
Kanamycin	Any	0	40	3	1	2.27	−0.041	0.790
	aacA-aphD	0			1	2.27		
Tetracycline	Any	6	32	2	4	22.73	0.558	<0.001
	tetK	6			4	22.73		
	tetM	0			1	2.27		
Vancomycin	Any	0	43	0	1	2.27	NT	NT
	vanA	0			1	2.27		
Fusidic Acid	Any	0	43	1	0		NT	NT
	fusB	0			0			
Mupirocin	Any	0	43	0	1	2.27	NT	NT
	ileS	0			1	2.27		
Rifampicin	Any	0	0	0	44	100	NT	NT
	rpoB	0			44	100		

[1] P⁺, phenotypic resistance; P⁻, phenotypic susceptibility; G⁺, resistant gene positive, G⁻, resistant gene negative.

[2] Association between resistant phenotypes and resistance genes. Values of P less than 0.05 were considered significant. NT correlation coefficients (r value) cannot be calculated (at least one variable is constant).

4 DISCUSSION

The resistance of *S. aureus* to antimicrobial agents is an increasing global problem. Determining the antimicrobial susceptibility profiles in a study collection is required not only for effective therapy but also for monitoring the spread of resistant strains in defined ecological niches (Hogan and Smith 2003; Coelho et al. 2009). In this study, the antimicrobial susceptibility profiles of S. aureus were determined and high levels of resistance to penicillin followed by erythromycin and tetracycline were detected. These data are similar to that contained in other reports (De Oliveira et al. 2000; Rajala-Schultz et al. 2004). In particular, more than half of the *S. aureus* isolates were resistant to penicillin (84.09%), a feature that was similar to other data previously reported in China (Shi et al. 2010). In contrast, this observation differed from trends being reported in other countries (Güler et al. 2005; Bengtsson et al. 2009). Most of isolates in this study were resistant to at least one antimicrobial compound and 9.09% (4/44) of the isolates in the collection were defined as being multidrug resistance according to the definition proposed by Magiorakos et al. (2012). It is tempting to speculate that these findings may arise from the frequent use of these antimicrobial agents in intramammary infections in China.

The presences of resistance-associated genes in S. aureus were detected in this study. All carried rifampicin-resistant gene rpoB. However, this gene was been shown to have little impact in terms of its contribution to relevant phenotypic resistance. Indeed, all isolates were sensitive to rifampicin. This can be attributed to the lack of expression of the resistance gene, which need to be activated (Hammad and Shimamoto 2014).

A high percentage of resistance genes associated with penicillin (blaZ), erythromycin (ermB, ermC) and tetracycline (tetK, tetM) were noted, and these most likely contribute to the high resistance rates recorded for these antimicrobials compounds. However, association analysis showed moderate correlation between geno- and phenotypic resistance for tetracycline alone. Frey et al. (2013) reported similar data in coagulase-negative staphylococci. Memon et al. (2013) also reported that high resistance rates against methicillin were found but no S. aureus isolate was positive for the mecA gene. In addition, genes aacA-aphD, mecA, ileS vanA, and fusB that conferred resistance to clinically relevant antimicrobials (gentamicin, tobramycin, kanamycin, methicillin, mupirocin, vancomycin and fusidic acid) were also detected. No associations were observed for these antimicrobials.

Resistance genes were detected in some susceptible isolates, while no resistance genes could be detected in some resistant isolates, which agreed with the results reported by Gao et al. (2012). In some isolates, phenotypic resistance may be caused by point mutations rather than gene acquisition. Additionally, except for the general resistance mechanisms, other factors such as biofilm formation may be the main resistance mechanism (Pantosti et al. 2007; Croes et al. 2009). Mechanisms of resistance to antibacterials are so complex that the presence or absence of a certain resistance gene does no certainly indicate that the particular isolate is resistant or sensitive to the corresponding antimicrobial agent (Gow et al. 2008).

5 CONCLUSION

Our study demonstrated that S. aureus from bovine mastitis in northwest China are generally resistant to many of the antimicrobial compounds commonly used for treatment of mastitis, especially penicillin. Susceptibility testing as part of the diagnosis is therefore recommended to guide the selection of the most appropriate chemotherapeutic agent. Our data also revealed that S. aureus isolates carry several genetic determinants. Although these determinants make little contribution to relevant resistance, they are potential threat to antimicrobial treatment.

ACKNOWLEDGEMENTS

This study was supported by the Central Public-Interest Scientific Institution Basal Research Fund, China (1610322015007), the Key Technology R&D Program of China during the 12th Five-Year Plan period (2012BAD12B03) and the Natural Science Foundation of Gansu Province, China (145RJYA311).

(Managing editor ZHANG Juan)

(发表于《Journal of Integrative Agriculture》, 院选SCI, IF: 0.724. 60345-7.)

Quantitative Structure Activity Relationship (QSAR) Studies on Nitazoxanide-based Analogues Against Clostridium Difficile in Vitro

ZHANG Han, LIU Xiwang, YANG Yajun, LI Jianyong*

(Key Laboratory of New Animal Drug Project, Gansu Province; Key Laboratory of Veterinary Pharmaceutical Development, Ministry of Agriculture; Lanzhou Institute of Husbandry and Pharmaceutical Sciences of CAAS, Lanzhou, 730050 China)

Abstract: Quantitative structure activity relationship (QSAR) has been established between the various physiochemical parameters of a series of nitazoxanide-based analogues and its antibacterial activity against Clostridium difficile. Genetic function approximation (GFA) and comparative molecular field analysis (CoMFA) techniques were used to identify the descriptors that have influence on biological activity. The most influencing molecular descriptors identified in 2D-QSAR include spatial, topological, and electronic descriptors, while electrostatic and stereoscopic fields were the most influencing molecular descriptors identified in 3D-QSAR. Statistical qualities (r^2, q^2) indicated the significance and predictability of the developed models. The study indicated that antibacterial activity of Clostridium difficile can be improved by increasing molecular connectivity index, local charge surface index, sharp index and decreasing molecular flexibility index.

Key words: Nitazoxanide-based analogues; Antibacterial activities; QSAR; Clostridium difficile

1 INTRODUCTION

Parasitic and bacterial infections were the major causes of morbidity and mortality in the world especially in developing countries. Diseases caused by intestinal parasites affected billions of people every year and there had been few drug innovations for treating theses infections in the past three decades. Fortunately, nitazoxanide (NTZ) named as 2-acetyloloxy-N-(5-nitro-2-thiazolyl) benzamide was developed as a promising compound to treat these diseases in the future. NTZ was first synthesized by Romark Laboratory in 1975 (Rossignol et al. 1975) and showed activities against Taenia saginata and Hymenolepis nana (Rossignol et al. 1984). Further studies reported that NTZ exhibited an unusual broad spectrum of activities against various parasites in

*Corresponding author, E-mail: lijy1971@163.com

human beings and animals (*Giardia lamblia*, *Cryptosporidium parvum*, *Entamoeba histolytica*, *Neospora caninum*, *Trichuris trichiura*, *Ascaris lumbricoides*, *Enterobius vermicularis*) (Rossignol et al. 1984; Rossignol et al. 2001; Adagu et al. 2002; Esposito et al. 2007; Elvia et al. 2003). In 1996, NTZ had been marketed in Latin America and then approved by US Food and Drug Administration (FDA) as an agent treating infections caused by parasites (Gilles et al. 2002; Anderson et al. 2007). NTZ also showed broad antimicrobial activities for anaerobic bacteria (*Clostridium difficile*, *Helicobacter pylori*) as well as notable anti-virus actions on influenza and hepatitis (Catherine et al. 2000; Guttner et al. 2003; Rossignol et al. 2009; Korba et al. 2008).

In contrast to metronidazole, NTZ has been shown nonmutagenic, which suggested that the two compounds have fundamental differences with the mode of action. In fact, recent studies revealed that NTZ inhibited pyruvate ferredoxin oxidoreductase (PFOR), a key enzyme of central intermediary metabolism in anaerobic organisms. NTZ appeared to interact with PFOR and the nitro group was not reduced. The mechanism was supposed that amide anion (NTZ^-) may couple directly with thiamine pyrophosphate (TPP), which was a cofactor of PFOR (Hoffman et al.2007). Recently, a number of nitazoxanide derivatives were synthesized and their activities were also investigated. The results demonstrated that the nitro group may be essential for the antimicrobial efficacy with a minimum inhibitory concentration of organisms (MIC_{90}) of 0.06-4.00mg·L^{-1} (Sisson et al. 2002; Glenn et al. 2006). But notably, specific replacements of the nitro group were also shown significant activities against intracellular parasites (Esposito et al. 2007) while invalidity for intestinal parasite Giardia lamblia. On the other hand, modifications of the salicylate moiety may reduce the activity even abrogate it (Muller et al. 2006). This indicated that other enzymes beside PFOR might also be relevant.

In order to prevent resistant emergence of NTZ and develop more effective drugs, a quantitative structure activity relationship (QSAR) study on nitazoxanide-based analogues was a useful tool. QSAR methodology is used to predict the activity of novel molecules by mathematical equations, which deduce the relationship (s) between a chemical structure and its biological activity (Kamalakaran et al. 2009). QSAR models are pointers to design effective drugs.

In this study, QSAR study has been investigated on 28 nitazoxanide-based analogues against Clostridium difficile in vitro. From the results obtained we established a functional predictive model, which can be further developed to aid in the design of novel derivatives with improved in vitro potency against Clostridium difficile and other anaerobic organisms including parasites.

2 MATERIALS AND METHODS

The antibacterial data of nitazoxanide-based analogues were used for QSAR study. These selected compounds should have enough variable grad in antibacterial activity and were divided into training set (23 compounds) and test set (5 compounds including 1a, 1, 8, 10, 17) as structural diversity with 80% rate. Moreover, the compounds in training set had enough quantity and included all structural information. The biological activity data were minimal inhibitory concentration (MIC) for Clostridium difficile in vitro and converted into pMIC by equation pMIC =-log

MIC. All compounds are shown in table 1 serially citing their structures and biological activities. All calculations were carried out with Discovery Studio 3.1 (DS3.1, Accelrys Software Inc., USA) QSAR software and default values were used in all parameters without specialized exception. The general structure of nitazoxanide-based analogues was followed as fig. 1.

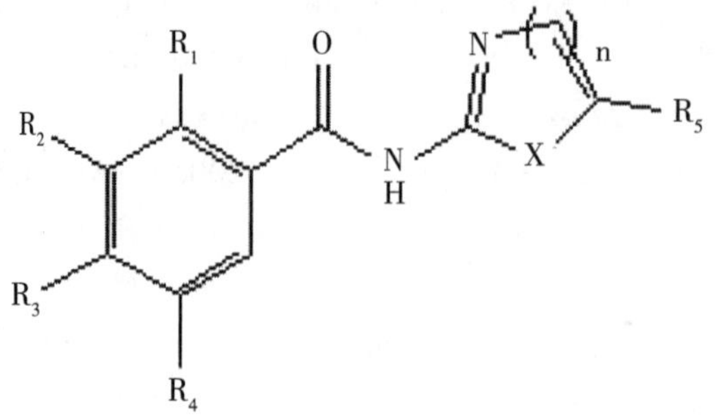

Fig. 1 The general structure of nitazoxanide-based analogues

2.1 2D-QSAR model

Initial conformation energy optimization 2D structures of all compounds were drawn with Chem Bio Draw 11.0 software and then imported into DS3.1 QSAR software. After pMICs were imported and molecular optimized parameters were set in Minimize Ligands, initial conformation energy optimization was carried out with second-generation force field MMFF.

2.2 Descriptors calculation

In the Calculate Molecular Properties, 98 Molecular descriptors for all compounds were selected. They included hydrophobic parameters, electronic parameters, space configuration parameters, the molecular character parameters, topological parameters and charge related parameters, such as, AlogP, Molecular Properties, Molecular Property Count, Surface Area and Volume, Topological Descriptors, Dipole, Jurs Descriptors, Molecular Properties, Principal Moments of Inertia, Shadow Indices, Surface Area and Volume.

2.3 2D-QSAR model establishment

Using genetic function approximation (GFA) method, the relationships between structure and biological activity of target compounds (training set compound) were established and analyzed. The generated model as planed contained 4-5 item parameters with the initial population numbers for 100, the iterative 5 000 times. Such parameters produced were the no cross validation correlation coefficient r^2, the adjusted coefficient of determination r^2 (adj), prediction correlation coefficient r^2 (pred), RMS Residual (RMS Residual Error), etc.

2.4 2D-QSAR model validation

In order to test the model's repeatability and accuracy and judge its predictive ability, active predictions of sample molecules in training set and test set were carried out by the GFA mode established with training set. In the DS 3.1 QSAR software, Calculate Molecular Properties module was run, and at the same time, test set and prediction set file data, GFATempModel-1 model were selected.

2.5 3D-QSAR model

Molecular optimization and overlay After energy optimization were carried out for threedimensional structure of small molecular compound in DS with molecular dynamics program such as Ligands Minimize, lowest energy conformation of each molecule was obtained. Through the most commonly used overlay method based on Molecular field, compound 1e with best activity was selected as template molecule. Then Molecular Overlay was started and at the same time all molecular force field with the same orientation was ensured.

2.6 3D-QSAR model establishment and validation

In the overlayed molecule surrounding, molecular field space range was defined and rectangular space was used in the software. The defined space evenly divided in accordance with 1.5 Å step length produced grid point. Such probe molecules as van der Waals carbon atom and H^+ were used to calculate the interaction of stereo energy and electrostatic energy between molecule and lattice point probe atom. After molecular field characteristics on lattice point were evaluated, molecular force field parameters (structural parameters) can be produced. The molecular fields of molecule at each grid point were calculated, and then the molecular field parameters as the independent variable and molecular activity value as the dependent variable. The model best principal component number n and cross validation correlation coefficient q^2 were determined by partial least squares (PLS) combined with interactive test. Then based on the best principal component, the relationship between structure and biological activity for goal compounds was established by PLS.

Table 1 Structures and activities (pMIC) of the nitazoxanide-based analogues.

Compound No.	Structure	MIC (μg/mL)	pMIC	Ref
1a* (NTZ)		0.25	0.602 06	Liu et al. 2014
1b		0.5	0.301 03	Liu et al. 2014
1c		1	0	Liu et al. 2014

(continued)

Compound No.	Structure	MIC (μg/mL)	pMIC	Ref
1d		0.5	0.301 03	Liu et al. 2014
1e		0.125	0.903 09	Liu et al. 2014
17d		4	−0.602 06	Liu et al. 2014
17e		2	−0.301 03	Liu et al. 2014
1*(TIZ)		0.5	0.301 03	Glenn et al. 2006
2		0.5	0.301 03	Glenn et al. 2006
3		0.25	0.602 06	Glenn et al. 2006
4		1.50	−0.176 09	Ballard et al. 2011

(continued)

Compound No.	Structure	MIC (μg/mL)	pMIC	Ref
5		0.88	0.055 517	Ballard et al. 2011
6		0.23	0.638 27	Ballard et al. 2011
7		0.16	0.795 88	Ballard et al. 2011
8*		0.49	0.309 80	Ballard et al. 2011
9		0.75	0.124 94	Ballard et al. 2011
10*		1.51	−0.178 98	Ballard et al. 2011
11		0.57	0.244 13	Ballard et al. 2011
12		0.41	0.387 22	Ballard et al. 2011

Compound No.	Structure	MIC (μg/mL)	pMIC	Ref
13		0.38	0.420 22	Ballard et al. 2011
14		0.74	0.130 77	Ballard et al. 2011
15		1.99	−0.298 85	Ballard et al. 2011
16		8	−0.903 09	Ballard et al. 2010
17*		7.99	−0.902 6	Ballard et al. 2010
18		8.01	−0.903 6	Ballard et al. 2010
19		1.52	−0.181 84	Ballard et al. 2010
20		0.99	0.004 365	Ballard et al. 2010
21		5.98	−0.776 7	Ballard et al. 2010

* Test set compounds.

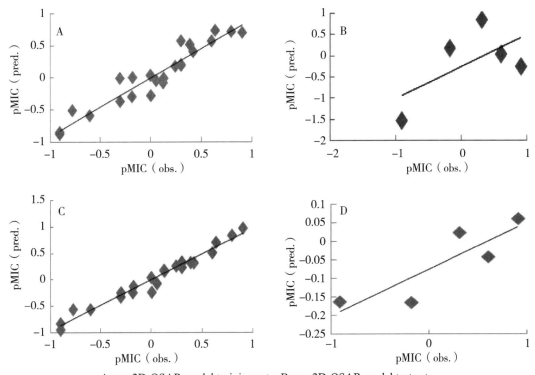

A was 2D-QSAR model training set, B was 2D-QSAR model test set,
C was 3D-QSAR model training set and D was 3D-QSAR model test set.

Fig. 2 Plots of predicted versus observed values for 2D-QSAR and 3D-QSAR model

Table 2 The active data and molecular descriptors of Nitazoxanide derivatives.

Compounds	pMIC	X_1	X_2	X_3	X_4
1a* (NTZ)	0.602 06	0.597 77	−1.419 91	3.751 11	4.477 58
1b	0.301 03	0.573 06	−1.387 43	2.949 6	3.580 02
1c	0	0.575 78	−1.516 76	2.949 6	3.580 02
1d	0.301 03	0.661 82	−1.732 5	3.410 1	4.001 86
1e	0.903 09	0.684 15	−1.694 19	3.609 66	4.001 86
17d	−0.602 06	0.491 31	−1.790 74	3.621 4	4.126 49
17e	−0.301 03	0.513 64	−1.713 25	3.830 57	4.126 49
1* (TIZ)	0.903 09	0.520 06	−1.511 9	2.762 1	3.417 9
2	0.301 03	0.686 72	−1.488 4	2.992 82	3.650 45
3	0.602 06	0.686 72	−1.489 5	2.992 82	3.650 45
4	−0.176 09	0.466 67	−1.337 13	2.711 21	3.216 31
5	0.055 5 17	0.529 67	−1.646 9	2.943 4	3.396 7
6	0.638 27	0.655 65	−1.396 87	3.107 79	3.656 28
7	0.795 88	0.684 15	−1.683 29	3.609 66	4.001 86

(continued)

Compounds	pMIC	X_1	X_2	X_3	X_4
8*	0.309 8	0.836 64	−1.784 09	3.796 52	4.444 97
9	0.124 94	0.497 76	−1.344 23	2.719 32	3.347 53
10*	−0.178 98	0.535 57	−1.140 07	3.071 12	3.799 93
11	0.244 13	0.442 26	−1.248 19	2.317 85	2.749 79
12	0.387 22	0.466 67	−1.092 26	2.317 85	2.749 79
13	0.420 22	0.560 12	−1.347 81	2.497 6	3.037 42
14	0.130 77	0.466 67	−1.330 18	2.497 6	3.037 42
15	−0.298 85	0.404 48	−1.117 52	4.378 29	5.001 98
16	−0.903 09	0.510 07	−1.237 46	4.370 88	5.463 79
17*	−0.902 6	0.522 46	−1.597 4	4.466 61	5.671 4
18	−0.903 6	0.584 65	−1.916 8	3.108 43	3.936 3
19	−0.181 84	0.802 13	−2.356 59	3.996 36	4.726 86
20	−0.776 7	0.678 1	−1.946 89	2.875 53	3.757 62
21	0.004 365	0.584 65	−1.622 98	2.702 79	3.462 17

* Test set compounds.

In the Create 3D QSAR Model modules, setting step length was 1.5A and high correlation value described was 0.9. After the establishment of 3D-QSAR model, at the same time test set data were input to validate model. The steric field and electrostatic field to activity contribution were intuitively reflected with three dimensional force field coefficient equipotential diagrams and other molecular modified information can be got.

3 RESULTS

3.1 2D-QSAR model

The structure and biological activity relationship model of target compounds was constructed by GFA in DS software. The optimal 2D-QSAR model with relevant activity data was finally determined after each parameter was investigated, which was expressed as regression equation (1) through GFA in training set molecule. pMIC=0.755 52+4.947 1*X_1+1.397*X_2+1.750 7*X_3−1.844 2*X_4 (model 1) n=23, r^2=0.905 2, r^2 (adj) = 0.884 2, r^2 (pred) = 0.840 2, RMS Residual Error =0.173 3, Friedman L.O.F.= 0.087 64, S.O.R p-value=1.65*$10^{-0.9}$.

X_1: molecular connectivity index, X_2: molecular local charge surface parameter, X_3: molecular shape index, X_4: molecular flexibility index.

3.2 3D-QSAR model

CoMFA method was chose for 3D-QSAR analysis in this study. Under unknowing receptor structure, by using two kinds of probe molecule (van der Waals C atom and H^+) to calculate electrostatic field and steric field and energy as a descriptor, CoMFA method studied the relationships between activity (y) and molecular field (x) and at the same time gave some suggestions for the existing molecular modification.

The results revealed that the non-cross validation correlation coefficient r^2 for the optimal model of 3D-QSAR was 0.967. The r^2 was greater than 0.8, which showed that the model had good correlation. In addition, its adjustment decision coefficient r^2 (adj) was 0.964, which indicated that the model had a strong predictability. RMS residual error was smaller as 0.090 05, indicating the model with good repeatability (table 4). The observed and predicted MICs of the compounds against difficult Clostridial were followed as table 2.

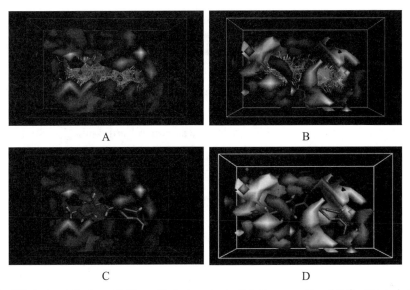

A and C for 3D electrostatic force field coefficient equipotential diagram, B and D for 3D steric force field coefficient equipotential diagrams, A and B for 3D-QSAR model equipotential diagrams of the electrostatic field and steric field, C and D for 3D equipotential diagram for the highest active compounds 1e.

Fig. 3 Steric and electrostatic equipotential diagrams

3.3 3D-QSAR model validation

By the best principal component number to get CoMFA model and with CoMFA model to forecast the molecular activity in the sample concentration for testing the model's repeatability and accuracy, the cross validation correlation coefficient for 0.744 ($q^2>0.5$) showed that its predictability was good. The correlation curve fig. for the predicted and observed value of the compound in training set and testing set was followed as fig. 3 (C, D). From fig. 3 (C, D), the linear relationship between predicted and observed value in model training set was good, and all data were concentrated near the trend line, which showed that this model had high repeatability and internal forecast ability. Therefore, 3D-QSAR model can be used to predict the compound activity with similar structure against C. difficile.

Equipotential diagram analysis of 3D-QSAR model The 3D force field coefficient equipotential diagrams by CoMFA method implied some important information for molecular modification.

The 3D equipotential diagrams of the electrostatic field and steric field were as shown in fig. 3. In the diagram A and C for 3D electrostatic force field coefficient equipotential diagram, red represented that the stronger the electro negativity of the area substituent, the better the compound activity. However, blue represented that the weaker the electro negativity of the area substituent, the better the compound activity. In the diagram B and D for 3D steric force field coefficient equipotential diagrams, yellow represented that the less the volume of the area substituent, the better the compound activity. However, green represented that the bigger the volume of the area substituent, the better the compound activity. The 3D equipotential diagram for the highest active compounds 1e was followed as fig. 3 (C and D).

4 DISCUSSION

In constructing a statistically significant predictive ability of the model, the model validation was indispensable. Non cross validation correlation coefficient r^2 was an important basis for testing the quality of QSAR model and the most commonly used internal inspection index, which can explain the deviation between molecular predicted value and observed value of training set. When r^2 was greater than 0.8 and the more close to 1, it was generally thought that the model has good predictability (Hansch and Verma, 2009). In the QSAR model, cross validation correlation coefficient q^2 was the statistical index of model predictability and usually used to judge the merits of the model, which meant that the higher the q^2, the stronger the predictability. When $q^2>0.3$, the established model in the 5% significant level has statistical significance, and while $q^2>0.5$, the established model has certain fitting ability and very remarkable statistical significance (Xu et al, 2004).

In constructing the 2D-QSAR model, 23 training set compounds and 98 Molecular descriptors such as hydrophobic parameters, electronic parameters, space configuration parameters, molecular character parameter, topological parameters and charge related parameters were used. Because of 23 compounds in training set, the QSAR model with 4–5 item parameters was constructed. After the initial population numbers was designed for 100 and the iterative 5 000 times, 10 GFA models were produced at the end of calculation. Compared with other models, r^2, r^2 (adj), r^2 (pred), RMS Residual Error, Friedman L.O.F. and S.O.R p-value for the selected optimal 2D-QSAR model has obvious improvement. From the formula (1), r^2 was 0.905 2 (>0.8), which indicated that this model could explain 90% of the variable. Adjustment coefficient r^2 (adj) and prediction correlation coefficient r^2 (pred) were more than 0.8, which indicated that the model has good internal forecast ability.

In order to validate the actual predictability of the 2D- QSAR model, 5 compounds in test set were used to forecast and evaluate the model. From predicted value, the model can predict accurately their activities. Moreover, the cross validation correlation coefficient q^2 for 0.371 (>0.3) indicated that the model credibility was more than 95%. Observed value of 28 compounds' activity and predicted value with GFA model were as shown in table 2, while linear relationship of observed value and predicted value in training set and test set were as shown in fig. 2 (A, B). The linear relationships between observed value and predicted value in in training set and test set were fitted well and their slopes were 0.905 and 0.751, which showed that GFA model selected had good internal predictability.

In constructing the 2D-QSAR model, over fitting phenomenon quite often happened. This usually resulted from too many parameters introduced, so only part parameters from the physical and chemical parameters can be selected. It is proposed that the ratio between the sample number and the number of parameters selected should be more than or equal to 5 in the less sample, while their ratio should be more than 2^n (n=parameter number taken) in the more sample, only so the appearance of over fitting phenomenon can be effectively avoided (Sharma et al, 2009). The model met the above principle, for example, the training set of sample number was for 23 and the model contained four parameters. In equation 1, all the parameters belonged to topological parameters and electrical parameters. Topological parameters reflected the characteristics of the molecular topology structure, generally including graphical description of molecular structure, the matrixing and quantification of graphics structure.

X_1 was molecular connectivity index, according to the atomic arrangement or connection order of each frame in the molecules to describe the nature of the molecular structure and with the reciprocal of square root for dot grade product to represent. It was the most representative topology parameter and reflected the molecular connectivity and branch nature. For the coefficient of the parameters, pMIC can be positive correlation with X_1, that was, the greater the molecular connectivity index, the better the compound activity, the smaller the MIC. From table 2, pMIC of compound 1e (0.684 15), 3 (0.686 72), 7 (0.684 15), 8 (0.836 64) with large X_1 were more than 0.390 8. Since X_1 coefficient was 4.947 1 and one of the biggest coefficient in equation 1, X_1 should be one of the most important parameters for activity. X_2 was molecular local charge surface parameter, combining molecular shape information with molecular charge information, and its numerical value was equal to polarizable surface area of all the negatively charged atomic solvent divided with all the negative charge, which was correlated to intestinal absorption of drug molecules. pMIC can be positive correlation with X_2, that was, the greater the molecular local charge surface parameter, the better the compound activity. For example, X_2 value of compound 9, 11, 12 were as followed: −1.344 23 < −1.248 19 <−1.092 26, and its activity also increased in turn. X_3 was molecular shape index and positive correlation with pMIC. Because only substituent (F, CF3) and different position differences in benzene ring of compound 1c (2.949 6), 1d (3.410 1), 1e (3.609 66) structure led to the shape index changes, the greater the X_3, the better the compound activity, so were compound 14, 15, 17. X_4 was molecular flexible index and negatively correlated with pMICs, which showed that the greater the X_4, the less the activity. For example, X_4 of compound 12, 14, 15, 16 were as followed: 2.749 79 < 3.037 42 < 5.001 98 <5.463 79, and the pMIC were as followed: 0.387 22 >0.130 77 >0.298 85 >0.903 09. In addition, X_4 of compound 1, 6, 1d, 8, 19, 17 were as followed: 3.417 9 <3.656 28 < 4.001 86 <4.444 97 <4.726 86 <5.671 4, and the pMIC were as followed: 0.903 09 >0.638 27 > 0.301 03 >0.309 8 >−0.181 84 >−0.902 6, which had the same rule.

3D-QSAR indirectly reflected nonbonding interaction characteristic between the ligand and receptor in the drug molecule action process, and had the rich connotation of physical chemistry.

Before constructing the 3D-QSAR model, different compounds were needed to molecule overlap. In the molecular field analysis, different molecular overlap method will lead to different calculation results, and it directly decided the credibility of the model. After comprehensive investigation of the molecular structure on this study system, the molecular field and static electric field were selected as overlap method. For model validation, when best principal component number for PLS was 3, maximum value for cross validation correlation coefficient q^2 was 0.744 and RMS error was minimum, which indicated that the model was highly credible and predictable.

From electrostatic field equipotential diagrams C, red area was distributed in the neighbouring of R_1, R_3 and R_4 and the surface of thiazole ring, and this indicated that electro negativity group in R_1, R_3 and R_4 can improve activity. For example, the pMIC of compound 1b greater than compound 5 was due to F as R_1 of compound 1b stronger electro negativity than H as R_1 of compound 5. The pMICs of compound 4, 6, 1e for 0.176 09, 0.638 27, 0.903 09 increased in turn. However, from their structure analysis, only a difference of R_3 for H, Cl, CF_3 in their structure and electro negativity increases successively showed that the activity of compounds increased with the enhancement of a substituent electro negativity in R3. The activity of compound 1b (R_1, R_3=F, R_2, R_4=H) was more than compound 1c (R_1, R_2=H, R_3, R_4=F) activity, and compound 1e (R_3=CF_3, R_1, R_2, R_4=H) activity was more than compound 1d (R_1, R_2=CF_3, R_3, R_4=H) activity. On the structure, the differences of compound 1b, 1c and 1d, 1e were F and CF_3 in the different position, respectively. This showed that electro negativity F as R1 had more biological activity than as R_4, and electro negativity CF_3 as R_3 had more biological activity than as R_1. It was concluded that the stronger electro negativity in R_3 can improve more biological activity than in R1 and at the same time it in R_1 was better than in R_4. Therefore, stronger electronegative substituent into R_3, R_1, R_4 in turn should be taken into account in designing ideal activity of thiazole nitazoxanide.

From steric field equipotential diagrams D, green area was mainly distributed in the neighbouring of R_1, R_4 and R_5, and this indicated that bigger volume than hydrogen atom in R_1, R_4 and R_5 can improve activity. For example, the only differences of compound 1a (R_1=OAc), 1 (R_1=OH), 4 (R_1=H) in structure were different substituent in R_1. As the substituent volume in R_1 became smaller, the pMIC of three compounds (0.602 06, 0.301 03, 0.176 09) also gradually became less and the activity weaker. Compound 7 (R_3=H) activity more than compound 8 (R_3=Cl) showed that larger volume Cl in R_3 can decrease the activity, and compound 1b (R_1=F, R_3=F) activity more than compound 5 (R_1=H, R_3=F), namely, was bulky F in the R_1 with improved activity.

In sum for steric field, stronger electro negativity and larger volume group in R_1, weaker electro negativity and smaller volume groups in R_2, stronger electro negativity and smaller volume group in R_3, can help to enhance compound activity against c. difficile. From table 1, H as R_1, R_2, R_3 and R_4 in benzene ring of compound 4 didn't enhance electro negativity in R_1, R_3 and R_4 position, and nor increased a steric hindrance in R_1 and R_4 position. Because of the worst accordance with CoMFA model, compound 4 was eventually one of 13 thiazole nitazoxanide derivatives with less activity.

5 CONCLUSION

The 2D and 3D QSAR for nitazoxanide-based analogues were studied with DS-QSAR software. Through the genetic function approximation method, a simple 2D- QSAR equation was generated: pMIC = 0.755 52+4.947 1 X_1+1.397 X_2+1.750 7 X_3−1.844 2 X_4 (model 1)

n = 23, r^2 = 0.905 2, r^2 (adj) = 0.884 2, r^2 (pred) = 0.840 2

The results showed that nitazoxanide-based analogues against C. difficile activity can be increased mainly through such as increase for the molecular connectivity index, increase for the molecular local charge surface parameters of the compound, increase for the compound molecular shape index and decrease for the compound molecular flexible index. The 3D model for r^2=0.967, q^2=0.744 was generated through CoMFA method. R_1, R_3 and R4 substituent were identified as the main factors for influencing compound activity by analyzing the equipotential diagram, and this was consistent with the results of 2D-QSAR model.

In a word, 2D and 3D-QSAR models with better predictability were constituted. They described the structure-activity relationship of nitazoxanide-based analogues and provided theoretical guidance for further designing new compounds with better activity.

（发表于《Pak J Pharm Sci》，SCI，IF：0.581．1 681−1 689.）

Evaluation of the Acute and Subchronic Toxicity of Ziwan Baibu Tang

XIN Ruihua[1], PENG Wenjing[1], LIU Xiaolei[1], LUO Yongjiang[1],
WANG Guibo[1], LUO Chaoying[1], XIE Jiasheng[1], LI Jinyu[1], LIANG Ge[2],
ZHENG Jifang[1*]

(1. Lanzhou Institute of Husbandry and Pharmaceutical Sciences of CAAS, Key Laboratory of New Animal Drug Project of Gansu Province, Engineering & Technology Research Center of Traditional Chinese Veterinary Medicine of Gansu Province, Key Laboratory of Veterinary Pharmaceutics Development, Ministry of Agriculture, Lanzhou 730050, China; 2. Sichuan Animal Science Academy, Chengdu 610066, China)

Abstract: [Background:] Ziwan Baibu Tang (ZBT) is used as a traditional Chinese prescription used in the clinic to cure a variety of respiratory diseases. The objective of this study was to evaluate the acute and subacute toxicity of ZBT. [Materials and Methods:] Fifty Kunming mice were assigned to 5 groups, and ZBT was administered at different concentrations. Symptoms, mortality and behavioural changes were recorded. Eighty Wistar rats were randomly allocated to 4 treatment groups or a control group for 4 weeks of treatment (half male and half female); general observations and weekly body weight (BW) gain were recorded. At the end of the 4-week experiment, the rats were sacrificed, and blood was collected for haematological and serum biochemical analyses. Samples from the primary organs were fixed in 10% buffered formalin for histopathological studies. [Results:] The calculated LD_{50} of acute oral toxicity was 319.16 g/kg BW. After the 4-week treatment, BW did not significantly differ in any of the dose groups, compared with the control group ($P>0.05$), and the rat blood red blood cell (RBC) and haemoglobin (HGB) levels in each dose group did not significantly differ ($P>0.05$). In addition, serum urea and creatine levels in each dose group did not significantly differ from those in the control group ($P>0.05$). [Conclusion:] These results indicate that ZBT does not possess toxic potential and that both the acute and subchronic toxicity in animals are very low.

Key words: Ziwan Baibu Tang; Acute; Subchronic; Toxicity; Herbal

1 INTRODUCTION

Herbs and herbal preparations have been used to treat disease throughout human history. According to World Health Organization (WHO) statistics, 70%–80% of the world's popula-

---

* Corresponding author, E-mail: xinruihuamys@126.com and zhengjifang100@126.com

tion employs plant-derived traditional treatment methods for health problems (Ahmad et al., 2006; Shirwaikar et al., 2009). Traditional Chinese medicine (TCM) has been used in China for thousands of years to diagnose and treat diseases (Chou et al., 2008). TCM theories are based on syndrome differentiation and holistic medicine. Chinese herbal medicine (CHM), the pharmaceutical and most important component of the TCM system, combines compounds in the form of processed natural products (Li et al., 2013) and is supported by extensive literature and clinical applications covering thousands of years (Xu et al., 2009). Acompound formulation is prescribed according to the principle, "Monarch, Minister, Assistant and Guide." Aformulation that contains more than two Chinese herbs conforms more closely to TCM theories and better reflects the characteristics of TCM than the administration of a single herb (Jiao et al., 2004). Studies of the effects of CHM formulas are attracting increasing attention globally (Shi et al., 2011).

Ziwan Baibu Tang (ZBT) is a traditional Chinese prescription that is now used in the clinic to cure a variety of respiratory diseases, such as pneumonia, respiratory tract infections, cough and phlegm (Cheng et al., 2014). ZBT consists of 7 crude drugs: Aster Tataricus, Stemona Japonica, Platycodon Grandiflorus, Cortex Mori, Herba Houttuyniae, Folium Eriobotryae and Glycyrrhiza. Aster Tataricus is the chief component in this prescription and is clinically used to eliminate phlegm and relieve cough in China. Aster Tataricus also possesses diuretic, antibacterial, antitumour, antiviral and anti-ulcer activities (Hou et al., 2006). Herba Houttuyniae is commonly used in TCM and provides antiviral, anti-inflammatory, antibacterial, and heat-clearing properties for the treatment of pneumonia and throat swelling; it has also been shown to increase the immune response in vivo and in vitro (Chen et al., 2014). Stemona Japonica (Zhu et al., 2010), Platycodon Grandiflorus (Song et al., 2006), Cortex Mori (Feng et al., 2004), Folium Eriobotryae (Ju et al., 2003) and Glycyrrhiza (Tian et al., 2006) are used traditionally as medicinal herbs for the treatment of respiratory tract infections in China, and several research studies have revealed that these herbs provide respiratory therapeutic effects (Chinese Pharmacopoeia Commission, 2010).

The safety of ZBT is generally assumed from its very long history of consumption in China. However, the consumption of plants and plant products whose content, toxicity profile and safe dose have not been determined may cause severe toxicity problems in humans and animals. Despite knowledge of the biological activities of ZBT, toxicological and side effect profiles have not been adequately documented. To provide scientific data for clinical drug safety, the present study investigated the acute and subchronic oral toxicity of ZBT by applying the recommended guidelines for safety or dose-dependent toxicity in mice and rats.

2 MATERIALS AND METHODS

2.1 Instruments and reagents

A haematology analyser (SYSMEX, pocH-100iV, Japan), thin semiautomatic microtome (Leica RM 2265), organization stand/dryer (Leica HI 1210), and automatic biochemistry analyser (Mindray Biomedical) were used. An aspartate aminotransferase kit (AST, 150211009), alanine aminotransferase kit (ALT, 150111013), albumin kits (ALB, 150909014), creatinine reagent kit (Creatinine, 151011012), urea kits (Urea, 151311013) and a creatine kinase kit (CK, 152511010) were provided by Mindray Biomedical Co., Ltd.

2.2 Preparation of ZBT extract

ZBT consists of 7 crude drugs; the plant components and their origins are listed in Table 1. Briefly, a mixture of 150 g of Aster Tataricus, 150 g of Stemona Japonica, 120 g of Platycodon Grandiflorus, 120 g of Cortex Mori, 180 g of Herba Houttuyniae, 150 g of Folium Eriobotryae and 130 g of Glycyrrhiza was macerated with 12 000 mL of distilled water for 2 h and decocted at 100℃ for 1.5 h. The filtrate was collected, and the residue was again decocted for 1.5 h in 10000 mL of distilled water. The extracts were combined and further condensed at 65℃ to obtain the aqueous extract of ZBT (1 000 mL).

Table 1 The composition of ZBT.

Species	Chinese name	Plant components	Origin	Grams(g)	%
Aster tataricus	Zi wan	Root, stem	Hebei, China	150	15
Stemona japonica	Bai bu	Root	Jiangxi, China	150	15
Platycodon grandiflorus	Jie geng	Root	Liaoning, China	120	12
Cortex mori	Sang bai pi	Rhizodermis	Henan, China	120	12
Herba houttuyniae	Yu xing cao	Whole herb	Sichuan, China	180	18
Folium eriobotryae	Pi pa ye	Leaves	Zhejiang, China	150	15
Glycyrrhiza	Gan cao	Root, stem	Xinjiang, China	130	13

2.3 Animals

Young mice weighing 18-22 g were purchased from the Animal Physiology Laboratory of Lanzhou University. Adult female and male Wistar rats (aged two months, weighing 181-199 and 185-205 g, resp.) were supplied by the production facility of the Animal Physiology Laboratory, Gansu University of Traditional Chinese Medicine (Gansu, China). The accession number was GTCM-150823. The animals were fed in a separate room with a barrier system under a controlled light-dark cycle (12-12 h, lights on 7: 00-19: 00), ventilation (air exchange rate of 18 cycles/h), temperature (23 ± 2℃) and relative humidity (55% ± 15%) during the study. The cages and chip bedding were changed twice each week (Wang et al., 2011). The experiments were initiated after acclimatizing the rats for 1 week. The study was performed in accordance with Veterinary Laboratory Biosafety Guidelines (The Chinese Ministry of Agriculture Guide, 2003).

2.4 Acute oral toxicity study in mice

Fifty healthy Kunming mice (equal numbers of males and females, 18-22 g) were randomly assigned to 5 groups. ZBT was dissolved at different concentrations in distilled water and administered to the Kunming mice via oral gavage at doses of 250.00, 300.00, 360.00, 432.00, and 518.40 g/kg body weight (BW) per day. All mice were fasted overnight to eliminate food from the gastrointestinal tract before dosing. After gavage, the mice were observed for 14 days to observe their symptoms, mortality, behavioural changes, skin, eyes, fur and somatic motor activity. Finally, the LD_{50} was calculated using Bliss software.

2.5 Subchronic oral toxicity test in rats

Eighty Wistar rats were assigned to 4 groups (high-dose group, medium-dose group, low-dose group, and control group; 20 rats in each group, half male and female). The treated groups received ZBT at a dose of approximately $10/LD_{50}$, $50/LD_{50}$ and $100/LD_{50}$ every day, and the control group received an equal dose of physiological saline between 9:00 a.m. and midnight every day for 4 weeks. The general observations for clinical signs included behavioural changes, signs of gross toxicity, fur condition and somatic motor activity. The animals were weighed on the first day of administration, and the individual BWs of the rats were recorded every subsequent week. The mean weekly BW gain was calculated for each sex and measured weekly. The rats were allowed unlimited access to food throughout the study.

At the end of the 4-week experiment, the animals were fasted for 12 h and then sacrificed by decapitation under anaesthesia with sodium pentobarbital (30 $mg.kg^{-1}$) intraperitoneally administered. Blood was collected in two tubes: tube 1, containing EDTA, was processed immediately for haematological parameters; tube 2, without additive, was centrifuged at 3 000 × g at 4℃ for 10 min to obtain serum (stored at −20℃ until analysis). The organs (heart, liver, spleen, lungs and kidneys) were weighed, and the absolute organ weights were converted to relative organ weights based on the organ-to-BW ratio. Organ samples were fixed in 10% formalin for histopathological examination.

2.6 Haematological and serum biochemical analysis

Haematology indexes were detected by pocH-100iV (haematology analyser and reagent) and included the white blood cell (WBC), red blood cell (RBC) and platelet (PLT) counts, as well as the haemoglobin (HGB) measurement. The following measurements were obtained: alanine aminotransferase (ALT), aspartate aminotransferase (AST), AST/ALT, albumin (ALB), creatine kinase (CK) and urea.

2.7 Histopathological studies

After weighing the collected organs, all samples were fixed in 10% buffered formalin. In the histological analysis, we assessed the tissue integrity of the organs and observed the presence of degeneration, necrosis, leukocyte infiltration, congestion, blood extravasation or fibrosis.

2.8 Statistical analysis

The LD_{50} was calculated using the Bliss method. The data for weekly BW, relative organ weights, haematology and serum biochemistry were evaluated by one-way ANOVA using SPSS 17.0 statistical software. All values are expressed as the mean ± SD. P values less than 0.05 were considered significant.

3 RESULTS

3.1 Acute oral toxicity

After administration of ZBT, the mice in the high-dose group lay quietly and rarely moved. In addition, the mice exhibited reduced spontaneous activity and slow movement. Few mice developed movement disorders, shortness of breath or obvious neurological symptoms. As the dose of ZBT increased, the symptoms intensified. Prior to death, the mice exhibited symptoms such as hind limb twitching, disordered hair coat and difficulty breathing. After 7 d, the surviving mice exhibited good appetite, improved vitality and normal movement. The hair coat of the surviving mice became smooth and shiny. Death mainly occurred within 1-2 h following oral gavage of ZBT. The mice were dissected and examined. The following major organ lesions were observed: duodenal haemorrhage, liver darkening and inconspicuous bleeding and bruising. No significant gross lesions were detected in other organs. As shown in Table 2, 0, 4, 9, 9 and 10 deaths were respectively observed in the 5 groups of mice within 7 d after administration of ZBT. The LD_{50} of ZBT in mice and the 95% confidence limits were calculated by the Bliss method using SPSS 13.0 software. The LD_{50} was 319.16 g/kg BW, and the 95% confidence interval was 270.31–363.68 g/kg BW.

Table 2 Mortality of the mice in the acute toxicity experiment.

Group	Dose (g/kg BW)	Number of mice	Mortality
1	250.00	10	0
2	300.00	10	4
3	360.00	10	9
4	432.00	10	9
5	518.40	10	10

3.2 Observation of gross toxicity and mortality

Experimental observations of the rats indicated that there were no abnormal changes in the experimental groups or control group in terms of food intake, water intake, mental state, respiratory condition, and coat gloss or body temperature. No rats died during the experimental period.

3.3 Body weight changes and organ weights

As shown in Tables 3 and 4, during the experimental period, there were no significant differences in BW between the control group and the groups of rats that received different doses of ZBT ($P>0.05$). Compared with the control group, female rats in the low-dose group exhibited significantly decreased heart weight/BW coefficient ($P<0.05$). By contrast, no significant differences in the heart weight/BW coefficient were observed between the rest of the groups and the control group ($P>0.05$). In addition, there were no significant differences between the control group and the various dose groups in the organ weight/BW coefficients of the liver, spleen, lung and kidney ($P>0.05$).

Table 3 Weekly body weight gain in subchronic oral administration test.

Groups	1 week	2 weeks	3 weeks	4 weeks
Female				
Vehicle	188.05 ± 7.52	203.55 ± 5.12	217.78 ± 6.77	232.07 ± 6.49
3 g/kg BW	188.10 ± 6.91	201.15 ± 5.35	215.50 ± 6.96	230.00 ± 7.22
15 g/kg BW	185.50 ± 8.63	200.75 ± 8.18	214.82 ± 9.50	230.20 ± 8.82
30 g/kg BW	182.65 ± 5.43	194.47 ± 5.79	205.44 ± 6.99	217.06 ± 7.11
Male				
Vehicle	202.87 ± 9.34	218.10 ± 8.43	235.11 ± 7.98	251.96 ± 8.08
3 g/kg BW	200.02 ± 5.63	216.23 ± 6.03	232.42 ± 6.45	247.10 ± 7.51
15 g/kg BW	199.84 ± 10.09	214.58 ± 9.79	229.33 ± 10.09	243.09 ± 9.89
30 g/kg BW	197.55 ± 6.67	209.41 ± 7.00	221.74 ± 7.16	232.90 ± 8.01

Values are means ± SEM for 20 rats in each group, half male and half female.

Table 4 Effects of subchronic oral administration on relative organ weights.

Group	Heart	Liver	Spleen	Lung	Kidney
Female					
Vehicle	0.68 ± 0.02	6.39 ± 0.66	0.37 ± 0.01	1.01 ± 0.03	0.51 ± 0.01
3 g/kg BW	0.62 ± 0.02	6.47 ± 0.83	0.35 ± 0.01	0.96 ± 0.03	0.48 ± 0.01
15 g/kg BW	0.62 ± 0.02	6.71 ± 0.91	0.34 ± 0.01	0.95 ± 0.03	0.48 ± 0.01
30 g/kg BW	0.58 ± 0.02*	6.25 ± 0.66	0.36 ± 0.01	0.96 ± 0.02	0.48 ± 0.01
Male					
Vehicle	0.69 ± 0.02	6.56 ± 0.83	0..41 ± 0.01	1.12 ± 0.04	0.55 ± 0.01
3 g/kg BW	0.68 ± 0.02	6.78 ± 0.75	0.40 ± 0.01	1.06 ± 0.03	0.54 ± 0.01
15 g/kg BW	0.69 ± 0.03	6.65 ± 0.88	0.39 ± 0.01	1.15 ± 0.04	0.53 ± 0.01
30 g/kg BW	0.69 ± 0.02	6.43 ± 0.56	0.40 ± 0.01	1.10 ± 0.03	0.53 ± 0.01

Values are means ± SEM for 20 rats in each group, half male and half female. Statistically significant compared to controls (*P<0.05).

3.4 Haematology

The haematological profiles of the treated and control groups are presented in Table 5. After continuous dosing for 4 weeks, WBC levels decreased significantly (P<0.05) in females in the high-dose group compared with the control group, and PLT levels significantly decreased (P<0.05) in both sexes in the high-dose group. No significant differences in RBC and HGB levels were observed in the dose groups (P>0.05).

Table 5 Haematological parameters in Wistar rats treated orally with ZBT for 4 weeks.

Group	WBC/($\times 10^9 \cdot L^{-1}$)	RBC/($\times 10^{12} \cdot L^{-1}$)	HGB/g·L^{-1}	PLT/($\times 10^9 \cdot L^{-1}$)
Female				
Vehicle	13.59 ± 2.89	8.41 ± 2.14	171.38 ± 23.04	420.88 ± 60.63
3 g/kg BW	12.53 ± 2.48	8.56 ± 2.13	172.69 ± 21.19	441.47 ± 65.21
15 g/kg BW	11.98 ± 3.99	8.24 ± 1.27	168.53 ± 43.90	524.56 ± 56.78
30 g/kg BW	10.43 ± 2.89*	8.60 ± 1.09	175.71 ± 19.73	596.50 ± 74.15*
Male				
Vehicle	10.73 ± 0.68	9.78 ± 1.12	177.64 ± 13.33	232.08 ± 56.28
3 g/kg BW	8.84 ± 0.43	8.34 ± 2.14	170.87 ± 24.02	257.87 ± 88.71
15 g/kg BW	10.69 ± 0.64	8.18 ± 1.16	166.00 ± 25.29	279.13 ± 46.42
30 g/kg BW	9.29 ± 0.64	7.86 ± 2.13	158.44 ± 21.32	384.27 ± 62.07*

Values are means ± SEM for 20 rats in each group, half male and half female. Statistically significant compared to controls *P<0.05, **P<0.01.

3.5 Serum chemistry

After 4 weeks of continuous administration of ZBT, no significant differences in urea levels were observed in the control group and the various groups of rats that received different doses of ZBT (P>0.05). In addition, there were no significant differences in serum creatinine levels between the control group and the low-dose group or between the control group and the moderate-dose group (P>0.05). Compared with the control group, female rats in the high-dose group exhibited significantly decreased serum ALB levels (P<0.05). Serum ALB levels were also markedly reduced in male rats in the moderate-dose group and high-dose group compared to the control group (P<0.05). No significant differences in serum AST activity were detected between the control group and various dose groups (P>0.05). Compared with the control group, male rats in the moderate-dose group exhibited significantly elevated CK activity (P<0.05). By contrast, there were no significant differences in CK activity between the control group and the low-dose or high-dose group (P>0.05). No dose-response relationship between serum CK activity and ZBT dose was observed (Table 6).

Table 6 Serum biochemistry in Wistar rats treated orally with ZBT for 4 weeks.

Group	AST (U/L)	AST/ALT	ALB (g/L)	Urea (mmol/L)	Creatinine (μmol/L)	CK (μmol/L)
Female						
Vehicle	318.50 ± 19.07	2.61 ± 0.15	34.04 ± 0.25	7.13 ± 0.28	73.83 ± 2.52	5 373.67 ± 198.71
3 g/kg BW	358.63 ± 16.52	3.06 ± 0.11	34.21 ± 0.29	6.99 ± 0.22	70.71 ± 1.21	4 570.75 ± 379.30
15 g/kg BW	241.88 ± 19.28	2.37 ± 0.15	33.59 ± 0.37	7.14 ± 0.14	70.88 ± 1.67	4 811.88 ± 394.28
30 g/kg BW	246.10 ± 11.29	2.69 ± 0.15	30.94 ± 0.26*	6.04 ± 0.29	62.20 ± 1.07	4 777.80 ± 296.45

(continued)

Group	AST (U/L)	AST/ALT	ALB (g/L)	Urea (mmol/L)	Creatinine (μmol/L)	CK (μmol/L)
			Male			
Vehicle	239.25 ± 7.17	2.22 ± 0.11	35.10 ± 0.14	6.67 ± 0.26	80.33 ± 1.65	3 704.44 ± 504.07
3 g/kg BW	268.14 ± 26.44	2.13 ± 0.17	34.34 ± 0.28	7.69 ± 0.27	72.88 ± 1.44	3 539.88 ± 497.87
15 g/kg BW	284.71 ± 23.26	2.25 ± 0.12	33.14 ± 0.33*	6.26 ± 0.26	69.29 ± 1.80	5 297.57 ± 306.58*
30 g/kg BW	199.50 ± 3.77	1.82 ± 0.20	33.93 ± 0.61*	6.54 ± 0.34	68.40 ± 2.40	3 260.50 ± 700.20

Values are means ± SEM for 20 rats in each group, half male and half female. Statistically significant compared to controls *P<0.05, **P<0.01.

3.6 Histopathology

In the histological investigation, organs including the heart, liver, spleen, lung and kidney exhibited no sign of pathological changes compared with the corresponding organs in the control group (Fig. 1).

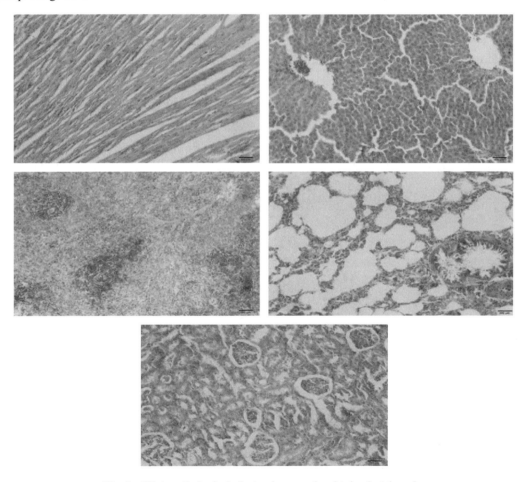

Fig. 1 Histopathological photomicrographs obtained at 4 weeks
Liver 200×, Heart 400×, Spleen 200××, Lung 200×, Kidney 200×

4 DISCUSSION

Safety studies of plants and plant products and establishing the effectiveness and safety of plants and plant products via scientific research are increasingly emphasized. The complexity of herbal preparations and their natural biological variations requires their safety, effectiveness and quality to be established (Shin et al., 2011; Lee et al., 2012). ZBT is a traditional Chinese prescription that is used in clinical practice. ZBT was developed by modifying Zi Wan San, an ancient medical prescription collected in "Tai Ping Sheng Hui Fang" (The Peaceful Holy Benevolent Prescriptions). ZBT consists of 7 Chinese herbs, including Aster Tataricus, Stemona Japonica, Platycodon Grandiflorus and Cortex Mori. Aster Tataricus relieves cough and asthma, clears away lung-heat, moistens intestines and relaxes bowels (Hou et al., 2006). Cortex Mori purges the lung of pathogenic fire, relieves the symptoms of asthma and induces diuresis to alleviate oedema (Feng et al., 2004). Herba Houttuyniae opens inhibited lung-energy, stops coughing and clears away heat and toxic substances (Chen et al., 2014). Glycyrrhiza uralensis expels phlegm, stops coughing, relieves acute symptoms, removes toxic substances and coordinates the effects of various drugs in the prescription (Tian et al., 2006). Therefore, ZBT produces effects such as relief of cough and asthma, heat clearing and detoxification. In addition, as natural medicines, TCMs possess unique advantages, including anti-stress activity, improvement of nonspecific immunity and regulation of metabolism. Combined with these unique advantages, ZBT exhibits good clinical effectiveness in the treatment of bronchial asthma, phlegm-heat panting and cough.

There were no significant differences in BW between rats in the control group and the groups exposed to different doses of ZBT during the experimental period ($P>0.05$). However, the BW of the rats in the various dose groups decreased in a dose-dependent manner, indicating that high doses of ZBT impact the appetite of rats. Increasing the dose of ZBT may have reduced food intake by the rats. Consequently, the control group exhibited greater weight gain, but the effect of ZBT on BW was not significant. Compared with the control group, female rats in the high-dose group exhibited a markedly reduced heart weight/BW coefficient ($P<0.05$). By contrast, no significant differences in the heart weight/BW coefficient were detected between the control group and the low- or moderate-dose group ($P>0.05$). The coefficient of organ weight/BW is a crucial detection index in the evaluation of drug safety and provides a quantitative reference for drug evaluation. However, it is impossible to identify changes in organs based on the coefficients alone (Yuan et al., 2003). Combining these coefficients with the results of a series of blood biochemical indices in the rats revealed no significant differences in AST and CK between the control group and various dose groups ($P>0.05$). Histopathological examination of heart tissue sections from the rats in the various dose groups revealed no abnormalities, indicating that ZBT had little effect on the hearts of female rats at doses of less than 30 g/kg BW.

The haematopoietic system can serve as an indicator of toxic effects because of its sensitivity to toxic compounds (Adeneve et al., 2006). Compared to the control group, the groups that received different doses of ZBT exhibited varying decreases in WBC levels in the blood. Among the ingredients of ZBT, the medicinal herb Aster tataricus contains shionone, a type of flavonoid. Shionone promotes S-phase (the phase of DNA synthesis) entry in mouse Kupffer cells and significantly increases the number of cells in the mitotic phase, indicating that shionone is capable of promoting the proliferation of liver Kupffer cells and activating the function of Kupffer cells. In addition, shionone significantly promotes the phagocytic activity of lung macrophages and peritoneal macrophages (Li et al., 1991). In the present study, no gross abnormalities were detected in various organs and tissues by histopathological examination. Therefore, ZBT does not appear to affect WBCs in the blood of rats. During the experimental period, blood PLT counts were significantly increased in both female rats and male rats exposed to high doses of ZBT compared with the control group. This result is consistent with findings by Chen et al. that adrenocortical hormones stimulate the haematopoietic activity of bone marrow and may increase the number of platelets (Chen et al., 2003). Compared to the control group, the male rats in the various dose groups exhibited significantly decreased blood RBC and HGB levels ($P>0.05$). One of the components of ZBT, Glycyrrhiza uralensis, contains enoxolone. The adrenocortical hormone-like activity of enoxolone likely initiates negative feedback regulation in vivo, thereby inhibiting the production of RBC and HGB. The specific mechanism remains to be further investigated. The results of the present study indicate that ZBT had little effect on the WBC, RBC, HGB and PLT levels in the blood of rats.

Most drugs cause liver damage during transformation in the liver. The extent of damage is related to the dosage and concentration of the drugs. Liver damage often leads to increased ALT and AST activity and elevated ALB content (Wang et al., 2010). However, because the increases in AST and ALT were not necessarily parallel and transaminase activity only reflects liver disease activity, the AST/ALT ratio is used as an index to evaluate liver health. The AST/ALT ratio is a highly sensitive index and accurately reflects the degree of liver damage and prognosis (Wang et al., 2001). Urea and creatinine are two indices commonly used to evaluate renal function. Under normal conditions, the production and excretion of urea and creatinine are in a state of dynamic equilibrium. However, when the kidney undergoes pathological changes, serum urea and creatinine levels significantly change. AST and CK are abundantly expressed in the myocardium. Under normal conditions, the activities of AST and CK are very low in the serum. When the myocardial tissue is injured, cell permeability increases. Large amounts of AST and CK enter the blood, resulting in significantly increased AST and CK activities. Serum AST and CK measurements exhibit high sensitivity and specificity and are recognized as the "gold standard" for the detection of cardiomyocyte damage (Meng et al., 2005).

Most drugs cause liver damage during transformation in the liver. The extent of damage is related to the dosage and concentration of the drugs. Liver damage often leads to increased ALT and AST activity and elevated ALB content (Wang et al., 2010). However, because the increases in AST and ALT were not necessarily parallel and transaminase activity only reflects liver disease activity, the AST/ALT ratio is used as an index to evaluate liver health. The AST/ALT ratio is a highly sensitive index and accurately reflects the degree of liver damage and prognosis (Wang et al., 2001). Urea and creatinine are two indices commonly used to evaluate renal function. Under normal conditions, the production and excretion of urea and creatinine are in a state of dynamic equilibrium. However, when the kidney undergoes pathological changes, serum urea and creatinine levels significantly change. AST and CK are abundantly expressed in the myocardium. Under normal conditions, the activities of AST and CK are very low in the serum. When the myocardial tissue is injured, cell permeability increases. Large amounts of AST and CK enter the blood, resulting in significantly increased AST and CK activities. Serum AST and CK measurements exhibit high sensitivity and specificity and are recognized as the "gold standard" for the detection of cardiomyocyte damage (Meng et al., 2005).

After administration of ZBT for 4 weeks, serum ALB levels did not significantly differ between the control group and female rats in the low-dose group or moderate-dose group ($P>0.05$). By contrast, serum ALB levels were significantly reduced in the high-dose group compared with the control group ($P<0.05$). However, there were no significant differences in the AST/ALT ratio between the control group and the various dose groups ($P>0.05$). These results indicate that ZBT did not cause liver damage in the female rats. Urea levels in the female and male rats that received various doses of ZBT did not significantly differ from those in the control group ($P>0.05$). In addition, creatinine levels did not significantly differ between the control group and the various dose groups ($P>0.05$). These results indicate that 4 weeks of continuous administration of ZBT had little effect on the kidney in rats. Compared with the control group, the high-dose group exhibited markedly reduced serum AST levels ($P<0.05$). However, CK content did not significantly differ among the groups ($P>0.05$). The combined experimental results indicate that ZBT did not induce myocardial injury in the female rats. Serum ALB levels indicated that compared to the control group, 4 weeks of continuous administration of ZBT did not cause liver damage in rats. CK levels were drastically increased in the moderate-dose group compared to the control group ($P<0.05$). However, there were no significant differences in serum AST activity between the control group and the low-, moderate- or high-dose groups ($P>0.05$). The above results suggest that ZBT had little effect on the rat myocardium.

5 CONCLUSION

The results of the present study demonstrate that the oral administration of ZBT has no significant impact on rat organs. After 4 weeks of continuous administration, the high-dose group (30 g/kg BW) was exposed to ZBT at a level far exceeding the recommended dosage in clinical applications. Therefore, ZBT should be considered an effective and highly safe Chinese herb medicine suitable for long-term clinical application.

Conflict of Interests: The authors of this paper declare no conflict of interest.

Authors' Contributions: Rui-Hua Xin and Wen-Jing Peng equally contributed to this work.

ACKNOWLEDGEMENTS

This work was financially supported by the Special Fund for Agro-Scientific Research in the Public Interest (No. 201303040-18).

(发表于《AFR J TRADIT COMPLEM》, SCI, IF: 0.553. 140–149.)

Evaluation of the Acute and Subchronic Toxicity of *Aster tataricus* L. F.

PENG Wenjing[1], XIN Ruihua[1], LUO Yongjiang[1], LIANG Ge[2],
REN Lihua[1], LIU Yan[1], WANG Guibo[1], ZHENG Jifang[1*]

(1. Lanzhou Institute of Husbandry and Pharmaceutical Sciences of CAAS, Key Laboratory of New Animal Drug Project of Gansu Province, Engineering & Technology Research Center of Traditional Chinese Veterinary Medicine of Gansu Province, Key Laboratory of Veterinary Pharmaceutics Development, Ministry of Agriculture, Lanzhou 730050, China; 2. Sichuan Animal Science Academy, Chengdu 610066, China)

Abstract: [Background:] *Aster tataricus* L. f. is used as a traditional Chinese drug to relieve cough and asthma symptoms and to eliminate phlegm. However, Aster tataricus L. f. possesses toxicity, and little systematic research has been conducted on its toxic effects in the laboratory. [Methods and Materials:] The acute group was administered 75% alcohol extract of *Aster tataricus* L. f. in a single dose. A subchronic toxicity study was performed via daily oral administration of *Aster tataricus* L. f. at a dose of 0.34 g/kg body weight in SD rats. The rats were divided into six groups: a petroleum ether extract (PEA) group, an ethyl acetate extract (EEA) group, an n-butyl alcohol extract (NEA) group, a remaining lower aqueous phases (REA) group, a 75% alcohol extract (AEA) group and a control group. Quantitative measurements of cytokines were obtained by fluorescence with a laser scanner using a Cy3 equivalent dye. [Results:] The LD_{50} of the 75% alcohol extract of Aster tataricus L. f. was 15.74 g/kg bw. In the subchronic toxicity study, no significant differences were observed among groups in relative organ weights, urine traits, liver antioxidase levels, or cytokine levels. However, significant sporadic differences were observed in body weight gains, haematology indices, biochemistry values, and histopathology features in PEA, EEA group. In addition, sporadic changes in other groups in measures such as WBC, MCHC, CK, ALP, AST, ALT, LDH, T-BIL, LDL-C, HDL-C, and TC were observed. [Conclusion:] The toxicity study showed that *Aster tataricus* L. f. can produce toxic effects, mainly on the liver; much less on the heart. The LD_{50} was 15.74 g/kg BW in mice, and the subchronic toxicity study, used a dosage of 0.34 g/kg/d.BW, showed that the toxic components of *Aster tataricus* L. f. were mainly concentrated in the petroleum ether fraction, followed by the ethyl acetate fraction, the n-butyl alcohol fraction, the lower aqueous phase and the 75% ethanol extracts.

Corresponding authors, E-mail: pengwenjingzy@163.com, xinruihuamys@126.com, zhengjifang100@126.com
Wen-jing Peng and Rui-hua Xin equally contributed to this work

Key words: *Aster tataricus* L. f.; Acute toxicity; Subchronic toxicity; Cytokines; Hepatic injury

Abbreviations: PEA, petroleum ether extract of *Aster tataricus* L. f.; EEA, ethyl acetate extract of *Aster tataricus* L. f.; NEA: n-butyl alcohol extract of *Aster tataricus* L. f.; REA: lower aqueous phases of *Aster tataricus* L. f.; AEA, 75% alcohol extract of *Aster tataricus* L. f.; WBC, white blood cell; RBC, red blood cell, PLT, platelet; HCT, haematocrit; MCV, mean corpuscular volume; HGB, haemoglobin; MCH, mean corpuscular haemoglobin; MCHC, mean corpuscular haemoglobin concentration; CREA, creatinine; LDH, lactate dehydrogenase; HDL-C, high-density lipoprotein cholesterol; LDL-C, low-density lipoprotein cholesterol; T-BIL, total bilirubin; ALT, alanine aminotransferase; ALP, alkaline phosphatase; AST, aspartate aminotransferase; TP, total protein; ALB, albumin; Glu, glucose; TC, total cholesterol; TG, triglycerides; CK, creatine kinase; GSH, Glutathione; MDA, malondialdehyde; T-SOD, total superoxide dismutase; TNF, tumour necrosis factor; IFN, interferon; MCP, monocyte chemotactic protein C.

1 INTRODUCTION

Aster tataricus L. f., a perennial herb of the Asteraceae (Morita et al., 1996), is widely distributed across Asia, Europe and North America. Majority of Aster species grow in low-lying wetlands, hills and low mountain meadows and marshes. There are an estimated 250 species of Aster worldwide, including 100 species that occur in China. A variety of compounds have been isolated from *Aster tataricus* L. f., including coumarins, flavonoids, anthraquinones (Sawai et al., 2011), sterols, peptides, terpenes (Akihisa et al., 1999), sesquiterpenes, astersaponin (Zhou et al., 2014), a triterpene and epifriedelinol (Fujioka et al., 1997; Shirota et al., 1994). A number of volatile oil constituents have been identified in *Aster tataricus* L. f. by gas chromatography (Tori et al., 2001).

However, in China, shionone is known as the main triterpenoid component of *Aster tataricus* L. f. The dried roots of this plant have been used in traditional Chinese medicine to eliminate phlegm and relieve cough (Sawai et al., 2011). *Aster tataricus* L. f. has been used as an expectorant, and it possesses diuretic, antibacterial, antitumour, antiviral and anti-ulcer activities (Morita et al., 1996; Shao et al., 1997; Wang et al., 1998). Yen et al. (1998) demonstrated that emodin, a component of *Aster tataricus* L. f., has antioxidant properties. Ng et al. (2003) showed that quercetin and kaempferol are effective antioxidants for inhibiting erythrocytic haemolysis and brain lipid peroxidation and that quercetin, kaempferol, scopoletin and emodin can inhibit the formation of superoxide free radicals. Additionally, terpene can cause cytotoxicity in tumour cells by inducing tumour cell apoptosis and DNA mutations (Zhou et al., 2014). In recent years, studies of the pharmacological activity of *Aster tataricus* have focused on cough relief and phlegm elimination or its anti-tumour effects, with few assessments of other pharmacological activities and no reports on liver toxicity associated with Aster. Considering its pharmacological activities and potential health benefits, and the lack of toxicological studies, there is a pressing need to clarify the toxicological profile of *Aster tataricus* L. f. Thus, we evaluated the LD50 and potential toxic effects of a 75% alcohol extract and four organic agentia extracts (petroleum ether extract, ethyl acetate extract, n-butyl alcohol extract, and aqueous phases) of *Aster tataricus* L. f. over 91 days to identify the extract with the greatest toxicity and determine the relationship between the toxicity and the duration of oral dosing. Haematology, biochemistry, urinalysis, liver biochemistry, histopathology, inflammation array, weight change and organ-to-body weight ratio were evaluated.

2 MATERIALS AND METHODS

2.1 Materials and chemicals

Urine was examined by using a KNF-100 Automated Urine Chemistry Analyser (Uritest Inc., China). An automatic biochemical analyser (BS-420) and biochemical kits were purchased from ShenZhen Mindray Bio-medical Electronics Co., Ltd. An automatic microplate reader was provided by Molecular Devices, Inc. (USA). Antioxidant enzyme kits were purchased from Nanjing Jiancheng Bioengineering Institute (Nanjing, China), and an inflammatory cytokines kit and reagents were provided by RayBiotech, Inc. (Guangzhou, China). Ethanol, petroleum ether, ethyl acetate extract, and n-butyl alcohol were purchased from Fuyu Chemical Co., Ltd. (Tianjin, China).

2.2 Animals

Young mice weighing 18-22 g were purchased from the Animal Physiology Laboratory of Lanzhou University, and a total of 36 male and 36 female 5-week-old Sprague-Dawley (SD) rats were supplied by the production facility of the Animal Physiology Laboratory, Gansu University of Traditional Chinese Medicine (Gansu, China). The accession number is GTCM-140823. The animals were fed in a separate room with a barrier system under a controlled light-dark cycle (12-12 h, lights on 7:00-19:00), ventilation (air exchange rate of 18 cycles/h), temperature (23 ± 2 ℃) and relative humidity ($55\% \pm 15\%$) during the study. The cages and chip bedding were changed twice per week. The experiments began after the rats had acclimatized for one week. This study was performed in accordance with the rulings of Gansu Experimental Animal Center (Gansu, China) approved by the Ministry of Health, P.R. China, in accordance with NIH guidelines.

2.3 Plant material

All the Aster tataricus root material used in this study were collected from The Yellow River Medicine Market (GanSu, China). Plant samples were identified by Professor Jifang Zheng at the Lanzhou Institute of Husbandry and Pharmaceutical Sciences of CAAS (Sample No. TONE-140512).

One thousand grams of *Aster tataricus* L. f. was weighed and added to water-diluted ethanol (purified water 25% v/v and ethanol 75% v/v) by three iterations of steam distillation over 2 hrs, filtered, and concentrated to remove the alcohol at 45 ℃ in a rotary evaporator. Next, an equal amount of ether petroleum (60-90 ℃) that had been extracted 3 times was added to the 75% ethanol extract using a merged extraction and concentrated to a paste with a weight of 6.885 g

(0.68% yield) to generate a petroleum ether extract of Aster tataricus L. f. (PEA) (Ye et al., 2014).

The lower aqueous phases were extracted 3 times with an equal amount of ethyl acetate using a merged extraction and concentrated to a paste with a weight of 13.529 g (1.35% yield) to generate an ethyl acetate extract of *Aster tataricus* L. f. (EEA). The remaining aqueous phases were extracted 3 times with equal n-butyl alcohol using a merged extraction and concentrated to a paste with a weight of 42.493 g (4.25% yield) to generate an n-butyl alcohol extract of *Aster tataricus* L. f. (NEA). The remaining lower aqueous phases were evaporated to a paste under reduced pressure to obtain 397.045 g (39.70% yield) to generate REA. (PEA, EEA, NEA, REA extract from AEA)

Aster tataricus L. f. (100 g) was extracted with 75% ethanol (400 mL of solvent ×3, 2 hrs per extraction) using a reflux unit, and the filtered solutions were combined and evaporated to a paste under reduced pressure at 40 ℃ to eliminate the alcohol, resulting in 48.002 g of alcohol extract (48.00% yield), or AEA. AEA was kept at -20℃ until use and suspended in distilled water.

2.4 Acute oral toxicity in mice

Sixty healthy Kunming mice (equal numbers of males and females, weighing 18-22 g) were randomly divided into 6 groups, with 5 males and 5 females in each group. AEA was dissolved to different concentrations in distilled water and administered to the Kunming mice via oral gavage at doses of 25.08, 20.81, 17.27, 14.34, 11.90 or 9.88 g/kg body weight per day at a dose ratio of 0.83 according to Research Methodology of Traditional Chinese Medicine Pharmacology (People's Medical Publishing House, 2006) (Chen, 2006). All of the mice were fasted overnight to eliminate food from the gastrointestinal tract before dosing. After gavage, the mice were observed for 14 days to observe symptoms, mortality, and changes in behaviour, skin, eyes, fur and somatic motor activity. After the experiment, all of the mice were euthanized, and their vital organs were individually observed for overt pathology by necropsy. Finally, the LD_{50} was calculated using SPSS 17.0 software.

3 SUBCHRONIC ORAL TOXICITY TEST

3.1 Experimental design

The five extracts were ground in Tween-80, dissolved in 0.5% water carboxymethyl cellulose Na and administered to the rats at a dose of 0.34 g original drug/kg/d.BW. Tween-80 and 0.5% water carboxymethyl cellulose Na were chosen because of their inert nature and the insolubility of PEA, EEA, and NEA. The rats were divided into 6 groups (PEA group, EEA group, NEA group, REA group, AEA group, and Control group). The treatment groups received Aster tataricus L. f. extract at a dose of approximately 0.34 g/kg body weight ($50/LD_{50}$), and the control group received an equal dose of 0.5% carboxymethyl cellulose Na. Both the treatment and control groups were dosed daily between 9:00 am and 12:00 am for 13 consecutive weeks.

Urine samples were collected for analysis at the end of the experiment. On the last day of the experiment, the animals were fasted overnight. Before collecting blood, the rats received a 3% sodium pentobarbital solution. Blood was collected on day 92 in 10% ethylenediaminetetraacetic acid (EDTA) tubes for haematology and in non-oxalate tubes for serum separation. The abdominal cavity was opened to dissect out and weigh the liver, lungs, kidneys, heart, spleen, thymus, testes and ovaries.

3.2 Observation of gross toxicity and urine examination

All the animals were observed twice daily for mortality. General observations were made for clinical signs including behavioural changes, signs of gross toxicity, fur condition and somatic motor activity, and these observations were performed once daily during the subchronic toxicity study.

On day 91, each rat was placed in a metabolic cage from 2:00-4:00 pm to collect urine. Urine colour, turbidity, pH, specific gravity (USG), protein, nitrite, bilirubin, urobilinogen, ketones, WBC, vitamin C and glucose levels were analysed using a KNF-100 Automated Urine Chemistry Analyser.

3.3 Body weight and food consumption

The animals were weighed on the first day of administration, and then individual body weights and food consumption were recorded every day. The mean weekly body weight gain and food consumption were calculated for each sex and measured weekly. The rats were allowed unlimited access to food throughout the study.

3.4 Macroscopic examination and relative organ weights

The rats were euthanized by administering sodium pentobarbital at dose of 30 mg/kg BW. The abdominal cavity of each rat was opened to observe the position, shape, size, colour, and consistency of the organs. Gross lesions were examined in all of the animals in all groups. The animals were eviscerated concomitantly. The liver, kidneys, spleen, lungs, heart, thymus, ovaries and testes were weighed. A portion of the liver (approximately 0.2-0.5 g) was immediately preserved at −80 ℃ for liver biochemical and cytokine analyses. The absolute organ weights were converted to relative organ weights based on the organ-to-body weight ratio.

3.5 Haematology and serum biochemistry

After the collection of blood in anticoagulant tubes on day 92, haematology indices were detected by using a pocH-100iV haematology analyser and reagents. The indices consisted of white blood cell (WBC), red blood cell (RBC), and platelet (PLT) counts; haematocrit (HCT); mean corpuscular volume (MCV); haemoglobin (HGB); mean corpuscular haemoglobin (MCH); and mean corpuscular haemoglobin concentration (MCHC).

Serum was isolated from non-anticoagulated tubes by centrifugation (Xiang Yi L-550, Changsha, China) at 3 000 r/min for 15 min. Measurements of the following were obtained: creatinine (CREA-j), lactate dehydrogenase (LDH), high-density lipoprotein cholesterol (HDL-C), low-density lipoprotein cholesterol (LDL-C), total bilirubin (T-BIL), alanine aminotransferase (ALT), alkaline phosphatase (ALP), aspartate aminotransferase

(AST), total protein (TP), albumin (ALB), urea, glucose (GLU), total cholesterol (TC), triglycerides (TG), and creatine kinase (CK).

3.6 Liver biochemical analysis

The supernatant was isolated from the liver-water suspension (volume ratio liver: water=1:9). GSH, MDA, and T-SOD tests were conducted with the liver supernatant according to the instructions provided with the kits.

3.7 Histopathology

After weighing the collected organs, all of the samples were fixed in 10% buffered formalin. The histological analysis aimed to assess the tissue integrity of the organs and to observe the presence of degeneration, necrosis, apoptosis, leukocyte infiltration, congestion, blood extravasation or fibrosis.

3.8 Quantitative measurements of liver inflammation using an array

Cytokines (IL-1α, IL-1β, IL-2, IL-4, IL-6, IL-10, IL-13, MCP-1, IFN γ, TNFα) were detected by fluorescence with a laser scanner using a Cy3 equivalent dye (Watanabe et al., 2005) between the control group, PEA and EEA group. Hepatic tissues were homogenized with 10 L protease inhibitors and 990 L 1X Cell Lysis Buffer and then centrifuged at 13,000 rpm for 20 min. The supernatants were used for the experiment. Cytokine standard dilutions were prepared before the test according to the kit instructions, and all of the kits and reagents were provided by RayBiotech, Inc., Guangzhou.

3.9 Statistical analysis

The LD_{50} values and data on weekly body weight, food consumption, relative organ weights, haematology, serum biochemistry, urinalysis and liver assessments were evaluated by one-way ANOVA using SPSS 17.0 statistical software. All of the values are expressed as the mean ± SD. P values less than 0.05 were considered significant.

4 RESULTS

Acute toxicity: All of the mice were observed for mortality and changes in behavioural patterns over 14 days. The mice became drowsy and depressed after dosing, and their fur lost its gloss. (Table 1) shows the group mortality rates. The LD_{50} of AEA in the mice was 15.74 g/kg bw based on calculations performed using Bliss in SPSS 17.0. The mice were convulsive before they died. The organs exhibited no macroscopic signs of toxicity, except in the liver. Significant pathological changes and lesions were observed in the liver.

Table 1 Mice mortality in the acute toxicity study.

Groups	Doses (g/kg)	Number of mice	Mortality	Survival rate (%)
1	25.08	10	10	0
2	20.81	10	8	20
3	17.27	10	5	50

				(continued)
Groups	Doses (g/kg)	Number of mice	Mortality	Survival rate (%)
4	14.34	10	5	50
5	11.90	10	2	80
6	9.88	10	0	100

5 SUBCHRONIC TOXICITY

5.1 Observations of gross toxicity and mortality

No deaths were observed throughout the experimental period. Compared with the solvent group, rats in the PEA, EEA, and NEA groups were drowsy and depressed, reacted slowly to outside stimulation, and displayed some fur loss beginning at 21 days after oral administration. No significant changes were observed in the remaining groups. These results suggest that the administration of PEA, EEA, and NEA in rats at levels up to 0.34 g/kg/day for 91 days have adverse effects on clinical observations.

5.2 Food consumption, body weight gain and relative organ weights

The gain in body weight was significantly higher ($P < 0.05$) in males of the petroleum ether extract group compared with the male controls at week 2 (no significant difference was observed in females). However, significant decreases in body weight gain relative to male controls were observed in males of the PEA group at weeks 7, 9, 11, and 12 ($P < 0.05$). In addition, almost all of the treatment groups lost weight sporadically from week 7 to week 12. Table 2 and Fig. 1 show the body weight gains and trends in weight gain in the rats. Food consumption is shown in Fig. 2. There were no significant differences in the weekly total food consumption between any of the extract groups and the control group throughout the experimental period. This result indicated that the *Aster tataricus* L. f. extracts affected rat weight gain in both sexes during the stable phase (weeks 7 to 12), especially in the groups that received the petroleum ether extract and the ethyl acetate extract of *Aster tataricus* L. f. No significant differences in the relative organ weights in either sex were observed between any of the extract groups and the control group (Table 3).

Table 2 Weekly body weight gain of rat treatment groups at 91 days.

Time (week)	Groups					
	Control group	PEA group	EEA group	NEA group	REA group	AEA group
	Male					
1	48.00 ± 7.94	53.00 ± 1.41	50.67 ± 9.02	46.50 ± 10.61	53.33 ± 7.51	55.50 ± 2.12
2	44.33 ± 6.03	53.50 ± 2.12*	49.67 ± 4.93	49.00 ± 1.41	48.00 ± 3.00	46.50 ± 0.71
3	42.33 ± 4.16	42.50 ± 3.54	41.67 ± 4.62	40.50 ± 3.54	43.33 ± 5.86	51.00 ± 2.83
4	32.30 ± 5.51	34.50 ± 2.12	38.33 ± 2.08	35.50 ± 3.54	35.00 ± 7.81	42.00 ± 1.41*

(continued)

Time (week)	Groups					
	Control group	PEA group	EEA group	NEA group	REA group	AEA group
	Male					
5	25.32 ± 5.51	25.00 ± 1.41	28.33 ± 5.51	33.50 ± 7.78	34.44 ± 2.08	30.00 ± 5.66
6	25.34 ± 7.64	23.50 ± 2.12	21.00 ± 5.00	18.50 ± 3.54	19.67 ± 6.35	20.00 ± 0.00
7	25.67 ± 4.93	14.00 ± 2.82*	18.00 ± 6.08	15.50 ± 0.71*	19.33 ± 5.04	17.50 ± 3.54
8	15.67 ± 1.53	10.00 ± 1.41	13.67 ± 3.79	16.50 ± 6.36	10.00 ± 2.65	9.50 ± 0.71
9	17.00 ± 5.29	6.50 ± 0.71*	14.00 ± 3.61	11.50 ± 6.36	20.00 ± 6.00	12.00 ± 2.83
10	9.00 ± 2.00	4.50 ± 0.71	7.67 ± 2.08	13.50 ± 2.12	18.00 ± 8.66*	14.00 ± 1.41
11	12.33 ± 5.03	20.00 ± 4.24*	16.33 ± 3.51	16.00 ± 0.00	23.67 ± 3.06**	19.00 ± 1.41
12	11.33 ± 1.53	4.50 ± 0.50*	5.33 ± 0.0.58*	16.50 ± 0.71	12.00 ± 5.29	13.50 ± 2.12
13	8.00 ± 1.02	9.00 ± 1.41	8.00 ± 2.00	5.50 ± 0.72	5.65 ± 1.52	6.50 ± 0.72
	Female					
1	38.50 ± 4.95	37.00 ± 2.65	26.50 ± 4.95*	38.67 ± 6.11	39.50 ± 0.71	33.00 ± 1.73
2	21.50 ± 4.95	24.33 ± 2.08	21.00 ± 1.41	26.67 ± 4.51	22.00 ± 5.66	17.33 ± 0.58
3	17.00 ± 0.00	15.00 ± 3.00	13.50 ± 3.54	15.67 ± 1.15	19.50 ± 2.12	21.33 ± 3.21
4	12.00 ± 2.83	18.33 ± 8.38	14.00 ± 4.24	14.33 ± 6.02	10.50 ± 4.95	12.67 ± 2.52
5	16.00 ± 1.41	15.32 ± 2.89	12.50 ± 4.95	17.67 ± 2.08	13.00 ± 1.41	11.00 ± 1.73
6	11.50 ± 0.71	12.00 ± 1.00	10.00 ± 1.41	13.67 ± 3.52	13.00 ± 2.83	10.00 ± 3.00
7	7.00 ± 1.41	6.33 ± 1.53	3.50 ± 0.71*	10.33 ± 1.53*	10.00 ± 1.41	9.67 ± 1.53
8	5.50 ± 0.71	2.33 ± 0.58**	2.00 ± 1.41**	−1.33 ± 0.58**	−1.50 ± 0.71**	3.33 ± 1.15**
9	6.50 ± 0.71	9.00 ± 3.00	2.50 ± 0.71	10.00 ± 3.61	3.50 ± 0.71	5.30 ± 1.00
10	9.50 ± 2.12	2.67 ± 1.15**	7.50 ± 2.12	6.67 ± 1.53	6.00 ± 1.41*	4.33 ± 1.15**
11	1.50 ± 0.71	7.33 ± 2.31**	1.50 ± 0.71	5.00 ± 1.00*	3.00 ± 1.41	3.67 ± 1.15
12	7.50 ± 2.12	0.67 ± 0.26**	5.50 ± 0.71*	1.33 ± 0.58**	4.00 ± 1.41**	0.33 ± 0.15**
13	2.50 ± 0.80	3.00 ± 1.00	1.75 ± 0.35	1.35 ± 0.58	3.50 ± 0.72	3.00 ± 1.00

Values are the mean ± SD for 6 rats in each group. * indicates a significant difference from controls ($P < 0.05$). ** indicates a highly significant difference from controls ($P < 0.01$).

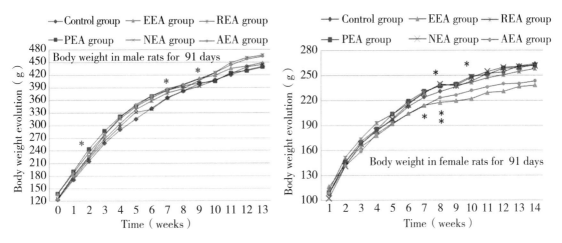

Fig. 1   Body weight changes in rats over 91 days.

The weekly average body weights (g) of male and female rats that received *Aster tataricus* L. f. extracts. * = statistically significant difference in body weight gain compared with the control group.

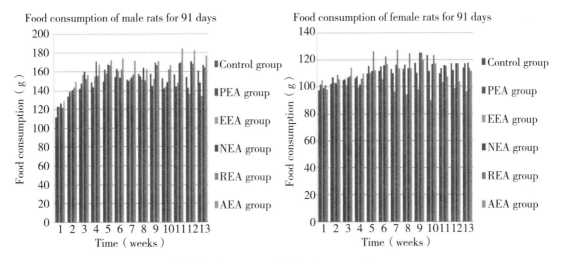

Fig. 2   Average weekly food consumption.

Values are presented as the mean ± SD (6 rats/sex/group). * = significant difference compared with the control group at P < 0.05.

Table 3   Effects of different *Aster tataricus* L. f. extracts on relative organ weights.

Parameters	Groups					
	Control group	PEA group	EEA group	NEA group	REA group	AEA group
	Male					
Liver	2.91 ± 0.06	2.68 ± 0.29	2.64 ± 0.28	2.79 ± 0.05	2.77 ± 0.05	2.73 ± 0.12
Heart	0.32 ± 0.02	0.36 ± 0.04	0.31 ± 0.03	0.35 ± 0.01	0.32 ± 0.02	0.33 ± 0.04
Spleen	0.17 ± 0.04	0.15 ± 0.02	0.16 ± 0.01	0.15 ± 0.03	0.17 ± 0.00	0.15 ± 0.03

(continued)

Parameters	Groups					
	Control group	PEA group	EEA group	NEA group	REA group	AEA group
Lungs	0.44 ± 0.02	0.45 ± 0.01	0.43 ± 0.06	0.49 ± 0.06	0.41 ± 0.00	0.47 ± 0.06
Kidneys	0.65 ± 0.08	0.59 ± 0.07	0.61 ± 0.00	0.67 ± 0.03	0.60 ± 0.02	0.70 ± 0.09
Thymus	0.10 ± 0.02	0.12 ± 0.06	0.11 ± 0.02	0.12 ± 0.02	0.10 ± 0.01	0.14 ± 0.03
Testes or ovaries	0.81 ± 0.03	0.87 ± 0.03	0.80 ± 0.06	0.77 ± 0.04	0.74 ± 0.05	0.79 ± 0.02
Female						
Liver	2.93 ± 0.18	2.68 ± 0.29	2.79 ± 0.13	3.05 ± 0.06	2.90 ± 0.03	3.01 ± 0.35
Heart	0.36 ± 0.04	0.37 ± 0.04	0.40 ± 0.03	0.41 ± 0.05	0.38 ± 0.03	0.41 ± 0.04
Spleen	0.19 ± 0.02	0.21 ± 0.03	0.22 ± 0.03	0.22 ± 0.03	0.21 ± 0.03	0.20 ± 0.01
Lungs	0.53 ± 0.05	0.55 ± 0.01	0.63 ± 0.03	0.55 ± 0.10	0.60 ± 0.04	0.58 ± 0.05
Kidneys	0.68 ± 0.03	0.64 ± 0.06	0.63 ± 0.03	0.69 ± 0.02	0.66 ± 0.03	0.68 ± 0.04
Thymus	0.16 ± 0.04	0.17 ± 0.00	0.15 ± 0.07	0.18 ± 0.01	0.14 ± 0.02	0.15 ± 0.01
Testes or ovaries	0.048 ± 0.01	0.046 ± 0.01	0.046 ± 0.01	0.046 ± 0.01	0.047 ± 0.00	0.050 ± 0.01

Values are the mean ± SD for 6 rats in each group. * indicates a significant difference from controls ($P < 0.05$). ** indicates a highly significant difference from controls ($P < 0.01$).

5.3 Urine examination

Following urine collection, the urine was observed to be luminous yellow in colour. The quantitative analyses of urine revealed no significant differences between any of the extract-treated groups and the control group in either sex. The urine parameters examined included pH (6.5-8.0), specific gravity (1.030), nitrite, glucose, protein, bilirubin, urobilinogen (3.3 μmol/L), ketones, WBC and vitamin C.

5.4 Haematology

No significant changes were observed in any haematological index except MCHC (Table 4). MCHC was decreased in females of the PEA, NEA, and REA groups relative to controls.

Table 4 Haematological parameters of Sprague-Dawley rats treated orally with *Aster tataricus* L. f. extracts.

Parameters	Groups					
	Control group	PEA group	EEA group	NEA group	REA group	AEA group
Males						
WBC (10^9/L)	9.70 ± 0.57	8.65 ± 2.19	11.37 ± 1.79	11.65 ± 0.07	11.13 ± 0.80	8.90 ± 0.14
RBC (10^{12}/L)	8.11 ± 0.39	8.59 ± 0.58	8.51 ± 0.53	8.50 ± 0.33	8.26 ± 0.35	8.38 ± 0.62

(continued)

Parameters	Groups					
	Control group	PEA group	EEA group	NEA group	REA group	AEA group
HGB (g/L)	175.67 ± 9.07	180.50 ± 4.95	180.00 ± 6.00	176.00 ± 5.66	175.67 ± 3.06	178.50 ± 4.95
HCT	0.46 ± 0.02	0.48 ± 0.02	0.49 ± 0.03	0.48 ± 0.01	0.48 ± 0.01	0.48 ± 0.03
MCV (fL)	57.33 ± 1.19	55.15 ± 0.07	55.17 ± 1.91	56.55 ± 3.04	57.70 ± 1.40	57.10 ± 1.13
MCH (pg)	21.67 ± 0.15	20.35 ± 0.49	20.33 ± 0.57	20.75 ± 1.48	21.27 ± 0.67	21.30 ± 0.99
MCHC (g/L)	378.00 ± 6.55	370.00 ± 9.90	374.67 ± 6.66	366.50 ± 6.36	368.67 ± 3.21	373.00 ± 9.90
PLT (10^9/L)	1 068.67 ± 24.01	1 138.00 ± 80.61	1 183.50 ± 61.52	1 052.00 ± 189.50	877.67 ± 32.53	1 056.00 ± 34.65
Females						
WBC (10^9/L)	6.83 ± 0.22	7.43 ± 0.55	5.75 ± 0.21	9.13 ± 2.93	9.15 ± 1.48	8.20 ± 0.20
RBC (10^{12}/L)	7.76 ± 0.24	8.19 ± 0.14	7.81 ± 0.20	8.26 ± 0.28	8.04 ± 0.76	8.09 ± 0.36
HGB (g/L)	172.60 ± 6.02	183.00 ± 8.19	167.00 ± 5.66	173.00 ± 3.61	169.50 ± 10.61	178.00 ± 10.15
HCT (%)	0.45 ± 0.01	0.47 ± 0.03	0.45 ± 0.01	0.47 ± 0.00	0.47 ± 0.03	0.47 ± 0.03
MCV (fL)	58.16 ± 0.80	60.40 ± 2.45	57.05 ± 0.07	56.70 ± 1.47	58.00 ± 1.56	58.20 ± 1.18
MCH (pg)	22.24 ± 0.11	22.33 ± 1.36	21.40 ± 0.14	21.27 ± 0.29	21.10 ± 0.71	22.00 ± 0.62
MCHC (g/L)	382.60 ± 6.88	369.67 ± 7.09*	375.00 ± 2.83	369.67 ± 5.86*	364.50 ± 2.12**	378.00 ± 13.45
PLT (10^9/L)	944.40 ± 78.74	1 021.67 ± 139.95	1 004.00 ± 83.44	1 029.00 ± 71.63	945.00 ± 4.24	914.33 ± 116.14

Values are the mean ± SD for 6 rats in each group. * indicates a significant difference from controls ($P < 0.05$).

** indicates a highly significant difference from controls ($P < 0.01$).

5.5 Serum chemistry

The results of biochemical analyses of blood are presented in Table 5. In male rats, the administration of each of the five extracts of *Aster tataricus* L. f. resulted in significant increases in CK, and similar trends were observed for the levels of ALT, AST, LDL-C, HDL-C, and TC in the PEA group. A significant increase in CK relative to controls was also noted in females in all of the tested groups. The plasma levels of ALT in female rats treated with PEA increased relative to the controls. Other parameters that showed relative increases were T-BIL (EEA and REA groups), LDH (PEA, EEA, NEA, and REA groups), and TC (AEA group). However, ALP was decreased in the PEA and AEA groups compared with the controls. The changes in ALT, AST, ALP, T-BIL, and CK indicate liver damage; other activities were not considered in any of the liver toxicological significance tests, although some irrelevant changes were observed.

Table 5 Serum biochemistry of SD rats treated orally with *Aster tataricus* L. f. extracts.

Indices	Groups					
	Control group	PEA group	EEA group	NEA group	REA group	AEA group
Male						
ALT (U/L)	46.67 ± 2.31	78.00 ± 26.87*	57.67 ± 14.05	53.50 ± 2.12	57.33 ± 10.79	47.00 ± 7.07
AST (U/L)	116.33 ± 15.95	175.50 ± 41.72*	142.00 ± 32.08	155.5 ± 3.54	147.67 ± 15.31	125.00 ± 42.23
ALP (U/L)	155.67 ± 13.43	149.50 ± 2.12	150.00 ± 19.52	124.00 ± 16.97	143.67 ± 29.54	132.50 ± 19.09
T-BIL (μmol/L)	6.05 ± 1.75	7.55 ± 1.93	5.71 ± 0.17	4.65 ± 0.01	5.51 ± 0.43	5.25 ± 0.66
TP (g/L)	67.00 ± 2.50	74.10 ± 2.62	69.27 ± 0.80	61.55 ± 4.74	64.60 ± 4.25	66.45 ± 3.04
ALB (g/L)	34.17 ± 2.59	37.30 ± 2.26	35.77 ± 0.46	33.15 ± 1.91	34.10 ± 1.65	34.35 ± 0.21
CK (U/L)	654.67 ± 54.00	1 958.50 ± 468.81**	1 745.00 ± 87.73**	1 281.00 ± 2.83**	1 982.33 ± 112.22**	1 884.00 ± 100.41**
LDL-C (mmol/L)	0.26 ± 0.04	0.42 ± 0.03**	0.28 ± 0.03	0.28 ± 0.04	0.25 ± 0.06	0.25 ± 0.05
LDH (U/L)	1 146.67 ± 381.34	1 458.00 ± 386.08	1 179.67 ± 310.78	1 641.00 ± 21.21	1 258.00 ± 416.45	1 274.00 ± 490.73
Urea (mmol/L)	6.57 ± 1.19	5.95 ± 2.33	6.10 ± 0.26	7.50 ± 0.14	7.83 ± 2.48	6.50 ± 1.84
Creatinine (μmol/L)	64.33 ± 8.62	74.50 ± 6.36	74.00 ± 5.29	62.00 ± 1.41	66.33 ± 3.79	60.50 ± 2.12
HDL-C (mmol/L)	1.07 ± 0.15	1.61 ± 0.24*	1.10 ± 0.05	1.12 ± 0.17	1.19 ± 0.22	1.02 ± 0.36
Glucose (mmol/L)	5.72 ± .83	5.29 ± 1.53	4.74 ± 0.34	6.10 ± 0.56	5.33 ± 0.59	5.37 ± 1.02
TC (mmol/L)	1.60 ± 0.30	2.30 ± 0.28*	1.63 ± 0.15	1.60 ± 0.28	1.67 ± 0.32	1.70 ± 0.14

(continued)

Indices	Groups					
	Control group	PEA group	EEA group	NEA group	REA group	AEA group
TG (mmol/L)	1.05 ± 0.13	1.15 ± 0.07	1.30 ± 0.49	0.75 ± 0.07	1.20 ± 0.18	1.21 ± 0.07
Female						
ALT (U/L)	45.50 ± 4.95	58.67 ± 2.52*	45.50 ± 2.12	50.33 ± 5.51	45.50 ± 3.54	47.33 ± 11.01
AST (U/L)	131.40 ± 32.74	141.00 ± 30.12	156.50 ± 24.75	152.67 ± 43.59	140.00 ± 36.77	116.33 ± 23.29
ALP (U/L)	112.50 ± 0.71	91.67 ± 4.93**	99.50 ± 0.71	109.33 ± 11.72	100.00 ± 1.41	89.00 ± 4.00**
T-BIL (μmol/L)	4.97 ± 1.31	5.48 ± 0.92	7.12 ± 0.14*	4.96 ± 0.66	7.53 ± 0.79*	5.19 ± 1.06
TP (g/L)	72.24 ± 7.50	71.47 ± 4.48	72.65 ± 1.77	67.53 ± 1.46	65.65 ± 3.75	72.50 ± 1.23
ALB (g/L)	37.70 ± 2.27	37.43 ± 0.91	36.80 ± 0.85	35.93 ± 0.64	34.85 ± 2.62	38.03 ± 1.36
CK (U/L)	1 544.50 ± 286.48	2 548.33 ± 458.26**	4 490.50 ± 441.94**	3 010.00 ± 284.96**	3 084.50 ± 665.39**	2 521.67 ± 487.26*
LDL-C (mmol/L)	0.21 ± 0.06	0.27 ± 0.04	0.26 ± 0.01	0.24 ± 0.05	0.27 ± 0.01	0.27 ± 0.07
LDH (U/L)	683.67 ± 32.32	1 206.67 ± 212.59**	1 347.50 ± 178.90**	1 265.67 ± 197.00**	1 120.00 ± 247.49*	978.67 ± 220.00
Urea (mmol/L)	7.22 ± 0.43	7.20 ± 1.57	6.65 ± 0.64	6.93 ± 2.14	7.95 ± 1.34	8.10 ± 1.84
Creatinine						
	70.40 ± 9.56	79.67 ± 9.61	72.50 ± 9.19	67.67 ± 7.51	68.00 ± 2.83	70.67 ± 2.52
(μmol/L)						
HDL-C (mmol/L)	1.81 ± 0.22	1.66 ± 0.42	1.72 ± 0.05	1.65 ± 0.17	1.77 ± 0.54	2.28 ± 0.19
Glucose (mmol/L)	5.85 ± 1.08	5.07 ± 0.80	4.76 ± 0.08	6.09 ± 0.55	5.56 ± 1.06	5.08 ± 0.57
TC (mmol/L)	2.24 ± 0.36	2.20 ± 0.46	2.40 ± 0.14	2.10 ± 0.20	2.40 ± 0.57	2.83 ± 0.31*
TG (mmol/L)	0.94 ± 0.39	1.13 ± 0.22	1.06 ± 0.17	0.76 ± 0.14	0.84 ± 0.23	0.59 ± 0.11

Values are the mean ± SD for 6 rats in each group. * indicates a significant difference from controls ($P < 0.05$). ** indicates a highly significant difference from controls ($P < 0.01$)

5.6 Liver biochemical analysis

Results of the liver biochemical analysis are summarized in Table 6. No significant changes were observed in glutathione (GSH), total superoxide dismutase (T-SOD), or malondialdehyde (MDA) activities in any of the groups compared with the control group.

Table 6 Liver biochemical indices during the 91-day subchronic oral administration test.

Indices	Solvent group	PEA group	EEA group	Groups NEA group	REA group	AEA group
			Male			
GSH	25.62 ± 0.98	26.96 ± 3.47	28.49 ± 3.73	28.99 ± 3.80	29.11 ± 3.50	30.71 ± 7.83
T-SOD	474.09 ± 24.50	541.95 ± 8.20	519.43 ± 55.29	491.70 ± 79.83	504.78 ± 63.11	539.40 ± 113.07
MDA	1.00 ± 0.25	1.04 ± 0.34	0.75 ± 0.10	0.90 ± 0.08	1.01 ± 0.16	0.83 ± 0.03
			Female			
GSH	23.77 ± 0.39	24.40 ± 4.75	26.62 ± 3.20	25.34 ± 3.09	26.48 ± 1.19	27.00 ± 1.71
T-SOD	550.76 ± 54.73	563.32 ± 21.43	572.90 ± 14.50	512.50 ± 36.71	497.35 ± 49.92	528.36 ± 62.98
MDA	1.11 ± 0.04	1.17 ± 0.05	1.19 ± 0.06	0.96 ± 0.06	1.17 ± 0.21	1.10 ± 0.03

Values are the mean ± SD for 6 rats in each group. * indicates a significant difference from controls ($P < 0.05$). ** indicates a highly significant difference from controls ($P < 0.01$).

5.7 Quantitative measurement of cytokines

There were no significant differences in cytokine measurements (Fig. 3). The alignment of the cytokine arrays is shown in Fig. 4A. Examples of the array blots after incubation with liver tissue samples (groups PEA and EEA) are shown in Fig. 4B. Twelve positively stained inflammation spots were clearly identified. All of the spots were slightly darker compared to those of the control rats, demonstrating that there were no significant changes in the levels of inflammatory cytokines.

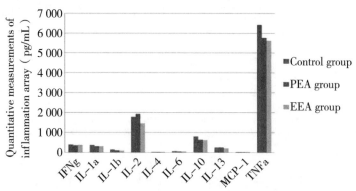

Fig. 3 Quantitative measurements of cytokines.

$*P < 0.05$, $**P < 0.01$ (extract groups vs. controls). Values are means ± SEM for 4 rats in each of the PEA, EEA and Control groups.

(A) The map of QAR-INF-1

	1	2	3	4	5	6	7	8
A	POS1	POS1	POS1	POS1	POS2	POS2	POS2	POS2
B	IFNg	IFNg	IFNg	IFNg	IL-1a	IL-1a	IL-1a	IL-1a
C	IL-1b	IL-1b	IL-1b	IL-1b	IL-2	IL-2	IL-2	IL-2
D	IL-4	IL-4	IL-4	IL-4	IL-6	IL-6	IL-6	IL-6
E	IL-10	IL-10	IL-10	IL-10	IL-13	IL-13	IL-13	IL-13
F	MCP-1	MCP-1	MCP-1	MCP-1	TNFa	TNFa	TNFa	TNFa

(B)   Control group   PEA group   EEA group

(A) The alignment of 12 cytokines in the Rat Inflammation Array.
(B) Examples of Rat Inflammation Array blots probed with liver samples from the PEA and EEA groups.

Fig. 4 Results of the rat inflammation array.

5.8 Histopathology

There were no significant macroscopic changes in observable toxicity in the main organs of the extract-treated rats at 91 days. However, the histopathological examination of organs from the extract-treated groups revealed sporadic lesions, slight congestion and hepatic cord disorders in the liver in the PEA and EEA groups (Fig. 5B, 5C). In addition, signs of swelling and dissolution of the cardiac muscle fibres were observed in myocardial cells (Fig. 5H) in the PEA group;

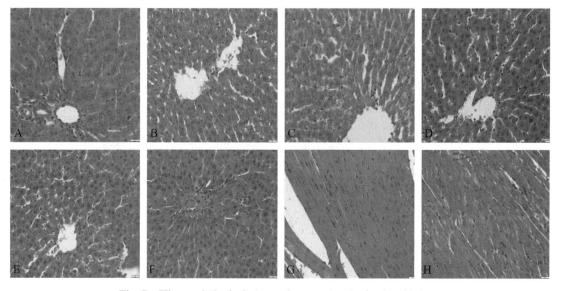

Fig. 5 Histopathological photomicrographs obtained at 91 days.

Photomicrographs of liver tissue from the control group (A) and B, C, D, E, F for PEA, EEA, NEA, REA, ANA group. Image G shows the condition of the heart in normal rats, and H shows that of the PEA group. All of the samples were stained with haematoxylin and eosin (40x).

however, no significant differences were observed between any of the remaining extract-treated groups compared with the controls. No changes in the other organs, including the kidneys, spleen, lungs, thymus, ovaries or testes, were detect

6 DISCUSSION

In recent years, there has been an increasing emphasis on the safety and efficacy of natural plant products and the elimination of side effects. The diversity and complexity of herbal components make it necessary to establish their safety, quality and effectiveness (Lee et al., 2012; Shin et al., 2011). We found that *Aster tataricus* L. f. extracts were toxic to animals and resulted in death during a pharmacologic study. Considering its lack of toxicological properties, to clarify the toxicological profile of *Aster tataricus* L. f., an acute toxicity test was conducted. The results demonstrated the toxic effects of *Aster tataricus* L. f. in mice. The LD_{50} of the 75% ethanol extract was greater than 15.74 g/kg BW in mice, and the detection of liver lesions indicated that *Aster tataricus* L. f. toxicity mainly affected the liver.

The preliminary hypothesis that *Aster tataricus* L. f. extracts can cause liver toxicity can be evaluated by examining the nidus in the liver. The lack of significant differences in the urine traits between each of the treatment groups and controls demonstrated that kidney function was normal in the extract-treated rats. No deaths related to extract administration were noted, but clinical signs such as unkempt fur, drowsiness and reduced appetite were observed in the treatment groups during the later period of the subchronic toxicity test compared with the control group. Increases or decreases in body weight usually reflect physiological changes, such as changes in liver function or hormones. The observed differences in weight gain indicate that these five extracts of *Aster tataricus* L. f., especially the petroleum ether and ethyl acetate extracts, can affect weight gain of rats, potentially via the accumulation of toxic effects.

The haematopoietic system can reflect toxic effects because of its sensitivity to toxic compounds (Adeneye et al., 2006). The significant decrease in MCHC observed may be due to diminished cell absorption caused by the accumulation of toxins. Iron deficiency can cause a decrease in MCHC values (Patel et al., 2008). Plasma ALT and AST levels are associated with changes in the permeability of the hepatocyte membrane or hepatic damage because soluble cytosolic enzymes are released into the blood when the permeability of the hepatocellular membrane is altered by either hepatocytic membrane injury or metabolic disturbances (Hor et al., 2011; Ihsan et al., 2010). Haematological analyses indicate that these phenomena generally result in the elevation of these enzymes in the blood. This type of release typically occurs from the apical side of the hepatocyte plasma membrane facing the canaliculi (Ramaiah, 2007). Cholestasis increases total serum bilirubin levels. Therefore, the significant increase in bilirubin observed in female rats confirmed the presence of disturbances in liver function because hyperbilirubinaemia caused by drug inhibition results in bilirubin transport into hepatocytes or the inhibition of bilirubin conjugation in the liver (Zucker et al., 2001). Lactate dehydrogenase (LDH) levels increased in the treated group, and this effect was very clear in females. Lactate dehydrogenase (LDH) leakage has been occasionally encountered in the literature as a potential marker of hepatocellular toxicity.

Alkaline phosphatase (ALP) is considered to be a cholestatic induction enzyme of hepatobiliary origin and shows minimal activity in normal hepatic tissue. ALP is generally used to detect impaired bile flow (cholestasis), but serum ALP activity typically increases rapidly in rats following a meal due to its high sensitivity, and thus, it cannot be reliably employed to detect cholestasis (Amacher, 2002). Considering the fasting employed during the last night of the test and the sensitivity of ALP to fasting, the decrease in ALP observed during the test period (a significant difference in female rats in the PEA and AEA groups) is not a good toxicity-related parameter. Increases in CK, LDH, and AST are considered to be indicators of myocardial damage. Nevertheless, the measurement of AST, CK, and LDH activities involves poor sensitivity and specificity in myocardial tissue, and therefore, the use of these measurements in clinical practice is limited. Currently, troponins are used as a replacement (Kulthinee et al., 2010; O'Brien, 2008; Walker, 2006). Hence, the observed increases in CK and LDH might only be a reference for myocardial damage, with greater cardiac toxicity observed in female compared to male rats. Serum creatine (Crea) and urea levels are used to assess renal toxicity (Kuroiwa et al., 2006), but no abnormities were observed in these activities in the present study. The observed increases in LDL-C, HDL-C and cholesterol in males in the PEA group suggest that petroleum ether extract of *Aster tataricus* L. f. affects lipid and carbohydrate metabolism in rats. HDL is known to be a strong inverse predictor of cardiovascular disease (Liju et al., 2013). Although some sporadic differences were observed, these differences were not considered to be related to toxicity because the observed changes were within the normal reference laboratory range (Petterino et al., 2006). Serum chemistry results showed that extracts of *Aster tataricus* L. f., especially petroleum ether extract, affect liver function and have some effect on the heart.

The production of excessive free radicals in liver tissue creates tissue injury, resulting in decreases in the levels of SOD and GSH. MDA is an important indicator of biological membrane system damage because it is a product of lipid peroxidation, and its content can reflect the degree of lipid peroxidation in the body and indirectly reflect the degree of cell injury. However, there were no significant changes in the three antioxidases. Furthermore, the results of the cytokine assessment revealed little effect of the petroleum ether extract and the ethyl acetate extract on inflammation, although these results may have been caused by inter-individual differences in the rats. The lack of obvious differences in these parameters indicated that liver toxicity was not severe at the dose of 0.34 g/kg/dBW. The histopathological results from the selected organs (heart, liver, lungs, spleen, thymus, testis, ovaries and kidneys) revealed normal architecture except for the liver in both sexes of group PEA and EEA. Therefore, these lesions can be considered to be spontaneous and/or incidental in nature but relevant to *Aster tataricus* L. f. treatment, especially in animals that receive petroleum ether extract or ethyl acetate extract of *Aster tataricus* L. f. (Lu et al., 2014).

The administration of low oral doses of *Aster tataricus* L. f. extracts to rats can cause mild liver injury, whereas the administration of high doses can cause acute liver toxicity or death. The results of our tests revealed that the petroleum ether and ethyl acetate extracts of *Aster tataricus* L. f. could cause liver toxicity after 91 days at a daily oral dose of 0.34 g/kg bw. The main components in petroleum ether and ethyl acetate extracts are shionone, friedelin, epifriedelinol, chrysophanol, emodin, quercetin, and luteolin (Ye, 2007). The pharmacologically active ingredients are also concentrated in these two polar segments; for example, shionone is the main active ingredient that eliminates phlegm and relieves cough (Qiny et al., 1984). An experimental study by Yanhua Lu et al. (1999) showed that petroleum ether and ethyl acetate extracts could eliminate phlegm. Emodin and quercetin have anti-inflammatory and bacteriostasis effects. Considering the toxicity of *Aster tataricus* L. f., it is important to consider this toxicity in clinical settings. Our future research will be focused on identifying drugs or other measures that can potentially inhibit the toxic effects of *Aster tataricus* L. f. to allow full benefit from its pharmacological effects.

7 CONCLUSION

Despite sporadic differences, the toxicity-related changes observed in this study were significant. The results of this study showed the toxic effects primarily affected the liver, with slight effects on the heart. The acute oral toxicity experiment showed that *Aster tataricus* L. f. is capable of toxic effects and resulted in an LD_{50} of 15.74 g/kg BW in mice. The subchronic experiment, conducted at a dose of 0.34 g/kg/d.BW, demonstrated that the toxic components of *Aster tataricus* L. f. were mainly concentrated in the petroleum ether fraction, followed by the ethyl acetate fraction, the n-butyl alcohol fraction, the lower aqueous phase and the 75% ethanol extract.

ACKNOWLEDGMENTS

This work was financially supported by the Special Fund for Agro-Scientific Research in the Public Interest (No. 201303040-18).

Conflict of Interest: The authors of this paper declare no conflicts of interest.

(发表于《Afr J Tradit Complement Altern Med.》, SCI, IF: 0.553. 38–53.)

Comparative Proteomic Analysis of Yak Follicular Fluid During Estrus

GUO Xian[*], PEI Jie, DING Xuezhi, CHU Min, BAO Pengjia,
WU Xiaoyun, LIANG Chunnian, YAN Ping[*]

(Key Laboratory of Yak Breeding Engineering of Gansu Province, Lanzhou Institute of Husbandry and Pharmaceutical Sciences, Chinese Academy of Agricultural Sciences, Lanzhou 730050, China)

Abstract: The breeding of yaks is highly seasonal, there are many crucial proteins involved in the reproduction control program, especially in follicular development. In order to isolate differential proteins between mature and immature follicular fluid (FF) of yak, the FF from yak follicle with different size were sampled respectively, and two-dimensional electrophoresis (2-DE) of the proteins were carried out. After silver staining, the Image Master 2D platinum software was used for protein analysis and matrix-assisted laser desorption ionization time of flight mass spectrometry (MALDI-TOF-MS) was performed for differential protein identification. The expression level of transferrin and ENOSF1 was determined by Western blotting for verification analysis. The results showed that 2-DE obtained an electrophoresis map of proteins from mature and immature yak follicular fluid with high resolution and repeatability. A comparison of protein profiles identified 12 differently expressed proteins, out of which 10 of them were upregulated while 2 were downregulated. Western blotting showed that the expression of transferrin and ENOSF1 was enhanced with follicular development. Both the obtained protein profiles and the differently expressed proteins identified in this study provided experimental data related to follicular development during yak breeding seasons. This study also laid the foundation for understanding the microenvironment during oocyte development.

Key words: Yak, Follicular fluid; Two-dimensional gel electrophoresis; Mass spectrometry; Western blot

1 INTRODUCTION

Yaks (Bos grunniens) are mainly distributed in the Qinghai-Tibetan Plateau and nearby regions; they provide meat, milk, wool and fuel for Tibetan and nomadic pastoralists. These animals are essential for agricultural production in high altitude environment regions. Their natural habitat ranges from between 2 500 to 6 000 m above sea level. Yaks are seasonally polyestrous, exhibiting cyclic ovarian activity and estrus from July to November (Sarkar and Prakash, 2005;

---

[*] Corresponding authors. Xian GUO. E-mail: guoxian@caas.cn. Ping YAN. E-mail: yanping@caas.cn.

Sarkar et al., 2008). The length of the estrous cycle lasts from 19 to 21 days. However, yaks live in an extremely harsh environment characterized by low temperature, low oxygen content, and high altitude; seasonal nutritional deficiencies seriously affect their reproductive efficiency (Liu et al., 2012). Unlike bovine females, yak cows rarely express estrus immediately after post-calving during the breeding season. The annual pregnancy rate of most yak breeds is 40%-60%, as only a small proportion of the cows return to estrus in the first breeding season after calving; most come into estrus in the second or third years under traditional management systems (Zi, 2003). These long postpartum anestrous intervals eventually result in poor reproductive efficiency.

Follicular fluid (FF) is a liquid in the ovarian follicle which fills the follicular antrum and surrounds the ovum. FF contains several blood plasma protein components derived from constituents that cross the blood-follicular barrier and several granulosa/theca proteins derived from granulosa and thecal cells (Nandedkar et al., 1992; Fortune, 1994). Thus, FF components may also provide information related to ovarian follicle development. Follicle development is closely associated with yak reproduction and FF provides a very important microenvironment for oocyte development. A previous analysis of FF protein composition aimed to identify molecules which may help follicular development and therefore assist in oocyte selection processes (Schweigert et al., 2006). The study of FF protein components have been made by detecting specific proteins and correlating them to estrus and oocyte quality. Recently, more complex molecular techniques have been introduced, allowing for simultaneous study of dozens of proteins and peptides in biological fluid, in a field known as "proteomics" (Revelli et al., 2009).

Protein composition is very complicated and dynamic, and is therefore a major problem in proteomic analysis of complex biological fluids; consequently, sensitive and high-resolution protein separation techniques are needed. The most popular FF proteonomic approach is presently based on two-dimensional gel electrophoresis (2-DE) followed by protein digestion and mass spectrometry (Görg et al., 2004). At the same time, Western blotting has been used to detect specific proteins in an extracted sample. Recently, easier techniques have been proposed and applied towards studying FF proteins based on protein prefractionation by isoelectric focusing as well as nanoliquid chromatography and mass spectrometry (Hanrieder et al., 2008). In order to improve the efficiency of yak breeding, comparative FF proteomics was applied to analyze different follicular diameters during estrus. This technical scheme involved 2-DE in combination with matrix-assisted laser desorption ionization time of flight mass spectrometry (MALDI-TOF-MS) and Western blotting.

2 MATERIALS AND METHODS

2.1 Collection of follicular fluid

Yak ovaries were obtained from a designed abattoir after slaughter in September 2014 in Qinghai Province of China. Sixty ovaries in total were collected from 30 non-pregnant yaks. The yaks were between 6-7 years of age, healthy, and free from any anatomical reproductive disorders;

they resided 3 200 m above sea level. Follicular fluid was directly aspirated from the ovaries by a sterile syringe and then divided into two groups according to follicular diameter (0–10 mm, > 10 mm), and mixed according to size. The FF was immediately centrifuged at 5 000 g for 15 min in order to remove blood cells and debris at room temperature. Then, the fluid supernatants were centrifuged a second time at 12, 000 g for 10 min to remove cells and fractionated at room temperature; the fluid supernatants were directly collected and frozen in liquid nitrogen for further analysis.

2.2 Protein extraction and quantitative analysis

Total FF protein extracts were prepared according to the phenol extraction method described by Zhou et al. (2011) with minor modifications. The most abundant serum proteins were eliminated using the ProteoExtract Albumin/IgG Removal Kit (Calbiochem, San Diego, CA, USA). Total protein concentrations were measured by the Bradford method (Bradford, 1976). Samples were prepared in triplicates and stored at −80 ℃ until use.

2.3 Two-dimensional gel electrophoresis (2-DE)

For 2-DE, 100 μg and 400 μg of proteins were loaded onto analytical and preparative gels, respectively. The Ettan IPGphor Isoelectric Focusing System (GE Amersham, Amersham, United Kingdom) and pH 3-10 immobilized pH gradient (IPG) strips (13 cm, nonlinear; GE Healthcare, Little Chalfont, United Kingdom) were used for isoelectric focusing (IEF). The IPG strips were rehydrated for 12 h in 250 μl of rehydration buffer containing the protein samples. The IEF was performed in four steps: 30 V for 12 h, 500 V for 1 h, 1 000 V for 1 h, and 8 000 V for 8 h. The gel strips were prepared for 15 min in equilibration buffer (50 mM Tris-HCl [pH 8.8], 6 M urea, 2% SDS, 30% glycerol, and 1% DTT). This step was repeated using the same buffer with 4% iodoacetamide in place of 1% DTT. The strips were then subjected to the 2-DE after transferring onto 12.5% SDS-polyacrylamide gels. Electrophoresis was then performed using the Hofer SE 600 system (GE Amersham, Amersham, United Kingdom) at 15 mA per gel for 30 min, next followed by 30 mA per gel, until the bromophenol blue reached the end of the gel. Three replicates were performed for each of the samples.

2.4 Gel Staining and image analysis

Protein spots in the analytical gels were visualized by silver staining. The preparative gels were stained by a modified silver staining method compatible with subsequent mass spectrometric analysis (Yan et al. 2000). The stained gels were scanned using an UMax Powerlook 2110XL (UMax, Taipei, Taiwan), and the image analysis was accomplished using ImageMaster 2D Platinum (Version 5.0, GE Amersham, Amersham, United Kingdom). Each paired spot was manually verified to ensure a high reproducibility level between the normalized spot gel volumes produced in triplicate data. Two spots were considered as significantly different if $P<0.05$ and proteins with a ⩾1.2x or greater overlap ratio threshold filtering were considered to be differentially expressed. Spots met the criteria were then selected and subjected to in-gel tryptic digestion. There were 30% ~ 35% higher cutoff points to be chosen for a silver stained gel.

2.5 Peptide identification

All the differentially expressed spots were selected and excised manually from the three preparative gels. Protein spots of interest were cut from the preparative gels, destained for 20 min in 30 mM potassium ferricyanide/100mM sodium thiosulfate (1 : 1 v/v) and washed with Milli-Q water until the gels were destained. The spots were incubated in 0.2 M NH_4HCO_3 for 20 min and then lyophilized. Each spot was digested overnight in 12.5 ng/μl trypsin in 25 mM NH_4HCO_3. The peptides were extracted three times with 60% acetonitrile (ACN) /0.1% trifluoroacetic acid (TFA). The extracts were pooled and dried completely by a vacuum centrifuge. The peptides were identified using MASCOT software (Matrix Science: Boston, MA, USA) by searching against the NCBI database (Bos Taurus). If the protein score C.I. was above 95% and protein score was above 60, then it was validated.

3 MALDI-TOF/TOF MS ANALYSIS

MS (Mass) and MS/MS data used for protein identification were obtained by using a MALDI-TOF-TOF instrument (4800 proteomics analyzer; Applied Biosystems, Waltham, MA, USA). Instrument parameters were set using 4000 Series Explorer software (Applied Biosystems, Waltham, MA, USA). The MS spectra were recorded in a reflector mode over a mass range from 800 to 4000 with a focus mass of 2000. MS used a CalMix5 standard to calibrate the instrument (ABI 4700 Calibration Mixture). For one main MS spectrum, 25 subspectra (with 125 shots per subspectrum) were accumulated using a random search pattern. For the MS calibration, autolysis peaks of trypsin ([M+H]+842.5100 and 2, 211.1046) were used as internal calibrates; up to 10 of the most intense ion signals were selected as MS/MS acquisition precursors, excluding the trypsin autolysis peaks and the matrix ion signals. In MS/MS positive ion mode, one main MS spectrum accumulated 50 subspectra (with 50 shots per subspectrum) using a random search pattern. Collision energy was 2 kV, the collision gas was air, and default calibration was set by using the Glu1-Fibrino-peptide B ([M+H]+1, 570.6696) which was spotted onto Cal 7 positions of the MALDI target. Combined peptide mass fingerprinting PMF and MS/MS queries were then performed by using the MASCOT search engine 2.2 (Matrix Science, Ltd., London, United Kingdom) embedded into GPS-Explorer Software 3.6 (Applied Biosystems, Waltham, MA, USA) on the NCBI database (Bos_taurus) with the following parameter settings: 100 ppm mass accuracy, trypsin cleavage with one missed cleavage allowed, carbamidomethylation set as fixed modification, methionine oxidation allowed as variable modification, and a MS/MS fragment tolerance set to 0.4 Da. A GPS Explorer protein confidence index ≥ 95% were then used for further manual validation.

3.1 Western blot analysis

Yak follicular fluid was homogenized for protein extraction in an ice-cold protein extraction buffer, which was divided into three groups according to follicular diameter (0-5 mm, 5-10 mm, >10 mm). Homogenates were centrifuged for 20 min at 13 000 g and supernatants were collected. The protein concentrations were determined by the Bradford method. Western blot analysis was performed as described in Moritz et al. (2014) with some modifications. About 300 μg of protein from each sample was denatured, electrophoresed, and transferred onto a polyvinylidene fluoride (PVDF) membrane. The membrane was blocked and blots were intubated in a specific antibody against ENOSF1 and transferrin, followed by secondary antibodies according to manufacturer. The NBT/BCIP system was used to evaluate the protein signal.

3.2 Immunoblotting

The proteins were separated by sodium dodecyl sulfate polyacrylamide gel electrophoresis (SDS-PAGE) on 10% polyacrylamide gels. Polypeptides were then transferred to PVDF membranes (0.22 μm, Amersham Life Science, Little Chalfon, UK) in a medium which consisted of 25 mM Tris-HCl (pH 8.3), 192 mM glycine, and 20% (v/v) methanol. After rinsing in the Tris-buffered saline (TBS) which contained 10 mM Tris-HCl (pH 7.5) and 150 mM NaCl, the blotted membranes were pre-incubated for 2 h in a blocking buffer containing 5% (w/v) non-fat milk dissolved in TBS supplemented by 0.05% (v/v) Tween-20 (TBST1). The components were then incubated by gentle shaking for 2 h at room temperature in the appropriate antibodies (anti-ENOSF1, rabbit polyclonal antibodies to ENOSF1 [ab182354; Abcam, Cambridge, MA, USA]) diluted 1 : 2 000 in the blocking buffer; anti-transferrin (sheep polyclonal antibodies to transferrin [ab112 892; Abcam, Cambridge, MA, USA] diluted 1 : 1 000 in the blocking buffer; anti-actin diluted 1 : 1 000 in the blocking buffer). Following extensive washes by TBST1, the membranes were then incubated with Goat-anti-Rabbit IgG conjugated with either horseradish peroxidase (1 : 5 000 diluted in TBST1) alone or Rabbit-anti-Mouse IgG conjugated with horseradish peroxidase (1 : 5 000 diluted in TBST1) at room temperature for 2 h. Afterwards, the substance was washed with TBST2 (50 mM Tris-HCl, pH 7.5, 150 mM NaCl, 0.1% [v/v] Tween-20) and TBS. Locations of antigenic proteins were visualized by scanning the membranes with the ImageQuant LAS 500 machine (GE Healthcare, Little Chalfont, United Kingdom).

3.3 Statistical analysis

The homogeneity of variance was checked by using Levene's test. If the data failed to pass the test, then a logarithmic transformation was used. The differences were evaluated by a one-way ANOVA followed by a Tukey's test using SPSS 13.0 (SPSS, Chicago, IL, USA). Significant differences between mature and immature follicular fluid were identified by a p-value of 0.05.

4 RESULTS

4.1 Comparative proteome analysis by 2-DE

Global protein components of yak follicular fluid were separated with high sensitivity and resolution by 2-DE; more than 100 spots were detected for each gel. Spots with a significant increase or decrease in relative abundance were considered to represent differently expressed proteins. After bioinformatic analysis, up to 12 differentially expressed proteins were obtained; 10 expressed proteins were found more abundantly in the mature follicular fluid, while 2 expressed proteins were upregulated in the immature follicular fluid (Fig. 1). Some of them were found to have the same isoelectric points (pI) and molecular weights (MW) as predicted in the database.

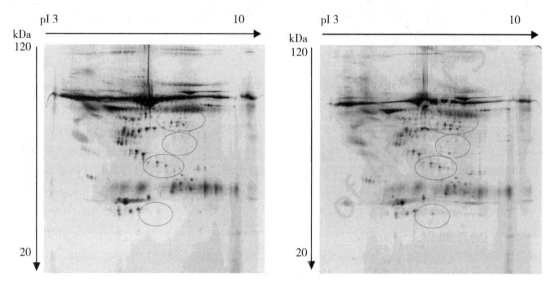

Fig. 1 Two-dimensional gel electrophoresis (2-DE) of mature (Left) and immature (Right) follicular fluid. Selected protein spots were numbered and collected for identification by MALDI-TOF-MS.

4.2 Identification of differently expressed proteins 179

Differently expressed proteins were picked from gels and identified using MS/MS. Fig. 2 shows typical examples of some differently expressed proteins identified by 2-DE. All data from the conventional gel-to-gel experiments are summarized in Table 1. Proteins with MW ranging from 20 to 120 kDa and pI between 3 and 10 were well-separated. Twelve significantly and consistently up- or downregulated proteins spots with a twofold change in volume intensity in the triplicate gels were trypsin-digested and analyzed by MALDI-TOF-MS/MS.

The following three criteria were used for the identification of specific proteins: (1) Several peptides were found to be specific for a given protein; (2) At least 12% of the amino acid sequence of protein was covered; (3) Observed MW and pI determined for a protein by 2-DE gel electrophoresis agreed with calculated values. Based on these criteria, 34 protein spots were successfully identified that corresponded to 14 proteins; all of these 12 proteins matched well with bovine proteins. Among these 12 proteins, 10 of them were upregulated in mature yak follicular fluid, including transferrin and ENOSF1. In comparison, the other 2 proteins were downregulated in immature yak follicular fluid.

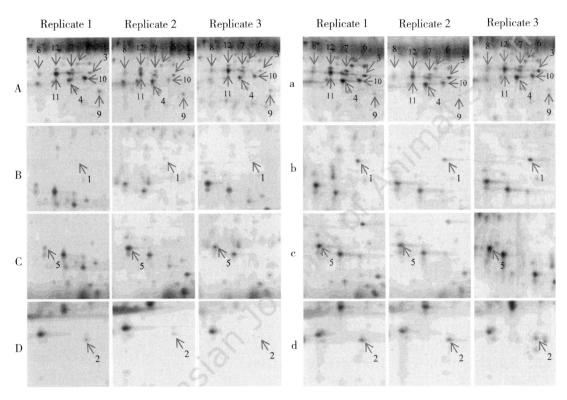

Fig. 2 Typical examples of some differently expressed protein spots with at least two-fold changes in mature (A, B, C, D) and immature (a, b, c, d) yak follicular fluid by 2-DE. Replicate 1, 2, and 3 images in the figure were triplicate gels.

Table 1 Matrix-assisted laser desorption/ionization time of flight mass spectrometry (MALDI-TOF-MS/MS) identification of differently expressed proteins in the yak follicular fluid.

NO.	Protein Name	Accession No.	Protein MW	Protein PI	Pep. Count	Protein Score	Protein Score C.I.%
1	serotransferrin precursor [Bos taurus]	gi|114326282	79 855.5	6.75	26	386	100
2	unnamed protein product [Bos taurus]	gi|110292444	71 264.2	5.82	30	600	100
3	serum albumin precursor [Bos taurus]	gi|30794280	71 274.2	5.82	28	220	100
4	Chain A, Crystal Structure Of Bovine Serum Albumin In Complex With 3, 5-Diiodosalicylic Acid [Bos taurus]	gi|529482051	68 415.6	5.6	16	692	100
5	mitochondrial enolase superfamily member 1 [Bos taurus]	gi|114052721	50 216.4	5.86	13	89	100
6	TPA: alpha-enolase [Bos taurus]	gi|296479148	47 596.5	6.37	15	111	100

(continued)

NO.	Protein Name	Accession No.	Protein MW	Protein PI	Pep. Count	Protein Score	Protein Score C.I.%
7	TPA: serum albumin precursor [Bos taurus]	gi|296486410	71 274.2	5.82	18	101	100
8	transferrin [Bos taurus]	gi|602117	79 869.5	6.75	16	79	99.852
9	PREDICTED: fibrinogen beta chain isoform X1 [Bos taurus]	gi|528981280	56 349.6	8.5	13	107	100
10	Deoxyribonuclease II beta [Bos taurus]	gi|109939926	38 864.2	9.06	8	69	98.594
11	PREDICTED: C-type lectin domain family 9 member A-like isoform X1 [Bos taurus]	gi|528950700	30 415.5	9.23	12	60	100
12	tetranectin precursor [Bos taurus]	gi|114051137	22 586.1	5.47	4	93	99.994

4.3 Analysis of differently expressed proteins

The function and molecular weight of 12 Bos taurus proteins were analyzed according to the NCBI database. From the candidates, ENOSF1 (enolase superfamily members) and transferrin were selected for Western blot analyses due to their correlation with animal reproduction. As shown in Fig. 3, the changes in expression of both selected proteins was consistent with the 2-DE and silver-staining results. These results demonstrated that the proteomic analysis of follicular fluid was a convincing assessment of reproduction response.

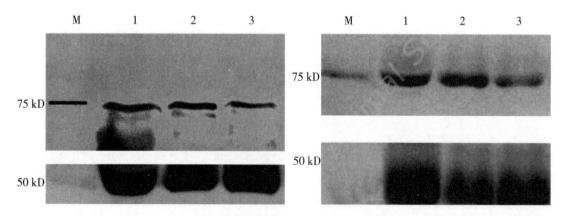

Figure 3 Antibody signals for transferrin and ENOSF1 obtained from there different follicular diameters
(0-5 mm, 5-10 mm, >10 mm)

Note: M: Marker, 1: Follicular diameter >10 mm, 2: Follicular diameter between 5 mm and 10 mm, 3: Follicular diameter <5 mm.

5 DISCUSSION

Yaks have seasonal polyestrous cycles with breeding that occurs from June to November; the largest proportion of female yaks come into estrus in the middle of this period (July to September) in most yak production areas (Zi, 2003; Sarkar and Prakash, 2005). Reproduction in yaks is low as a result of seasonal breeding, delayed puberty, and lower estrus frequency. Therefore, numerous recent studies have focused on reproduction and physiology in yaks, as well as assisted reproductive techniques (Sarkar et al., 2008; Guo et al., 2012). Follicular development is a very complicated process; it plays a role in oocyte maturation and the ovulation mechanism during yak breeding season.

Proteomics is a significant focus of life science studies in the post-genomic era. Proteomics-based studies examine the overall protein expression in tissues or cells, increasing the possibility of answering more complex biological questions. In addition, mass spectrometry techniques have enabled scientists to carry out molecular analysis of peptides and proteins in a sensitive, rapid, and specific manner; they have fundamentally transformed the biological and biochemical study of proteomics (Han et al., 2008). Therefore, an overall analysis of proteomics combined with spectrometry can alleviate challenges in assessing the complex FF composition (Upadhyay et al., 2013). Proteomic-based analysis of FF protein changes during yak follicular development helps to identify proteins that either possibly participate or are even key to follicular development and oocyte maturation; it also helps to determine potential regulatory mechanisms and yak seasonal breeding at a proteome level. Thus, proteomic-based analysis provides new insights for improving yak ovarian reserve capacity and reproductive performance.

Proteins associated with follicle and oocyte maturation and development can be identified by using the proteomic approaches comparing protein profiles in different developmental stages of FF. Spitzer et al. (1996) carried out 2-DE to compare protein profiles between mature and immature human follicles; their results indicated different FF protein contents in different developmental stages could be used as follicular maturation biomarkers. Liu et al. (2007) performed MALDI-TOF-MS to analyze specific FF peptides of different developmental stages by hydrolyzing FF proteins and analyzing final products using MALDI-FOL-MS. A comparison of peaks obtained from samples at different development stages identified 5 specific peaks with a 100% occurrence in mature oocyte FF and 25 specific peaks with a 100% occurrence in immature oocyte FF. By searching the protein database, numerous proteins associated with oocyte maturation were identified. In this study, in order to investigate proteomic changes occurring in follicular fluid between mature and immature yak oocyte follicles, FF proteomes were analyzed using the 2-DE and MALDI-TOF-MS approaches. In total, 34 different expressed protein spots were obtained, 12 of which were successfully identified to correspond with bovine proteins. The 12 proteins selected spots were identified as unique proteins, 9 of which were validated (protein score >100). A comparison of protein profiles identified 12 differently expressed proteins, out of which 10 were upregulated while 2 were downregulated. Functional analyses revealed that most of up- or downregulated proteins were both involved in metabolism, binding, signal transduction, cell developmental pathways and cell structure. The study provides a solid basis for futher research into the process that regulates follicular development and oocyte maturation in yak during estrus.

Western blotting is also called immunoblotting, a type of immunochemical technique used to detect a protein immobilized on a matrix. This study used Western blotting to detect transferrin and ENOSF1 expression. Our results suggested transferrin and ENOSF1 expression were enhanced with follicular development (0-5 mm, 5-10 mm, >10 mm).

5.1 ENOSF1

The enolase superfamily (ENOSF), named after the enolase enzyme of glycolysis, is used as a model of protein superfamily evolution (Gerlt et al., 2005). The enolase superfamily members share a common ENOSF fold and catalyze a common half reaction: they all abstract protons that are adjacent to carboxyl groups from numerous substrates. The ENOSF1 gene was originally identified as a naturally occurring human TYMS gene antisense transcript (Dolnick, 1993) and codes for two proteins (rTSα and rTSβ) through alternative RNA splicing (Dolnick, 1993). The ENOSF1 gene function appears primarily to regulate TYMS locus expression both via the antisense transcript and through encoded proteins (Dolnick et al., 2003). Evolutionarily conserved acidic residues, in loops located at the end of two β-sheets lining the ENOSF fold C-terminal barrel, coordinate with an essential magnesium; they are shared by all ENOSFs. Within the superfamily, different ENOSF families are distinguished by the identity of the third magnesium ligand and the different general acid/base catalytic residue combinations at the ends of the remaining β sheets in the C-terminal barrel (Gerlt et al., 2005). In addition, Finckbeiner (2011) reported ENOSF1 function in vertebrates, showing that ENOSF1 is required for embryonic development. The increased apoptosis following ENOSF1b knockdown suggests a potential survival advantage for increased ENOSF1β expression in human cancers. Our study showed that the ENOSF1 expression level in mature yak FF was higher than in immature yak FF. The protein level increased along with follicle development, indicating ENOSF1 may play a regulatory role and/or can be considered a biomarker for yak follicular development. However, its possible role in a regulating mechanism requires further investigation.

5.2 Transferrin

Transferrin is the main iron transport protein found in the circulation. The transferrin saturation level in the blood is an important indicator of iron status (Kovac et al., 2009). Transferrin is can bind two Fe^{3+} ions with very high affinity and donate iron to cells throughout the body via transferrin receptor 1 (TfR1) (Kovac et al., 2009). The crystal structure of the single transferrin polypeptide chain (which is comprised of 680-690 amino acid residues) has been determined in both the diferric (Bailey et al., 1988) and iron-free (apo-transferrin) forms (Wally et al., 2006). The chain is folded into two lobes (the N-lobe and C-lobe) derived from the N-terminal and C-terminal halves of the protein, respectively. The two lobes share 60% homology and are presumed to arise from gene duplication and fusion (Park et al., 1985). Each lobe is folded into two subdomains which come together to form a cleft, providing a binding site for one ferric ion (Baker et al., 2003). In the serum-free culturing of follicular granulosa cells, transferrin is often added as a growth factor into the medium to stimulate the cell growth and development (Orly et al., 1980). In this study, MALDI-TOF-MS-based isolation and identification of immature and mature yak FF found that transferrin was significantly expressed in mature FF; the protein level increased along with yak follicle development. This result indicated transferrin may significantly stimulate follicular maturation and therefore be a potential biomarker for yak follicular maturation.

6 CONCLUSION

In the breeding season, expression of different proteins in yak follicular fluid may vary over different developmental stages. This study identified 12 differently expressed proteins, of which 10 exhibited upregulated expression and 2 had downregulated expression during follicular maturation. Western blotting showed that transferrin and ENOSF1 expression were enhanced with follicular development. These differently expressed proteins may regulate and influence follicular maturation according to their level of function.

CONFLIST OF INTEREST

We certify that there is no conflict of interest with any financial organization regarding the material discussed in the manuscript.

ACKNOWLEDGEMENTS

The work was supported or partly supported by grants from National Natural Science Foundation of China (31301976), Science and Technology Support Projects in Gansu Province (1504NKCA052), China Agriculture Research System (CARS-38), National Science and Technology Support Project (2013BAD16B09), and the Innovation Project of Chinese Academy of Agricultural Sciences (CAAS-ASTIP-2014-LIHPS-01).

（发表于《Asian Australas. J. Anim. Sci.》，SCI，IF：0.541. 1-7.）

Syntheses, Crystal Structures and Antibacterial Evaluation of Two New Pleuromutilin Derivatives[①]

SHANG Ruofeng[1], XU Shuijin[2], YI Yunpeng[1], AI Xin[1], LIANG Jianping[1*]

(1. Key Laboratory of New Animal Drug Project of Gansu Province/Key Laboratory of Veterinary Pharmaceutical Development, Ministry of Agriculture/Lanzhou Institute of Husbandry and Pharmaceutical Sciences of CAAS, Lanzhou 730050, China; 2. YanchengYouhua Pharmaceutical & Chemical Technology Co., Ltd., Yancheng 224555, China)

Abstract: Two new pleuromutilin derivatives, 14-O-[(4-amino-6-methoxyl-pyrimidine-2-yl)-thioacetyl] mutilin (4) and 14-O-[4-amino-1-methyl-6-oxo-1,6-dihydropyrimidin-2-yl) thioacetyl] mutilin (5), were synthesized and structurally characterized by IR, NMR spectra, HRMS and single-crystal X-ray diffraction. These compounds contain a 5-6-8 tricyclic carbon skeleton and a pyrimidine ring. Compound 4 is in the monoclinic system, space group P12]1 with a = 10.251 7 (4), b = 12.565 5 (4), c = 10.343 5 (4) Å, V = 1 315.69 (8) Å^3, Z = 2, D_c = 1.309 g/cm^3, F (000) = 558, μ= 0.166 mm^{-1}, S = 1.047, R = 0.045 7 and wR = 0.093 4 for 4721 unique reflections (R_{int} = 0.032 2) with I > 2σ (I). Compound 5 belongs to the orthorhombic system, space group P2$_1$2$_1$2$_1$ with a = 7.366 7 (4), b = 13.999 0 (7), c = 29.043 4 (13) A, V = 2 995.1 (2) Å^3, Z = 4, D_c = 1.250 g/cm^3, F (000) =1 216, μ= 0.153 mm^{-1}, S = 1.031, R = 0.054 5 and wR = 0.098 2 for 5242 unique rejections (R_{int} = 0.047 6) with I > 2σ (I)). The in vitro antibacterial activity study showed the title compounds 4 and 5 displayed slightly less activity against methicillin-resistant Staphylococcus aureus (MRSA) and methicillin-resistant Staphylococcus epidermidis (MRSE), and lower potent against Escherichia coli (E. coli) and Bacillus subtilis (B. subtilis) when compared to those of tiamulin fumarate.

Key words: Pleuromutilin; Single-crystal structure; Antibacterial activities

1 INTRODUCTION

Pleuromutilin (1), a natural product composed of a 5-6-8 tricyclic carbon skeleton[1, 2], was first isolated in a crystalline form from cultures of two basidiomycetes species, Pleurotus mutilus and P. passeckerianus, in 1951[3]. The pleuromutilin derivatives inhibit bacterial protein synthesis via a specific interaction with the 23S rRNA of 50S bacterial ribosome subunit[4, 5]. Further studies have shown that the domain V of 23S rRNA at the peptidyl transferase center (PTC) is the pleuro- mutilin derivatives binding site, in which the tricyclic core of the pleuromutilin positions

---

* Corresponding author. Tel: 0931-2115287, E-mail: liangjp1963@163.com
DOI: 10.14102/j.cnki.0254-5861.2011-0896

in a pocket close to the A-tRNA binding site, whereas the C-14 extension points toward the P-tRNA binding site[6, 7]. These compounds have received much investigative attention due to their unique antibacterial mechanism, potent pharmacodynamic properties, and their lack of target-specific crossresistance to other antibiotics[4, 8]. Modification of pleuromutilin has led to three drugs: tiamulin, valnemulin, and retapamulin. The first two drugs are used as veterinary drugs[9, 10], while retapamulin is the first pleuromutilin drug approved for human use in 2007[11, 12].

Our previous work proposed heterocyclic rings bearing polar groups at the C14 side chain of pleuromutilin derivatives raise their antibacterial activity[13]. The compounds with pyrimidine ring were widely found in ature products, such as nucleotides, thiamine (vitaminB1) and alloxan[14]. The pyrimidine system also turned out to be an important pharmacophore, and is found in many synthetic compounds such as antibacterial drug, trimethoprim, barbiturates, the HIV drug, and zido- vudine[14]. Herein, we prepared the title compounds with pyrimidine ring, 14-O-[(4-amino-6-methoxyl-pyrimidine-2-yl) thioacetyl] mutilin (4) and 14-O- [4-amino- 1-methyl-6-oxo-1, 6-dihydropyrimidin-2-yl) - thioacetyl] mutilin (5) as shown in Scheme 1. Furthermore, their single-crystal structures and antibacterial activities are also reported.

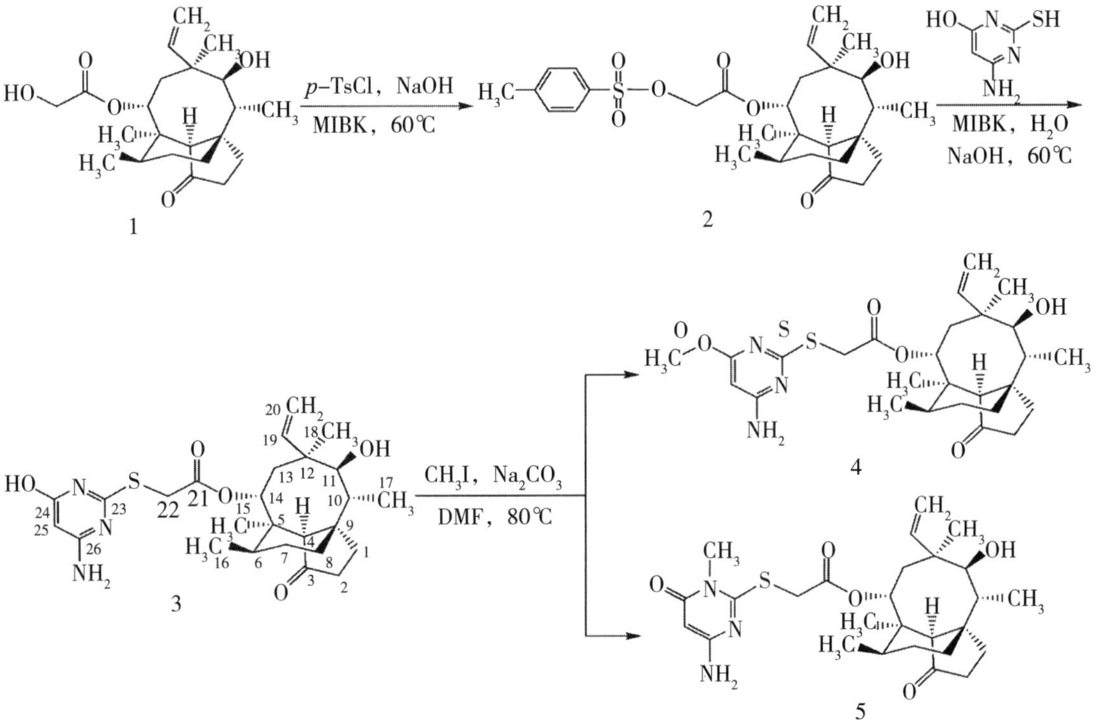

Fig. 1 Synthetic route of the title compound

2 EXPERIMENTAL

2.1 Reagents and instruments

All reagents obtained from commercial sources were of AR grade and used without further purification. IR spectra were obtained on a Thermo Nicolet NEXUS-670 spectrometer and recorded as KBr thin film, and the absorptions are reported in cm^{-1}. NMR spectra were recorded on a Bruker-400 MHz spectrometer in appropriate solvents. Chemical shifts (δ) in ^{1}H NMR were expressed in parts per million (ppm) relative to the tetramethylsi- lane. ^{13}C NMR spectra were recorded on 100 MHz spectrometers. High-resolution mass spectra (HRMS) were obtained with a Bruker Daltonics APEX II 47e mass spectrometer equipped with an electrospray ion source. The single-crystal structures of the title compounds were determined on an Agilent SuperNova X-diffractometer. All reactions were monitored by thin-layer chromatography (TLC) on 0.2 mm thick silica gel GF254 pre-coated plates. After elution, plates were visualized under UV illumination at 254 nm for UV active materials. Column chromatography was carried out on silica gel (200 ~ 300 mesh). The products were eluted in appropriate solvent mixture under air pressure.

Concentration and evaporation of the solvent after reaction or extraction were carried out on a rotary evaporator.

2.2 Synthesis and characterization of the title compound

2.2.1 Synthesis of intermediate 2

A 10 M NaOH (5 mL, 50 mmol) was added dropwise to a mixture of pleuromutilin (7.57 g, 20 mmol) and p-toluenesulfonyl chloride (4.2 g, 22 mmol) in water (5 mL) and t-butyl methyl ether (20 mL). The result mixture was stirred under reflux for 1 h followed by dilution with water (50 mL) and stirred under an ice bath for 15 min. The result suspension was washed with water (50 mL) and cold t-butyl methyl ether (20 mL) 3 times. The crude production was filtered to give 9.8 g compound 2 as a white powder. It was directly used in the next step without further purification[13]. Yield: 93%, m. p.: 147 ~ 148 ℃. IR (KBr): 3446 (OH, s), 2924 (CH_2, s), 2863 (CH_3, s), 1732 (C=O, vs), 1597 (C=C-C=, m), 1456 (CH_2, s), 1371 (CH_3, s), 1297 (S=O, s), 1233 (S=O, s), 1117 (C=O, s), 1035 (C-OH, s), 832 (C=C-C=, m), 664 (CH_2, m) cm^{-1}. ^{1}H NMR (400 MHz, CDCl$_3$) δ 7.80 ~ 7.82 (d, 2H, J = 4.0 Hz), 7.35 ~ 7.37 (d, 2H, J = 4.0 Hz), 6.43 (q, 1H, J = 17.2 Hz, 10.8 Hz, C_{19}-H), 5.75 ~ 5.78 (d, 1H, J = 4.2 Hz, C_{14}-H), 5.31 ~ 5.34 (d, 1H, J = 6.4 Hz, C_{20}-H), 5.17 ~ 5.21 (d, 1H, J = 8.8 Hz, C_{20}-H), 4.48 (s, 2H, C_{22}-H), 3.34 (d, 1H, J = 6.4 Hz, C_{11}-H), 2.45 (s, 3H), 2.21 ~ 2.29 (m, 3H), 2.01 ~ 2.08 (m, 3H), 1.63 ~ 1.65 (dd, 2H, J_1 = 10Hz, J_2 = 7.2 Hz), 1.46 ~ 1.50 (m, 5H), 1.41 ~ 1.44 (m, 1H), 1.33 ~ 1.36 (m, 1H), 1.22 ~ 1.26 (s, 5H), 1.11 ~ 1.15 (m, 1H, C_6-H), 0.87 (d, 3H, J = 6.8Hz, C_{16}-H) 0.63 (d, 3H, J = 6.8 Hz, C_{17}-H). ^{13}C NMR (100 MHz, CDCl$_3$) δ 216.7 (C_3), 164.8 (C_{21}), 145.2 (C_{19}), 138.6, 132.5, 129.9, 127.9, 117.2 (C_{20}), 74.4 (C_{14}), 70.2 (C_{11}), 64.9 (C_{22}), 57.9 (C_4), 45.3, 44.4, 43.9, 41.7, 36.4, 35.9, 34.3, 30.2, 26.7, 26.3, 24.7, 21.6, 16.4 (C_{16}), 14.6 (C_{15}), 11.4 (C_{17}). HRMS (ESI): [M + Na]$^+$ calcd. for $C_{29}H_{40}O_7S$, 533.2501. Found, 533.2507.

2.2.2 Synthesis of intermediate 3

To a solution of 4-amino-6-hydroxy-2-mercapto-pyrimidine monohydrate (1.65 g, 10 mmol) in 20 mL methanol, 10 M NaOH (1.1 mL, 11 mmol) was added and stirred for 30 min. A solution of compound 2 (5.33, 10 mmol) in 20 mL of DCM was added dropwise to the reaction mixture. The mixture was stirred for 40 h at room temperature and evaporated under reduced pressure to dryness. The crude product was extracted with a solution of ethyl acetate (60 mL) and water (20 mL) and treated with saturated $NaHCO_3$. The target compound 3 was then precipitated from the solution and purified by flash silica column chromatography (ethyl acetate: ethanol 20 : 1 v/v) to yield 3.93 g (78%), m. p.: 204~206 ℃. IR (KBr) 3368 (NH, m), 2930 (CH_2, s), 1729 (C=O, s), 1629 (C=O, vs), 1575 (C=C-C=, s), 1542 (C=C-C=, s), 1457 (CH_2, s), 1286 (OH, s), 1117 (C=O, m), 1019 (C-O, m), 981 (C-H, m), 809 (C-H, m) cm^{-1}. ^{1}H-NMR (400 MHz, $CDCl_3$) δ 6.47 (dd, J = 17.4, 11.1 Hz, 1H, C_{19}-H), 5.74 (d, J = 8.3 Hz, 1H, C_{14}-H), 5.24~5.13 (m, 2H, C_{20}-H), 4.79 (s, 2H, C_{26}-NH_2), 4.11 (q, J = 7.1 Hz, 1H), C_{25}-H, 3.85 (d, J = 16.5 Hz, 1H, C_{22}-H), 3.69 (d, J = 16.5 Hz, 1H, C_{22}-H), 3.35 (d, J = 6.2 Hz, 1H, C_{11}-H), 2.29~1.98 (m, 6H), 1.75 (d, J = 14.1 Hz, 1H), 1.63 (dd, J = 21.3, 11.1 Hz, 2H), 1.56~1.18 (m, 9H), 1.10 (d, J = 20.2 Hz, 4H), 0.86 (d, J = 6.8 Hz, 3H), C_{16}-H, 0.71 (d, J = 6.9 Hz, 3H, C17-H). ^{13}C NMR (100 MHz, $CDCl_3$) δ 216.9 (C_3), 166.9 (C_{21}), 166.0 (C_{24}), 162.7 (C_{26}), 160.3 (C_{23}), 139.2 (C_{19}), 117.2 (C_{20}), 84.0 (C_{25}), 74.6 (C_{14}), 70.1 (C_{11}), 60.4 (C_{22}), 58.1 (C_4), 45.4, 43.9, 41.9, 36.7, 36.0, 33.3, 30.4, 26.9, 26.3, 24.8, 21.0, 16.8 (C_{16}), 14.9 (C_{15}), 11.5 (C_{17}). HRMS (ESI): $[M+H]^+$ calcd. for $C_{26}H_{37}N_3O_5S$ 504.2526. Found 504.2520.

2.2.3 Syntheses of compounds 4 and 5

Methyl iodide (0.71 g, 5.0 mmol) was added dropwise into a suspension of compound 3 (2.51 g, 5.0 mmol) in anhydrous DMF (20 mL) and K_2CO_3 (2.07 g, 15 mmol). The reaction mixture was stirred at 80~85 ℃ for 4 h. After the mixture was cooled, the inorganic material was taken off by filtration and the solvent was removed in vacuum. The solid residue, consisting of a mixture of 4 and 5 isomers, was obtained and purified using silica column chromatography (petroleum ether: ethyl acetate = 1 : 15, v/v) to yield 0.75 g (29%) and 1.58 g (61%) of compounds 4 and 5, respectively. Compound 4: m. p.: 183~185 ℃. IR (KBr) 3374 (NH, m), 2928 (CH_2, s), 1732 (C=O, s), 1626 (C=O, vs), 1585 (C=C-C=, s), 1548 (C=C-C=, s), 1390 (OH, s), 1307 (C-O-C, s), 1273 (OH, s), 1212 (O-C=O, s), 1152 (C-O, s), 1117 (C=O, m), 1048 (C-O-C, s), 1017 (C-O, m), 982 (C-H, m), 917 (C-H, m) cm^{-1}. ^{1}H NMR (400 MHz, DMSO-d6) δ 6.67 (s, 2H, C_{26}-NH_2), 6.11 (dd, J = 17.7, 11.2 Hz, 1H, C_{19}-H), 5.52 (d, J = 8.2 Hz, 1H, C_{14}-H), 5.02 (dd, J = 23.6, 14.4 Hz, 2H, C_{20}-H), 4.49 (d, J = 6.0 Hz, 1H, C_{25}-H), 3.85 (m, 2H, C_{22}-H), 3.75 (s, 3H, O-CH_3), 3.41 (t, J = 5.7 Hz, 1H C_{11}-H), 2.39 (s, 1H), 2.32~1.92 (m, 5H), 1.63 (s, 2H), 1.49~1.15 (m, 9H), 1.02 (s, 5H), 0.80 (d, J = 6.8 Hz, 3H, C_{16}-H), 0.58 (d, J = 6.8 Hz, 3H, C_{17}-H). ^{13}C NMR (100 MHz, DMSO-d6) δ217.6 (C_3), 169.5 (C_{21}), 168.3 (C_{24}), 168.1 (C_{26}), 165.6 (C_{23}), 141.3 (C_{19}), 115.7 (C_{20}), 82.0 (C_{25}), 73.1 (C_{14}), 70.2 (C_{11}), 60.2 (C_{22}), 57.7 (C4), 53.7, 45.4, 44.5, 42.0, 36.8, 34.5, 33.7, 30.6, 28.9, 27.1, 24.9, 21.2, 16.6 (C_{16}), 15.0 (C_{15}), 12.0 (C_{17}). HRMS (ESI): $[M+H]^+$ calcd. for

$C_{27}H_{39}N_3O_5S$ 518.268 3. Found 518.268 1.

Compound 5: m. p.: 218~221 ℃. IR (KBr) 3422 (NH, m), 2929 (CH_2, s), 1730 (C=O, s), 1630 (C=O, vs), 1509 (N-H, s), 1457 (C-N, s), 1414 (OH, s), 1282 (OH, s), 1154 (C-O, s), 1117 (C=O, m), 1094 (C-N, s), 807 (C-H, m) cm^{-1}. ^{1}H NMR (400 MHz, DMSO-d6) δ 6.28 (s, 2H, C_{26}-NH_2), 6.10 (dd, J = 17.8, 11.2 Hz, 1H, C_{19}-H), 5.52 (d, J = 8.1 Hz, 1H, C_{14}-H), 5.05 (dd, J = 27.5, 14.5 Hz, 2H, C_{20}-H), 4.52 (d, J = 5.9 Hz, 1H, C_{25}-H)), 4.07~3.96 (m, 2H, C_{22}-H), 3.42 (d, J = 5.4 Hz, 1H, C_{11}-H), 3.34 (s, 3H, O-CH_3), 2.40 (s, 1H), 2.22~1.93 (m, 5H), 1.62 (dd, J - 24.8, 12.6 Hz, 2H), 1.52~1.13 (m, 9H), 1.05 (s, 4H), 0.82 (d, J = 6.8 Hz, 3H, C_{16}-H), 0.61 (d, J = 6.8 Hz, 3H, C17-H). ^{13}C NMR (100 MHz, DMSO-d6) δ217.6 (C_3), 170.8 (C_{21}), 166.8, (C_{24}) 161.9 (C_{26}), 161.5 C_{23}), 141.2 (C_{19}), 115.9 (C_{20}), 81.2, (C_{25}) 73.1 (C_{14}), 70.8 (C_{11}), 60.2 (C_{22}), 57.7 (C_4), 45.4, 44.6, 43.9, 42.0, 36.8, 34.6, 30.6, 29.1, 27.1, 24.9, 21.2, 16.6 (C_{16}), 14.9 (C_{15}), 12.0 (C_{17}); HRMS (ESI): $[M + H]^+$ calcd. for $C_{27}H_{39}N_3O_5S$ 518.268 3. Found 518.269 6.

2.3 Crystal data and structure determination

The crystals of the title compounds 4 and 5 were obtained from the mixture of dichloromethane and ethanol with 1 : 1 (v/v) for about twenty days at room temperature. Two colorless single crystals with dimensions of 0.32mm × 0.27mm × 0.25mm and 0.31mm × 0 25mm × 0 24mm, respectively were selected and mounted in air onto thin glass fibers. X-ray intensity data were collected at 294.58 (10) and 173.00 (10) K, respectively on an Agilent SuperNova-CCD diffractometer equipped with a mirror-monochromatic MoKα (λ = 0.710 73 Å) radiation. A total of reflections of compound 4 were collected in the range of 3.54≤θ≤26.01° (index ranges: −12≤h≤12, −15≤k≤15, −12≤l≤12) by using an ω-2θ scan mode with 4721 independent ones (R_{int} = 0.032 2), of which 4010 with I > 2σ (I) were considered as observed and used in the succeeding refinements. A total of 8695 reflections for crystal 5 including 5242 independent ones (R_{int} =0.047 6) were collected in the range of 3.43≤θ≤26.02° (index ranges: −8≤h≤9, −9≤k≤17, −20 ≤l≤35) by using the same scan mode as compound 4.

The structures were solved by direct methods using ShelXS program[15] and refined with ShelXL program[16] by full-matrix least-squares techniques on F^2. The non-hydrogen atoms were refined aniso-tropically, and hydrogen atoms were determined with theoretical calculations. For compound 4, the full-matrix least-squares refinement gave the final R =0.0457, wR = 0.0934 (w = 1/[$σ^2$ (F_O^2) + (0.379P)^2 + 0.252 0P], where P = (F_O^2 + 2_C^2)/3), (Δ/σ)$_{max}$ = 0.000, S = 1.047, (Δρ)$_{max}$ = 0.170 and (Δρ)$_{min}$ = −0.318 e/Å^3. For compound 5, the final R = 0.054 5, wR = 0.098 2 (w = 1/[$σ^2$ (F_O^2) + (0.370P)^2 + 0.0000P], where P = (F_O^2 + $2F_C^2$)/3), (Δ/σ)$_{max}$ = 0.000, S = 1.031, (Δρ)$_{max}$ = 0.217, and (Δρ)$_{min}$ = −0.218 e/Å^3.

2.4 In vitro antibacterial activities

Compounds 3, 4 and 5 were further tested for their minimum inhibitory concentrations (MIC) and Oxford cup assay against methicillin-resistant Staphylococcus aureus ATCC 29213

(MRSA), methicillin-resistant Staphylococus epidermidis ATCC 35984 (MRSE), Bacillus subtilis ATCC 11778 (B. subtilis), and Escherichia coli ATCC 25922 (E. coli) according to the literature[13]. Tia- mulin fumarate was co-assayed as a reference drug.

3 RESULTS AND DISCUSSION

The title compounds 4 and 5 were prepared according to Scheme 1. Almost all pleuromutilin derivatives were synthesized from 22-O-tosylpleuromutilin (2) which was obtained by the reaction of pleuro- mutilin and p-toluenesulfonyl chloride to activate the 22-hydroxyl of pleuromutilin. The intermediate 3 was then obtained in 78% yield by the nucleophilic attack of 4-amino-6-hydroxy-2-mercaptopyrimidine monohydrate on compound 2 under alkaline conditions. The title compounds 4 and 5 were simultaneously obtained as isomers (1.0 : 2.1 ratio) from 3 in the presence of K_2CO_3 because of the keto-enol equilibrium of the pyrimidinone scaffold[17]. The IR, ^{1}H NMR, ^{13}C NMR and H RMS for the obtained compounds are all in good agreement with the assumed structures. Two single crystals of compounds 4 and 5 were cultured for X-ray diffraction analysis to confirm their configurations. The selected bond lengths and bond angles for 4 and 5 are shown in Tables 1 and 2, respectively. And the hydrogen bonding parameters for these two compounds are shown in Table 3.

Table 1 Selected bond lengths (Å) and bond angles (°) for compound 4

Bond	Dist.	Bond	Dist.	Bond	Dist.
S(1)-C(10)	1.796(3)	O(5)-C(22)	1.202(3)	C(3)-C(17)	1.516(5)
S(1)-C(11)	1.763(3)	N(1)-C(11)	1.329(4)	C(5)-C(19)	1.533(4)
O(1)-C(1)	1.473(3)	N(1)-C(14)	1.336(3)	C(6)-C(23)	1.536(4)
O(1)-C(9)	1.324(3)	N(2)-C(11)	1.318(4)	C(7)-C(22)	1.543(4)
O(2)-C(9)	1.208(3)	N(2)-C(12)	1.357(4)	C(8)-C(25)	1.550(4)
O(3)-C(14)	1.347(3)	N(3)-C(12)	1.363(4)	C(8)-C(27)	1.545(4)
O(3)-C(15)	1.440(4)	C(1)-C(8)	1.560(4)	C(9)-C(10)	1.505(4)
O(4)-C(4)	1.431(3)	C(3)-C(16)	1.533(4)	C(13)-C(14)	1.367(4)
Angle	(°)	Angle	(°)	Angle	(°)
C(11)-S(1)-C(10)	100.82(14)	O(2)-C(9)-O(1)	125.3(3)	N(1)-C(11)-S(1)	118.1(2)
C(9)-O(1)-C(1)	119.8(2)	O(2)-C(9)-C(10)	122.6(3)	O(3)-C(14)-C(13)	117.6(3)
C(14)-O(3)-C(15)	117.5(2)	O(1)-C(9)-C(10)	112.1(3)	N(1)-C(14)-O(3)	118.4(3)
C(11)-N(1)-C(14)	114.3(3)	C(9)-C(10)-S(1)	117.7(2)	N(1)-C(14)-C(13)	124.0(3)
C(11)-N(2)-C(12)	116.1(2)	N(1)-C(11)-S(1)	118.1(2)	O(3)-C(14)-C(13)	117.6(3)

321

Bond	Dist.	Bond	Dist.	Bond	Dist.
O(1)-C(1)-C(2)	106.3(2)	N(2)-C(11)-S(1)	113.8(2)	O(5)-C(22)-C(7)	128.5(3)
O(1)-C(1)-C(8)	104.42(19)	N(2)-C(11)-N(1)	128.1(3)	O(5)-C(22)-C(21)	123.3(3)
O(4)-C(4)-C(3)	109.1(2)	N(2)-C(12)-N(3)	116.5(3)	O(5)-C(22)-C(7)	128.5(3)
O(4)-C(4)-C(5)	112.6(2)	N(2)-C(12)-C(13)	120.8(3)	O(5)-C(22)-C(21)	123.3(3)
O(1)-C(9)-C(10)	112.1(3)	N(3)-C(12)-C(13)	122.7(3)		

Table 2 Selected bond lengths (Å) and bond angles (°) for compound 5

Bond	Dist.	Bond	Dist.	Bond	Dist.
S(1)-C(22)	1.811(3)	N(1)-C(23)	1.356(4)	C(1)-C(11)	1.545(5)
S(1)-C(23)	1.760(3)	N(1)-C(26)	1.418(4)	C(4)-C(15)	1.516(5)
O(1)-C(5)	1.435(4)	N(1)-C(27)	1.474(4)	C(4)-C(17)	1.536(5)
O(2)-C(12)	1.204(4)	N(2)-C(23)	1.298(4)	C(11)-C(19)	1.536(4)
O(3)-C(2)	1.478(3)	N(2)-C(24)	1.379(4)	C(12)-C(13)	1.517(5)
O(3)-C(21)	1.339(4)	N(3)-C(24)	1.339(4)	C(21)-C(22)	1.508(4)
O(4)-C(21)	1.203(4)	O(6)-C(28)	1.423(5)	C(24)-C(25)	1.372(5)
O(5)-C(26)	1.245(4)	C(1)-C(8)	1.550(4)	C(28)-C(29)	1.482(6)
Angle	(°)	Angle	(°)	Angle	(°)
C(23)-S(1)-C(22)	99.35(16)	O(3)-C(2)-C(3)	106.1(2)	N(2)-C(23)-S(1)	120.1(3)
C(5)-O(1)-H(1)	109.5	O(2)-C(12)-C(8)	128.6(3)	N(2)-C(23)-N(1)	125.6(3)
C(21)-O(3)-C(2)	117.1(2)	O(2)-C(12)-C(13)	123.5(4)	N(3)-C(24)-N(2)	114.9(3)
C(23)-N(1)-C(26)	120.1(3)	O(3)-C(21)-C(22)	109.5(3)	N(3)-C(24)-C(25)	122.8(3)
C(23)-N(1)-C(27)	122.1(3)	O(4)-C(21)-O(3)	125.8(3)	C(25)-C(24)-N(2)	122.3(3)
C(26)-N(1)-C(27)	117.8(3)	O(4)-C(21)-C(22)	124.7(3)	O(5)-C(26)-N(1)	118.2(3)
C(23)-N(2)-C(24)	116.0(3)	C(21)-C(22)-S(1)	106.6(2)	O(5)-C(26)-C(25)	127.1(3)
O(3)-C(2)-C(1)	108.1(2)	N(1)-C(23)-S(1)	114.3(2)	C(25)-C(26)-N(1)	114.7(3)

Table 3 Hydrogen bond lengths (Å) and bond angles (°) for compounds 4 and 5

Compound	D-H-A	d(D-H)	d(H···A)	d(D-A)	∠DHA
	N(2)-H(2)···O(4) [a]	0.86	2.16	2.935(3)	150
	N(3)-H(3A)···O(2) [b]	0.86	2.31	3.121(3)	157

(continued)

Compound	D-H-A	d(D-H)	d(H···A)	d(D-A)	∠DHA
4	O(3)-H(3B)···N(4) (c)	0.86	2.22	3.039(3)	158
	O(4)-H(4)···N(2) (d)	0.82	2.26	2.935(3)	140
	C(16)-H(16B)···O(4)	0.96	2.42	2.779(4)	102
	C(27)-H(27B)···O(5)	0.96	2.21	2.936(4)	131
	N(3)-H(3A)···O(5) (e)	0.90	1.99	2.864(4)	162
	N(3)-H(3B)···O(1) (f)	0.98	1.94	2.917(4)	172
	O(6)-H(6A)···O(5) (g)	0.80	1.92	2.717(4)	179
	O(1)-H(1)···O(6)	0.84	1.91	2.718(4)	163
	C(2)-H(2)···O(4)	1.00	2.29	2.738(4)	106
	C(20)-H(20A)···O(2)	0.98	2.34	2.925(5)	118
	C(20)-H(20C)···O(3)	0.98	2.33	2.784(4)	107
	C(27)-H(27B)···O(5)	0.98	2.25	2.701(4)	107

Symmetry codes: (a) 1+x, y, z; (b) -x, -1/2+y, 1-z; (c) 1+x, y, z; (d) -1+x, y, z; (e) -1+x, y, z; (f) -3/2-x, -1-y, 1/2+z; (g) -1/2-x, -1-y, -1/2+z

The crystal structures of compounds 4 and 5 in Fig. 1 show that the two molecules consist of a 5-6-8 tricyclic carbon skeleton and a pyrimidine ring, respectively, in which all bond lengths are in normal ranges. The five-membered rings in 4 and 5 (C(7), C(22), C(21), C(20), C(6) for compound 4 and C(7), C(8), C(12), C(13), C(14) for compound 5) are not planar. Both of the eight-membered rings in two molecules exhibit a boat conformation and the six-membered rings exhibit a chair conformation.

The side chain of compound 4 exhibits a zig-zag conformation with the dihedral angle of 69.72(19)° between the plane formed by O(1), C(9) and C(10) and another plane generated by C(10), S(1) and C(11) atoms. The side chain of compound 5 also exhibits the zig-zag conformation but the dihedral angle between planes O(3)-C(21)-C(22) and C(22)-S(1)-C(23) is 15.91(195)°. For the ester groups on the side chains, the bond lengths of C=O (O(2)-C(9) in 4 and O(4)-C(21) in 5 are shorter than that of C-O (O(1)-C(9) in 4 and O(3)-C(21) in 5, which may be caused by the conjugation with carbonyl (C=O). The pyrimidine ring and the plane formed by N(2), C(4) and S(2) atoms in compound 4 is slightly nonplanar with the dihedral angle of 1.094(148)°.

The molecules of compound 4 are linked together by four noticeable intermolecular hydrogen bonds, among which two intermolecular hydrogen bonds are formed between C atoms of the eight-membered ring and O atoms of hydroxyl and carbonyl groups, respectively. These intermolecular contacts link the molecules into a three-dimensional network (Fig. 2A). One CH_3CH_2OH molecule cocrystallizes in the crystal structure of compound 5. There are three intermolecular hydrogen bonds, N(3)-H(3A)···O(5), N(3)-H(3B)···O(1), and O(6)-H

(6A)…O(5), which are formed among two title molecules and alcohol. These intermolecular hydrogen bonds link the molecules to give a three-dimensional network and play key roles in stabilizing the crystal packing (Fig. 2B).

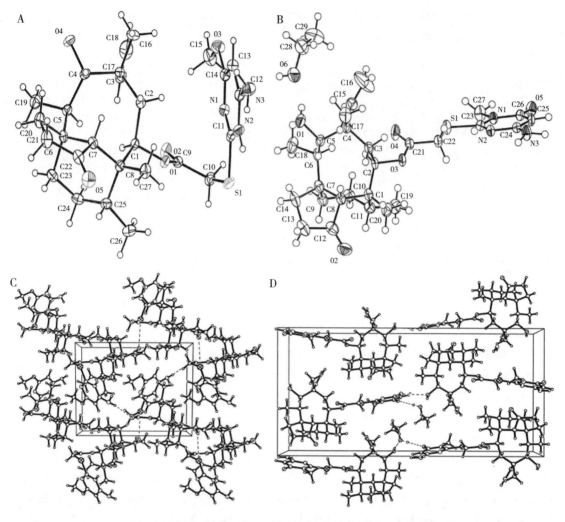

Fig. 2 Molecular packing of compounds 4 (A) and 5 (B)

The synthesized pleuromutilin derivatives 3, 4 and 5 along with tiamulin fumarate used as reference drug were screened for their in vitro antibacterial activity against MRSA, MRSE, E. coli and B. subtilis using the agar dilution method for MIC and Oxford cup assay for zones of inhibition. The antibacterial activities are reported in Table 4. All screened compounds showed potent activity against MRSA and MRSE, but slightly less potent against E. coli and B. subtilis. Compound 3 showed the most antibacterial activities which were superior to or comparable with that of tiamulin fumarate. Compounds 4 and 5 showed slightly less activity against MRSA and MRSE and lower potent against E. coli and B. subtilis when compared to that of tiamulin fumarate.

Table 4 MIC and zones of inhibition of compounds 3, 4, 5 and tiamulin
fumarate for MRSA, MRSE, *E. coli* and *S. subtilis*.

Compound	MIC (μg/mL)				Zone of inhibition (mm)							
	MRSA	MRSE	*E. coli*	*S. subtilis*	MRSA (μg/mL) 320	160	MRSE (μg/mL) 320	160	*E. coli* (μg/mL) 320	160	*S. subtilis* (μg/mL) 320	160
3	0.25	0.125	16	32	17.92	16.82	18.22	16.97	15.25	13.46	13.75	11.19
4	1	0.5	32	16	17.28	15.70	17.53	16.19	13.67	12.06	14.95	13.64
5	2	1	32	16	16.85	15.22	16.95	16.03	13.21	11.73	14.55	13.32
Tiamulin fumarate	0.5	0.25	8	32	17.44	16.35	15.84	14.08	15.97	13.85	13.72	12.17

（发表于《CHINESE J STRUC CHEM》，SCI，IF：0.538. 529–536.）

Flavonoids and Phenolics from the Flowers of *Limonium aureum*[*]

LIU Yu[*], SHANG Ruofeng, CHENG Fusheng, WANG Xuehong,
HAO Baocheng, LIANG Jianping

(Key Laboratory of New Animal Drug Project, Gansu Province, Key Laboratory of Veterinary Pharmaceutical Development, Ministry of Agriculture, Lanzhou Institute of Husbandry and Pharmaceutical Sciences of CAAS, 730050, Lanzhou, China)

Plants of the Limonium aureum (L.) Hill ex Kuntze (Plumbaginaceae) are rich sources of flavonoids and phenolics [1, 2]. Plants of the Limonium aureum were reported to possess blood invigorating and activating properties and hemostatic and anticancer activity [3].

We studied the flavonoid and phenolic compositions of the flowers of Limonium aureum collected from the north of China, in September 2010 (Gansu region, PRC).

Air-dried flowers (1.0 kg) were exhaustively extracted with ethanol. The extract was evaporated to a thick consistency and diluted with 400 mL water. The flavonoids and phenolics were extracted with ethyl acetate and chromatographed over polyamide, silica gel, and Sephadex LH-20 in our continuing chemical examination of this plant. The compounds were isolated from the flowers of Limonium aureum and identified by UV, mass, and NMR spectra and comparison with authentic samples.

Eriodictyol, white needle crystals, mp 266~267℃. UV (MeOH, λ_{max}, nm): 286, 321. ^{1}H NMR and ^{13}CNMR confirmed the identity of the compound as eriodictyol [4].

Luteolin, mp 330~332℃. UV (MeOH, λ_{max}, nm): 255, 270, and 350. MS (EI): 287 (18), 286 (100), 258 (12), 153 (20), 134 (6), 129 (10), 62 (21), and 44 (47). Authentic samples were used to identify luteolin [5].

Apigenin, mp 346~347℃. UV (MeOH, λ_{max}, nm): 267, 334. MS (EI): 270 (100), 242 (13), 153 (15), 152 (10), 121 (11), 96 (3), 69 (4) and 43 (7). Authentic samples were used to identify apigenin [6].

5,7-Dihydroxychromone, white needle crystals, mp 245~247℃. ^{1}H NMR (400 MHz, DMSO-d_6, δ, ppm, J/Hz): 6.07 (1H, d, J = 6, H-6), 6.13 (2H, d, J = 2.0, H-3), 6.24 (2H, d, J = 2.4, H-8), 7.78 (1H, d, J = 6, H-2), 10.30 (1H, s, H-7), 12.50 (1H, s, H-5). ^{13}C NMR (100 MHz, DMSO-d_6, δ, ppm): 93.7 (C-8), 98.9 (C-6), 105.0 (C-10), 110.3 (C-3), 155.6 (C-2), 157.6 (C-4), 161.5 (C-5), 164.0 (C-7),

---

[*] Corresponding author, E-mail: liuyu8108@163.com.

181.0（C-4），characterized as 5, 7-dihydroxychromone [7].

Quercetin 3-O-β-D-xyloside, yellow powder, mp 202～204 ℃. ^{1}H NMR（400 MHz, DMSO-d$_6$, δ, ppm, J/Hz）: 2.95-3.44（sugar H）, 5.35（1H, d, J = 7.0, H-1″）, 6.19（1H, br.s, H-6）, 6.39（1H, br.s, H-8）, 6.85（1H, d, J = 8.5, H-5′）, 7.54（1H, dd, J = 2.0, 8.5, H-6′）, 7.56（1H, d, J = 2.0, H-2′）^{13}C NMR（100 MHz, DMSO-d$_6$, δ, ppm）: 66.0（C-5″）, 70.3（C-4″）, 74.6（C-2″）, 76.9（C-3″）, 94.1（C-8）, 99.2（C-6）, 102.3（C-1″）, 104.4（C-10）, 116.3（C-2′）, 116.6（C-5′）, 121.0（C-1′）, 121.9（C-6′）, 133.2（C-3）, 146.3（C-3′）, 150.1（C-4′）, 156.1（C-2）, 156.1（C-9）, 162.6（C-5）, 164.2（C-7）, 177.2（C-4）, characterized as quercetin-3-O-β-D-xyloside [8].

Myricetin 3-β-D-xylopyranoside, mp 186～187℃. UV spectrum（MeOH, λ$_{max}$, nm）: 256, 284, 360. ^{1}H NMR（400 MHz, CD$_3$OD, δ, ppm, J/Hz）: 3.50（1H, dd, J = 3, H-5″）, 3.67（1H, dd, J = 3, H-4″）, 3.80-3.87（2H, m, H-2″, 3″）, 5.19（1H, d, J = 5.8, H-1″）, 6.22（1H, d, J = 2）, 6.41（1H, d, J = 2）, 7.33（2H, s）; acid hydrolysis produced myricetin and xylose, characterized as myricetin 3-O-β-D-xylopyranoside [9].

This research was supported by the Natural Science Foundation of Gansu Province, China（No. 1010RJZA004）, Central level public welfare scientific research institutes funds for Basic scientific research（No. 1610322013006）, and "Five-Year" plan of national science and technology projects in rural areas（No. 2011AA10A214）.

（发表于《Chemistry of Natural Compounds》，SCI，IF：0.509. 130–131.）

The Role of Porcine Reproductive and Respiratory Syndrome Virus as a Risk Factor in the Outbreak of Porcine Epidemic Diarrhea in Immunized Swine Herds

HUANG Meizhou[*], WANG Hui[*], WANG Shengyi,
CUI Dongan, TUO Xin, LIU Yongming[**]

(Engineering and Technology Research Center of Traditional Chinese Veterinary Medicine of Gansu Province, Key Lab of New Animal Drug Project of Gansu Province, Key Lab of Veterinary Pharmaceutical Development of Ministry of Agriculture, Lanzhou Institute of Husbandry and Pharmaceutical Sciences of Chinese Academy of Agricultural Sciences, Lanzhou, Gansu, China)

Abstract: The aim of the study was to investigate the relation between porcine epidemic diarrhea (PED) and porcine reproductive and respiratory syndrome virus (PRRSV). Here, we constructed a dynamic clinical experimental model in which sows were infected with PRRSV. The stability of the experimental model was surveyed by detecting the levels of antibodies against classical swine fever virus (CSFV) or PRRSV, respectively. In both group A (infected group) and group B (healthy group), the CSFV antibody level was not significantly different ($P > 0.05$). In group B, the PRRSV antibody level was also not significantly different ($P > 0.05$) at the three times of testing. The incidence of porcine epidemic diarrhea virus (PEDV) in piglets and PEDV-carrying rate of sows was detected by real-time fluorescent RT-PCR. In group A, PEDV infection of piglets and the PEDV-carrying rate of sows were significantly higher than in group B ($P < 0.01$) and the PEDV strains were within the same group according to the phylogenetic analysis based on the complete S gene. The results suggested that PRRSV was a risk factor in the outbreak of PED in immunized swine herds. Besides that, the experimental model was stable and no interference factors affected the results of the study.

Key words: Porcine; Diarrhea; Lactation; Piglet; Virus; Immunity

1 INTRODUCTION

Porcine epidemic diarrhea (PED) is considered to be a devastating disease for producers, and the spread of infection is very rapid (1). Clinical symptoms include severe enteritis, vomiting, and watery diarrhea, with high infectivity and lethality in piglets, which causes great financial losses (2, 3). Therefore, intensive swine commercial farms vaccinate with a combination of killed

---

[*] These authors contributed equally to this work.
[**] Corresponding author, E-mail: myslym@sina.com

or attenuated vaccines against transmissible gastroenteritis virus (TGEV) and PEDV infection (4-6). Since the outbreak of PEDV in 2010, it has rapidly spread throughout China (7, 8). Although most large-scale pig farms vaccinate according to a proper immunization schedule, PEDV still emerges in immunized swine herds (4).

Porcine epidemic diarrhea virus (PEDV) is an enveloped, single-stranded, positive-sense RNA virus that mutates easily. Molecular epidemiology studies have shown that virus mutation is one of the causes of PEDV outbreaks (9-11). Some intensive swine commercial farms have used autogenous vaccines that complied with basic quality and safety requirements, but failed to control outbreaks of PEDV (12-14). Porcine reproductive and respiratory syndrome (PRRS) is considered to be one of the most important diseases affecting pigs. This is mainly due to its impact on production, especially as the virus is putatively immunosuppressive and concurrent diseases are common (15). Animals known to have been infected with PRRSV have a noticeable increase in the number of secondary viral and bacterial infections (15-18). Piglets are protected against PEDV by specific IgG antibodies from the colostrum and milk of immune sows until they are 4 to 13 days old. The duration of immunity depends on the maternal antibody titer (3, 19). If a sow's immune system was suppressed, the piglet's antibody level of PEDV would be decreased. Moreover, PEDV mainly infects piglets at 3 to 7 days of age. Hence, whether PRRSV-infected sows are a risk factor for PED occurrence in immunized swine herds has not been documented. Here, we provide proof to verify this hypothesis by constructing a model of PRRSV infected-sows and by detecting the PEDV-positive piglets and the rate of PEDV carrier sows in the model.

2 MATERIALS AND METHODS

2.1 Study subjects

The trial was conducted between April 2014 and January 2015 in the Gansu Province of Northwest China. There was a persistent outbreak of PEDV in the area during that time period. The study included a total of 2 intensive swine commercial farms with 840 ternary breeding sows. They were managed under an intensive husbandry system with similar health, nutrition, and husbandry practices. Following the protocols approved by the Institutional Animal Care and Use Committee of Lanzhou Institute of Husbandry and Pharmaceutical Sciences of the Chinese Academy of Agricultural Sciences (animal use permit: SCXK20008-0010), the animals utilized in these experiments were treated humanely and with respect for the alleviation of suffering.

2.2 Experimental protocol

The experiment consisted of five parts (Fig. 1). The first part involved constructing the experimental model by enrollment and exclusion. The experimental model was divided into two groups. The sows were enrolled into group A (PRRSV-infected group) according to the following enrollment criteria. The sows were infected with PRRSV but did not show any clinical symptoms and had a good mental and physical state and appetite. The sows were confirmed not to carry any other contagions, such as classical swine fever virus (CSFV) or porcine circovirus (PCV), and there was a clinical history of veterinary quarantine and clinical records. The sows were enrolled into group B (healthy group) according to the same enrollment criteria as above except that the sows were negative for PRRSV. If sows were negative for PRRSV in group A, they were excluded and their piglets were not included in the final results. In group B, if sows were positive for PRRSV, they were excluded from the study. Additionally, seriously ill sows were excluded from the study. All sows had similar parities (3 to 5 births) in both group A and group B.

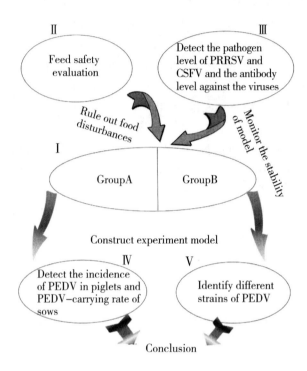

Fig. 1 The framework of experiment.

The second part was feed safety evaluation, including the detection of PEDV and basic nutritional composition analysis, which was used to rule out food disturbances in the experiment because previous research has documented that contaminated feed is a risk factor for PEDV (20).

The third part was monitoring the stability of the experiment model by detecting pathogen levels of PRRSV and CSFV and the antibody levels against the viruses. The experiment was designed by randomized complete blocks and differentiated stages by block. The different stages of sows and piglets were divided into three blocks: gestation sows, lactation sows, and 1- to 7-day-old suckling piglets.

The fourth part involved detecting the incidence of PEDV in piglets and the PEDV-carrying rate of sows. The fifth part was identifying different strains of PEDV based on the complete S gene, which confirmed the PEDV strains at the two farms during the experiment.

2.3 Sample collection

To evaluate feed safety, the feed of pregnant and lactating sows was collected from the two farms every 3 months during the experiment.

Blood samples were collected from the anterior vena cava at the two farms every 3 months from April 2014 to January 2015. The blood was held at room temperature for 2 h and then centrifuged at 3 000 × g for 10 min. The serum was transferred to new centrifuge tubes and stored at −20 ℃ prior to detecting PRRSV and the antibody levels against CSFV and PRRSV. Fecal scores were used to.

identify diarrhea in piglets following the following fecal scoring system: 1 = hard, dry pellet; 2 = firm, formed stool; 3 = soft, moist stool that retains shape; 4 = soft, unformed stool that assumes the shape of the container; 5 = watery liquid that can be poured. The piglets were considered to have diarrhea when the fecal scores were at level 4 or 5 (21, 22). The feces of diarrheal piglets were collected by squeezing the abdomen, and sows' feces were collected by rectal swab.

2.4 Feed safety evaluation

Feed samples were evaluated using PEDV-TGEV-PRV triple real-time fluorescent RT-PCR kits (Anheal Laboratories, China). Specific steps were carried out in accordance with the manufacturer's protocols. The feeds of the sows during pregnancy and lactation were sent to the Center for Quality Supervising, Inspecting, and Testing of Animal Fiber, Fur, and Leather Products of the Ministry of Agriculture, P.R. China, to analyze the basic nutrients.

2.5 Experimental model surveillance by detecting CSFV and PRRSV antibody levels

During the experiment, blood samples were collected once every 3 months. Ten blood samples were taken at random from each block of the experimental model each time. The blood samples were sent to a laboratory for testing. The antibody levels against CSFV and PRRSV were detected by ELISA kits (IDEXX, USA). Samples were tested in duplicate; the specific steps were carried out in accordance with the manufacturer's protocols. The results were determined using an enzyme standard instrument (MDC Spectramax M2e, USA).

2.6 Detection of CSFV and PRRSV

We used CSFV real-time fluorescent RT-PCR kits and PRRSV real-time fluorescent RT-PCR kits (Anheal Laboratories, China) to detect the virus in the serum samples of piglets and sows, respectively. Real-time fluorescent RT-PCR protocols are shown in Table 1. Results were detected using real-time fluorescent RT-PCR instrument (Bio-Rad CFX96, Germany).

2.7 Detection of the incidence of PEDV in piglets and the PEDV-carrying rate of sows

We used PEDV-TGEV-PRV triple real-time fluorescent RT-PCR kits (Anheal Laboratories, China) to detect the virus in the fecal samples of piglets and sows. Real-time fluorescent RT-PCR protocols are shown in Table 1. Results were determined using a real-time fluorescent RT-PCR instrument (Bio-Rad CFX96, Germany).

Table 1 Real-time fluorescent quantitative RT-PCR protocols.

Items	Step 1 Extract RNA template	Step 2 Reaction system	Step 3 Reaction conditions	Step 4 Reporter dye	Step 5 Results analysis
Detection of CSFV	Extract virus RNA from samples according to manufacturers protocols	Add RNase-free dH2O, RT-PCR buffer, enzyme mix TaqMan probe, and RNA template according to manufacturer's protocols	The amplification was performed as follows: one cycle at 42 ℃ for 5 min and 95 ℃ for 10 s, followed by 40 cycles at 95 ℃ for 5 s, 60 ℃ for 35 s, and 72 ℃ for 5 min	FAM	If positive control shows specific amplification curve and negative control has no Ct value. $0 < $ Ct value $\leqslant 30$ was recognized as a positive sample.
Detection of PRRSV					
Detection of PEDV					

2.8 Identification of different strains of PEDV

2.8.1 Designing the primer

To obtain the complete S gene sequence of PEDV, primers were designed based on the sequence of a reference PEDV strain (Accession Number: AF353511.1). Forward Primer: 5'-ATGAGGTCTTTAATTTACTTCTGGTTG-3'; Reverse Primer: 5'-TCACTG CACGTG-GACCTTT-3'.

2.8.2 PEDV RNA extraction

The fecal pretreatment protocol was based on previous literature reports. Each sample was diluted with phosphate-buffered saline to make a 10% (v/v) suspension. The suspensions were vortexed for 1 min and clarified by centrifugation for 10 min at 5 000 × g. The supernatants were collected for extraction of PEDV RNA and dissolved in 50 μL of RNase-free dH$_2$O, as described in the TaKaRa miniBEST viral RNA/DNA Extraction Kit (TaKaRa, China). Samples were stored at −80 ℃.

2.8.3 RT-PCR

The S gene of PEDV was amplified by RT-PCR using the PrimeScript one-step RT-PCR kit. The amplification was performed as follows: one cycle at 50 ℃ for 30 min, then 94 ℃ for 2 min, followed by 30 cycles at 94 ℃ for 30 s, 56 ℃ for 30 s, and 72 ℃ for 5 min. Samples were stored at 4 ℃.

2.8.4 Sequencing of RT-PCR products

The RT-PCR products were identified by electrophoresis on 0.5% agarose gel. The positive results were sent to the Beijing Genomics Institute for sequencing. All sequencing reactions were performed in duplicate. The size of the RT-PCR products was 4 152 bp.

2.8.5 Sequence analysis of the S gene

The nucleotide sequences of the S gene from the strains were aligned and analyzed by MEGA 6.0 software. The CV777 strains were used for sequence alignment and analysis with the PEDV strains.

2.8.6 Phylogenetic analysis of PEDV

Phylogenetic trees were constructed using Molecular Evolutionary Genetics Analysis (MEGA 6.0) software with the neighbor-joining method based on the sequence of the S gene.

2.9 Statistical analysis

Antibody levels against CSFV and PRRSV were expressed as means and standard deviations ($\pm$ SD). Statistical analysis was performed with one-way analysis of variance (ANOVA). Table analysis was used to detect the incidence of PEDV in piglets and the PEDV-carrying rate of sows. All statistical analyses used SAS software (version 9.2, USA), and $P < 0.05$ was considered significant.

3 RESULTS

3.1 Construction of the experimental model

We used PRRSV real-time fluorescent RT-PCR kits (Anheal Laboratories, China) to detect PRRSV in the serum of sows. Each sow received a clinical examination, including identification of pathological conditions, rectal temperature, and mental and physical state. During the experimental period, 183 sows were enrolled in the experimental model and divided into two groups (group A, n = 81; group B, n = 102) according to the enrollment and exclusion criteria. In addition, there were 1723 suckling piglets (group A, n = 726; group B, n = 997) in the experimental model. All suckling piglets were assured colostrum intake from their mothers within 1 to 4 h after birth, and the nipples of the sows were disinfected prior to suckling.

3.2 Feed safety evaluation

The basic nutrition of the sows' feed during pregnancy and lactation agreed with the national standards (Table 2). The results of the PEDV-TGEV-PRV triple real-time fluorescent RT-PCR showed that all feed samples were negative for PEDV (Ct = 0) (Table 3).

Table 2　The basic nutrition of the sow feed during pregnancy and lactation.

Indexes	DM(%) GB/T 6432—1994	EE(g/kg) GB/T 6433—2006	CF(g/kg) GB/T 6434—2006	Moisture(%) GB/T 6435—2006	Ca(%) GB/T 6435—2006	TP(%) GB/T 6435—2006	Ash(%) GB/T 6435—2006
Pregnant sows' feed	14.28 ± 0.50	19 ± 1	59 ± 2	12.89 ± 0.53	4.25 ± 0.17	0.39 ± 0.09	5.31 ± 0.26
Lactating sows' feed	17.13 ± 0.32	43 ± 1	45 ± 3	12.34 ± 0.69	4.23 ± 0.35	0.43 ± 0.03	5.13 ± 0.18

Values are mean ± standard deviation. DM: Crude protein; EE: crude fat; CF: crude fiber; TP: total phosphorus.

Table 3 The results of PEDV detection in fecal and feed samples by PEDV-TGEV-PRV triple real-time fluorescent RT-PCR.

Indexes	Group A	P (Ct value)	P/T*	Group B	P (Ct value)	P/T*
Positive rate of PEDV in feed	0%	0 (0)	0/12	0%	0 (0)	0/12
Positive rate of PEDV in piglet feces	76.4%**	204 (12.6 ~ 28.7)	204/267	53.0%	108 (17.3 ~ 27.9)	108/198
Positive rate of PEDV in sow feces	40%**	8 (18.5 ~ 29.6)	8/20	15%	3 (14.3 ~ 23.6)	3/20

P: Positive samples; P/T*: positive samples/ total samples. 0 < Ct value ≤ 30 was recognized as a positive sample. **: P < 0.01.

3.3 Detection of CSFV and PRRSV antibody levels

The levels of antibodies against CSFV or PRRSV were detected by ELISA kits (Figure 2). In both group A and group B, the level of antibodies against CSFV was not significantly different at the three times of testing ($P > 0.05$). In group B, the level of antibodies against PRRSV was also not significantly different ($P > 0.05$). However, the level of antibodies against PRRSV in group A was significantly different ($P < 0.01$) between the first time point and the latter two time points in the block of suckling piglets. In group A, the level of antibodies against PRRSV was significantly higher than in group B in all blocks ($P < 0.01$). The level of antibodies against CSFV in group A was significantly lower than in group B in the block of lactating sows and suckling piglets ($P < 0.01$).

3.4 Detection of CSFV and PRRSV

The results of the detection of PRRSV and CSFV are shown in Table 4. The sows and piglets in group B were CSFV- and PRRSV-negative. The sows and piglets in group A were CSFV-negative. The sows in group A were PRRSV-positive, while the piglets in group A were PRRSV-negative.

3.5 Detection of the incidence of PEDV in piglets and the PEDV-carrying rate of sows

There were 465 piglets with diarrhea (group A, n = 267; group B, n = 198). In group A, the diarrhea rate of the piglets was significantly higher than in group B (group A was 36.78%, group B was 19.8%, $P < 0.01$). Analysis of the cycle threshold (Ct) values of the PEDV real-time RT-PCR-positive samples indicated that all of the samples with Ct values below 30 were PEDV-positive. The fecal samples of 312 piglets were PEDV-positive (group A, n = 204; group B, n = 108). In group A, PEDV infection was significantly higher than in group B ($P < 0.01$). Eleven samples (group A, n = 8; group B, n = 3) out of 40 sow fecal samples (group A, n = 20; group B, n = 20) were PEDV-positive. The rate of PEDV carrier sows in group A was significantly higher than in group B ($P < 0.01$) (Fig. 3; Table 3).

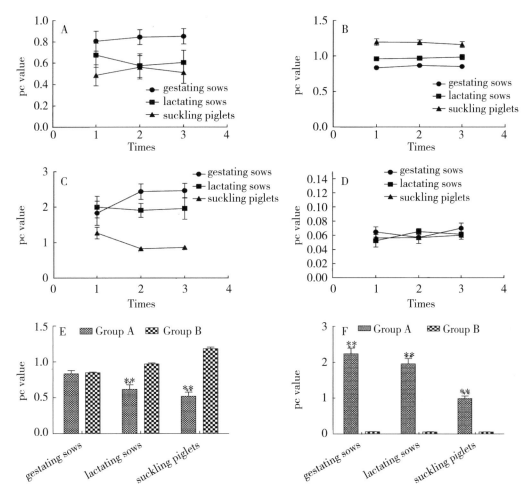

Fig. 2 The fluctuation of CSFV and PRRSV antibody levels during different periods from April 2014 to January 2015.

Note: A is the antibody level against CSFV in group A, and B is the antibody level against CSFV in group B. C is the antibody level against PRRSV in group A. D is the antibody level against PRRSV in group B. E is the comparative antibody levels against CSFV in swine in farm A and farm B. F shows the comparison of antibody levels against PRRSV in swine in group A and group B. Values are represented as mean ± standard deviation and the treatments in one experiment were repeated twice. **: P < 0.01.

3.6 Identification of different strains of PEDV

During the experiment, we extracted 24 PEDV RNA samples from PEDV-positive fecal samples. The 24 RNA samples were converted to cDNA samples by reverse-transcription. The 10 cDNA samples were successfully sequenced. The nucleotide sequences of the 10 Cdna samples had 99% sequence identity. More precisely, the nucleotide sequences of the 10 cDNA samples had 98% sequence identity with CV777 (Accession No: JN599150.1). We chose 2 cDNA sequences from a different group (Accession No: KR902706 and KR902707) together with the PEDV CV777 strain (Accession No: JN599150.1) to perform a phylogenetic analysis. The results showed that all of the sequences fell into two groups (Figure 4). The two PEDV field isolates were within one group.

4 DISCUSSION

An experimental model, in which sows were infected with PRRSV, was constructed based on enrollment and exclusion criteria. The selection of the appropriate experimental model for the research was complex due to animal welfare considerations and the changeable clinical environment. The selection of a suitable experimental model largely depends on whether it is stable and meets the specific research demands in question (23). Many static experimental models can be built to manually control influencing factors and infection, which allows scientists to evaluate research factors (24, 25). The static experimental models have shortcomings in emerging infectious disease research. The animals are euthanized after infection in these models. The outrage of society reminds investigators to adhere to high standards of humane animal usage (25). To monitor the stability of the experimental model, we chose to survey the antibody levels against PRRSV and CSFV. The use of laboratory data for passive disease surveillance is limited by its lack of ability to identify disease outbreaks, reemerging diseases, or novel pathogens (26, 27). Monitoring the test results of commonly used first-order tests for a known disease may be a unique form of syndromic data collection for the timely identification of novel disease outbreaks in swine populations (28). In both group A and group B, the level of anti-CSFV antibodies was not significantly different at the three time points ($P > 0.05$). The results suggested that the experimental model was stable. The piglets had antibodies against CSFV from the colostrum and milk of immune sows until they were first vaccinated. If a sow's immune system was suppressed, the piglet's antibody levels against CSFV would decrease. The antibody level against CSFV in group A was significantly lower than that of group B in the block of lactating sows and suckling piglets ($P < 0.01$), which showed that PRRSV may suppress the host's immune system. PRRSV has a tropism for immune cells and has been shown to suppress the host's immune system (15-18). PEDV-contaminated feed may be a risk factor for PEDV transmission, which may lead to PEDV occurrence (19). Therefore, we performed a feed safety evaluation. The results of the feed safety evaluation showed that all of the feed samples were negative for PEDV ($Ct = 0$), and the basic nutrition of the feed met country standards. The results indicated that the feed was not related to the outbreak of PEDV in the experimental model. Moreover, many researchers have documented a remarkable increase in PEDV outbreaks attributable to the emergence of new strains (14, 29). The PEDV S glycoprotein plays a pivotal role in regulating interactions with specific host cell receptor glycoproteins to mediate viral entry, which could be a primary target for the development of effective vaccines against PEDV (30). The mutation of the PEDV S gene is possible because of the PEDV outbreak. This paper showed that the PEDV strains had high similarity and were within one group, based on the construction of phylogenetic trees using the complete S gene sequence from the experimental model. Therefore, we deduced that the outbreak of PEDV was not related to PEDV mutation in the experimental model. The incidence of PEDV in piglets and the PEDV-carrying rate in sows were significantly increased in group A. Therefore, we deduced that the sows were infected with PRRSV, which led to immunosuppression in swine in group A. If the immunity of the sows and piglets was inhibited, there would be a high risk of infection by secondary viruses and bacteria. The rate of PEDV-positive sows was significantly increased, and although few sows showed any clinical signs, they could excrete the virus in their feces, which could infect piglets by the fecal-oral route (1, 19). Therefore, we propose that PRRSV infection might be related to an outbreak of PEDV in immunized swine herds.

Table 4 PRRSV and CSFV levels in serum samples using PRRSV real-time fluorescent RT-PCR and CSFV real-time fluorescent RT-PCR, respectively.

Indexes	Group A	P(Ct value)	P/T*	Group B	P(Ct value)	P/T*
Positive rate of PRRSV in piglet serum	0%	0(0)	0/30	0%	0(0)	0/30
Positive rate of PRRSV in sow serum	100%	30(9.67~28.7)	30/30	0%	0(0)	0/30
Positive rate of CSFV in sow serum	0%	0(0)	0/30	0%	0(0)	0/30
Positive rate of CSFV in piglet serum	0%	0(0)	0/30	0%	0(0)	0/30

P: Positive samples; P/T*: positive samples/ total samples. 0 < Ct value ≤ 30 was recognized as a positive sample.

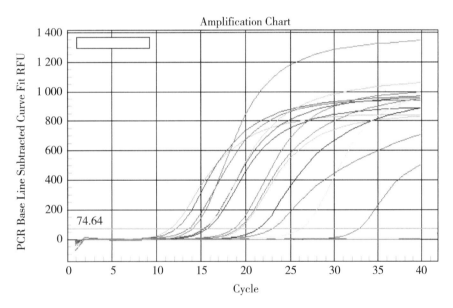

Fig. 3 Real-time fluorescent quantitative RT-PCR amplification curve of PEDV.

Fig. 4 The relationship between the PEDV field isolates of the experimental model and PEDV CV777 stain (Accession No: JN599150.1) based on the full-length S gene.

In conclusion, many risk factors can cause an outbreak of PEDV, such as virus mutation and feed contamination. The results of this study showed that PRRSV was a risk factor in the outbreak of porcine epidemic diarrhea in immunized swine herds.

Acknowledgments

The project was funded by the Special Fund for Agro-Scientific Research in the Public Interest (No. 201303040-18) and the National Key Technology Research and Development Program of the Ministry of Science and Technology of China (No. 2012BAD12B03). The authors would like to thank the Key Laboratory of Veterinary Pharmaceutical Development of the Ministry of Agriculture, Key Laboratory of New Animal Drug Project of Gansu Province, and Engineering and Technology Research Center of Traditional Chinese Veterinary Medicine of Gansu Province personnel for their contribution to the project.

（发表于《Turkish Journal of Veterinary and Animal Sciences》，SCI，IF：0.242. doi:10.3906/vet-1508-43.）

Crystal Structure of 14-[(1-benzyloxycarbonyl-amino-2-methylpropan-2-yl)sulfanyl] Acetate Mutilin, C₃₄H₄₉NO₆S

SHANG Ruofeng[1*], YI Yunpeng[2], LIANG Jianping[2]

(1. Lanzhou Institute of Husbandry and Pharmaceutical Sciences of CAAS - Key Laboratory of New Animal Drug Project of Gansu Province; and Key Laboratory of Veterinary Pharmaceutical Development, Ministry of Agriculture lanzhou, Gansu Province 730050; 2. Lanzhou Institute of Husbandry and Pharmaceutical Sciences of CAAS - Key Laboratory of New Animal Drug Project of Gansu Province/Key Laboratory of Veterinary Pharmaceutical Development, Ministry of Agriculture Lanzhou, Gansu730050, China)

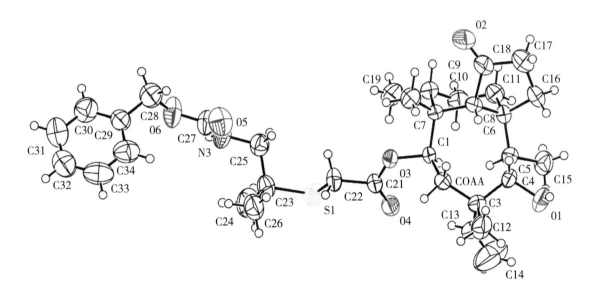

DOI 10.1515/ncrs-2015-0142

Received July 3, 2015; accepted January 13, 2016; available online February 16, 2016

Abstract

$C_{34}H_{49}NO_6S$, monoclinic, $P2_1$ (no. 4), a = 11.711 2(8) Å, b = 10.013 5(5) Å, c = 14.641 1(10) Å, V = 1 664.4 Å³, Z = 2, $R_{gt}(F) = 0.065\ 9$, $wR_{ref}(F^2) = 0.126\ 9$, T = 293 K.

CCDC no.: 1446995

* Corresponding author, E-mail: shangrf1974@163.com

Table 1 Data collection and handling.

Crystal	Colorless block, size 0.25 mm × 0.34 mm × 0.37 mm
Wavelength:	Mo Kα radiation (0.710 73 Å)
μ:	1.40 cm^{-1}
Diffractometer, scan mode:	SuperNova (Dual, Cu at zero, Eos) ω scans
2θ$_{max}$:	52.04°
N(hkl)$_{measured}$, N(hkl)$_{unique}$:	6782, 5443
N(poram) refined:	386
Programs:	OLEX2 [6], SHELXS [7]

Table 2 Fractional atomic coordinates and isotropic or equivalent isotropic displacement parameters (Å^2).

Atom	Site	x	y	z	U$_{iso}$
H(1)	2a	1.082 1	0.221 4	0.121 4	0.103
H(1A)	2a	0.835 9	0.518 1	0.273 3	0.046
H(22A)	2a	0.741 9	0.595 2	0.525 3	0.069
H(22B)	2a	0.708 4	0.444 5	0.534 0	0.069
H(3)	2a	0.497 4	0.818 3	0.704 0	0.077
H(0AA)	2a	0.905 2	0.310 1	0.375 3	0.056
H(0AB)	2a	1.026 9	0.365 5	0.368 1	0.056
H(4)	2a	1.082 7	0.365 6	0.217 8	0.060
H(5)	2a	0.876 2	0.491 9	0.150 4	0.056
H(8)	2a	1.110 6	0.511 5	0.302 1	0.053
H(9)	2a	0.987 5	0.827 6	0.328 9	0.058
H(10A)	2a	0.812 3	0.696 5	0.191 4	0.071
H(10B)	2a	0.840 0	0.849 5	0.190 5	0.071
H(11A)	2a	1.024 3	0.805 5	0.164 7	0.071
H(11B)	2a	0.927 8	0.736 1	0.085 8	0.071
H(12A)	2a	1.072 4	0.153 5	0.305 2	0.117
H(12B)	2a	0.980 3	0.084 5	0.222 8	0.117
H(12C)	2a	0.951 6	0.104 3	0.321 1	0.117
H(13)	2a	0.753 3	0.314 6	0.218 9	0.082
H(14A)	2a	0.800 5	0.138 6	0.096 0	0.193
H(14B)	2a	0.674 0	0.194 3	0.103 2	0.193
H(15A)	2a	0.999 4	0.462 9	0.008 7	0.103

(continued)

Atom	Site	x	y	z	U_{iso}
H(15B)	2a	0.894 3	0.562 6	0.002 8	0.103
H(15C)	2a	0.872 1	0.408 2	0.002 9	0.103
H(16A)	2a	1.174 1	0.511 7	0.141 6	0.086
H(16B)	2a	1.132 9	0.641 5	0.081 9	0.086
H(17A)	2a	1.309 3	0.639 5	0.233 5	0.114
H(17B)	2a	1.244 4	0.771 8	0.190 5	0.114
H(19A)	2a	0.763 3	0.744 1	0.347 0	0.115
H(19B)	2a	0.808 9	0.891 8	0.352 1	0.115
H(19C)	2a	0.862 6	0.791 5	0.433 5	0.115
H(20A)	2a	1.001 7	0.642 0	0.489 2	0.096
H(20B)	2a	1.113 4	0.695 7	0.460 1	0.096
H(20C)	2a	1.094 5	0.541 6	0.468 0	0.096
H(26A)	2a	0.512 9	0.407 4	0.625 0	0.125
H(26B)	2a	0.538 6	0.507 0	0.710 2	0.125
H(26C)	2a	0.638 6	0.470 3	0.660 4	0.125
H(24A)	2a	0.365 3	0.706 1	0.527 6	0.148
H(24B)	2a	0.365 4	0.636 0	0.623 4	0.148
H(24C)	2a	0.345 4	0.551 5	0.530 6	0.148
H(25A)	2a	0.670 6	0.711 7	0.631 3	0.071
H(25B)	2a	0.567 7	0.799 4	0.572 7	0.071
H(28A)	2a	0.680 6	0.944 9	0.969 1	0.101
H(28B)	2a	0.751 5	0.817 6	0.953 0	0.101
H(30)	2a	0.572 7	0.9368	1.082 1	0.098
H(31)	2a	0.502 5	0.816 9	1.194 9	0.115
H(32)	2a	0.509 5	0.588 9	1.194 9	0.110
H(33)	2a	0.586 0	0.477 7	1.087 1	0.111
H(34)	2a	0.654 1	0.594 4	0.976 6	0.097

Table 3 Atomic displacement parameters (A^2).

Atom	Site	x	y	z	U_{11}	U_{22}	U_{33}	U_{12}	U_{13}	U_{23}
S(1)	2a	0.541 6 (2)	0.566 3 (2)	0.466 5 (1)	0.056 (1)	0.069 (1)	0.060 (1)	0.008 (1)	0.016 (1)	−0.009 (1)
O(1)	2a	1.016 2 (5)	0.254 9 (5)	0.106 9 (4)	0.089 (4)	0.064 (3)	0.058 (3)	0.013 (3)	0.027 (3)	−0.016 (3)

(continued)

Atom	Site	x	y	z	U_{11}	U_{22}	U_{33}	U_{12}	U_{13}	U_{23}
O(2)	2a	1.2135(5)	0.7759(6)	0.3619(4)	0.089(5)	0.103(4)	0.084(5)	−0.049(4)	0.028(4)	−0.025(4)
O(3)	2a	0.8428(4)	0.5179(4)	0.4134(3)	0.050(3)	0.054(3)	0.037(3)	−0.006(2)	0.016(2)	−0.008(2)
O(4)	2a	0.6660(5)	0.4461(5)	0.3312(4)	0.057(3)	0.091(4)	0.047(3)	−0.027(3)	0.017(3)	−0.018(3)
O(5)	2a	0.7335(5)	0.7133(6)	0.8062(4)	0.047(3)	0.112(5)	0.087(4)	0.018(4)	0.011(3)	0.009(4)
O(6)	2a	0.5985(5)	0.8391(6)	0.8549(4)	0.066(4)	0.143(5)	0.050(4)	0.013(4)	0.015(3)	−0.014(4)
C(1)	2a	0.8962(6)	0.5045(6)	0.3322(4)	0.041(4)	0.042(4)	0.039(4)	−0.006(3)	0.020(3)	−0.006(3)
C(22)	2a	0.6945(6)	0.5220(7)	0.4930(5)	0.054(5)	0.077(5)	0.049(5)	0.006(4)	0.027(4)	−0.003(4)
N(3)	2a	0.5642(5)	0.7806(6)	0.7067(4)	0.055(4)	0.086(4)	0.049(4)	0.026(4)	0.009(3)	−0.006(3)
C(OAA)	2a	0.9441(6)	0.3617(6)	0.3357(4)	0.057(5)	0.041(4)	0.043(4)	−0.004(4)	0.017(4)	0.002(3)
C(3)	2a	0.9345(6)	0.2806(6)	0.2439(5)	0.061(5)	0.038(4)	0.054(5)	−0.003(4)	0.017(4)	−0.003(3)
C(4)	2a	1.0034(7)	0.3468(6)	0.1791(5)	0.069(5)	0.045(4)	0.039(4)	−0.003(4)	0.021(4)	−0.010(3)
C(5)	2a	0.9521(6)	0.4799(6)	0.1346(4)	0.057(5)	0.046(4)	0.039(4)	−0.006(4)	0.014(4)	−0.004(3)
C(6)	2a	1.0294(6)	0.6013(6)	0.1760(5)	0.051(4)	0.050(4)	0.040(4)	−0.005(4)	0.020(3)	−0.002(3)
C(3)	2a	1.0764(6)	0.6003(6)	0.2857(5)	0.044(4)	0.041(4)	0.048(4)	−0.010(3)	0.013(3)	0.000(3)
C(7)	2a	0.9884(6)	0.6217(6)	0.3470(5)	0.047(4)	0.040(4)	0.038(4)	−0.012(3)	0.013(3)	−0.007(3)
C(9)	2a	0.9260(6)	0.7590(6)	0.3147(5)	0.052(4)	0.044(4)	0.050(5)	−0.005(4)	0.013(4)	−0.003(3)
C(10)	2a	0.8746(7)	0.7624(7)	0.2081(5)	0.073(6)	0.040(4)	0.067(6)	0.005(4)	0.021(5)	0.001(4)
C(11)	2a	0.9662(7)	0.7345(7)	0.1526(5)	0.082(6)	0.049(4)	0.048(5)	−0.008(4)	0.020(4)	0.008(4)
C(12)	2a	0.9900(8)	0.1426(6)	0.2764(6)	0.122(8)	0.042(4)	0.079(6)	0.019(5)	0.042(6)	0.005(4)
C(13)	2a	0.8046(8)	0.2635(7)	0.1939(6)	0.077(6)	0.057(5)	0.072(6)	−0.020(5)	0.019(5)	−0.021(4)
C(14)	2a	0.756(1)	0.193(1)	0.1252(9)	0.12(1)	0.16(1)	0.20(2)	−0.011(9)	0.01(1)	−0.10(1)

(continued)

Atom	Site	x	y	z	U_{11}	U_{22}	U_{33}	U_{12}	U_{13}	U_{23}
C(15)	2a	0.927 1(7)	0.478 2(8)	0.027 3(5)	0.090(6)	0.073(5)	0.043(5)	0.003(5)	0.018(4)	0.006(4)
C(16)	2a	1.145 3(7)	0.602 1(8)	0.144 0(5)	0.074(6)	0.085(6)	0.068(6)	−0.023(5)	0.041(5)	−0.003(5)
C(17)	2a	1.233 7(8)	0.684(1)	0.215 7(7)	0.074(6)	0.112(8)	0.114(9)	−0.035(6)	0.050(6)	−0.026(6)
C(18)	2a	1.180 2(7)	0.696 0(9)	0.298 0(6)	0.062(5)	0.075(6)	0.070(6)	−0.021(5)	0.024(5)	−0.006(5)
C(19)	2a	0.831 2(7)	0.800 5(7)	0.366 8(6)	0.089(7)	0.056(5)	0.091(7)	0.008(5)	0.034(6)	−0.016(4)
C(20)	2a	1.055 8(6)	0.625 6(8)	0.450 8(4)	0.063(5)	0.081(5)	0.042(5)	−0.017(5)	0.003(4)	−0.007(4)
C(21)	2a	0.728 9(7)	0.490 9(7)	0.401 8(5)	0.056(5)	0.051(4)	0.051(5)	−0.003(4)	0.028(4)	−0.003(4)
C(23)	2a	0.518 4(7)	0.603 8(7)	0.582 4(5)	0.049(5)	0.062(5)	0.063(5)	0.017(4)	0.025(4)	0.008(4)
C(26)	2a	0.555 6(8)	0.486 0(8)	0.651 0(6)	0.115(8)	0.066(5)	0.089(7)	0.010(6)	0.059(6)	0.012(5)
C(24)	2a	0.386 5(7)	0.626(1)	0.564 3(7)	0.060(6)	0.132(8)	0.113(8)	0.007(6)	0.037(6)	−0.035(7)
C(25)	2a	0.587 0(7)	0.730 2(7)	0.620 4(5)	0.071(5)	0.055(4)	0.054(5)	0.008(4)	0.020(4)	0.006(4)
C(27)	2a	0.641 4(7)	0.771 8(9)	0.790 9(6)	0.052(5)	0.089(6)	0.052(5)	0.009(5)	0.013(4)	−0.008(5)
C(28)	2a	0.673 3(8)	0.852(1)	0.951 1(6)	0.079(7)	0.116(7)	0.054(6)	−0.026(6)	0.007(5)	−0.013(5)
C(29)	2a	0.622 2(7)	0.775 6(9)	1.018 7(6)	0.059(5)	0.080(6)	0.046(5)	−0.016(5)	0.009(4)	−0.002(4)
C(30)	2a	0.575 9(8)	0.844 0(9)	1.083 2(7)	0.091(8)	0.075(6)	0.079(7)	0.007(5)	0.020(6)	0.002(5)
C(31)	2a	0.533 3(9)	0.772(1)	1.150 7(7)	0.092(8)	0.114(9)	0.090(8)	0.009(7)	0.040(6)	0.003(7)
C(32)	2a	0.537 7(9)	0.637(1)	1.150 6(7)	0.081(8)	0.111(9)	0.086(8)	−0.007(7)	0.026(6)	0.007(7)
C(33)	2a	0.583 0(8)	0.571(1)	1.086 4(8)	0.069(6)	0.079(6)	0.114(9)	−0.009(6)	−0.004(6)	0.009(7)
C(34)	2a	0.624 2(8)	0.641(1)	1.020 7(7)	0.060(6)	0.077(7)	0.100(8)	−0.004(5)	0.009(5)	−0.005(6)

The crystal structure is shown in the figure. Tables 1-3 contain details of the measurement method and a list of the atoms including atomic coordinates and displacement parameters.

SOURCE OF MATERIAL

Benzyl carbonochloridate (0.56 g, 3.3 mmol) was added drop-wise to a solution of 14-O-[(1-amino-2-methylpropane-2-yl)thioacetyl] mutilin (1.39 g, 3. mmol) and triethylamine (0.81 g, 8 mmol) in dichloromethane (60 mL) and was stirred at 0 ℃ or room temperature for 4.5 h. Then the reaction mixture was washed with saturated aqueous NH4Cl and water and dried with anhydrous Na_2SO_4 overnight. The solvent was evaporated in vacuum and the residue was chromatographed on silica gel (petroleum ether: ethyl acetate 2 : 1 v/v) to afford a pure product (1.40 g, yield: 78°; m.p. 60 ~ 63 ℃). The crystals were obtained by slow evaporation from an dichloromethane solution at room temperature.

DISCUSSION

The title compound was synthesized to be used as an antibiotic candidate against some Gram-positive bacteria and mycoplasmas. The starting material was pleuromutilin which was first isolated in a crystalline form from cultures of two species of basidiomycetes, Pleurotus mutilus and P.passeckerianus in 1951 [1]. The pleuromutilin class has a unique mode of action, which involve inhibition of protein synthesis by primarily inhibiting ribosomal peptidyl transferase center (PTC) [2], and thus display high antibacterial activities with no target-specific cross-resistance to other antibiotics [3]. The molecular modifications of pleuromutilin led to three drugs: tiamulin, valnemulin, and retapamulin. The first two drugs were used as veterinary medicines [4]. The retapamulin became the first pleuromutilin approved for human use in 2007 [5].

The crystal structure of the title compound is built up of $C_{33}H_{48}N_2O_6S$ molecules containing a 5-6-8 tricyclic carbon skeleton and a phenyl group, in which all bond lengths are in normal ranges. The five-membered ring (C6-C16-C17-C18-C8) is not planar and the six-membered ring (C6-C11-C10-C9-C7-C8) exhibits a chair conformation. The eight-membered ring (from C1 to C8) exhibits a boat conformation. The synthesized side chain of pleuromutilin exhibits zig-zag conformation. The torsion angles are 67.54°, 172.51°, 107.39° and -112.90° for C22-S1-C23-C25, S1-C23-C25-N3, C23-C25-N3-C27 and C27-O6-C28-C29, respectively. Weak NH...O and OH...O hydrogen bonds are present.

Acknowledgements: This work was supported by grants from Basic Scientific Research Funds in Central Agricultural

Scientific Research Institutions (No. 1610322014003) and The Agricultural Science and Technology Innovation Program (CAAS-ASTIP-2014-LIHPS-04).

（发表于《Z. Kristallogr. NCS》，SCI，IF：0.122．465–467.）

The Complete Mitochondrial Genome of *Ovis ammon darwini* (Artiodactyla: Bovidae)

MAO Hongxia[1], LIU Hanli[1], MA Guilin[1],
YANG Qin[1], GUO Xian[2], ZHONG Lamaocao[3]

(1. Animal Husbandry Science and Research Institute of Gannan Tibet Autonomous Prefecture in China, Hezuo, China; 2. Lanzhou Institute of Husbandry and Pharmaceutical Sciences, Chinese Academy of Agricultural Sciences, Lanzhou, China; 3. Animal Disease Prevention and Control Center of Gannan Tibet Autonomous Prefecture in China, Hezuo, China)

Abstract: *Ovis ammon darwini*: has been listed as a nearthreatened species by IUCN-World Conservation Union due to over-hunting and poaching, competition with domestic livestock, and habitat destruction. Study on molecular biology research provides the scientific basis for the protection and sustainable utilization of key endangered species. Mitochondrial genome is very useful for researches in ecology, systematics and conservation biology. In this study, the complete mitochondrial genome sequence of *O. a. darwini* was determined by next-generation sequencing data, which is 16,618 bp in length and contains 13 protein-coding genes (PCGs), 2 rRNAs genes, 22 tRNAs genes and a non-coding control region. Base composition of genome is A (33.7%), C (25.8%), G (13.1%), T (27.4%) with an A + T content of 61.1%. Phylogenetic analysis suggested that the systematic status of *O. a. darwini* was more closed related to *O. a. hodgsoni* clustered with *O. ammon*. The mitogenome of *O. a. darwini* offered significant information for molecular genetic research of *O. ammon*.

Key words: *Ovis ammon darwini*; Mitochondrial genome; Illumina sequencing

The argali (species *Ovis ammon*) is the world's largest sheep within Order Artiodactyla, Family Bovidae. This species is considered an endangered due to habitat loss from overgrazing of domestic sheep and sport hunting for meat and hair. Developing reasonable conservation and utilization measures for *O. a. darwini* requires a deeper understanding of molecular genetics. Mitochondrial DNA, as an effective genetic maker, has been successfully applied to the study of population genetics, molecular phylogenetics and evolutionary studies. In this research, we assembled and characterized the complete mitochondrial genome sequence of *O. a. darwini*, which serves as powerful molecular tools to provide deeper insights into the detailed history of argali evolution.

The total DNA was extracted from the muscle tissue of *O. a. darwini* collected from Maqu county (Gansu, China; 34° 08′N, 101° 72′E). A genomic library with insert fragment of 180 bp was prepared, tagged and subjected to NGS on Illumina HiSeq 2500 Sequencing System (Illumina, CA,

USA) according to the standard operating protocol. In all, 14, 985, 138 raw paired reads with average length of 125 bases was obtained. After quality assessment and data filtering with CLC Genomics Workbench v85 (CLC, Bio, Arhus, Demark), high-quality clean reads was used to assemble the mitochondrial genome using the program MITObim v1.8 (Hahn et al. 2013), with that of Marco Polo wild sheep (GenBank: KT781689) as initial reference. A total of 2638 reads mapped to reference mitochondrial genome with average reads depth of 20.7 ×. Following the assembly, the mitochondrial genome was annotated using GENEIOUS R9 (Biomatters Ltd, Aukland, New Zealand). In brief, protein-coding genes, transfer RNA (tRNAs) genes, ribosomal RNAs (rRNAs) genes and control region (CR) were identified and annotated based on alignment with counterparts from the mitochondrial genome of known Bovidae species (GenBank: KT781689). Simultaneously, tRNAs genes were proofreaded and amended using MITOS Web Server (Bernt et al. 2013).

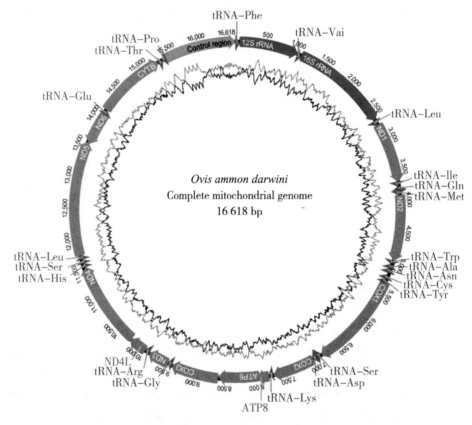

Fig. 1 The structure of *Ovis ammon darwini* mitochondrial genome

Our research findings revealed that the circular genome is 16618 bp, which consists of 13 protein-coding genes (PCGs), 2 rRNAs genes, 22 tRNAs genes and a non-coding control region (Fig. 1), showing that the gene composition and arrangement are similar to reported O.ammon (Jiang et al. 2013; Meadows et al. 2011). The contents of A, C, T and G in mitochondrial genome were 33.7, 25.8, 13.1 and 27.4%, respectively. An overall A + T content of whole mitochondrial genome is 61.1%. The sequence was deposited in GenBank (GenBank: KX609626). To validate this novel mitogenome, each protein-coding genes was detected by using amino acid sequences translated from corresponding genes and abnormal stop codons and frameshift mutations were not found. In addition, we validated the status of *O. a. darwini* by investigating its phylogenetic position within Caprinae (Avise et al. 1987; Dutton et al. 1999). A phylogenetics analysis was conducted on 27 mitochondrial genomes from Caprinae and an entire mitogenome of Bos javanicus as an outgroup. The neighbor-joining and maximum likelihood tree were constructed in Mega 5 (Tamura et al. 2011), with node supports estimated from 1000 replicate searches. Both NJ and ML trees showed identical topology and major internal nodes were well supported by bootstrap values (Fig. 2). Phylogenetic analyses indicated the presence of two distinct clades in subfamily Caprinae and several monophyletic clades, as shown in previous studies (Jiang et al. 2013). Interestingly, For clade of *O. ammon*, *O. a. darwini* and *O. a. hodgsoni* formed a separate subclade and these two subspecies showed closer relationship within *O. ammon*. The compete mitochondrial genome of *O. a. darwini* provide essential information for understanding phylogenetic relationships of *O. ammon* mitochondrial genome and will be useful for the conservation genetics of this near-threatened species.

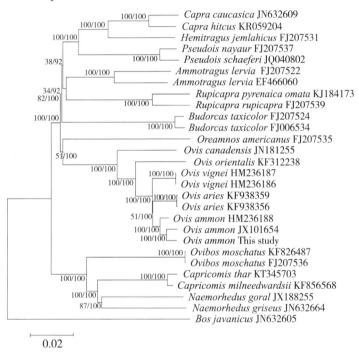

Fig. 2 Phylogeny of 27 mitochondrial genomes from Caprinae with Bos javanicus as an outgroup based on the neighbor-joining (NJ) and maximum likelihood (ML) analysis.

Note: The bootstrap values were based on 1000 resamplings

Acknowledgments: The work was supported or partly supported by grants from The Great Project of Science and Technology of Gansu Province in China (Contract No. 1102NKDA027), National Natural Science Foundation of China (31301976) and Science and Technology Support Projects in Gansu Province (1504NKCA052).

(发表于《Conservation Genet Resour》, SCI, IF: 0.446.)

荷斯坦奶牛乳腺组织冻存及乳腺上皮细胞原代培养技术改进

林杰，王旭荣，王磊，张景艳，王学智，孟嘉仁，杨志强[*]，李建喜[*]

（中国农业科学院兰州畜牧与兽药研究所，
甘肃省中兽药工程技术研究中心，兰州 730050）

摘　要：为了改进奶牛乳腺上皮细胞原代培养技术和探究乳腺组织冻存方法，选用24月龄无泌乳史荷斯坦奶牛，采用组织块种植法，通过侧置和倒置两种方式，分别从新鲜和冻存（8个月）的乳腺组织中分离培养奶牛乳腺上皮细胞。结果表明，侧置处理能使乳腺细胞爬出时间缩短1~2d，并改善组织块贴壁速度和效果；回收于前次培养后的乳腺组织块再分离培养，贴壁第2天即有乳腺上皮细胞爬出；一步冷冻（-80℃）液氮保存较大奶牛乳腺组织块，冻存液A（FBS：DMSO：DMEM/F12=7：2：1）冻存的组织块存活率为50%，冻存液B（FBS：DMEM/F12：DMSO：HEPES/丙酮酸钠贮存液=10：7：2：1）冻存的组织块存活率为56%。可见，作者建立的侧置处理法和组织块回收法可以用于奶牛乳腺上皮细胞的分离培养，可以缩短组织块种植法实验周期，冻存液中加入缓冲液HEPES/丙酮酸钠后有利于组织块生物活性保持。

关键词：乳腺上皮细胞；侧置处理；组织冻存；原代培养；奶牛

Cryopreservation of the Mammary Gland and Improvement on the Culture of the Primary Epithelial Cells in Holstein Dairy Cows

LIN Jie, WANG Xurong, WANG Lei, ZHANG Jingyan, WANG Xuezhi,
MENG Jiaren, YANG Zhiqiang[*], LI Jianxi[*]

(Engineering & Technology Research Center of Traditional Chinese Veterinary Medicine of Gansu Province, Lanzhou Institute of Husbandry and Pharmaceutical Science of CAAS, Lanzhou 730050, China)

基金项目：国家奶牛产业技术体系（CARS37）；公益性行业专项（20130304）
作者简介：林杰（1991—），女，满族，河北唐山人，硕士生，主要从事奶牛疾病研究，Tel：0931-2115262，E-mail：qielinjie2009@163.com
[*]通讯作者：杨志强，研究员，主要从事临床兽医学与奶牛疾病防治研究，E-mail：zhiqyang2006@163.com；李建喜，研究员，主要从事兽医药理毒理与中兽医学研究与应用，E-mail：lzjianxil@163.com

Abstract: The present study was performed to improve the process of tissue explant method and explore the cryopreservation method of bovine mammary gland. The mammary epithelial cells were isolated and cultured from the fresh and cryopreserved (8mouths) mammary gland that collected from a 24months' Holstein dairy cow without history of lactation by tissue explant method including the side-place and inversion pattern respectively. Side-place method could shorten 1—2 days for appearance of cell, and improve the speed and adherence effect of tissue pieces, and the cells could climb out on the second day from the tissue pieces that were cultured and moved another flask. The survival rate was 50% for the tissue mass by liquid nitrogen storage followed by one-step cryopreservation (-80℃) with freeze-stored solution FBS:DMSO:DMEM/F12=7:2:1while the survival rate was 56% with freeze-stored solution FBS:DMEM/F12:DMSO:HEPES/sodium pyruvate=10:7:2:1. The side-place method and tissue pieces recycling method were established for mammary epithelial cells culture in the presented study, which could shorten the cycle of tissue explant method. And the buffer HEPES/sodium pyruvate was benefit for maintaining the tissue bioactivity.

Key words: Mammary epithelial cells; Side-place method; Tissue cryopreservation; Primary cultural; Dairy cow

体外培养的奶牛乳腺上皮细胞适宜建立细胞模型，是开展生理、病理、药理等研究的重要实验材料。国外已有成熟永生化奶牛乳腺上皮细胞系，如MAC-T、BME-UV1[1-2]。国内也有永生化奶牛乳腺细胞系研究的报道，如朱佳杰、李吉霞、上官陶、代瑞等通过转染hTert、SV40T等获得永生化乳腺上皮细胞[3-6]。但是，目前仍没有成熟的商品化奶牛乳腺上皮细胞系，多数实验室只能通过原代培养获取奶牛乳腺上皮细胞。通常获取原代奶牛乳腺上皮细胞的方法有以下4种：组织块种植法、酶消化法、机械剪切法和乳汁分离法，其中以组织块种植法和酶消化法最为常用。由于组织块种植法培养周期较长，酶消化法和机械剪切法会对细胞产生一定损伤，乳汁分离法细胞得率较低，故需要不断完善和改进奶牛乳腺上皮细胞原代培养技术。本研究对组织块种植法分离乳腺上皮细胞的关键技术步骤进行改进，提出了侧置处理法和组织块回收培养法。该方法改善了组织块种植法培养奶牛乳腺上皮细胞过程中的组织块贴壁效果，缩短了试验周期。同时，为了解决用于细胞分离培养过程中奶牛乳腺组织获取成本高的问题，笔者对奶牛乳腺组织块冻存技术进行了探索，成功建立了冻存乳腺组织分离培养奶牛乳腺上皮细胞的方法。

1 材料与方法

1.1 材料

1.1.1 动物

24月龄健康荷斯坦奶牛，无泌乳史，购自甘肃省荷斯坦奶牛繁育中心，运送至兰州海石湾屠宰场宰杀。

1.1.2 主要试剂

DMEM/F12（Gibco）、胎牛血清（FBS）（四季青）、青霉素钠（160万单位）（哈药集团有限公司兽药厂）、硫酸链霉素（100万单位）（浙江金力制药有限公司）、两性霉素B（Biosharp）、鼠尾胶原蛋白（gibco）、DMSO（Vetec）、角蛋白18单克隆抗体（Abcom）、FITC标记的兔抗鼠IgG（Abcom）。

1.1.3 主要溶液配制

DMEM/F12培养液：pH为7.0～7.2，0.22μm微孔滤膜过滤。完全培养液：DMEM/F12基础培养液+20%FBS+500U·mL^{-1}青霉素+500U·mL^{-1}链霉素+5μg·mL^{-1}两性霉素B。

D-Hank's溶液：pH为7.0～7.2，0.22μm微孔滤膜过滤。应用液：D-Hank's溶液+500U·mL^{-1}青霉素+500U·mL^{-1}链霉素+5μg·mL^{-1}两性霉素B。

胰酶消化液：准确称取20.0mg EDTA-2Na、250mg胰蛋白酶，D-Hank's液溶解，定容至100mL，0.22μm微孔滤膜过滤，于-20℃保存备用。

冻存液A：按FBS∶DMSO∶DMEM/F12=7∶2∶1比例配置（DMSO、DMEM/F12均用0.22μm微孔滤膜过滤除菌），现配现用。

冻存液B：FBS∶DMEM/F12∶DMSO∶HEPES/丙酮酸钠贮存液=10∶7∶2∶1比例配置（贮存液中HEPES 0.8mol·L^{-1}、丙酮酸钠0.02mol·L^{-1}，DMEM/F12、DMSO均用0.22μm微孔滤膜过滤除菌），现配现用。

1.1.4 主要仪器

细胞培养箱（HERACELL150i），倒置显微镜及相机（Nikon eclicse TS100-F；Nikon DS-U2；SAMSUNG WB150F），荧光显微镜（OLYMPUS IX71 DP72），扫描电镜（JEOL JSM-6510）。

1.2 方法

1.2.1 乳腺组织取样

试验奶牛屠宰后，乳房表面用清水充分冲洗，75%酒精消毒，高压灭菌解剖刀剖离整个乳房，置于75%酒精处理过的采样箱中，1h内带回实验室，再用75%酒精消毒组织表面，避开脂肪组织、结缔组织和血管，剪取约2.5cm^{3}的乳腺实质，75%酒精浸泡3min后转移入细胞间。

1.2.2 乳腺上皮细胞分离培养及纯化

用D-Hank's液浸洗组织4次以去除酒精，剔除外部泛白组织，将乳腺实质剪碎至1mm^{3}小块。D-Hank's液清洗后均匀接种于铺满鼠尾胶原蛋白的细胞培养瓶，分别取不同培养瓶侧置和倒置（图1），培养箱中分别静置30min和1h，吸除瓶底液体，缓慢加入完全培养液至刚好淹没组织，继续培养。待细胞铺满瓶底后，移除组织块，并将组织块回收再接种培养。待细胞铺满瓶底约80%时，差时胰酶消化法纯化奶牛乳腺上皮细胞。

1.2.3 奶牛乳腺上皮细胞形态学观察

采用倒置显微镜和扫描电镜观察细胞形态特点。扫描电镜样品制备如下：参照刘莉莉[7]方法制作细胞爬片，待细胞融合至70%～80%，D-Hank's液清洗后用2.5%戊二醛4℃下固定1h；弃去戊二醛，D-Hank's液和蒸馏水充分清洗后，依次用40%、50%、60%、70%、80%、90%、95%、100%、100%酒精逐级脱水，自然干燥后喷晶观察。

1.2.4 奶牛乳腺上皮细胞鉴定

参照刘莉莉[7]试验方法制作细胞爬片。对角蛋白18进行免疫荧光标记，用PI染料核染，荧光显微镜下观察[8]，鉴定上皮细胞并判断其纯度。

A. 侧置处理法；B. 倒置处理法
A. Side-place method；B. Inversion method

图1　培养瓶摆放方式
Fig. 1　Placement of cell culture flask

1.2.5　奶牛乳腺上皮细胞传代

当细胞铺满瓶底约80%时，胰酶消化，显微镜下观察到多数细胞回缩、变圆、细胞间隙扩大时，终止消化，反复吹打后，以$1\times10^5 \cdot mL^{-1}$细胞密度传代培养。

1.2.6　奶牛乳腺组织冻存

将乳腺实质剪至$4\sim5mm^3$的小块，每5mL冻存管中放入$2\sim6$块组织，分别加入冻存液A和B。纱布包裹冻存管，于-80℃冻存24h后转移至液氮中保存。

1.2.7　乳腺组织复苏及细胞培养

8个月后冻存管37℃水浴快速解冻，D-Hank's液充分清洗组织块后，按照新鲜组织处理方法分离培养奶牛乳腺上皮细胞。接种组织块时每30块为一组，每组3个重复，连续观察10d，对有细胞爬出的组织块进行计数，并计算其平均值，比较A、B冻存液保存的奶牛乳腺组织爬出细胞的活力。

2　结果

2.1　奶牛乳腺上皮细胞分离培养

试验奶牛乳腺组织脂肪含量较高（脂肪颗粒如图2A所示），清洗小组织块过程中易漂浮、接种后不易贴壁。相同时间内侧置处理组织贴壁效果优于倒置处理组织，30min即可达到较好贴壁效果。侧置处理的组织块培养至第4—5天可见细胞爬出，倒置处理的组织块培养至第6天可见细胞爬出。鹅卵石样乳腺上皮细胞后于成纤维细胞爬出，两者间有明显界限（图2A）；部分组织可直接爬出鹅卵石样上皮细胞，无成纤维细胞爬出（图3）。回收的组织接种后第2天即有细胞爬出，且鹅卵石样细胞比例较高。

2.2　奶牛乳腺上皮细胞纯化

成纤维细胞消化$30\sim60s$，乳腺上皮细胞消化$2\sim3.5min$，纯化$2\sim3$次即可得到较纯的奶牛乳腺上皮细胞。纯化过程中，对乳腺上皮细胞进行一次瓶内消化平铺，可解决细胞团块中心部位密度过大的问题（图2B）。

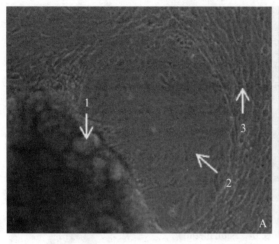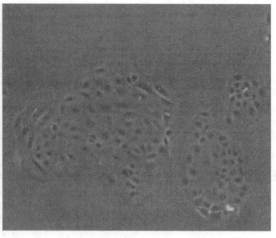

A. 纯化前的细胞团，1为脂肪颗粒，2为纯化前的奶牛乳腺上皮细胞，
3为纯化前的纤维细胞；B. 纯化后的奶牛乳腺上皮细胞团块

A. Cells after purification, 1 indicated fat particles, 2 indicated the mammary epithelial cells before purification,
3 indicated Fibroblasts before purification; B. Bovine mammary epithelial cells after purification

图2　细胞原代培养（Nikon，10×）
Fig. 2　Cell primary culture（Nikon，10×）

2.3　奶牛乳腺上皮细胞形态学特征

奶牛乳腺上皮细胞呈单层生长，鹅卵石样聚集，呈岛屿状生长，有少部分细胞呈多角型、摊鸡蛋样和梭形等，大小不一（图4A），但其与成纤维细胞的形态差异极大（图4C、D）。乳腺上皮细胞的细胞核呈圆形或椭圆形，细胞间可见上皮细胞特有的拉网结构，存在接触抑制现象。细胞传至20代后，扁平圆饼状极性细胞数量增多，也有三角形、长形、多边形极性细胞（图4B）。

在扫面电镜下细胞轮廓清晰，大小不一。鹅卵石样乳腺上皮细胞的细胞核呈椭圆形，边缘不光滑，摊鸡蛋样细胞中央有皱褶，表面较光滑，细胞体积较大（图4E、F）。电镜扫描下新鲜组织培养的上皮细胞和冻存组织培养的细胞形态基本相同。

2.4　奶牛乳腺上皮细胞的鉴定

用荧光标记第5代乳腺上皮细胞角蛋白18，荧光显微镜下可观察到大量阳性细胞，且其细胞核位置与细胞核染色位置基本重合，表明本研究成功获得纯化的奶牛乳腺上皮细胞（图5A、B、C）。

2.5　奶牛乳腺上皮细胞稳定性

侧置与倒置处理得到的细胞生长速度和形态学无明显差异，传至第10代的细胞形态和生长速度基本一致；20代后细胞生长速度虽无明显变化，但形状不规则的极化细胞逐渐增多（图2D、E、F）。

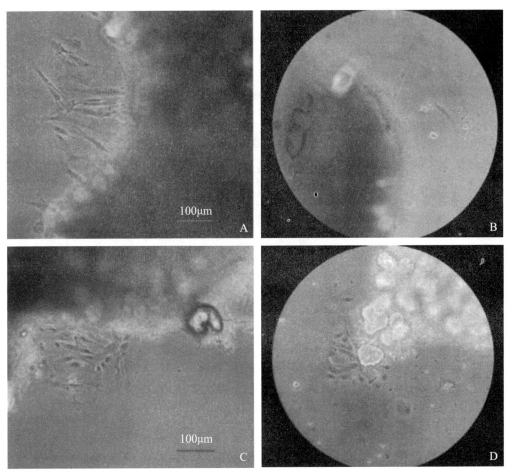

A.冻存组织培养第5天爬出的成纤维细胞（Nikon）；B.新鲜组织培养第5天爬出的成纤维细胞
（SAMSUNG）；C.冻存组织培养第5天爬出的鹅卵石样上皮细胞（Nikon）；
D.新鲜组织培养第5天爬出的鹅卵石样上皮细胞（SAMSUNG）

A. The Fibroblasts climbed out from the cryopreserved tissue on the fifth day（Nikon）；B. The fibroblasts climbed out from the fresh tissue on the fifth day（SAMSUNG）；C. The oval morphology of mammary epithelial cells climbed out from the cryopreserved tissue on the fifth day（Nikon）；D. The oval morphology of mammary epithelial cells climbed out from the fresh tissue on the fifth day（SAMSUNG）

图3　形态学观察比较（10×）

Fig. 3　Comparison of morphology（10×）

2.6　冻存奶牛乳腺组织活性与培养细胞形态观察

分别取冻存液A和B中的组织块进行分离培养，每天爬出细胞的组织块数见表1。来自于冻存液A的组织块存活率为50%，来自冻存液B的组织块存活率为56%。结果显示冻存液B对组织块的保存效果较好。

形态学观察结果表明，冻存组织在第4天有细胞爬出，且其与新鲜组织爬出的细胞形态无明显差别；部分组织块也可直接爬出鹅卵石样奶牛乳腺上皮细胞（图3）。鉴定纯化结果显示，乳腺上皮细胞有角蛋白18表达。

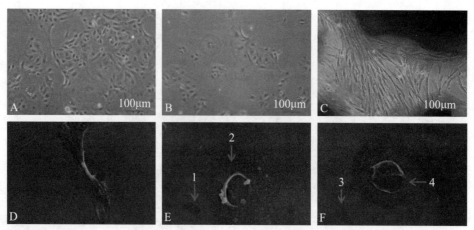

A.20代内乳腺上皮细胞，多呈鹅卵石样（Nikon，10×）；B.20代后扁平圆饼状极性细胞增多（Nikon，10×）；C.倒置显微镜下的成纤维细胞（10×）；D.扫描电镜下的成纤维细胞（1 000×）；E.新鲜组织培养的鹅卵石样（1）和摊鸡蛋样（2）上皮细胞（扫描电镜图，1 000×）；F.冻存组织培养的鹅卵石样（3）和摊鸡蛋样（4）上皮细胞（扫描电镜图，1 000×）

A. The mammary epithelial cells showed oval morphology within 20generations（Nikon，10×）；B. Polarized morphology cells increased gradually after 20generations（Nikon，10×）；C. The fibroblasts under the inverted microscope（10×）；D. The fibroblasts under the scanning electron microscope（1 000×）；E. The oval morphology cells(1)and the pancake morphology mammary epithelial cells(2)cultured from fresh tissue(scanning electron microscope，1 000×）；F. The oval morphology（3）and the pancake morphology mammary epithelial cells（4）cultured from cryopreserved tissue（scanning electron microscope，1 000×）

图4　奶牛乳腺上皮细胞形态学观察
Fig. 4　The morphology of bovine mammary epithelial cell

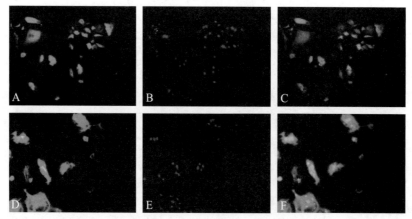

A. 第5代细胞，细胞角蛋白18荧光标记；B. 第5代细胞，细胞核染色；C.A图与B图的合并图；
D. 第24代细胞，细胞角蛋白18荧光标记；E. 第24代细胞，细胞核染色；F.D图与E图的合并图

A. Green fluorescence labeling of cytokeratin 18 on the 2th generation cells；B. Red fluorescent labeling of nucleus on the 2 th generation cell；C. Merged figure of A and B；D. Green fluorescence labeling of cytokeratin 18 on the 24 th generation cells；E. Red fluorescent labeling of nucleus on the 24 th generation cell；F. Merged figure of D and E

图5　免疫荧光鉴定角蛋白18（20×）
Fig. 5　The immuneofluorescence assay of cytokeratin 18（20×）

表 1 组织块存活率
Table 1 Survival rate of tissue piece

冻存液 Stored solution		时间/d Days							平均值 Average	存活率/% Survival rate
		4	5	6	7	8	9	10		
冻存液 A Stored solution A	A1	4	+7	+3	−1	+1	0	0	15	50
	A2	6	+6	+2	+2	0	0	0		
	A3	2	+6	+3	+1	+1	0	0		
冻存液 B Stored solution B	B1	4	+7	+3	+2	0	0	0	16.7	56
	B2	4	+6	+3	+2	+1	0	0		
	B3	5	+7	+4	−1	+1	0	0		

3 讨论

3.1 组织材料的选择

哺乳动物乳腺腺泡通常到妊娠中晚期或泌乳期才能充分发育，泌乳后期又开始萎缩，是一个循环的组织重塑过程[9]。一般认为，取自妊娠中晚期或泌乳期的乳腺组织较易获得乳腺上皮细胞，因此利用健康的4～7岁泌乳高峰期或泌乳中后期的奶牛乳腺组织，培养乳腺上皮细胞报道较多[1, 10-12]。也有研究表明，非泌乳期乳腺组织也可用于乳腺上皮细胞的分离培养。如M. Jedrzejczak等和G. Rauner等报道，利用小母牛乳腺组织分离、培养乳腺上皮细胞的效率高，获得的原代细胞活性好[13-14]。本试验所选奶牛（24月龄）已达到性成熟，无泌乳史，可避免泌乳期乳腺组织初乳不易清洗的问题，获得的乳腺上皮细胞生长速度快，可隔天传代。试验结果提示，腺泡未得到充分发育的奶牛乳腺组织也可用于乳腺上皮细胞的分离培养。

3.2 组织材料的处理

细菌污染是困扰奶牛乳腺上皮细胞原代培养的主要问题，其主要原因有两方面：一是奶牛患隐性乳房炎，其乳腺组织带菌，一般难控制、不易消除；二是人为污染。本研究通过3个环节克服以上问题：（1）选择无泌乳史奶牛，患隐性乳房炎概率极低，可以保证乳腺组织不带菌；（2）注意无菌转移，组织块短时间浸于75%酒精中转移至无菌操作台，可保证组织表面无菌；（3）提高培养液和D-Hank's中抗生素浓度，高于报道用量5倍，待细胞爬出后，再逐渐降低抗生素浓度，可保证组织块处理和初期培养过程中对可能存在细菌产生及时有效的抑制作用。

3.3 侧置处理的优势

本试验中采用侧置处理组织块中乳腺上皮细胞爬出时间为4～5 d，倒置处理组织块中乳腺上皮细胞爬出时间为6～7 d，侧置比倒置缩短了1～2 d。分析认为，侧置处理能够尽快沥干组织块表面液体，这有利于组织块的快速贴壁，能更好地保持组织活力。乳腺上皮细胞免疫荧光鉴定和传代试验结果表明，两种方法得到的乳腺上皮细胞特异性角蛋白18表达水平和增殖效果基本一致。可见，侧置处理较常用的倒置[15]处理优势在于缩短组织块种植法培养乳腺上皮细胞的周期。

3.4 形态学特征

据报道，体外培养的上皮细胞有3种类型：分泌上皮细胞、导管上皮细胞、肌上皮细胞。传统的差速贴壁法和差时消化法很难将这3种上皮细胞分开，需要通过密度梯度离心的方法才能将其分开[16-17]。本试验荧光鉴定结果显示，分离获得的乳腺上皮细胞均能表达上皮细胞特有的角蛋白18，但其形态不一，推测可能有以上3种类型上皮细胞。扫描电镜下分离到的乳腺上皮细胞呈鹅卵石样，这与刘明江观察到的结果[18]类似。但本试验观察到的摊鸡蛋样乳腺上皮细胞还未见报道。

3.5 奶牛乳腺组织冻存法

组织冻存技术目前在遗传育种、生物医疗领域已得到广泛应用，不仅可以保存稀缺组织，同时可以减少某些研究需要反复取材培养的麻烦。本研究采用冻存较大组织块（$4\sim5mm^3$）的方法保存奶牛乳腺组织，经-80℃一步冷冻过夜后于液氮中保存8个月，冻存液A保存的乳腺组织成活率为50%，冻存液B保存的乳腺组织成活率为56%。本方法虽比孙苏军等[19]采用小组织块（d<0.3mm）和梯度冻存法保存的山羊乳腺组织块复苏后成活率（89.8%）低，但可大大简化组织块剪切、清洗、冻存和复苏等操作过程，且冻存液中加入缓冲液HEPES/丙酮酸钠后有利于组织块生物活性的保持，能够满足细胞分离培养的需要。

4 结论

在利用组织块种植法培养奶牛乳腺上皮细胞过程中，应用侧置处理法与组织块回收法，从无泌乳史奶牛乳腺组织中成功分离培养出了奶牛乳腺上皮细胞；在乳腺组织冻存过程中，发现冻存液中加入缓冲液HEPES/丙酮酸钠后有利于组织块生物活性的保持，采用一步冻存较大奶牛乳腺组织块仍可保持其活性，能满足乳腺上皮细胞分离培养需要。

（编辑　白永平）

（发表于《畜牧兽医学报》，一级学报，1 067-1 074.）

6株牛源副乳房链球菌的分离和鉴定

李新圃，罗金印，杨 峰，王旭荣，李宏胜*

（中国农业科学院兰州畜牧与兽药研究所/农业部兽用药物创制重点实验室/
甘肃省新兽药工程重点实验室，甘肃兰州 730050）

摘 要：从奶牛临床型乳房炎乳汁中分离出6株细菌，通过生化试验、16SrDNA鉴定、小鼠致病性和抗生素耐药性检测，结果显示，这6株细菌为革兰阳性链球菌，与副乳房链球菌（Spu）同源性为99%；在血琼脂平面上产生α溶血，过氧化氢酶反应阴性，CAMP试验阴性；对小鼠有不同程度的致病性，半数致死量（LD_{50}）为$5.6 \times 10^7 \sim 5.2 \times 10^8$CFU；对四环素耐药，对氨苄西林、氯霉素、红霉素、卡那霉素、氧氟沙星和万古霉素敏感。本试验检出并报道牛源Spu，研究数据将有助于进一步开展Spu与奶牛乳房炎之间的相关性研究。

关键词：奶牛；副乳房链球菌；分离；鉴定

Isolation and identification of six *Streptococcus parauberis* from cow

LI Xinpu, LUO Jinyin, YANG Feng, WANG Xurong, LI Hongsheng*

（Lanzhou Institute of Animal Science and Veterinary Pharmaceutics of CAAS/Key Laboratory of Veterinary Drug Discovery of Agricultural·ministry/Key Laboratory of New Animal Drug Project of Gansu province, Lanzhou 730050, China）

Abstract: Six bacteria were isolated from milk of cow with clinical mastitis, and biochemical tests, 16SrDNA identification, mouse pathogenicity and antibiotic resistance were carried out. The result showed the 6isolates gram-positive *Streptococcus* shared 99% homology with *Streptococcus parauberis*, which had biochemical reaction of α-hemolysis, catalase-negative, and CAMP-negative, displayed varying degrees of pathogenicity to mice with LD_{50} of $5.6 \times 10^7 -$

基金项目："十二五"国家科技支撑计划资助项目（2012BAD12B03）；甘肃省自然科学基金资助项目（1308RJZA119）；甘肃省农业科技创新资助项目（GNCX-2013-59）；甘肃省科技支撑计划资助项目（144NKCA240）
作者简介：李新圃（1962—），女，副研究员，博士。
*通讯作者，E-mail: myslhs@163.com
DOI: 10.16303/j.cnki.1002-4545.2016.10.12

5.2×10^8 CFU, were resistance to tetracycline but sensitive to ampicillin, chloramphenicol, erythromycin, kanamycin, ofloxacin and vancomycin. It was the first time to detect *S. parauberis* from cow in domestic, and the data contribute to further understand the correlation between *S. parauberis* and cow mastitis.

Key words: Cow; *Streptococcus parauberis*; Isolation; Identification

副乳房炎链球菌（*Streptococcus parauberis*，Spu）是一种环境性致病菌，能够引发奶牛乳房炎。因为常规微生物和生化鉴定试验很难将其与乳房链球菌（*Streptococcus uberis*，Su）相区分，所以早期有关奶牛乳房炎病原菌的研究报道并未涉及Spu。1990年，英国的WILLIAMS等[1]首次采用DNA技术对SuⅠ型和Ⅱ型的16SrRNA核酸序列进行了反转录酶分析，发现它们具有不同系统发育，处于不同种状态，因此提议将SuⅡ型命名为Spu。随着分子生物学技术的发展应用，DNA和PCR技术已被成功用于Spu的分离鉴定[2-3]。国外已有人从奶牛乳房炎乳汁中分离鉴定出Spu[4]，但国内还没有人对牛源Spu进行研究报道[5-6]。2014年，在陕甘宁等地部分奶牛场采集了304份临床型乳房炎乳样，在进行细菌分布动态监测时，从中发现6株Spu。为了解Spu与奶牛乳房炎的相关性，本研究对该菌进行了生化试验、16SrDNA鉴定，致病性和耐药性研究。

1 材料与方法

1.1 主要仪器及试剂

PCR扩增仪（ABI），凝胶成像仪（BOILAD），显微镜（Nikon），电泳仪（北京六一仪器公司），培养箱（上海一恒科学仪器有限公司）。16SrDNA PCR试剂盒（TaKaRa），细菌DNA提取试剂盒（OMEGA），DL2000 DNA Marker（TaKaRa）。

药敏纸片青霉素10U、氨苄西林10μg、氯霉素30μg、红霉素15μg、卡那霉素30μg、氧氟沙星5μg、链霉素10μg、四环素30μg、万古霉素30μg，肺炎链球菌标准株ATCC49619（ATCC：美国模式培养物集存库），以及培养基MHA（水解酪蛋白琼脂）购自杭州微生物试剂有限公司。增菌肉汤、改良肉汤、琼脂血平面等自制。

1.2 实验小鼠

清洁级昆明小鼠，雌雄各半，6~8周，体质量16~20g，由兰州大学GLP实验室提供。

1.3 奶牛乳样

临床型乳房炎乳样，采自陕西、甘肃、宁夏、山西等地奶牛场。用温水清洗患病乳区，酒精棉消毒乳头，弃去头2把乳，将乳汁挤在灭菌试管中，远途乳样保存在琼脂斜面上，冷藏保存，并尽快进行细菌分离鉴定。

1.4 细菌分离

挑取均匀乳样（或琼脂斜面上的菌落），划分在血平面上，37℃孵育24h。从血平面上挑选单个菌落，接种到增菌肉汤中，37℃孵育24h。进行载玻片涂片、革兰染色、显微镜观察。对分离纯化的细菌，进行CAMP和过氧化氢酶试验。分离纯化菌株可加入脱脂乳中冷冻保存。

1.5 16SrDNA鉴定

采用DNA提取试剂盒提取细菌DNA，使用16SrDNA PCR试剂盒对DNA模板进行PCR。对PCR扩增产物进行10g/L琼脂糖凝胶电泳和凝胶成像，将符合测序标准的PCR产物纯化，并送北京六合华大基因科技股份有限公司进行基因测序。使用PubMed数据库，对所测得序列片段进行Blast（http://www.ncbi.nlm.nih.gov/pubmed/）搜索比对，确定细菌种属。

1.6 试验菌悬液

取分离纯化菌株，接种到改良肉汤中，在37℃培养24h，使细菌浓度达到5麦氏浊度以上。采用平板菌落计数法测定细菌浊度，计算每毫升菌悬液中菌落形成单位（CFU/mL），用改良肉汤稀释细菌到试验浊度。

1.7 分离菌对小鼠的致病性

参考研究资料[7-8]，每株菌设立4个试验组和1个对照组，每组5只小鼠，试验组分别按0.2，0.4，0.6和0.8mL的注射量，于腹腔注射1.2×10^9 CFU/mL的菌悬液，对照组注射0.8mL的改良肉汤，找出试验菌对小鼠0%和100%致死量的估计值。在这2个估计值之间，制备4个不同浓度的菌悬液，每株菌设立4个试验组和1个对照组，每组5只小鼠，试验组分别腹腔注射0.5mL不同浓度的菌悬液，对照组注射改良肉汤，使用序贯法[9]计算小鼠的半数致死量（LD_{50}）。试验小鼠观察1~15d，记录小鼠的病理变化及死亡情况，对发病死亡小鼠进行剖解，观察内脏器官的病理变化，同时挑取少许肝、脾组织，划线在血平面上，37℃孵育24h。观察血平面上细菌生长情况，若有细菌生长，并且形态特征和DNA鉴定符合试验菌，则说明试验菌具有致病性。

1.8 分离菌的耐药性

参照我国卫生行业标准WS/T-125—1999[10]和美国临床和实验室标准协会标准CLSL 2013[11]，以ATCC49619（链球菌药敏试验质控菌株）作为质控菌株，吸取0.5麦氏浊度的试验菌悬液200μL，均匀涂布在直径90mm的MHA培养基（含5%羊血）上，放置5片抗菌素药敏纸片，37℃孵育24h，测定抑菌圈直径，判断耐药、敏感和中敏。每株菌进行3次平行试验，结果取其平均值。

2 结果

2.1 细菌生化形态

6株Spu，在改良肉汤中生长良好，革兰染色阳性，镜检菌体呈圆形、短链、无荚膜（图1），血琼脂平面上生长的菌落细小、灰白、光滑、突起，呈现不同程度的α溶血（图2）；CAMP试验阴性（图3），过氧化氢酶反应阴性。

2.2 16SrDNA鉴定结果

结果见图4，图5。使用试剂盒提供的Forward Primer/Reverse Primer1扩增引物，获得500bp左右的DNA扩增片段，测序结果通过Blast搜索比对，6株分离菌株与Spu JJI51（基因登记号FJ009631）、DSM6631（基因登记号TN630843）、349（基因登记号AY942570）、332（基因登记号AY942572）、328（基因登记号AY942571）和F64（基因登记号KP137349）等的同源性为99%。

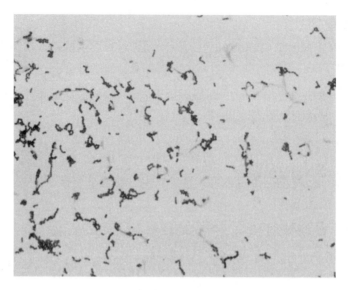

图1 分离菌株的染色形态

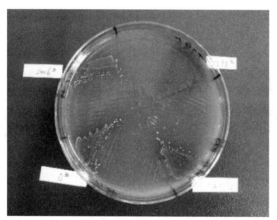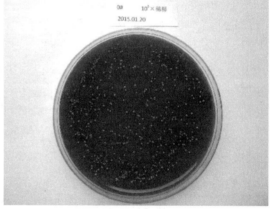

图2 分离菌株在血平面上生长状态

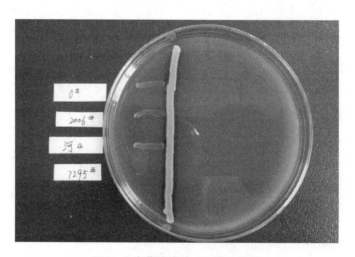

图3 分离菌株的CAMP试验结果

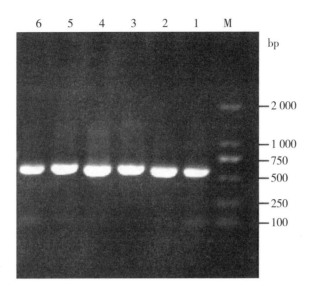

图4 分离菌株16SrDNA PCR电泳图 M. DL2000 DNA Marker；1~6.6株Spu分离菌

ATTTTGATTCTGGCTACACGACGCTGGCGGCGTGCCTAAT-
ACATGCAAGTAGAACGCTGAAGACTGGTGCTTGCACTAG-
TCAGATGAGTTGCGAACGGGTGAGTAACGCGTAGGTAAC-
CTACCTTATAGCGGGGGATAACTATTGGAAACGATAGCT-
AATACCGCATGACAATTAAGTACTCATGTACTAAATTTA-
AAAGGAGCAATTGCTTCACTATGAGATGGACCTGCGTTG-
TATTAGCTAGTTGGTGAGGTAACGGCTCACCAAGGCTAC-
GATACATAGCCGACCTGAGAGGGTGATCGGCCACACTGG-
GACTGAGACACGGCCCAGACTCCTACGGGAGGCAGCAGT-
AGGGAATCTTCGGCAATGGGGGCAACCCTGACCGAGCAA-
CGCCGCGTGAGTGAAGAAGGTTTTCGGATCGTAAAGCTC-
TGTTGTTAGAGAAGAACGGTAATGGGAGTGGAAAATCCA-
TTACGTGACGGTAACTAACCAGAAAGGGACGGCTAACTA-
CGTGCCAGCCGCCGCGGTAATACGTCGTGACTGGGAAACC-
CCTTGGGGA(557 bp)

图5 分离菌株16S rDNA测序结果

2.3 小鼠致病性

试验组小鼠腹腔注射$6.4 \times 10^7 \sim 5.9 \times 10^8$ CFU的Spu，出现不同程度的精神萎靡，不思饮食，站立不稳，嗜睡昏迷，感染后1~3d开始出现死亡，死亡情况持续10d左右，之后未死亡的小鼠开始恢复。对照组小鼠，生长发育正常。剖解死亡鼠，发现肺脏充血，肝脏发暗有白斑，肠管水肿，腹腔积水。取死亡小鼠的肝、脾组织，接种到血平面上，长出单一细菌，检测结果为Spu。Spu对小鼠的DL_{50}为$5.6 \times 10^7 \sim 5.2 \times 10^8$ CFU。

2.4 抗生素耐药性

结果见表1。6株Spu对四环素耐药，对青、链霉素既有耐药株，也有敏感株，对氨苄西林中度敏感，对氯霉素、红霉素、卡那霉素、氧氟沙星和万古霉素敏感。

表1 药敏试验结果

抗生素	菌株数（占试验菌株的百分比，%）		
	耐药	中敏	敏感
青霉素	4(66.7)	2(33.3)	0(0.0)
氨苄西林	0(0.0)	4(66.7)	2(33.3)
氯霉素	0(0.0)	0(0.0)	6(100.0)
万古霉素	0(0.0)	0(0.0)	6(100.0)
卡那霉素	0(0.0)	0(0.0)	6(100.0)
四环素	6(100.0)	0(0.0)	0(0.0)
氧氟沙星	0(0.0)	0(0.0)	6(100.0)
红霉素	0(0.0)	0(0.0)	6(100.0)
链霉素	4(6.3.7)	2(33.3)	0(0.0)

3 讨论

微生物和生化试验是分离与鉴定细菌的常规方法，主要是利用细菌及其产物的生化反应，将肉眼无法观察到的细菌转化成肉眼可见的实物和颜色变化，其过程受到很多不确定因素的影响，如细菌在保存过程可能产生的变化，不同批次培养基和生化试验管之间的活性差异，不同观察者判断颜色和形态的主观意识差别，因此很难通过微生物和生化试验鉴别出亲缘关系很近的Spu和Su。有研究报道，在现有常规生化试验条件下，Spu与Su的生化反应区别仅在于前者不产生β-D-葡萄糖苷酸酶，后者产生[12]。但是利用微生物和生化试验，可以缩小菌群范围，便于Spu和Su的分离鉴定。本研究分离鉴定出6株Spu，与Su同样属于非β溶血性、过氧化氢酶反应阴性、CAMP试验阴性的革兰阳性链球菌。但如果将链球菌的范围扩大为球菌，具有以上生化反应特性的菌群可以扩大到包括七叶苷阳性链球菌、肠球菌、乳球菌和空气菌等[13]。

微生物和生化试验鉴定很容易造成同属细菌之间的误判。由于同属不同种的细菌，致病性和药物敏感度往往存在较大的差别[14]，菌种误判可能造成细菌致病性被忽视以及药敏试验结果可信度降低的不良后果。因此，采用分子生物学鉴定技术，准确区分细菌种属，不仅有助于细菌本身的分离鉴定研究，而且有助于对细菌引发疾病的有效诊断和治疗。随着分子生物学鉴定技术的发展及测试成本的降低，该技术将会逐渐替代生化试验成为鉴别细菌的主要方法，而生化试验则会成为以了解细菌生化特性为主的辅助性鉴定法。研究资料显示，利用分子生物学方法，德国已从131株Su中鉴别出1株Spu[12]，荷兰从83株Su中鉴别出2株Spu，澳大利亚从27株中Su鉴别出6株Spu[15]，芬兰从137株Su中鉴别出2株Spu[4]，国内至今无人进行相关的研究报道。

由于常规微生物和生化试验很难区分Spu和Su，加上Spu的检出率较低，近年来有关牛源Spu的研究报道很少。2014年期间，从陕西、甘肃、宁夏、山西等地采集了304份临床型乳房炎乳样，从中检测出包括27个种属的227株细菌。其中检出率较高的致病菌有无乳链球菌27.6%，金黄色葡萄球菌7.0%，大肠杆菌4.7%。并且从中发现6株Spu，检出率为2.3%，这是国内首次从奶牛乳房炎乳汁中分离鉴定出Spu。而Su仅检出1株，检出率为0.4%。6株Spu对小鼠均有一定的致病性，对四环素和青、链霉素存在不同程度的耐药性。本试验中，Spu和Su的检出比例是6∶1，与国外研究资料显示的比例（0.008~0.128）完全相反，说明国内已经出现Spu引发的奶牛乳房炎。Spu与奶牛乳房炎之间存在的关系，是否与Su一致，还是存在某些差别？还需要进一步研究观察。

（发表于《中国兽医学报》，一级学报，1 710–1 713.）

苦马豆素的来源、药理作用及检测方法研究进展

黄 鑫，梁剑平，高旭东，郝宝成*

（中国农业科学院兰州畜牧与兽药研究所，农业部兽用药物创制重点实验室，
甘肃省新兽药工程重点实验室，甘肃兰州 730050）

摘 要：吲哚里西啶类生物碱苦马豆素是疯草及其内寄生真菌的代谢产物，其有良好的抗病毒、抗肿瘤和提高机体免疫力的作用，具有巨大的药用潜力。作者简要阐述了苦马豆素的药理作用、检测方法、药用价值和研发潜力，并对其来源及疯草内生真菌合成苦马豆素的相关研究进展进行总结，以期丰富苦马豆素的来源，为其在科研和医疗上的广泛应用提供保障。

关键词：苦马豆素；来源；药理作用；检测方法；内生菌合成

Research Advance on Sources, Pharmacological Effects and Detection Methods of Swainsonine

HUANG Xin, LIANG Jianping, GAO Xudong, HAO Baocheng*

(Key Laboratory of New Animal Drug Project of Gansu Province/Key Laboratory of Veterinary Pharmaceutical Development of Ministry of Agriculture, Lanzhou Institute of Husbandry and Pharmaceutical Sciences of CAAS, Lanzhou 730050, China)

Abstract: Swainsonine, one of indole quinolizidine alkaloids, is a metabolite of locoweed and its parasitic fungi. The swainsonine has good anti-virus, anti-tumor effects and can enhance the role of immunity, has great potential for medicine development. This paper briefly expounds pharmacology, detection method, medicinal value and development potential of swansonine. And the research progress of its source and locoweed endophytic fungus synthesizing spherosin were summarized to enrich swainsonine's source and provide the supply of abundant raw materials for

---

基金项目：中国农业科学院科技创新工程兽用天然药物创新工程（CAAS-ASTIP-2014-LIHPS-04）
作者简介：黄鑫（1992—），男，甘肃清水人，硕士生，主要从事新型兽用天然药物的研究，E-mail: gsau520@sina.cn
*通讯作者：郝宝成（1983—），男，博士，助理研究员，主要从事兽用天然药物的研究与创制研究，E-mail: haobaocheng@sina.cn

its scientific and medical application.

Key words: Swainsonine; Source; Pharmacologic action; Detection method; Endophytic fungi synthesis

苦马豆素（Swainsonine，SW）是一种吲哚里西啶类生物碱，是从部分疯草植物及其内生真菌中提取出来的具有抗病毒、抗肿瘤和提高免疫力等生物活性的生物碱，也是引起家畜疯草中毒的主要原因。自20世纪70年代首次分离获得苦马豆素纯品以来，其功能和药理作用受到人们的广泛关注。据相关报道，直接投喂或者注射SW均能引起山羊、绵羊、家兔和大鼠等实验动物出现相似的病理症状，出现的病理症状均是由SW竞争性抑制哺乳动物细胞内溶酶体α-甘露糖苷酶和高尔基体甘露糖苷酶Ⅱ的活性引起的，进而使动物体中低聚糖代谢紊乱、糖蛋白合成受阻、组织细胞空泡化，相关组织器官的功能紊乱甚至丧失，最终导致动物的中毒或死亡[1-4]。

此外，SW还具有良好的抗肿瘤、抗病毒和提高免疫力等药理作用。陈基萍等[5]研究发现苦马豆素还有一定的抑菌作用。但由于苦马豆素来源有限、人工合成困难、提取效率低、真菌发酵技术不成熟及市场价格昂贵等原因，限制了其在抗病毒、抗肿瘤等领域的应用与发展[6-7]。

1 苦马豆素的来源

苦马豆素的结构为吲哚里西啶环1，2，8位上各连有一个活性羟基基团，故被称为1，2，8-三羟基八氢吲哚里西啶或吲哚里西啶三醇。SW具有良好的抗病毒、抗肿瘤和免疫调节作用，因此备受相关领域的关注。但是，SW的来源一直是限制其在各领域发展的重要因素。SW的主要来源大概分为3种：化学合成、植物中提取和内生真菌发酵。

1.1 化学合成苦马豆素

由于SW具有良好的生物活性而天然来源含量很低，因此引起化学家对其实验室合成的极大兴趣。自从1984年对SW进行首次全合成以来，研究人员已经发现了四十多种化学合成路线[8]，但由于其化学结构是由两个环共同分享一个碳原子和一个氮原子，且环上含有3个羟基基团，构成了4个手性碳原子，所以对其进行手性全合成难度很大。将现有的合成方法总结归纳，可分为以下四类：

（1）利用烯烃复分解反应合成SW的中间体吲哚里西啶环，该类反应的关键在于将含有杂原子氮和氧的环状手性烯烃化合物经过钌催化烯烃复分解反应，合成SW的中间体吲哚里西啶环。典型反应是N. Buschmann等[9]以具有手性的噁唑烷酮类化合物作为合成原料，经过水解、酰胺烷基化、TBDM-SOTf保护、钌催化烯烃复分解反应得到吡咯衍生物，再经过亲核取代、脱烯丙基甲酸酯后得到不饱和的吲哚里西啶环，此反应共经过12步，其产率达到40%，因其应用了烯烃复分解反应产率较高，适用于工业大规模生产SW。此外，C. W. G. Au等[10]利用分子内的烷基化反应合成六氢吡啶衍生物，并选择性地获得C-8手性构型，然后经过钌催化烯烃复分解反应合成SW中间体不饱和吲哚里西啶，此反应共经过10步，但其产率仅为12%。I. Déchamps等[11]以L-吡咯氨酸为原料经14步反应，总收率为14%。

（2）经1，3-偶极环加成反应或者双羟基化不对称反应合成SW的关键中间体二环内酰胺类化合物，因烯烃双羟基化反应具有所需反应条件温和，对温度、水和氧气等条件要求不严格，并且大多数烯烃能获得高产率、高ee（对映体超量）值的光学活性邻二醇等优点，所以在SW合成过程中双羟基化反应起到了重要作用；此外，1，3-偶极环加成反应是合成SW中间体二环内酰胺类化合物的重要方法之一[12]。

（3）不对称反应和还原氨基化反应是合成SW的重要方法。M. Trajkovic等[13]用L-脯氨酸催化二噁烷酮衍生物发生不对称性羟醛缩合反应（还原胺化反应），形成具有手性的杂环化合物，再经过8步反应合成SW，总产率为24%。不对称氧化反应体系的组成简单、合成容易、原料经济、底物具有手性选择性；还原氨基化反应条件温和、产率较高、合成方便，是合成烯胺六元环的重要方法。

（4）H. Guo等[14]将丁内酯和2-锂呋喃经羟基保护、Noyori不对称还原、Achmatowicz反应、Boc保护及其酯化反应等，终得到吡喃酮类化合物，再经糖基化反应、二羟基化反应、氧化还原及叠氮化反应，最终成功合成SW。P. Rajasekaran等[15]将抗坏血酸衍生的烯丙醇进行不同的取代，然后在加热和钯催化的条件下发生Overman重排反应获得SW。

SW的化学合成路径已经超过了四十多条，其合成步骤8～20步不等，但由于SW结构具有两个手性碳原子，使其合成受到限制，而且合成的产物纯度较差，所以给SW化学合成带来了巨大的困难和挑战。如果找到一种化学合成方法，利用简单易得的合成原料，操作条件简单，能选择性的构建手性原子，这将给SW的高产率低成本合成提供一条捷径，这也是化学科学家一直追求的。V. Dhand等[8]试图找到一种独特的方法，其不需要手性分子作为合成原料，由5-氯戊醇为底物，以有机催化和不对称R-氯化为基础的反应可以控制反应基团选择性构成手性中心，由于其操作简单，且原料容易获得，所以这个反应过程可能会应用于一系列吲哚里西啶、吡咯环和类吡咯环等天然产物的合成。

1.2 植物中提取苦马豆素

SW是一种具有神经毒性的生物碱，因其最初在苦马豆（*Swainsona canescens*）中被发现，因此命名为苦马豆素；其也是豆科棘豆属和黄芪属有毒植物的主要毒性成分和动物疯草中毒的根本原因，故又称疯草毒素。此外，有报道称分布于地球南半球的某些锦葵科（Malvaceae）黄花稔属、旋花科（Con-volvulaceae）的植物中也含有SW[16-19]。

自20世纪70年代末，澳洲科学家Colegate首次从灰苦马豆（*Swainsona canescens*）中分离获得SW以来，国内外学者先后从斑荚黄芪（*Astragalus membranaceus* Bunge）、黄花棘豆（*Oxytropis ochrocephala*）、变异黄芪（*Astragalus variabilis* Bunge）、茎直黄芪（*Astragalus strictus* Grah. Ex Bend.）、绢毛棘豆、甘肃棘豆（*Oxytropis kansuensis*）、急弯棘豆（*Oxytropis deflexa* DC.）、宽苞棘豆（*Oxytropis latibracteata* Jurtz）、冰川棘豆（*Oxytropis glacialis* Benth）等种属的植物中提取出了SW。A. D. Vallotton等[20]在实验室中移植了3种密柔毛黄芪属（*Astragalus mollissimus*）的植物，试验发现这3种密柔毛黄芪属的植物都能产生SW，但其产SW的能力并不相同。B. C. Hao等[21]采用纤维素酶法提取西藏茎直黄芪中SW，并用气相色谱检测其含量，试验中采用单因素设计和正交试验等方法确定了提取SW的最佳条件：在粉碎筛为40目，料液比为1∶40，温度为50℃的条件下加入3.5%的纤维素酶酶解3h，其提取效率最高，此种方法可以更加高效快速、方便节能地提取茎直黄芪中的SW。宋岩岩等[22]将急弯棘豆植物的草粉经过酸化碱化处理后，分别用氯仿、乙酸乙酯和正丁醇等萃取，然后将萃取液在薄层板上展开，用欧式液（Ehrlich's）显色后分析，结果显示萃取物中含有SW，这说明急弯棘豆属于疯草类植物。五爪吉祥草[*Reineckia carnea*（Andr.）Kunth]

属于旋花科植物[22]，D. Cook等[23]在检测巴西草原上五爪吉祥草时发现其中含有SW，其含量是随着季节的变化而变化。S. Takeda等[24]在研究中发现，蒙古西部的许多山羊、绵羊、牛和马因摄入小花棘豆而出现共济失调、全身麻痹、肌肉震颤等神经症状，因此他们将干燥的小花棘豆植物处理后进行化学分析，发现这些植物中含有0.02%～0.05%的SW。

虽然从上述植物中可以提取分离到SW，但此类植物分布面积广泛，且大多生长在寒冷或干旱的环境恶劣地区，从植物中大量提取SW会对草原草场造成不可恢复性的破坏。而且从植物组织中提取SW过程十分复杂，提取所用溶剂不仅污染环境，且对原料利用率较低，因此在植物中提取SW的方法，环保性和可行性较差。

1.3 内生真菌发酵

1.3.1 疯草内生真菌

内生真菌（Endophytic fungi）是指某种真菌短暂或者长期地寄生于活的植物组织内部，但对寄生植物不会造成明显的病害症状，这种内生真菌包括病原真菌、腐生真菌和菌根菌[25]。1982年K. Braun[26]首次从绿僵菌（*M. anisopliae*）中分离获得SW，开创了从内生真菌中获得苦马豆素的先例。因此，部分研究人员认为疯草内生真菌与SW之间也可能存在一定联系。K. Braun从3种疯草植物中分离得到内生真菌，经过形态学鉴定均属于Undifilum属真菌，并将分离到的真菌在体外培养后，在代谢产物中检测到了SW，这是首次在人工体外培养的条件下从疯草内生菌中获得SW。据相关文献研究，金龟子绿僵菌（*Anisopliae*）、豆类丝核菌属（*Leguminicola*）和埃里格孢属（*Undifilum*）等主要真菌都能产SW。截至目前，从各种疯草中分离出的内生真菌均为埃里格孢属真菌（*Undifilum*）[27]。D. Cook等[28]为了分析SW、疯草植物和内生真菌之间的关系，调查了豆科中黄芪属、棘豆属和苦马豆属等能产生SW的3个属的植物，结果如表1所示。

K. Braun等[25]为了了解内生真菌与SW的关系，采集了11种有毒疯草植株，并从其中的8种有毒疯草植株的花、叶、茎和种子中分离出真菌，经检测均能产生SW，在进行体外培养时发现，其具有很厚的横向隔，在培养基中生长速度十分缓慢，偶尔会产生分生孢子。将体外培养的18株菌株的菌丝体采用PCR技术扩增出菌丝体B-微管蛋白编码区域rDNA的ITS序列，并对其进行序列分析，分析结果显示不同宿主植物、寄生在同一植株不同部位的所有的内生菌ITS序列基本相同，结果提示内生菌是通过植物种子垂直传播。形态学和ITS区序列的分析结果表明，内生菌与埃里格孢属真菌的关系最为密切。B. M. Pryor等[29]通过观察发现，疯草棘豆属产SW的内生真菌在形态结构上与埃里格孢属菌真菌（*Undifilum*）非常相似。结合形态学与线粒体小亚基rDNA基因分析，B. M. Pryor将棘豆属疯草中产SW的内生真菌划分为子囊菌纲（Ascomycetes）假球壳目（Pleosporales）假球壳科（Pleosporaceae）埃里格孢菌属（*Undifilum*）。D. Cook等[30]在南美洲的旋花科植物五爪吉祥草中发现了一种产生SW的内生真菌，体外培养并提取DNA，在对其ITS1F和ITS4基因序列进行测序时发现其为五爪吉祥草内共生菌，并在马铃薯琼脂培养基（PDA）上能够缓慢生长。用杀菌剂处理五爪吉祥草种子的试验证明该内生真菌通过植物种子进行垂直传播。

1.3.2 疯草内生真菌合成苦马豆素

SW是从部分疯草植物内生真菌的次级代谢产物中也可以获得的具有生物活性的生物碱SW。有学者报道，疯草中SW的含量与疯草感染内生真菌的数量具有较高的相关性[31]。E. Oldrup等[32]在一种疯草植物中发现了内生真菌*Undifilum oxytropis*，在利用PCR定量技术测定疯草植物中所含内生真菌的浓度时发现，植物中所含内生真菌的浓度与其所含SW的含量呈正相关；在没有检测到SW的疯草植株中也没有检测到 Undifilum oxytropis 内生真菌的存在。结果表明，疯草植物中SW是由植株中的共生内生菌合成产生。D. Cook等[33]指出，内生真菌的基因型可能是影响SW在不同植株中含量不同的原因。其在同一生长条件下种植同种绢毛棘豆植株，在这些植株中检测SW的含量时发现，在同一生长条件下不同的植株之间SW的含量是有差异的，这说明了不同植株所侵染的内生真菌的基因型不同，其产生苦马豆素的能力不同。

表1 黄芪属、棘豆属和苦马豆属疯草植物中产苦马豆素内生菌种类及检测方法 [28]

Table 1 Species of the swainsonine Fungal endophyte in *Astragalus*, *Oxytropis* and *Swainsona* plant and the detection methods of these endophytes

疯草种类 Locoweed	内生菌种类 Endophyte	苦马豆素检测方法 Detection method of the Swainsonine	真菌检测方法 Detection method of fungi	
			分离培养 Isolation and cultivation	PCR
Astragalus allochrous	*Undifilum* sp.	酶法		
Astragalus asymmetricus		TLC		
Astragalus didymocarpus		TLC		
Astragalus lentiginosus	*Undifilum fulvum*	TLC、LC-MS	×	×
Astragalus drummondii		酶法		
Astragalus purshii				
Astragalus amphioxys		LC-MS		×
Astragalus variabilis	*Undifilum oxytropis*	GC-MS、LC-MS	×	×
Astragalus emoryanus		TLC、MS、酶法		
Astragalus bisulcatus		TLC、酶法		
Astragalus strictus	*Undifilum oxytropis*	GC-MS	×	
Astragalus missouriensis		酶法		
Astragalus flavus		酶法		
Astragalus humistratus		酶法		
Astragalus lonchocarpus		酶法		
Astragalus mollissimus	*Undifilum cinerum*	TLC、LOMS、酶法	×	×
Astragalus nothoxys				
Astragalus oxyphysus		TLC		
Astragalus pycnostachyus		TLC		
Astragalus pehuenches		TLC		

(continued)

疯草种类 Locoweed	内生菌种类 Endophyte	苦马豆素检测方法 Detection method of the Swainsonine	真菌检测方法 Detection method of fungi	
			分离培养 Isolation and cultivation	PCR
Astragalus praelongus		TLC/酶法		
Astragalus pubentissimus	Undifilum sp.	LC-MS	×	×
Astragalus wootoni	Undifilum sp.	TLC、LC-MS	×	×
Astragalus tephrodes		酶法		
Astragalus thurberi		酶法		
Oxytropis glacialis	Undifilum oxytropis	GC-MS	×	
Oxytropis kansuensis	Undifilum oxytropis	GC-MS	×	
Oxytropis besseyi				
Oxytropis glabra	Undifilum oxytropis	GC-MS、LC-MS	×	×
Oxytropis campestris				
Oxytropis sericea	Undifilum oxytropis	TLC、LC-MS、酶法		
Oxytropis lambertii	Undifilum oxytropis	TLC、GC-MS、LC-MS、酶法	×	×
Oxytropis ochrocephala	Undifilum oxytropis	GC-MS	×	
Oxytrofiis sericopetala	Undifilum oxytropis	GC-MS	×	×
Swainsona greyana		GC-FID		
Swainsona canescens	Undifilum sp.	GC-FID、LC-MS	×	×
Swainsona galegifolia		GC-FID		

"×"表示首次检测到此种真菌的检测方法

× means the method detected the fungi first time

近年来,疯草内生真菌合成SW的学说已经被大多数学者所认同。为了提高疯草内生真菌产生SW的产量,张蕾蕾等[34]研究不同环境因素对内生真菌Undifilum oxytropis产生SW的能力及生长的影响,设计了不同培养条件对疯草内生真菌生长的影响,结果表明,不同培养条件对其真菌的生长和SW的产量均有影响,且U. oxytropis真菌具有自我调节适应环境的能力,偏酸或偏碱条件均可促进SW合成。L-哌可酸、L-赖氨酸和α-酮戊二酸等可能是合成SW的前体物质,所以均会对SW的合成产生影响,且各物质在培养基中的浓度与SW的产率密切相关。郭伟等[35]设计了对产生SW真菌埃里格孢菌FEL3菌株的发酵培养条件优化试验,最后通过测定菌丝中SW的含量确定最佳培养条件:碳源为燕麦片、氮源为豆粉、无机盐为$CuSO_4 \cdot 5H_2O$、培养基初始pH为6.5~7.0、接种量为体积分数的8%时,埃里格孢菌FEL3菌丝和菌液中SW浓度都显著高于其他组。陈基萍等[36]对产生SW的内生菌菌株*Aspergillus ustus*的培养发酵条件进行优化筛选，结果表明，菌株*Aspergillus ustus*在豆粕培养基，硝酸钾、麦芽糖分别为16、80mg·L^{-1}，pH为6.0，温度为35℃的条件下培养时生长状态最佳，且SW的产量显著高于其他组。真菌原生质体的制备是研究真菌遗传转化和基因组学的重要工具，疯草内生真菌原生质体可以进一步确定疯草内生真菌合成SW的相关基因功能，阐明分子调控机制，提高疯草内生真菌合成产生SW的产量。张蕾蕾等[37]对产SW内生真菌Undifilum oxytropis进行酶解，把菌体的细胞壁用溶壁酶溶解后将原生质体释放，然后把原生质体接种于PDA培养基中培养30d，最后经过对原生质体再生菌丝中SW的含量测定，结果显示再生菌丝中SW的含量略高于野生菌株的菌丝。

2 苦马豆素的药理作用

SW具有良好的抗病毒、抗肿瘤和提高机体免疫力作用，并能够很好地抑制肿瘤细胞的转移和增长。

2.1 抗病毒作用

SW是内生真菌的次级代谢产物，具有较好的抗病毒效果。Y. Tanaka等[38]研究证明SW对3型人副流感病毒（HPIV3）具有一定的抑制作用，其作用机制是SW通过对内质网中α-甘露糖苷酶抑制导致内质网功能丧失，阻止了HPIV3表面蛋白血凝素—神经氨酸酶（HN）和融合（F）蛋白的表达，降低病毒侵染细胞的能力，从而达到抗病毒的效果。刘文明等[39]使用SW治疗鸡马立克病（MD）的试验表明，其可以维持鸡体内的白细胞、血清磷酸酶（AKP）和乳酸脱氢酶（LDH）水平，发挥其免疫调节作用，同时SW能有效抑制MD肿瘤的形成和转移，具有良好的抗肿瘤作用。吴延磊等[40]研究金龟子绿僵菌次级代谢产物SW对鸡新城疫病毒（NDV）疫苗的影响，试验中用14日龄雏鸡建立新城疫病毒模型，将其分SW高、中、低剂量组和对照组，再使用新城疫病毒疫苗免疫治疗，结果显示，SW治疗组雏鸡体内的抗新城疫病毒抗体显著高于对照组，这说明SW与新城疫病毒疫苗联用可增强免疫效果，并对新城疫病毒有一定的抑制作用。郝宝成等[41]发现在一定剂量范围内，体外SW能有效地抑制牛病毒性腹泻病毒（BVDV）的增殖和感染，并推测其能直接灭活游离的牛病毒性腹泻病毒。

2.2 抗肿瘤作用

SW能够促进肿瘤细胞凋亡，以达到治疗癌症的效果。目前，国外已进入药物Ⅲ期临床阶段，P. E. Goss等[42]通过静脉注射SW的方式为肿瘤患者治疗，在6周后肿瘤患者的肿瘤减小了一半，其中2位淋巴癌患者在静脉注射SW一周后，临床症状有了明显减轻。J. Hamaguchi等[43]研究发现，SW能阻止大肠癌细胞转移扩散，降低其对抗癌药物5-氟尿嘧啶（5-fluorouracil）的耐受性，能提高大肠癌的化疗效果。此外，F. M. Santos等[44]发现SW对艾氏腹水癌（*Ehrlich ascites carcinoma*，EAC）的治疗药物顺铂（cisplatin）有增效作用。N. You等[45]使用SW对HepG2、SMCC7721、Huh7、MHCC97-H和HL-7702等人肝癌细胞进行治疗的试验发现，其对人肝癌细胞具有良好的抑制作用，但对人体正常的肝细胞并无损伤。由于MHCC97-H肝癌细胞具有高度的增殖性和致癌性，研究者选择使用其作为肝癌细胞模型，在对其用SW治疗时发现，其能显著抑制MHCC97-H肝癌细胞的生长，且在G0/G1细胞周期诱导癌细胞凋亡。苦马豆素是如何诱导癌症细胞的凋亡呢？N. You发现SW能使肝癌细胞在

G0/G1期的细胞周期蛋白（D1和E）减少，而细胞周期蛋白依赖性激酶（Cdk2的和Cdk4的）CDK抑制剂p21和p27蛋白增加，从而导致癌症细胞凋亡。SW能增强MHCC97-H肝癌细胞中Bax基因表达的上调和Bcl-2基因表达的下调，继而导致其细胞凋亡。此外，SW还能通过限制紫杉醇诱导的核因子κB（NF-κB）的积累，使其在体内外的细胞毒性作用增强，这说明SW可以直接抑制肝癌细胞的生长，减轻紫杉醇治疗肝癌细胞时的过度反应。D. Singh等[46]用SW对HL-60白血病细胞进行治疗时发现其能减少细胞周期蛋白（D1和E）表达，增加细胞周期蛋白依赖性激酶CDK抑制剂p21和p27蛋白的表达，从而达到对HL-60白血病细胞的治疗效果。可见，SW具有巨大的药用价值和科研潜力，但是长期使用又会对机体造成一定的损伤，如何将其制成高效、安全的抗癌药物将成为当前及今后研究的热点。

2.3　调节机体免疫力

苦马豆素不仅具有良好的抗癌、抗病毒作用，同时还是一种具有较强活性的免疫调节剂。O. A. Oredipe等[47]研究发现SW能够刺激骨髓细胞增殖，提高骨髓细胞转化为淋巴细胞的能力，增强机体的免疫系统功能。张志敏等[48]研究发现SW能激活小鼠腹腔免疫系统生成巨噬细胞，提高一氧化氮合酶的活性和TNF-α的水平，提高机体免疫系统功能。试验还表明高剂量的SW能抑制巨噬细胞的形成。因此，SW可以通过双向调节腹腔巨噬细胞发挥免疫作用。

淋巴细胞是体现机体免疫系统功能的表征之一，T淋巴细胞是免疫系统中最为多样和多效的一种免疫细胞。T细胞按其表面表达CD分子的不同一般分为CD_4^+T淋巴细胞和CD_8^+T淋巴细胞两大类。CD_4^+T淋巴细胞在机体免疫应答反应中主要起免疫调节作用，通过此作用，使机体针对不同抗原的免疫应答朝着定向类型发展，达到完全清除体内病原体的目的；CD_8^+T淋巴细胞主要通过细胞毒作用发挥作用，杀死体内被病毒（细菌）感染的细胞、胞内寄生菌感染的细胞和肿瘤细胞等，并将其吞噬消化。张建军等[49]给小鼠灌服一定剂量的SW，利用流式细胞仪分析血液中T淋巴细胞的种类和数量，发现适当剂量SW给药的小鼠体内CD_4^+/CD_8^+的比值比正常数值显著提高。这说明，低剂量SW能提高小鼠的免疫能力，而中、高剂量SW只能在短时间内提高小鼠的免疫机能，高剂量长时间给药反而会降低小鼠的免疫能力。

3　苦马豆素的检测方法

根据文献报道，现有的检测SW方法主要有薄层色谱法（thin layer chromatography，TLC）、气相色谱法（gas chromatography，GC）、气相—质谱联用法（GC-MS）、高效液相色谱法（High Performance Liquid Chromatography，HPLC）、液相—质谱联用法（LC-MS）、α-甘露糖苷酶活性抑制分析法（α-man-nosidase activity inhibition analysis，α-MIA）等[50-51]。R. J. Molyneux等[52]采用薄层色谱法检测SW，其建立了SW的Ehrlich's试剂显色方法，该方法不仅具有特异性，而且灵敏度较高、应用广泛、操作方便、经济实用，但检测时易受硅胶板薄厚、硅胶粒度、板面平整度及边缘效应等因素影响。气相色谱法（GC）和气质联用法（GC-MS），采用气相色谱法检测SW前，需将样品制备为容易气化并且稳定的衍生物，比薄层色谱方法操作复杂，但检出率较高、灵敏度较高。气质联用检测SW主要是用于定性分析，可用于检测SW是否存在[53]。在检测SW的方法中，LC-MS和HPLC法应用广泛，其优点是迅速、连续、高效、灵敏。HPLC法需根据检测物质的性质选择适当的检测器，目前常用检测器为DAD、MS、ELSD和PAD等。LC-MS法检测SW时，对于样品的处理要求不高，且具有极高的灵敏度，但是仪器成本过高。由于SW阳离子与甘露糖苷水解过程中形成的甘露糖阳离子的半椅式空间结构非常相似，SW竞争性抑制α-甘露糖苷酶活性，α-MIA法就是按这一原理间接分析SW的含量。该法对检测样品的要求较低，方法简单快捷，适用于大量样品的检测，但是准确度较差，检测范围较窄。此外还有荧光光谱法、离子抑制色谱法等[54-55]，各种检测方法对照如表2所示。

表2 各种检测方法的比较
Table 2 Comparison of the detection methods of swainsonine

检测方法 Method	适用范围 Sample	样品条件要求 Requirement	最低检出限/(mg·g^{-1}) Lowest detection limit	文献来源 Reference
薄层色谱	各种样品	不严格	0.50×10^{-3}	[56]
气相质谱联用	提取物	需制备衍生物	0.10×10^{-9}	[57]
液相质谱联用	提取物	需制备衍生物	0.10×10^{-9}	[58]
酶分析法	血清、乳样品	pH、温度限制	0.50×10^{-3}	[59]
荧先先请	食糜样品、提取物	需除杂	0.10×10^{-8}	[54]
离子抑制色谱	提取物	不严格	0.29×10^{-4}	[55]

4 结论与展望

SW具有良好的抗病毒、抗肿瘤和提高免疫力等药用作用和科研价值，其发展前景广阔。近年来，国内外学者对其抗病毒作用研究主要集中在3型人副流感病毒[38]、牛腹泻病毒[41]、鸡马立克病毒[39-40]等研究上，并且证明SW能通过抑制内质网和高尔基体中α-甘露糖苷酶的活性，阻止病毒自身蛋白质的合成，并发挥免疫调节作用，提高机体的防御系统来实现抗病毒作用；但其抗病毒作用机制阐述并不是十分清楚，适用范围也有待进一步发掘。SW抗肿瘤效果显著，受到国内外学者的广泛关注，在美国SW已经进入抗肿瘤药物Ⅲ期临床阶段[61]，但在国内SW抗肿瘤研究属于起步阶段，其抗肿瘤的作用机制尚不明确，在细胞水平上的作用靶点和代谢通路等相关研究鲜有报道。虽然SW在抗病毒、抗肿瘤方面效果显著，但其在植物体内的含量较低，加之提取工艺落后，提取效率较低，人工合成的成本过高，产率较低，导致SW的价格高昂、使用范围有限[61]。SW的来源将成为限制其发展和应用的最大难题，因此拓宽其新的来源将成为今后研究的主要方向。目前，许多研究人员通过真菌发酵方法来解决其来源短缺问题，疯草内生真菌、金龟子绿僵菌和豆类丝核真菌均能合成SW，但是SW是如何在这些菌中合成代谢的，其合成机制与代谢途径也不完全清楚。

近年来，合成生物学的概念逐渐被引入到天然产物学领域，已经应用于青蒿素、紫杉醇等多种天然药物的合成[62]。将合成生物学、基因工程学、蛋白质组与代谢组学等多种方法相结合，研究这些真菌合成SW的代谢途径，找到与合成SW相关的关键基因。一方面，可利用基因工程技术、蛋白质组学技术和微生物诱变技术等，将关键基因导入高表达目标载体，改造出一株生长周期短、产量高的菌株，以便SW的大规模工业生产与应用，解决其来源问题。另一方面，可以利用基因敲除技术、RNA干扰等，将内生菌中合成SW的关键基因敲除，或转录后沉默与其相关的基因表达，培育营养价值高且无毒"疯草"新品种。

（编辑 白永平）

（发表于《畜牧兽医学报》，一级学报，1 075–1 085.）

牦牛胎儿皮肤毛囊的形态发生及E钙黏蛋白的表达和定位

佘平昌[1,2]，梁春年[1,2]，裴 杰[1,2]，褚 敏[1,2]，郭 宪[1,2]，阎 萍[1,2]*

(1.中国农业科学院兰州畜牧与兽药研究所，兰州 730050；
2.甘肃省牦牛繁育工程重点实验室，兰州 730050)

摘 要：研究了牦牛毛囊的形态发生过程及胎儿毛囊发生的起始部位，分析了E-cadherin（CDH1）对毛囊发育的作用。采取牦牛胚胎头部皮肤制备组织切片，观察毛囊的形态发生过程；利用免疫组织化学技术，定位E-cadherin蛋白在毛囊中的表达；使用qRT-PCR方法，比较不同胎龄胎儿头部皮肤中E-cadherin mRNA转录水平。结果表明，毛囊黑色素颗粒在牦牛胎儿皮肤中不同部位先后聚集，牦牛胎儿头部皮肤在胎龄60~70d时初级毛囊开始形成毛芽，胎龄130d时形成毛球结构。次级毛囊在胎龄80~90d时从初级毛囊中分化出来；并且毛囊的发育可能最先是从头部开始的，其中唇部、眉毛、睫毛、角缘发育最明显；E-cadherin蛋白定位于头部皮肤的表皮、真皮及毛囊中，在表皮中表达较高，在毛囊中呈中等阳性表达；E-cadherin mRNA在牦牛不同胎龄皮肤毛囊中相对转录水平整体呈上升趋势，而在90d时转录量较低，显著低于70d、120d和130d时（$P<0.02$）。牦牛头部毛囊在60~70d时开始形成毛芽，牦牛毛囊发育的起始部位可能是从头部开始的。E-cadherin基因可能参与了牦牛毛囊的发育。

关键词：牦牛；毛囊；E-cadherin；形态发生；qRT-PCR；免疫组织化学

Fetal Skin Hair Follicle Morphogenesis and E-Cadherin Expression of the Yak

SHE Pingchang[1,2], LIANG Chunnian[1,2], PEI Jie[1,2],
CHU Min[1,2], GUO Xian[1,2], YAN Ping[1,2]*

(1. Lanzhou Institute of Husbandry and Pharmaceutical Sciences of CAAS, Lanzhou 730050, China; 2. Key Laboratory of Yak Breeding Engineering of Gansu Province, Lanzhou 730050, China)

基金项目：牦牛遗传资源与育种（CAAS-ASTIP-2014-LIHPS-01）；科技部科技支撑计划（2012BAD13B05）
作者简介：佘平昌（1989—），男，云南普洱人，硕士生，主要从事家畜分子生物学研究，E-mail：sheemail@163.com
*通信作者：阎萍，研究员，博士生导师，主要从事动物遗传育种与繁殖研究，E-mail：pingyanlz@163.com

Abstract: The objective of this study was to observe yak hair follicle morphogenesis and fetal hair follicle original sites, study the importance of E-cadherin (CDH1) in the development of hair follicles. Skin tissue sections of the Yak embryo head were observed, and the hair follicle morphogenesis was studied. The melanin granules of fetal skin at different month and the development of hair follicle original sites were detected. The expression of E-cadherin protein was analyzed by using immunohistochemistry. The mRNA transcription levels of E-cadherin gene in different month fetal head skin was tested by using qRT-PCR method. Results showed that the fetal head skin begins to form primary hair buds in embryonic age of 60~70 days, and form the hair bulbs structure in the embryonic age of 130 days. Secondary hair follicles differentiate out from primary hair follicles in gestational age of 80~90 days; The parts of the fetal skin melanin gathering situations in different periods were observed, the hair follicles differentiation was started from head, of which lower, eyebrows, eyelashes, and horn site were the most obvious; E-cadherin protein was located on the epidermis, dermis and hair follicles, and showed high expression in epidermis, and moderate positive expression in hair follicles. E-cadherin gene mRNA transcription were uptrend in different embryonic age of yak skin hair follicle, but was lower at the embryonic age of 90days, it was obviously below those at 70 days, 120 days and 130days ($P<0.02$). Yak head hair follicle is developed in the embryonic age of 60~70 days to form hair buds, and hair follicle development is likely forms from head. E-cadherin gene is most likely involved in the development of yak hair follicles.

Key words: Yak; Hair Follicle; E-cadherin; Morphogenesis; qRT-PCR; Immunohistochemistry

  牦牛（*Bos grunniens*）是青藏高原及毗邻山区的特色宝贵特殊畜种，为牧民提供肉、奶、毛、役力、燃料等生产生活资料，对改善牧民生活水平具有重要的意义。为了适应高寒的生存环境，牦牛身上的被毛较为厚密，按长短分为短毛和长毛，裙毛为牦牛特有的长毛，可长达50cm，被广泛运用于假发生产；按粗细可分为粗毛和绒毛，能够提供较好的纺用材料和工业材料[1]。基于此，作者对牦牛毛囊的形态发生及毛囊发生的调控因子进行研究，了解牦牛毛囊发育规律及相关调控因子，拟为牦牛的分子育种和利用牦牛毛发资源提供科学依据。人类的毛囊发育最先是从头部开始的[2]，并沿着头部到臀部这条轴线在电磁效应的推动下波浪式展开[3]。胚胎期毛囊干细胞在上皮细胞、间质和神经嵴相互作用下开始分化[4-5]，在早期人类和小鼠的研究中将其划分为8个阶段[6-7]，分为基板前期、基板期、毛芽期、毛钉期、毛囊期几个时期。很多因子参与了毛囊的形成和发展，β-catenin、Wnt信号通路具有重要作用[8-9]，骨形态发生蛋白（BMPs）对皮肤附属物及细胞的增殖也具有调控作用[10]。E钙黏蛋白（E-cadherin）是钙黏蛋白超家族的一员，存在于胚胎和正常组织中，通过介导同种细胞黏附而发挥作用。其介导的信号途径主要包括，Wnt信号转导途径、Rho-GTP激酶系统及与受体型酪氨酸激酶的互作。E-cadherin对哺乳动物胚胎期上皮的维护和功能具有重要作用，是多细胞上皮发育的关键因素[11]。同时它还参与细胞的维护和转移，E-cadherin功能的丢失可能导致癌症的发生[12]。在毛囊的相关研究中，E-cadherin是毛囊发育期黑色素形成的关键因素之一，其与钙黏蛋白家族的P-cadherin、H-cadherin共同完成黑色素细胞与角质细胞的粘连[13]，同时E-cadherin还具有粘连表皮和毛囊并参与毛囊更新的作用[14]。目前对人类、小鼠和羊等的毛囊研究较多[15-18]，而对牦牛毛囊形态发生、发生部位和发育调控因子的研究很少。作者通过组织切片观察牦牛毛囊形态发生及E-cadherin的表达定位，为探究牦牛毛囊发育规律及发育调控因子提供理论基础，从而促进牦牛分子育种、品种资源的开发和保护。

1 材料与方法

在国家大通种牛场屠宰场牦牛的屠宰期中采集胎龄各个时期胎儿的额部皮肤样品，液氮保存，并对其全身进行拍照，测量其顶臀长，根据胎儿顶臀长与胎龄的线性关系推算胎龄[19]。后期试验在中国农业科学院兰州畜牧与兽药研究所牦牛繁育工程重点实验室进行。

1.1 毛囊发育形态学观察

皮肤样品修整至长宽厚1.5cm×1.5cm×3mm大小后，滤纸吸干水分，OCT包埋剂（日本樱花）包埋，-80℃ 30min。冰冻切片机（莱卡CM1950）25℃，10μm切片。AAF固定液（乙醇85%，冰乙酸5%，甲醛10%）固定5min，纯净水洗10s，苏木素（碧云天）染色5min，水洗6min，分化5s，返蓝，梯度乙醇脱水2min，伊红（碧云天）染色50s，梯度乙醇涮洗10s，显微镜（莱卡）观察。

1.2 E-cadherin蛋白的表达定位

冰冻组织切片方法同上，冰冻切片于4℃丙酮中固定30min，3% H_2O_2甲醇溶液温室处理30min，蒸馏水洗涤3次×5min，滴加5% BSA封闭液，室温作用20min后甩去多余液体。滴加兔抗E-cadherin多克隆抗体（博士德产品），湿盒内4℃过夜，PBS冲洗2次，每次3min，滴加羊抗兔IgG，37℃孵育20min后PBS冲洗5次×3min，滴加SABC，37℃作用20min后PBS冲洗4次×5min，DAB显色，苏木素复染，观察，拍照，记录结果。

1.3 E-cadherin mRNA 的 qRT-PCR

1.3.1 RNA的提取及cDNA的合成

Trizol法提取不同胎龄牦牛头部皮肤的总RNA，按照TaKaRa公司试剂盒Prime Script RT reagent Kit with gDNA Eraser（Perfect Real Time）说明书合成cDNA。

1.3.2 荧光定量PCR

参照K. Nganvongpanit等[20]的研究设计E-cadherin基因的上下游引物，F：5′-GTA-CACCTTCATCGTCCAGAGCTAA-3′，R：5′-GCTCTTCAATGGCTTGTCCATTTGA-3′。根据GenBank中黄牛GAPDH基因序列（EU192062.1）用Primer Premier 5.0软件设计上下游引物F：5′-CACCCTCAAGATTGTCAGC-3′，R：5′-TCATAAGTCCCTCCACGAT-3′。引物由上海英骏生物公司合成。产物经2%琼脂糖凝胶电泳检测。

定量PCR（CFX96）反应使用25μL体系：SYBR®Premix Ex Taq™ Ⅱ（Tli RNaseH Plus）12.5μL，上下游引物各1μL，cDNA 2μL，dH_2O 8.5μL。荧光定量PCR的反应程序：95℃预变性30s；95℃变性5s，60℃退火30s，40个循环；之后进行熔解曲线分析，程序：65～95℃逐渐升温，每0.5℃读板一次。每个时期3个个体重复，每个样本重复3次。反应结束后以熔解曲线来判定qRT-PCR反应的特异性；获得每个样品的Ct值，运用$2^{-\Delta\Delta Ct}$法计算相对转录水平。

1.4 图像数据处理

莱卡显微镜观察，PS图像标记处理，Excel统计及SAS进行One-way ANOVA分析，用Origin作图。

2 结果

2.1 毛囊发育起始部位的观察

对于有色毛来说，黑色素的合成只发生在毛囊形成期，黑色素的多少及分布决定了毛发的颜色及分布。通过对比发现，在牦牛胎龄70d的时候可见眉毛处有黑色素颗粒细胞（MC）的聚集（图1-A-a），在胎龄90d的时候观察到牦牛的眉毛及睫毛出现MC聚集的情况（图1-B-a、图1-B-b），胎龄105d时可以明显的看到胎儿嘴部周围及额头有明显变黑的MC聚集毛囊发育的迹象（图1-C）。胎龄140d时已经能够非常明显的看到胎儿的头部已经全面有毛囊的发育（图1-D），通过细节（图1-E）发现其中胎儿的下唇部已经有毛孔的出现（图1-E-c）。同时发现胎儿角部周围的毛囊聚集特别明显（图1-C-e图1-D-e），说明牛角的形成可能与周围毛囊的作用有关系。

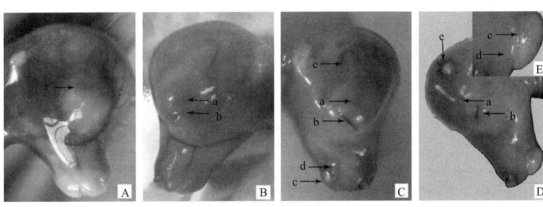

A.胚胎70d胎儿头部；B.胚胎90d胎儿头部；C.胚胎105d胎儿头部；D.胚胎140d胎儿头部；E.唇部放大观察；a.眉毛；b.睫毛；c.下唇；d.上唇；e.角基

A. Head at 70 days of fetal period; B. Head at 90 days of fetal period; C. Head at 105 days of fetal period; D. Head at 140 days of fetal period; E. Lip enlarged view; a. Eyebrow; b. Eyelash; c. Lower lip; d. Upper lip; e. Horn site

图1 牦牛胚胎毛囊发育起始部位的观察

Fig. 1 Observation of Yak hair follicle original sites

2.2 毛囊的形态发生

毛囊的形成是表皮细胞和真皮细胞之间通过一系列的信号分子来促进细胞群体的增殖分化完成的。通过观察牦牛额头皮肤的组织切片发现，在胎龄55d之前看到表皮结构较薄，但已经形成，没有出现细胞的聚集（图2-A）。胎龄60~70d时，表皮增厚，表皮周围开始出现大量细胞聚集，开始形成毛芽（图2-B-c）。胎龄75~80d时上皮成角质细胞大量增殖，毛芽继续向真皮深入，并形成较宽阔的圆柱形（图2-C），圆柱形末端形成一个圆形真皮毛乳头（图2-C-d）。胎龄80~90d时毛囊逐渐向真皮深入，变长变粗，已经延长成一个实心立体柱状结构（图2-D），由围绕中心轴排列的多层成角质细胞组成，毛囊侧面形成一个膨大部，为次级毛囊原细胞（图2-D-e）。发育到胎龄95～105d时表皮增厚，毛囊顶端与表皮下层形成伞状连接（图2-E），在初级毛囊的侧面可以观察到次级毛囊毛芽的出现（图2-E-f）。胎龄110～120d，毛乳头被毛母质包围形成近似球形（图2-F-g）。胎龄130d时毛囊继续发育，毛乳头被毛母质包围（图2-G-h、i），形成毛球，同时可见皮脂腺的形成（图2-G-j）。

次级毛囊的观察。次级毛囊（SF）是绒毛初始的基本机构，它的发育相对初级毛囊（PF）晚，SF的变化过程与PF的变化过程基本相同，它的分化是由PF靠近皮肤的一端分化出来的。胎龄85d时当PF发育到毛钉期时，毛囊靠近皮肤的一端出现膨大部（图2-D-e），这个膨大部就是SF的毛芽，而后和PF以同样的发育顺序向真皮深入，当胎龄130d时，SF已经发育形成真皮毛乳头（图2-G-f），随后继续发育形成完整的毛囊结构。

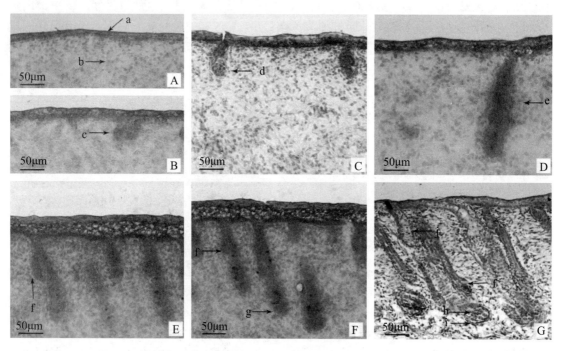

A. 胚胎55d头部皮肤；B. 胚胎65d头部皮肤；C. 胚胎75d头部皮肤；D. 胚胎85d头部皮肤；E. 胚胎100d头部皮肤；F. 胚胎115d头部皮肤；G. 胚胎130d头部皮肤；a. 表皮；b. 真皮；c. 毛芽前体；d. 真皮毛乳头；e. 次级毛囊原细胞；f. 次级毛囊；g. 毛球；h. 毛乳头；i. 毛母质；j. 皮脂腺

A. Head skin at 55 days of fetal period；B. Head skin at 62days of fetal period；C. Head skin at 75 days of fetal period；D. Head skin at 85 days of fetal period；E. Head skin at 100 days of fetal period；F. Head skin at 115 days of fetal period；G. Head skin at 130days of fetal period；a. Epidermis；b. Dermis；c. Precursor of hair germ；d. Dermal papilla；e. The hair germ of original secondary follicles；f. Secondary hair follicle；g. Hair bulb；h. Dermal papilla；i. Hair matrix；j. Sebaceous gland

图2　牦牛毛囊形态发生过程（HE）

Fig. 2　Changeable process of hair follicle morphogenesis of the Yak in the fetal period（HE）

2.3 E-cadherin 蛋白的表达定位

为了观察E-cadherin蛋白的表达位置，选取胎龄120d毛囊发育较明显的皮肤组织做免疫组化试验。根据染色程度，可将E-cadherin染色结果分为4个等级：阴性：无着色；弱阳性：染色弱，呈淡黄色；中等阳性：中等染色，呈黄褐色；强阳性：染色强，呈棕褐色。结果显示，E-cadherin在毛囊、表皮及真皮层中都有表达（图3-B、C）。E-cadherin在皮下组织无着色为阴性（图3-B-c），毛囊中E-cadherin阳性产物呈黄褐色（图3-B-b），为中等阳性，而毛囊中的毛球部位比毛球以上的毛囊部位具有更多的阳性产物。而在真皮层部分呈淡黄色（图3-B-d），为弱阳性。在皮肤的表皮层有较多的阳性产物的存在，呈现棕褐色（图3-C-a），为强阳性。

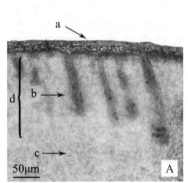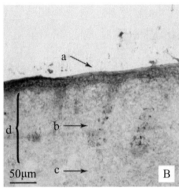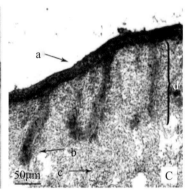

A.HE 染色；B.E-cadherin 的免疫组织化学；C.E-cadherin 的免疫组织化学苏木精复染；
a. 表皮；b. 毛囊；c. 皮下组织；d. 真皮

A. HE staining; B. Immunohistochemical staining of E-cadherin; C. Immunohistochemical staining of E-cadherin and hematoxylin counterstain; a. Epidermis; b. Hair follicle; c. Subcutaneous tissue; d. Dermis

图3 E-cadherin在牦牛胚胎120 d皮肤中的免疫组化染色
Fig. 3 The immunohistochemical stains of E-cadherin in the Yak skin at 120 days of fetal period

2.4 不同时期 E-cadherin mRNA 转录水平变化

为了观察E-cadherin mRNA在毛囊发育各个时期的相对转录水平，选择了胎龄70d、90d、120d和130d样本做荧光定量试验。半定量RT-PCR扩增，2%琼脂糖凝胶电泳检测发现E-cadherin mRNA在头部皮肤中有表达。熔解曲线分析发现，E-cadherin基因的PCR产物均呈较为锐利的单一峰（图4-B），排除了形成引物二聚体和非特异性产物对结果带来影响的可能，同时说明引物有很好的特异性。通过扩增曲线可知，E-cadherin基因及GADPH内参的扩增效果较好，内参最先达到荧光阈值（图4-A）。检测结果发现，E-cadherin mRNA在牦牛头部皮肤的相对转录量整体呈上升趋势（图5），而在胎龄90d时转录量最低，显著低于70、120和130d（$P<0.05$），极显著低于130d（$P<0.01$）；胎龄70d转录量也较低，显著低于130d（$P<0.05$）。

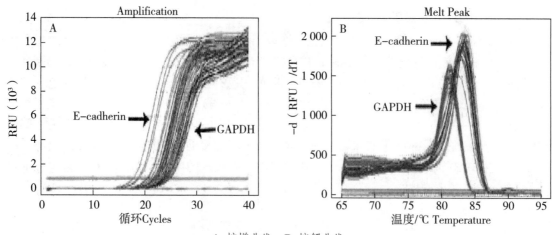

A. 扩增曲线；B. 熔解曲线

A. Amplification plots；B. Melt curve

图4　E-cadherin mRNA qRT-PCR的扩增曲线和熔解曲线

Fig. 4　qRT-PCR amplification plots and melt curve of E-cadherin mRNA transcription

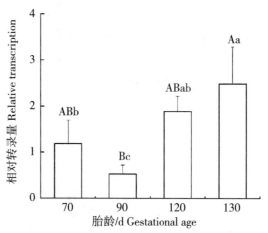

不同小写字母表示差异显著（P<0.05），不同大写字母表示差异极显著（P<0.01）

Lowercase letter indicates P<0.05，capital letter indicates P<0.01

图5　牦牛毛囊不同发育阶段E-cadherin mRNA的转录水平

Fig. 5　Analysis of the relative transcription of E-cadherin in different developmental stages of hair follicle

3　讨论

关于毛囊发育部位顺序的研究发现，牦牛毛囊的发育首先是从头部开始的，其中以下唇部、眉毛、睫毛、角缘发育最明显。对于有色毛发来说，黑色素合成只发生于毛囊生长期，毛球部的黑色素细胞合成黑色素颗粒，转运到毛干皮质细胞。黑色素颗粒的多少、颗粒性状以及分布决定毛发的颜色[21]。K. A. Holbrook等[2]在人类的研究中发现，人类的毛囊发育始于头部，最先从眉毛、睫毛及下唇缘开始逐渐扩展到整个胚胎。在这一点上牦牛和人类是类似的。但是为什么毛囊的发育是从头部开始的，又是怎么从头部开始的呢？毛囊的发育沿着头部到臀部这条轴线展开[3]，头部的毛囊在基底细胞的诱导下形成并极化形成电磁感应现象，推动毛囊发育波浪式展开。毛囊发育部位的研究对于探索毛囊发育规律及动物毛发资源的利用都具有很大的意义，本研究中通过观察有色毛囊黑色素颗粒的聚集形式来观察毛囊发育顺序，但是这种观察方法是有局限性的不能准确定位各部位发育的先后。

关于毛囊的形态发生的研究发现，牦牛胚胎额头皮肤毛囊在胎龄60~70d的时候开始发育，胎龄130d时发育形成毛球结构，次级毛囊在胎龄80~90d时从初级毛囊中分化出来。动物毛囊的发育是由胚胎干细胞在经历一系列刺激的作用下分化完成的[4, 22]，根据小鼠模型可分为基板期、毛芽期、毛钉期及毛囊期几个时期[23]。在草食动物中，M. H. Hardy等[24]在1922年对绵羊的毛囊发育做了系统的分析。本研究中，牦牛的额头皮肤在胎龄55d前观察不到毛囊发育的迹象，此阶段可能是毛囊发育的准备期，细胞接受相关刺激朝着一定的方向进化。胎龄60~70d时，毛芽开始出现，与内蒙古的绒山羊相比要晚[25]，这可能与物种的差异性、妊娠期及动物的耐低温都有一定的关系。而后毛芽继续发育在胎龄130d时已经基本形成完整的毛囊结构。次级毛囊的发育在初级毛囊毛钉时期开始，次级毛囊是牦牛等耐寒动物御寒的基础，也是产毛量多少的基础。次级毛囊的形成是在初级毛囊的一侧支生出来的，时期大约在胎龄85d左右。本研究发现头部的毛囊毛芽是从胎龄60~70d的时候开始的，说明牦牛毛囊的发育比较早，而躯干及四肢毛囊的发育可能会比额头要晚5~20d，而形成这一差异的调控因子需进行深入研究。

关于E-cadherin蛋白表达和定位的研究发现，E-cadherin表达于牦牛额头皮肤毛囊中的表皮、真皮及毛囊中，在真皮中表达较弱，而在表皮及毛囊中表达较高。E-cadherin mRNA在牦牛头部皮肤的相对转录水平整体呈上升趋势，而在90d时转录量较低，显著低于70d、120d和130d（$P<0.05$），极显著低于130d（$P<0.01$）；胎龄70d转录量也较低，显著低于130d（$P<0.05$）。钙黏蛋白超家族是一种钙离子依赖性跨膜糖蛋白。通过介导同种细胞间发生黏附而发挥作用。存在于胚胎组织和正常组织的上皮细胞中。钙黏蛋白（E-cadherin）主要通过3条介导信号通路：一是Wnt信号转导途径，主要是通过Beta-catenin的作用；二是Rho-GTP激酶系统通过Rho蛋白的竞争来调节E-cadherin的黏附活力；三是通过与受体型酪氨酸激酶的作用来调节细胞黏附，E-cadherin对毛囊黑色素形成、毛囊与表皮的黏附都具有重要作用[11-14]。J. Kishimoto等[26]发现E-cadherin通过Wnt信号通路对哺乳动物胚胎期毛囊的发育具有调控作用。同时E-cadherin可以促进细胞间的黏附聚集以及维持细胞极性、上皮细胞结构的完整性，并参与细胞发育调节组织的发育及形成，抑制细胞迁移，维护正常细胞间的选接和极性，而且对细胞发育、生长、肿瘤转移、细胞运动起主要作用，这可能是导致E-cadherin在表皮中较高表达的原因。S. Müller-Röver等[27]研究发现在小鼠脱毛后毛发再生的过程中E-cadherin基因转录量较低，推测E-cadherin基因对毛囊发育具有负调节作用；本研究中在胎龄90d时E-cadherin基因转录量显著低于70d、120d和130d，可能是E-cadherin基因对毛囊发育具有调节作用。K. Nganvongpanit等[20]发现E-cadherin在黄牛的毛囊发育中有表达，D. L. Gay等[28]也在人类的毛囊中发现了它的表达，同时它又是一种信号转导因子[29]；毛囊的发育是由上皮细胞及位于下层的间充质细胞相互作用刺激分化的，同时E-cadherin又在细胞黏附及黑色素生成中具有重要作用。由此推测，E-cadherin可能参与牦牛毛囊的发育。

4 结论

牦牛胎儿头部皮肤在胎龄60~70d时毛囊开始形成毛芽,胎龄130d时形成毛球结构,次级毛囊在胎龄80~90d时从初级毛囊中分化出来。毛囊的发育可能是从头部开始的。E-cadherin蛋白表达于牦牛额头皮肤毛囊中的表皮、真皮及毛囊中,在表皮中表达较高,在毛囊中呈中等阳性表达。E-cadherin mRNA在牦牛头部皮肤的转录水平整体呈上升趋势,而在90d时转录量较低,显著低于70d、120d和130d。推测E-cadherin基因可能参与了牦牛毛囊的发育。

(编辑 白永平)
(发表于《畜牧兽医学报》,院选中文,397–403.)

五氯柳胺口服混悬剂的制备及其含量测定

张吉丽，李　冰，司鸿飞，郭　肖，朱　阵，尚晓飞，周绪正，张继瑜*

（中国农业科学院兰州畜牧与兽药研究所，农业部兽药创制重点实验室，兰州　730050）

摘　要：研究的目的是制备五氯柳胺混悬剂，研究其理化性质并建立含量测定方法。通过单因素筛选及二次回归正交旋转设计，以沉降体积比和再分散性为考察因素，优化处方制备五氯柳胺混悬剂，用透射电子显微镜、激光粒度测定仪对其形态和粒径进行测定，考察制剂的稳定性，建立测定含量的高效液相色谱方法。五氯柳胺混悬剂每100mL中含五氯柳胺3.2g，卡波姆974p 0.2g，十二烷基硫酸钠0.3g，对羟基苯甲酸甲酯0.02g，焦亚硫酸钠0.4g。混悬剂沉降体积比为0.999，再分散性良好。五氯柳胺在18~100μg·mL^{-1}的质量浓度范围内与峰面积呈良好的线性关系。制备的五氯柳胺混悬剂工艺简单，物理性状良好，稳定性和分散性好，质量可控。

关键词：五氯柳胺混悬剂；二次回归正交旋转设计；含量测定：HPLC

Preparation and Content Measurement of Oxyclozanide Oral Suspension

ZHANG Jili, LI Bing, SI Hongfei, GUO Xiao, ZHU Zhen, SHANG Xiaofei, ZHOU Xuzheng, ZHANG Jiyu*

（Key Laboratory of Veterinary Pharmaceutical Development of Ministry of Agriculture, Lanzhou Institute of Husbandry and Pharmaceutical Sciences, Chinese Academy of Agricultural Sciences, Lanzhou 730050, China）

Abstract: The study aimed to prepare oxyclozanide oral suspension and evaluate its physical and chemical properties, and establish an HPLC method for the determination of oxyclozanide. The sedimentation volume ratio and redispersibility were used as variable factors,

---

基金项目：国家科技支撑计划项目（2015BAD1101）；国家现代农业产业技术体系专项（CAR-38）
作者简介：张吉丽（1992—），女，黑龙江大庆人，硕士生，主要从事动物抗寄生虫药研究，E-mail：zhangjlzjl@sina.com
* 通信作者：张继瑜（1967—），E-mail：infzjy@sina.com

the single factor experiment and the quadratic regression orthogonal rotational combination design were carried out to optimize the prescription. Then, the shape, size of oxyclozanide suspension was measured by transmission electron microscopylaser and particle size analyzer, and the content of oxyclozanide in suspension was determined by HPLC. The result showed that the suspension (100 mL) was consisted of 3.2 g oxyclozanide, 0.2 g carbomer 974 p, 0.3 g sodium dodecyl sulfate sodium salt, 0.02 g methyl 4-hydroxybenzoate and 0.4 g sodium pyrosulfite. With good redispersibility, the sedimentation volume ratio of oxyclozanide suspension was 0.999. Excellent line relationship was obtained in the range of 18 ~ 100 μg · mL^{-1} (r^2=0.999 3). The oxyclozanide suspension was successfully prepared with simple preparation process, good stability and well quality-controlled product.

Key words: Oxyclozanide suspension; Quadratic regression orthogonal rotational combination design; Determination of content; HPLC

肝片吸虫病（fascioliasis hepatica）是由肝片吸虫引起的食源性人畜共患病，该病遍及欧洲、亚洲、非洲、美洲和大洋洲等区域[1]，引起动物消瘦、贫血、使役能力下降和生产性能降低，严重阻碍了畜牧养殖业的健康发展，同时对食品安全、公共卫生安全和环境生态安全构成了严重威胁，造成了严重的经济损失[2-3]。

临床对肝片吸虫病的防治主要依靠化学药物。在疫苗预防方面，虽然肝片吸虫重组卡介苗的研制取得了明显进展，但距离批量化生产和大面积推广还有很大差距。常用于防治肝片吸虫感染的化学药物有三氯苯达唑（triclabendazole）、阿苯达唑（albendazole）、氯舒隆（clorsulon）和五氯柳胺（oxyclozanide）等[4-5]，其中三氯苯达唑、阿苯达唑和氯舒隆等由于在临床长期大面积使用已产生了耐药性，导致临床疗效大大降低[6-7]。五氯柳胺作为水杨酰苯胺类化学合成抗蠕虫药物，主要用于防治牛、羊等动物的吸虫病、绦虫病和蛔虫病，对线虫也有一定的治疗效果，其通过抑制虫体线粒体中氧化磷酸化过程从而阻碍三磷酸腺苷的生成，使虫体因能量耗尽而死亡[8]。该药的驱虫效率高、毒性小、安全范围大。国外大量的临床应用证实了五氯柳胺对蠕虫感染具有良好的治疗效果，特别是对肝片吸虫的感染效果十分突出[9-11]。至今为止，该药在欧、美和非洲等地区仍广泛应用，且未出现耐药性。

五氯柳胺在国内尚未开发为兽药产品，基于其优良的药效性能，为满足我国对兽医吸虫病防控的巨大临床需求，作者开展了抗牛、羊吸虫病五氯柳胺新制剂的研制。本研究通过单因素筛选以及二次回归正交旋转设计方法建立和优化制剂处方，制备了具有自主知识产权的五氯柳胺口服混悬剂，并对其物理性状、再分散性、沉降体积比等药学特性进行了研究，建立了药物有效成分含量测定的高效液相色谱方法，为五氯柳胺新药开发和临床应用奠定基础。

1 材料与方法

1.1 主要材料

1.1.1 主要仪器

ME403型电子天平（梅特勒—托利多仪器有限公司）；KQ-600DE数控超声清洗器（昆山市超声仪器限公司）；D2 010W电动搅拌器（上海梅颖浦仪器仪表制造有限公司）；

UPT-Ⅱ-40L优普超纯水制造系统（上海优普实业有限公司）；FE20实验室pH计（梅特勒—托利多仪器有限公司）；ZetasizerNano ZS ZEN 3600激光动态散射仪（英国Malvern-instrument公司）；TECNAL G2型场发射高分辨透射电子显微镜（美国FEI公司）；Waters2695高效液相色谱仪（美国Waters公司）；Waters2489紫外检测器（美国Waters公司）；WD-A药物稳定性检测仪（天津药典标准仪器有限公司）；针筒式微孔滤膜过滤器（孔径0.22μm，广州优瓦仪器有限公司）；Hypersil ODS色谱柱（4.6mm×150mm，5μm，大连依利特分析仪器有限公司）。

1.1.2 主要试剂

五氯柳胺原料药（常州亚邦齐晖医药化工有限公司，批号：RD1506001，含量：99%以上）；五氯柳胺标准品（美国TSZ公司，批号：T121-043）；十二烷基硫酸钠（化学纯，国药集团化学试剂有限公司）；卡波姆974p（上海阿拉丁生化科技股份有限公司）；对羟基苯甲酸甲酯（化学纯，国药集团化学试剂有限公司）；焦亚硫酸钠（分析纯，国药集团化学试剂有限公司）；甲醇（色谱纯，美国费舍尔公司）；磷酸（色谱纯，Sigma公司）；超纯水和纯水为作者实验室自制。

1.2 试验方法

1.2.1 混悬剂的制备

将五氯柳胺、焦亚硫酸钠、十二烷基硫酸钠、卡波姆974p等原料药和辅料分别用研钵研磨，过筛，备用。将卡波姆974p、焦亚硫酸钠等辅料成分分别缓慢加入装有蒸馏水的烧杯中，调动搅拌机转速，搅拌1.5h，得溶液A；另将十二烷基硫酸钠和五氯柳胺加入盛有蒸馏水的烧杯中，混匀，得到溶液B；将溶液B缓慢加入到正在搅拌的溶液A中，加蒸馏水定容；搅拌1.5h，即得[12]。

1.2.2 评价指标及测定

依据《中国兽药典》2010年版对口服混悬剂的质量评价要求，以沉降体积比和再分散性作为主要评价指标[13]并进行测定。

1.2.3 辅料及其用量筛选

主要辅料中，选择卡波姆974p作为助悬剂，十二烷基硫酸钠作为润湿剂，焦亚硫酸钠作为抗氧化剂，通过单因素试验确定各种辅料的最佳用量范围。采用SAS软件中"GLM"程序分析辅料各成分的显著性。

以单因素试验中显著性因素作为主要考察指标，选用二次正交回归旋转设计对处方进行优化，各因素水平见表1，并通过SAS软件进行试验设计。

以沉降体积比及再分散性作为评价指标，计算Y值进行综合评价。计算方法如下：

沉降体积比（F）测定：将各处方混悬剂置于50mL的具塞试管中，放置，记录初始高度H_0，分别在3h、1d、3d和7d进行观察，记录沉降后体积H，以H/H_0为沉降体积比数值。

再分散性次数（N）测定：将各种处方混悬剂的具塞试管倒转180°，测定恢复到初制混悬状态时所需翻转的次数即为再分散性次数。

表1　二次回归正交旋转设计各因素水平

Table 1　The levels of factors in the quadratic regression orthogonal rotational combination design

因素Factors	水平Levels				
卡波姆974p Carbomer 974p	0.1	0.18	0.3	0.42	0.5
十二烷基硫酸纳 Sodium dodecyl sulfate	0.1	0.18	0.3	0.42	0.5
焦亚硫酸纳 Sodium pyrosulfite	0.01	0.21	0.505	0.8	1.0

Y值计算方程：

$$Y = F + \frac{1}{N}$$

Y为沉降体积比与再分散性倒数之和，值越大则混悬剂效果越好。

1.2.4　验证性试验

按照筛选的最优处方制备3批五氯柳胺混悬剂样品，测定粒径大小及分布、再分散性、沉降体积比、外观颜色和pH等指标。

（1）粒径大小及分布：测定方法如下。

粒径大小测定：采用激光动态散射仪对五氯柳胺混悬剂进行粒径的测定。将制备好的混悬剂进行100倍稀释，超声5min，在激光粒度仪中进样检测，每个样品测定3次，记录平均粒径的大小及分布范围。

形态分布测定：将3批样品用超纯水稀释10倍，负染后，自然挥干，置于透射电镜下观察，在50和100nm标尺下检测样品的形态分布。

（2）再分散性及沉降体积比：测定方法同"1.2.3"中所述。

（3）外观颜色和pH：观察并记录五氯柳胺混悬剂的外观色泽，用酸度计进行pH测定。

1.2.5　含量测定方法

（1）色谱分析条件：色谱柱为Hypersil ODS（150mm×4.6mm，5μm）；流动相为甲醇：0.1%磷酸水（80∶20）；流速：1.0mL·min^{-1}；检测波长：300nm；柱温：25℃；进样量：20μL。

（2）方法专属性试验：将甲醇配制的五氯柳胺标准品溶液，五氯柳胺混悬剂供试品溶液和空白混悬剂样品溶液，分别经0.22μm微孔滤器过滤，注入液相色谱仪，依次进样测定。

（3）线性考察：精密称取五氯柳胺标准品10.00mg，置于100mL容量瓶中，用甲醇溶解并定容，配制成100μg·mL^{-1}的五氯柳胺标准品母液，待用。采用梯度稀释法，取标准品母液配制成质量浓度为100、60、45、36、18μg·mL^{-1}的标准品溶液，精确吸取20μL进样，记录峰面积。以五氯柳胺标准品的峰面积为纵坐标Y，以五氯柳胺标准品的质量浓度（C）为横坐标，绘制峰面积（Y）–质量浓度（C）标准曲线，计算标准曲线的回归方程[14-15]。

（4）回收率试验：精密称取五氯柳胺标准品，用甲醇溶解，配制成30（低）、60（中）、80μg·mL^{-1}（高）3个质量浓度的对照品溶液，每份溶液分别进样3次，测定每个样品的实测质量浓度。根据实测质量浓度与理论质量浓度的比值，计算回收率及相对标准偏差（RSD）。

（5）精密度试验：精密称取五氯柳胺标准品，用甲醇配制成30（低）、60（中）、80μg·mL^{-1}（高）3个质量浓度的对照品溶液，分别考察日内精密度和日间精密度。

（6）重复性试验：按供试品溶液制备方法，配制供试品溶液2份，并按上述色谱条件，精密吸取各样品溶液10μL进样，记录峰面积，计算含量及其RSD值[16]。

（7）样品含量测定：分别精密吸取3批五氯柳胺混悬剂样品（含五氯柳胺32mg·mL^{-1}）各0.2mL置于100mL的容量瓶中，用甲醇稀释并定容，经0.22μm微孔滤器过滤，注入液相色谱仪，记录峰面积，计算含量及RSD值。

1.2.6 混悬剂的稳定性考察

为了对混悬剂的稳定性进行评价，试验中进行了为期半年的短期稳定性考察和相关试验，内容包括加速试验和强光、高温条件下的影响因素试验。

（1）加速试验：将3批样品在市售包装条件下，放置于恒温恒湿试验箱中，在温度40℃±2℃、相对湿度75%±5%的条件下进行6个月的稳定性考察，分别于第0个月、第1个月、第2个月、第3个月、第6个月进行外观性状、再分散性和沉降体积比考察，测定pH及含量。

（2）影响因素试验：包括强光照射试验和高温试验。

强光照射试验：将3批样品置于照度4 500lx±500lx的光照箱中放置10d，分别于第0天、第5天和第10天取样分析。与第0天比较，观察外观性状，考察沉降体积比、再分散性、pH等指标的变化，并进行含量测定。

高温试验：将3批样品在60℃下放置10d，分别于第0天、第5天和第10天取样分析，与第0天比较，检测指标同上所述。

2 结果

2.1 二次回归正交旋转设计试验

通过单因素筛选各辅料的最佳用量范围，应用SAS软件对各因素的显著性进行分析，结果表明，卡波姆974p、十二烷基硫酸钠及焦亚硫酸钠三种辅料对五氯柳胺混悬剂的稳定性影响极显著。

根据对沉降体积比与再分散性指标的测定，计算Y值的评价结果见表2。二次回归正交旋转设计试验的结果如图1所示，其中，图A为当十二烷基硫酸钠固定时，卡波姆974p与焦亚硫酸钠对混悬剂稳定性的影响；图B为当焦亚硫酸钠固定时，卡波姆974p与十二烷基硫酸钠对混悬剂稳定性的影响；图C为卡波姆974p固定时，焦亚硫酸钠与十二烷基硫酸钠对混悬剂稳定性的影响。

表 2 二次回归旋转正交设计试验结果

Table 2 The result of the quadratic regression orthogonal rotational combination design test

试验次数 Test times	因素 Factors			Y值 The value of Y
	卡波姆974p Carbomer	十二烷基硫酸钠 Sodium dodccyl sulfate	焦亚硫酸钠 Sodium pyrosulfite	
1	0.180	0.180	0.210	1.027
2	0.180	0.180	0.800	0.986
3	0.180	0.420	0.210	1.579
4	0.180	0.420	0.800	1.495
5	0.420	0.180	0.210	0.839
6	0.420	0.180	0.800	1.455

(continued)

试验次数 Test times	因素 Factors			Y值 The value of Y
	卡波姆974p Carbomer	十二烷基硫酸钠 Sodium dodccyl sulfate	焦亚硫酸纳 Sodium pyrosulfite	
7	0.420	0.420	0.210	1.999
8	0.420	0.420	0.800	1.999
9	0.100	0.300	0.505	1.323
10	0.500	a 300	0.505	1.735
11	0.300	0.100	0.505	0.912
12	0.300	0.500	0.505	1.486
13	0.300	0.300	0.010	1.500
14	0.300	0.300	L 000	1.582
15	0.300	0.300	0.505	1.590
16	0.300	0.300	0.505	1.350
17	0.300	0.300	0.505	1.480
18	0.300	0.300	0.505	1.331
19	0.300	0.300	0.505	1.596
20	0.300	0.300	0.505	1.112
21	0.300	a 300	0.505	1.577
22	0.300	0.300	0.505	1.484
23	0.300	a 300	0.505	1.053
24	0.300	0.300	0.505	1.572
25	0.300	a 300	0.505	1.041

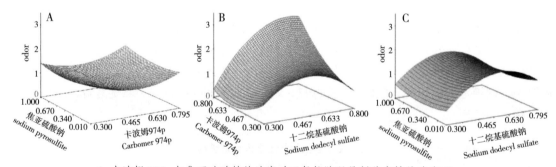

A. 卡波姆974p与焦亚硫酸钠的浓度对五氯柳胺混悬剂稳定性的响应面；
B. 卡波姆974p与十二烷基硫酸钠的浓度对五氯柳胺混悬剂稳定性的响应面；
C. 十二烷基硫酸钠与焦亚硫酸钠的浓度对五氯柳胺混悬剂稳定性的响应面。

A. Response surface of carbomer 974p and sodium pyrosulfite on the stable of oxyclozanide suspension; B. Response surface of carbomer 974p and sodium dodecyl sulfate concentration on the stable of oxyclozanide suspension; C. Response surface of sodium pyrosulfite and sodium dodecyl sulfate concentration on the stable of oxyclozanide suspension

图1 SAS分析二次回归旋转正交设计的响应面结果

Fig. 1 The response surface of the quadratic regression orthogonal rotational combination design test by SAS analysis

通过对响应面的SAS软件分析得到一个驻点如表3所示，由于此驻点是一个鞍点，因此作者对其进行了岭分析，找到了符合实际情况的最大值即：Y=1.999 943，此时，卡波姆974p最优值为2.0mg·mL^{-1}，十二烷基硫酸钠最优值为3.0mg·mL^{-1}，焦亚硫酸钠最优值为4.0mg·mL^{-1}。

试验得到的最优处方：以100mL计，当卡波姆974p为0.2g、焦亚硫酸钠为0.4g、十二烷基硫酸钠为0.3g、对羟基苯甲酸甲酯为0.02g时，沉降体积比最大为0.999 8，接近于1，再分散性最好。

2.2 验证试验

2.2.1 粒径大小及分布结果

根据试验优化后得到的处方，分别配制3批五氯柳胺的混悬剂，测定粒径大小及分布如图2所示，A、B、C为3批样品的粒径分布图，粒径大小分别为791nm、767nm和713nm。

在透射电镜下，对混悬剂的稀释液进行检查，图3表示在标尺为50nm和100nm时的粒径分布，可见粒径分布较集中，大小适宜。

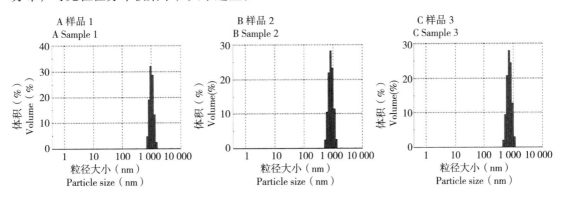

图2　混悬剂粒径分布

Fig. 2　The particle distribution of suspension

表3　SAS分析二次回归旋转正交设计的驻点结果

Table 3　The stationary point in the quadratic regression orthogonal rotational combination design test by SAS analysis

因素 Factor	临界值Critical value	
	编码Coded	非编码Uncoded
卡波姆974p Carbomer 974p	−0.962 517	−0.949 736
十二坑基硫酸纳Sodium dodecyl sulfate	0.261 698	0.691 937
焦亚硫酸纳Sodium pyrosulfite	0.114 565	0.192 698
在驻点的顿测值为1.476 144 Predicted value at stationary point：1.476 144		

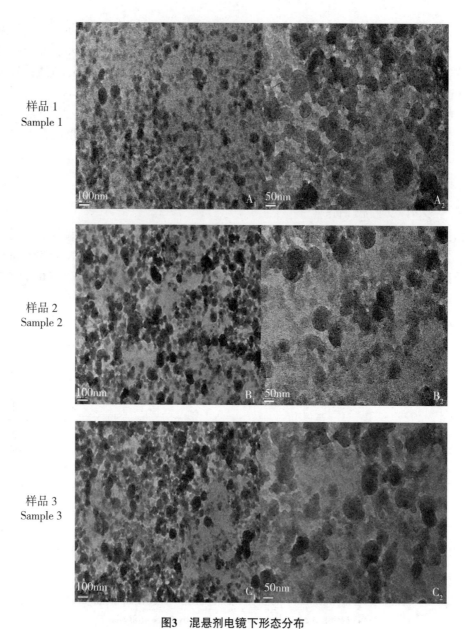

图3 混悬剂电镜下形态分布

Fig. 3 The electron micrograph of suspension morphology distribution

2.2.2 再分散性及沉降体积比测定结果

3批样品放置一个月后，测定沉降体积比及再分散性，结果如表4所示，沉降体积比均值为0.999 3，再分散性较好。

2.2.3 外观颜色观察和pH测定结果

本试验中制得的五氯柳胺混悬剂为乳白色均匀、浑浊、无丁达尔效应的混悬液，静止较长时间可能出现分层现象，上清液无色透明，下层为乳白色的沉降物，但轻轻振摇后重新恢复均匀体系。测定3批样品的pH分别为5.98、6.03和6.08，平均pH为6.03。

表 4 沉降体积比及再分散性结果

Table 4　The result of sedmientation rate and redispersion

样品 The samples	沉降体积比 Sedmientation rate	再分散性振摇次数 The shake frequency of redispersion
1	0.999	2
2	0.999	1
3	1	1

由以上结果可知，优化后的处方粒径大小适宜，粒径均匀，沉降体积比较大，再分散性良好，pH值较为稳定，符合2010年版《中国兽药典》中关于混悬剂的质量标准要求。

2.3 含量测定

2.3.1 专属性考察

由图4结果可见，空白的混悬剂溶剂对五氯柳胺的含量测定无干扰，该方法专属性强。

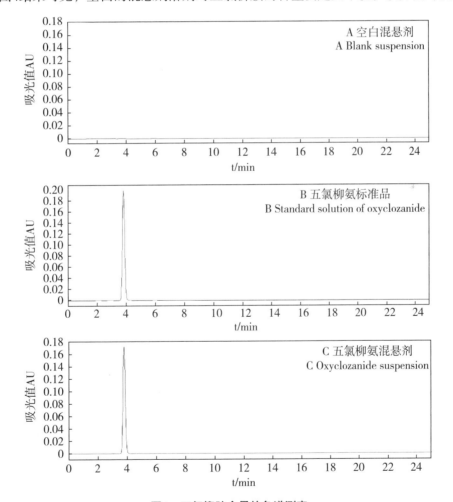

图4　五氯柳胺含量的色谱测定

Fig. 4　The content measurementchromatogram of oxyclozanide

2.3.2 线性考察

得到的标准曲线见图5，线性回归方程为：$y=24\,942x-113\,147$（$r^2=0.999\,3$），表明五氯柳胺在18~100μg·mL^{-1}的质量浓度范围内与对应峰面积的线性关系良好。

2.3.3 回收率与精密度

五氯柳胺的平均回收率为99.37%±0.14%，RSD为0.14%（表5），表明该方法准确度较高。

五氯柳胺的日内精密度的RSD值为0.37%，日间精密度的RSD值为0.64%，表明该方法的精密度良好。

2.3.4 重复性

取同批五氯柳胺混悬剂样品5份，测定五氯柳胺混悬剂的含量RSD值为0.67%，表明该方法的重复性良好。

2.3.5 样品含量测定

将3批样品稀释液分别进样20μL，测定3次，记录峰面积并计算实测值与理论值的比值，3批样品的平均含量为标示量的96.52%±0.47%，RSD为0.47%，如表6。

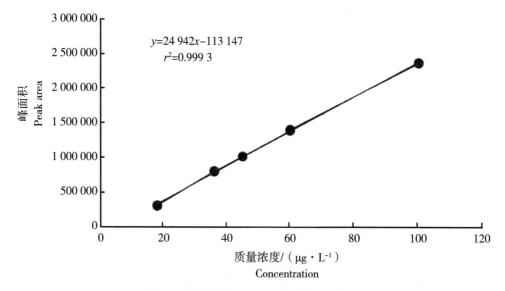

图5 五氯柳胺峰面积—质量浓度标准曲线

Fig. 5 The peak area—concentration standard curve of oxyzanide

表5 五氯柳胺混悬剂回收率试验结果（n=9）

Table 5 The recovery test results of oxyclozanide oral suspension

样品质量浓度(μg·mL^{-1}) The sample concentration	实测质量浓度(μg·mL^{-1}) The actual concentration	回收率/% The recovery rate	平均回收率 The average recovery rate	RSD/%
30	29.35	97.82		
60	59.71	99.51	99.37±0.15	0.15
80	80.61	100.77		

表 6 含量测定结果

Table 6 The result of content determination

样品 The sample	质量浓度/(μg·mL^{-1}) Concentration	标示量含量/% Labeled amount of content	平均标示量含量($\bar{x} \pm s$)/% The average labeled amount of content	RSD/%
1	61.72	96.44		
	61.38	95.90		
	61.41	95.95		
2	62.10	97.04		
	62.13	97.08	96.52 ± 0.47	0.47
	61.89	96.70		
3	61.49	96.08		
	61.85	96.64		
	61.99	96.86		

2.4 稳定性试验

2.4.1 加速试验

在加速条件下，检测五氯柳胺混悬剂三批样品的外观性状、含量、pH值、再分散性和沉降体积比都无明显变化。说明在市售包装下的五氯柳胺混悬剂在加速条件下性质稳定。检测结果如表7所示。

2.4.2 影响因素试验

在高温、强光照射条件下，五氯柳胺混悬剂的外观性状、含量、pH值、再分散性和沉降体积比几乎没有变化，说明五氯柳胺混悬剂对热、光很稳定。检测结果如表8所示。

表 7 加速试验结果

Table 7 The result of accelerated test

项目 Project	常态下各项指标 The indicators under normal condition	第0个月 Month 0	第1个月 Month 1	第2个月 Month 2	第3个月 Month 3	第6个月 Month 6
外观性状 Appearance	乳白色、均匀、浑浊，静止后可能出现分层，上层为澄清透明，下层为乳白色沉降物	同上所述，出现分层	同上所述，出现分层	同上所述，出现分层	同上所述，出现分层	同上所述，出现分层
pH	5.5 ~ 6.3	6.01 5.93 5.88	5.99 5.86 5.90	6.02 5.91 5.95	6.05 5.88 6.03	6.03 5.92 5.99
沉降体积比 Sedimentation volume ratio	0.9以上	0.999 1 1	0.998 0.999 0.998	0.983 0.992 0.989	0.979 0.982 0.981	0.973 0.979 0.974

(continued)

项目 Project	常态下各项指标 The indicators under normal condition	第0个月 Month 0	第1个月 Month 1	第2个月 Month 2	第3个月 Month 3	第6个月 Month 6
再分散性 Redispers-ibility	轻轻振摇即恢复均匀状态	1 1 1	1 1 1	1 1 1	2 1 1	2 2 2
含量 Content	95%~105%	97.08% 96.73% 96.89%	97.10% 96.07% 96.14%	96.92% 96.11% 95.97%	96.57% 95.83% 95.68%	96.63% 95.67% 95.49%

表8 高温、强光照射试验结果

Table 8　The result of high temperatures test and strong light test

项目 Project	常态下各项目指标 The indicators under normal condition	高温试验 High temperatures test			强光照射试验 Strong light test		
		第0天 Day 0	第5天 Day 5	第10天 Day 10	第0天 Day 0	第5天 Day 5	第10天 Day 10
外观性状 Appearance	乳白色、均匀、浑浊.静止后可能出现分层，上层为澄清透明，下层为乳白色沉降物	同上所述，出现分层	同上所述，出现分层	同上所述，出现分层	同上所述，出现分层	同上所述，出现分层	同上所述，出现分层
pH pH value	5.5~6.3	5.98 5.88 6.03	5.94 5.85 5.99	5.97 5.83 6.01	5.88 5.99 6.09	5.93 5.95 6.03	5.87 5.98 6.07
沉降体积比 Sedimentation volume ratio	0.9以上	1 1 1	0.999 0.999 0.998	0.997 0.998 0.998	1 1 1	0.999 0.998 0.999	0.998 0.995 0.997
再分散性 Redispers-ibility	轻轻振摇即恢复均匀状态	1 1 1	1 1 1	1 1 1	1 1 1	1 1 1	1 1 1
含量 Content	95%~105%	96.75% 96.35% 97.39%	96.69% 96.28% 97.04%	96.59% 96.31% 97.17%	95.67% 96.85% 96.39%	95.98% 96.43% 96.47%	95.73% 96.52% 96.36%

3　讨论

据报道，在研药物中难溶性药物比例高达70%以上，因其溶解性大大限制了药物的开发及应用。在药物制剂过程中，可以将不溶性药物考虑制成混悬剂，使不溶性药物粉粒分散在液体分散媒中形成的不均匀分散体系。混悬剂制备方法多样，生产过程简单，工业化大生产可行性强[17-19]。本研究制备的五氯柳胺口服混悬剂，由于药物粒径小，颗粒生物黏附性提高，与目前市售的抗寄生虫药物口服制剂如片剂、胶囊剂相比，五氯柳胺口服混悬剂的药物吸收良好，颗粒分布均匀，在胃肠道中降低了食物对药物吸收的影响，提高了难溶性药物的口服生物利用度，且具有分布面积大、吸收快、用量精准、给药方便等优点，该剂型在兽医临床上备受青睐。

通过筛选，最佳处方工艺中各辅料的用量均符合《药用辅料手册》中关于辅料用量的相关规定[20]以及《兽药研究技术指导原则汇编》中关于兽用化学药物制剂研究基本技术指导原则[21]。在处方工艺中，为了增加混悬剂的物理稳定性，制备时需加入能使混悬剂稳定的稳定剂，包括助悬剂、润湿剂等。本试验中选择卡波姆974p作为助悬剂，使混悬剂的黏度适中，有助于药物均匀分散，不易沉降和絮凝[22]，且制成的混悬剂为假塑性流体[23]，静止时黏度较大，倒出时粘度较小，有利于临床用药，增加了五氯柳胺的动力学稳定性。本试验选用十二烷基硫酸钠作为润湿剂，它是一种阴离子型表面活性剂，它的加入改变了五氯柳胺与水相之间的表面张力，分散相的分散度大，使之形成均匀稳定的分散体系。

在制剂制备过程中，考虑到混悬剂具有动力学不稳定性，粒径的大小及分布直接影响其沉降速度，因此，本实验采用研磨法，该法适用范围广，制备过程简便，制备的混悬剂颗粒分布窄，混悬性能良好。采用该方法制备的五氯柳胺混悬剂外观呈乳白色，均匀，浑浊，且无丁达尔效应，其沉降体积比、再分散性、粒径分布、pH值及含量测定等结果，均符合我国关于兽药混悬剂的相关要求。

本研究提供了一种制备工艺简单、性质稳定、质量可控、成本低廉、投药方便的用于防治牛、羊肝片吸虫病的五氯柳胺口服混悬剂，弥补了国内五氯柳胺制剂的空白。在动物抗寄生虫药物混悬剂的研制与应用方面，由于近年来我国畜牧养殖业迅速向规模化方向发展，临床对药物剂型的需求也在发生变化，传统的片剂剂型由于给药难度大，人力耗费以及剂量准确性的等因素的影响，在临床使用中带来了一定的局限性。本研究对满足兽医临床对五氯柳胺的需求、提升我国兽药制剂水平和保障畜牧业健康发展具有重要意义。

（编辑　白永平）

（发表于《畜牧兽医学报》，院选中文，2 115–2 125.）

牦牛角性状候选基因的筛选

佘平昌[1,2]，吴晓云[1,2]，梁春年[1,2]，褚敏[1,2]，丁学智[1,2]，阎萍[1,2]*

(1.中国农业科学院兰州畜牧与兽药研究所，兰州 730050；
2.甘肃省牦牛繁育工程重点实验室，兰州 730050)

摘 要：本研究旨在筛选牦牛角性状候选基因，为角发育机制的研究和无角牦牛的培育提供科学依据。采集牦牛角发育初期胎龄90d的有角和无角牛的角基及额部皮肤组织，利用实时荧光定量PCR技术（qRT-PCR），比较牛角性状候选基因OLIG1、OLIG2、IFNAR1、IFNAR2、C1H21orf62、GCFC1、IFNGR2、SYNJ1、IL10RB、GART、RXFP2、FOXL2、TWIST1、TWIST2、ZEB2、E-cadherin和P-cadherin mRNA在角基间及皮肤间的相对表达量。结果显示，候选基因OLIG1、IFNAR1、C1H21orf62、SYNJ1、IL10 RB、GART、RXFP2、TWIST2和E-cadherin在有角和无角牦牛角基间和皮肤间转录量差异不显著（P>0.05）。IFNGR2、FOXL2、TWIST1和P-cadherin在有角和无角牦牛角基间转录量差异显著（P<0.05）。IFNAR2和GCFC1在有角和无角牦牛角基间和皮肤间转录量差异都显著（P<0.05）。OLIG2和ZEB2在有角和无角牦牛角基间转录量差异极显著（P<0.01）。推测OLIG2和FOXL2可能是牛角性状的关键候选基因，同时ZEB2、TWIST1和IFNGR2基因可能对牛角的形成具有重要作用，而P-cadherin基因可能参与了牛角的发育。

关键词：牛角；无角性状；牦牛；候选基因；qRT-PCR

Screening of Candidate Genes for Horn Trait of Yak

SHE Pingchang[1,2], WU Xiaoyun[1,2], LIANG Chunnian[1,2],
CHU Min[1,2], DING Xuezhi[1,2], YAN Ping[1,2]*

(1. Lanzhou Institute of Husbandry and Pharmaceutical Sciences of Chinese Academy of Agricultural Sciences, Lanzhou 730050, China; 2. Key Laboratory of Yak Breeding Engineering of Gansu Province, Lanzhou 730050, China)

Abstract: This study screened the candidate genes of horn trait which provides scientific basis for the horn developmental mechanism and the polled yak breeding. The horn bud and forehead

---

基金项目：牦牛遗传资源与育种（CAAS-ASTIP-2014-LIHPS-01）；科技部科技支撑计划（2012BAD13B05）
作者简介：佘平昌（1989—），男，云南普洱人，硕士生，主要从事家畜分子生物学研究，E-mail：sheemail@163.com
*通信作者：阎萍，研究员，博士生导师，主要从事动物遗传育种与繁殖研究，E-mail：pingyanlz@163.com

skin tissue from horn and polled yaks with 90dgestational age were collected, and the mRNA different expression of OLIG1, OLIG2, IFNAR1, IFNAR2, C1H21orf62, GCFC1, IFNGR2, SYNJ1, IL10RB, GART, RXFP2, FOXL2, TWIST1, TWIST2, ZEB2, E-cadherin, P-cadherin genes were analyzed using real-time quantitative PCR (qRT-PCR). The results showed that expressions of OLIG1, IFNAR1, C1H21orf62, SYNJ1, IL10RB, GART, RXFP2, TWIST2and E-cadherin genes showed no significant differences in horn and skin between horned and polled yaks (P>0.05). There were significant difference at transcript levels of IFNGR2, FOXL2, TWIST1and Pcadherin genes in horn buds between horned and polled yaks (P<0.05). IFNAR2and GCFC1 had significant transcription difference in the expression of horn bud and skin between horned and polled yaks (P<0.05). OLIG2and ZEB2had highly significant transcription difference in horn bud expression between horned and polled yaks (P<0.01). We concluded that OLIG2and FOXL2 might be the key candidate genes. ZEB2, TWIST1and IFNGR-2might also involve in the formation of horn. P-cadherin might participate in the development of horn.

Key words: Horn; Polled trait; Yak; Candidate genes; qRT-PCR

牛角是牛科动物特有的组织结构。在男耕女织的古代原始生产过程中，牛角是进行耕地过程中牵牛的基础，所以牛角在原始耕作时代中是一种生产力。但是进入集约规模化饲养后，牛群常发生争斗及伤害饲养人员的行为，造成了动物的伤害及养殖场的混乱，牛角变成了一种制约家畜集约化和设施养殖的性状。所以，很多养牛场将犊牛进行去角[1]，然而去角增加了养殖成本，同时容易引起犊牛的应激反应，是违背动物福利的行为。动物福利问题已经在发达国家引起高度重视，欧盟启动了关于禁止动物阉割和去角的ALCASDE工程[2]，而德国的动物福利组织也联合政府发表申明将积极推进无角公牛精液的使用。因此，无角牛品种的培育是未来牛育种的主要方向之一。

牛角根据生物学分类属于洞角，由角质鞘和骨心组成。牛角的发育较早，60d胎龄的胎牛就细微可见角的形成，90d胎龄时发育更加明显，基本可以通过肉眼判定角的有无，在角的生长部位可见真皮乳头状的凸起形成角窦[3]，同时出现表皮增厚的现象[4]，角窦和皮肤间通过真皮乳头进行信号传导[5]。但是只有在出生时才可见螺纹状角的出现[6]，出生后6月龄才开始出现角的空腔结构[3]。组织学观察发现胚胎期牛角与皮肤间的差异主要有4点：腺导管真皮细胞的集群不同、囊泡化的角质细胞、没有毛囊的发育[7]、真皮中出现大量的神经束[8]，推测头皮层的神经纤维对牛角形成具有触动作用，所以神经发育相关基因可能会在角基部位高表达。在牛角性状候选基因的研究中，根据目前的研究进展，很多研究将无角突变定位在1号染色体（BTA1，Bos taurus autosome1）1.265~2.198M上1M左右的一段序列内，包括C1H21orf62、OLIG1、OLIG2、SYNJ1、GCFC4、GART、IL10RB、IFNAR1、IFNAR2、IFNGR2等基因[9-14]。而不在该序列内的一些基因也可能对牛角形成和发育具有重要作用，FOXL2和RXFP2基因在牛角形成和发育过程中高度表达[7,15]，而ZEB2、TWIST1和TWIST2是畸角牛的候选基因[16-18]，E-cadherin和P-cadherin对胚胎分化及表皮、毛囊、神经等的形成具有重要作用[19-21]。牛角是一种质量性状（Qualitative trait），其遗传方式可能为多基因遗传（Polygenic trait）。目前对牛角性状相关基因还没有进行精确定位，还需进一步研究。本研究通过荧光定量PCR（Real-time quantitative PCR，qRT-PCR）方法分析牛角性状候选基因mRNA的表达水平，初步揭示牛角发育和形成的分子机制，以期为无角牦牛品种的培育提供科学依据和理论支撑。

1 材料与方法

在青海省大通牦牛种牛场采集90d胎龄的有角和无角牦牛的角基部及额部皮肤样品，液氮保存，并对其角基部及全身进行拍照，测量其顶臀长，根据胎儿顶臀长与胎龄的线性关系推算胎龄[22]。

1.1 RNA 提取及 cDNA 合成

采用Trizol（Invitrogen）法提取有角和无角牦牛的角基及额部皮肤组织的总RNA，按照TaKaRa公司试剂盒Prime Script RT reagent Kit with gDNA Eraser（Perfect Real Time）说明书合成cDNA。

1.2 荧光定量 PCR

使用Premier 5.0设计候选基因mRNA荧光定量PCR引物（表1），选择GAPDH作为内参基因。引物由上海英骏生物公司合成，产物经2%琼脂糖凝胶电泳检测。

定量PCR（CFX96）反应使用25μL体系：SYBR® Premix Ex Taq^{TM} Ⅱ（Tli RNaseH Plus）12.5μL，上下游引物各1μL，cDNA 2μL，ddH$_2$O 8.5μL。荧光定量PCR的反应程序：95℃预变性30s；95℃变性5s，53.5℃退火30s，40个循环；之后进行熔解曲线分析，反应程序：65℃逐渐升温至95℃，每升温0.5℃读板1次。每个组织4个个体重复，每个样本3次试验重复。反应结束后以熔解曲线来判定qRT-PCR反应的特异性；获得每个样品的Ct值，运用$2^{-\Delta\Delta Ct}$法计算相对转录水平。

1.3 图像数据处理

本试验采用Excel 2013对数据进行统计，SPSS 17.0进行one-way ANOVA分析，Origin7.5作柱状图。

2 结果

2.1 BTA1 上候选区域内基因的相对表达量分析

为了分析候选基因的转录情况，本研究对比候选基因在有角牛和无角牛的角基间及额部皮肤间的相对转录水平，据此分析该基因是否参与了角的形成和发育（图1）。OLIG1、OLIG2、IFNAR1、IFNAR2、C1H21orf62、GCFC1、IFNGR2、SYNJ1、IL10RB和GART是1号染色体上1M左右片段上的基因。研究发现，OLIG1、IFNAR1、C1H21orf62、SYNJ1、IL10RB、GART、PXFP2、TWIST2和E-cadherin在有角和无角牦牛的角基间及皮肤间的转录量差异都不显著（P>0.05）；OLIG2在有角牛角基的转录量极显著高于无角牛角基（P<0.01），而在皮肤间差异不显著；IFNAR2在有角牛的角基和皮肤的转录量都要显著高于无角牛（P<0.05）；相反，GCFC1在有角牛的角基和皮肤的转录量都要显著低于无角牛（P<0.05）。IFNGR2在无角牛角基中的转录量显著高于有角牛（P<0.05），而在皮肤间的转录差异不显著（P>0.05）。

表1 荧光定量PCR引物序列
Table 1　Primer sequences used for the real-time quantitative PCR

基因名称 Gene name	上游引物序列(5'-3') Forward primer	下游引物序列(5'-3') Reverse primer
OLIG1	GTGGCGATCTTGGAGAGC	GAGAGGAAGCGGATGCAC
OLIG2	CCGGCATCTACCAGTACATTATAGC	GCACTCGTTGAGGCTGAGGT
IFNAR1	TCTTGACAGACTCACTGCTGCT	GAAGCAAATCTGAAGCCTGAA
IFNAR2	CTCAAGGGCTCTCACACACA	ACTTCCGAGGAGGACAGTGA
ClH21or/62	ACCAGCTGTCAAGCCTGAGT	CAGAAGGAGGTGCCATTCTC
GCFC1	GATGATGAGAAACGCCGCAT	AGGCATTGTCTGGTAGGTGT
IFNGR2	CTTGGACCAGCACTTTGGTT	CTCGCCAAAGGATGACACTT
SYNJ1	TGGTACGGTGTTGGTCTCAA	ATTCAAGGCAGAGCTTCCCT
IIAORB	TCTGTGATGACGCTCAGTTTG	TTTATGAGCCACCCTCATCAC
GART	GCCCGAGTACTTGTCATTGG	CCTGGGGTAACCAACACTTG
RXFP2	AGAACCCCACAATCCAGATG	TGACCCTGGAGAAGTTCCTG
FOXL2	CCGGCATCTACXAGTACATTATAGC	GCACTCGTTGAGGCTGAGGT
TWIST1	TGTGAGTCAGTTTGATCCCAAT	GGCATCATTATGGACTTTCTCC
TWIST2	TTCAACTGGACCAAGGCTCT	CCAGGCTAACTTGAGACAGTCA
ZEB2	ACACGGGTTCTGAAACTGATGA	GGCACGCTAGCTGGACTTCT
E-cadherin	GTACACCTTCATCGTCCAGAGCTAA	GCTCTTCAATGGCTTGTCCATTTGA
P-cadherin	GCAGCCTGATACGGTAGAGC	AGATCGAACCGGGTACTGG
GAPDH	ATGCAACCTGGTGATAAAAGAG	GAGGGCGCTAAGAGAAGAGTCAT

2.2 其他重要候选基因的相对表达量分析

RXFP2、FOXL2、TWIST1、TWIST2和ZEB2也是重要的角性状候选基因。研究发现，RXFP2和TWIST2在有角和无角牦牛的角基间和皮肤间转录量差异都不显著（$P>0.05$）；FOXL2在有角牛角基的转录量显著高于无角牛（$P<0.05$），而在皮肤间差异不显著；TWIST1却在无角牛角基的转录量显著高于有角牛（$P<0.05$），皮肤间的转录量差异不显著。ZEB2在有角牛角基的转录量极显著低于无角牛（$P<0.01$），而在皮肤间同样差异不显著（$P>0.05$）。

2.3 角发育相关基因的相对表达量分析

关于E-cadherin和P-cadherin与角性状的相关研究报道较少，但E-cadherin是表皮及毛囊的发育关键基因，P-cadherin的表达对表皮发育及神经分化具有重要作用，而据研究进展可知，角的发育是角基表皮分化的过程，中间也伴随着毛囊的发育，引入分析为试验提供参考。研究发现，E-cadherin在角基间和皮肤间的转录差异都不显著（$P>0.05$），而P-cadherin在有角牛角基中的转录量显著高于无角牛（$P<0.05$），皮肤间的转录量差异不显著（$P>0.05$）。

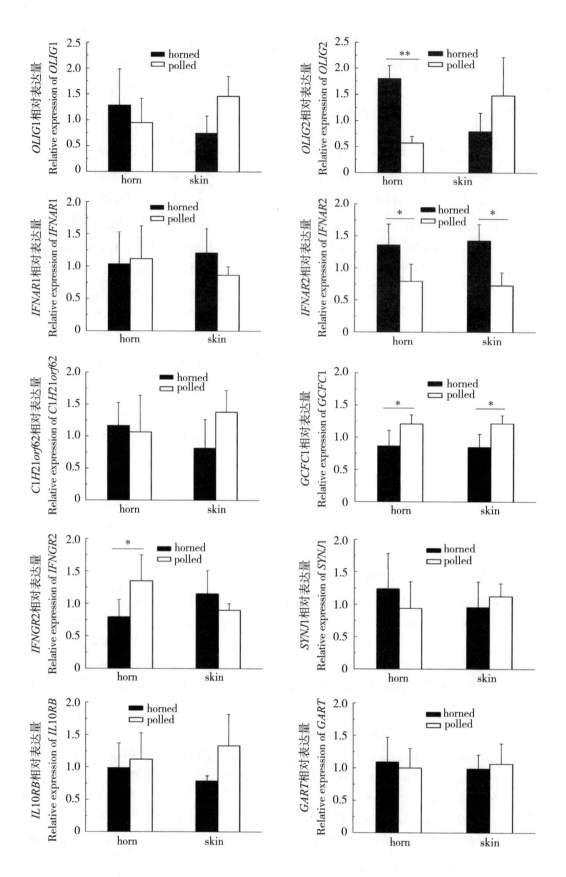

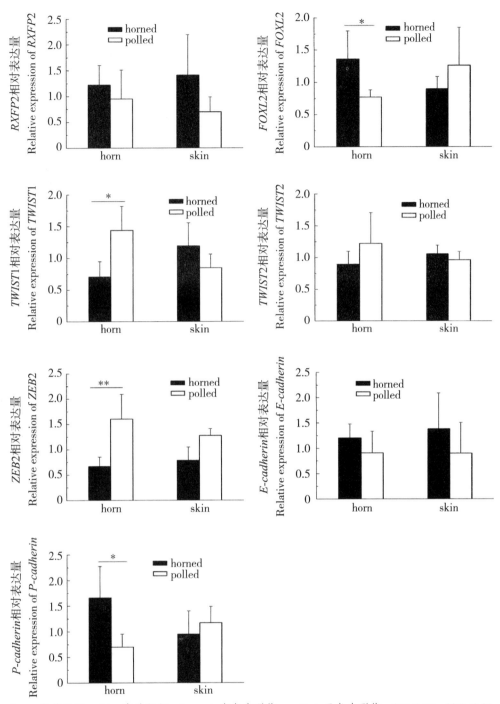

horn. 角基组织；skin. 皮肤组织；horned. 有角牛群体；polled. 无角牛群体；*.P<0.05；**.P<0.01
horn. Horn bud; skin. Forehead skin; horned. Horned yaks; polled. Polled yaks; *. P<0.05; **. P<0.01

图1 候选基因在有角和无角牦牛的角基间和皮肤间的相对转录水平

Fig. 1 Relative expression analyses of candidate genes in horn bud and skin tissues between horned and polled yaks by qRT-PCR

3 讨论

本研究对BAT1上定位的无角性状基因座区域内的候选基因的表达量进行了分析。研究结果显示，OLIG1、IFNAR1、C1H21orf62、SYNJ1、IL10RB和GART在有角和无角牦牛的角基间及皮肤间的转录量差异不显著（P>0.05）；OLIG2、IFNAR2、GCFC1和IFNGR2的mRNA转录量在角基间具有显著（P<0.05）或极显著（P<0.01）差异。候选基因中，OLIG2为少突胶质细胞转录因子，其主要参与少突胶质细胞的形成分化及髓鞘的形成，同时对哺乳动物的组织分化起重要调节作用[23]。本研究发现，OLIG2在角基间转录差异极显著，研究结果与在荷斯坦牛[7]和西门塔尔牛[15]中的研究结果一致。近期研究发现，OLIG2参与神经分化，推测该基因对牛角形成的作用机制可能与神经调控有关，说明OLIG2对角的发育有重要作用，因此将该基因作为荷斯坦奶牛、西门塔尔牛及牦牛无角性状的候选基因[7, 13, 15]。IFNAR2和GCFC1基因在角基间和皮肤间转录差异均显著，说明IFNAR2和GCFC1的表达模式不具有角组织特异性，IFNAR2和GCFC1可能对表皮的发育过程具有关键作用。IFNGR2是参与T细胞活动的一个细胞因子，在无角荷斯坦牛中该基因存在突变位点AC000128:g.1390292G>A[24]，但是S. Rothammer等[12]的研究又否定了该突变与无角性状的相关性，本研究发现，有角牛中IFNGR2基因的表达量显著低于无角牛，所以还不能排除IFNGR2可能为角性状候选基因。

本研究发现，其他候选基因RXFP2、TWIST2和E-cadherin在有角和无角牦牛的角基间和皮肤间转录量差异不显著（P>0.05）；FOXL2、TWIST1、ZEB2和P-cadherin的mRNA转录量在角基间具有显著（P<0.05）或极显著（P<0.01）差异。候选基因中，FOXL2又名叉头框因子2，是叉头框转录因子家族成员之一，对哺乳动物早期胚胎的发育和生殖系统功能维持具有重要作用[25]，本研究发现，FOXL2在角基间的表达量存在显著差异，在西门塔尔牛有角牛角基中该基因的表达量也显著高于无角牛[15]，同时FOXL2基因是羊角性状的主效基因之一[26]，所以FOXL2可能参与了牛角的形成和发育。TWIST1[16]和ZEB2[18]都是目前研究畸角牛的关键基因，TWIST1是在胚胎发育的一个转录因子，对诱导细胞移动及组织朔形具有重要作用，参与骨骼、肌肉和神经等的形成。ZEB2属于E盒结合锌指蛋白家族，在胚胎发育过程中对神经系统的发育具有重要作用。而牦牛中没有畸角牛，但是本研究发现其在角基中表达量差异显著，推测TWIST1和ZEB2可能与牛角的形成或发育具有调控作用，但是其功能仍需进一步明确。P-cadherin基因为钙黏蛋白家族成员，在胚胎发育过程中对表皮及神经的发育发挥着重要的作用，而角的发育为表皮增厚的过程，本研究中P-cadherin在角基间转录差异显著，推测该基因可能参与了牛角的发育。

本研究中，候选基因OLIG1、IFNAR1、C1H21orf62、SYNJ1、GART和RXFP2虽然在有角和无角牦牛的角基间和皮肤间的转录均不存在显著差异（P>0.05），但是根据本研究的内容不能完全排除这些基因可能参与角的形成或发育的可能。而IL10RB、TWIST2和E-cadherin基因的相关研究较少，且本试验结果显示这些基因在有角和无角牛角基及皮肤间的表达量都不存在差异，故推测IL10RB、TWIST2和E-cadherin基因可能不是牛角早期形成和发育的关键基因。

4 结论

牦牛角形状候选基因的表达研究发现，候选基因OLIG1、IFNAR1、C1H21orf62、

SYNJ1、IL10RB、GART、RXFP2、TWIST2和E-cadherin在有角和无角牦牛的角基和皮肤间相对转录水平差异不显著（P>0.05）。IFNGR2、FOXL2、TWIST1和P-cadherin基因在角基间转录量差异显著（P<0.05）。IFNAR2和GCFC1基因在角基和皮肤间转录量差异都显著（P<0.05）。OLIG2和ZEB2基因在角基间转录量差异极显著（P<0.01）。结合前期研究推测OLIG2和FOXL2基因可能是牛角性状的关键候选基因，同时ZEB2、TWIST1和IFNGR2基因可能对牛角的形成具有重要作用，而P-cadherin基因可能参与了牛角的发育。

（编辑　郭云雁）

（发表于《畜牧兽医学报》，院选中文，1 147-1 153.）

基于GC-MS技术的蹄叶炎奶牛血浆代谢谱分析

李亚娟[1,2]，王东升[1]，张世栋[1]，严作廷[1]，
杨志强[1]，杜玉兰[2]，董书伟[1]，何宝祥[2]

（1.中国农业科学院兰州畜牧与兽药研究所/农业部兽用药物创制重点实验室/甘肃省新兽药工程重点实验，兰州　730050；2.广西大学动物科技学院，南宁　530005）

摘　要：〔目的〕利用气相色谱/质谱联用技术（gas chromatograph tandem mass spectrometer technology，GC-MS）对奶牛血浆进行代谢组学分析，以明确奶牛发生蹄叶炎时血浆中代谢物的变化，进一步揭示其发病机制。〔方法〕根据蹄叶炎奶牛的临床症状和活动量情况，选取蹄叶炎患病组S和健康对照组C奶牛各10头，采集血浆样品，经衍生化处理后，利用GC-MS方法对两组奶牛的血浆进行全部代谢物检测。用SIMCA-P 11.5软件进行主成分分析（principal component analysis，PCA）和偏最小二乘法分析（supervised partial least squares-discriminant analysis，PLS-DA），对蹄叶炎组奶牛和健康组奶牛检测数据进行模式识别分析，筛选差异代谢物，并利用生物信息学方法进行代谢通路富集分析。为进一步验证代谢组学的试验结果，检测了与差异代谢通路相关的脂质代谢指标和抗氧化能力指标。〔结果〕利用GC-MS方法，在奶牛血浆中共检测到242种代谢物，通过多元统计分析结合t-检验筛选出37种差异代谢物（VIP>1，P<0.05），其中在蹄叶炎奶牛血浆中3种物质含量上调（FC>1），分别是氨基氧乙酸、油酸、乳糖，34种物质含量下调（FC<1），包括脂肪酸、氨基酸等。经代谢通路富集分析，发现变化显著（P<0.05）的代谢通路包括脂肪酸生物合成通路，甘氨酸/丝氨酸/苏氨酸代谢通路，不饱和脂肪酸生物合成通路，嘌呤代谢通路，甲烷代谢通路，苯丙氨酸代谢通路和氰基氨基酸代谢通路。在验证试验中，发现两组奶牛血浆中脂质代谢相关指标差异显著，且蹄叶炎患病牛血浆抗氧化能力显著低于健康牛，与代谢组学结果一致。〔结论〕利用GC-MS技术分析了蹄叶炎患病牛与健康奶牛血浆代谢谱的变化，发现两组间代谢物存在显著差异，并筛选出了37种差异代谢物，而这些差异分子可能成为奶牛蹄叶炎早期诊断或群体监测的潜在生物标记物。本结果全面揭示了奶牛蹄叶炎发生后血浆代谢物的代谢规律，为进一步阐明其发病机制提供理论依据。

关键词：代谢组学；蹄叶炎；气相色谱质谱联用技术；奶牛

基金项目：国家自然科学青年基金项目（31302156）；甘肃省青年科技基金计划1506RJYA145；中央级基本科研业务费项目（1610322014012）

联系方式：李亚娟，E-mail：18719826933@163.com。通信作者董书伟，E-mail：dongshuwei@caas.cn。通信作者何宝祥，E-mail：hebaox@gxu.edu.cn

Plasma Metabolic Profiling Analysis of Dairy Cows with Laminitis Based on GC-MS

LI Yajuan[1,2], WANG Dongsheng[1], ZHANG Shidong[1], YAN Zuoting[1],
YANG Zhiqiang[1], DU Yulan[2], DONG Shuwei[1], HE Baoxiang[2]

(1. Key Laboratory of Veterinary Pharmaceutical Development, Ministry of Agriculture/Key Laboratory of New Animal Drug Project, Gansu Province/Lanzhou Institute of Husbandry and Pharmaceutical Sciences of Chinese Academy of Agricultural Sciences, Lanzhou 730050;
2. College of Animal Science and Technology, Guangxi University, Nanning 530005)

Abstract: 〔Objective〕 In order to discover the difference between the cows suffered with laminitis and healthy cows, metabonomics analysis based on gas chromatograph tandem mass spectrometer technology (GC-MS) was utilized in the study. 〔Method〕 Twenty plasma samples were collected from cows with laminitis and healthy groups. Plasma metabolic profiling was obtained by GC-MS after derivatization. Multivariate statistics including principal component analysis (PCA) and supervised partial least squares-discriminant analysis (PLS-DA) were conducted to screen the changed metabolites. Bioinformatics was analyzed to identify the significantly changed metabolic pathways. 〔Result〕 A total of 242 variables were detected and 37 compounds were significantly different between healthy and laminitis-suffered cows of which 3 were up-regulated including aminooxyacetic acid, lactose and oleic acid and 34 were down-regulated in laminitis group involved amino acid and fatty acid. The main changed metabolic pathways were fatty acid biosynthesis, Gly, Ser, Thr metabolism, biosynthesis of unsaturated fatty acids, purine metabolism, methane metabolism, cyanoamino acid metabolism and phenylalanine metabolism. Results of the the experiment of validation showed that the indexes related to fatty acid metabolism and antioxidation were significantly different between healthy cows and sick cows. 〔Conclusion〕 Based on GC-MS, this study revealed the changed regularity of metabolism in plasma with laminitis, and the discovered different metabolites would be potential biomarkers in occurrence and development process of laminitis, which can lay a theoretical foundation for further understanding of the pathogenesis, early diagnosis and therapy of laminitis in dairy cows.

Key words: Metabonomics; Laminitis; Gas chromatograph tandem mass spectrometer technology; Dairy cow

引言

【研究意义】随着奶牛养殖业的发展,奶牛蹄病已成为继乳房炎和繁殖疾病之后危害奶牛健康的重要疾病，而蹄叶炎是最为常见的一种蹄病[1]。蹄叶炎是蹄壁真皮的乳头层和血管层发生的弥漫性、浆液性和无菌性炎症[2]，可导致蹄变形、蹄底溃疡病及白线病等多种蹄病，引起奶牛疼痛不安，严重影响了动物福利，造成奶牛生产性能明显下降，带来了巨大的经济损失，制约了奶业的健康发展。因此，开展对奶牛蹄叶炎病因、发病机理、诊断和防治等方面的研究，对防治奶牛蹄叶炎、提高奶牛生产性能具有重要的实用价值和现实意义。【前人研究进展】有人认为奶牛蹄叶炎是机体全身代谢紊乱的局部表现，也与营养、遗传和管理等综合因素有关，还可继发于瘤胃酸中毒、乳房炎、胎衣不下、酮病和子宫内膜炎等[3-4]。如当奶牛摄入过多精料，发生瘤胃酸中毒时，瘤胃内微生物异常发酵，释放出内毒素、组胺等有害物质，随血液循环作用于蹄部，引发蹄叶炎，而乳酸、内毒素和组织胺正是引发该病的主要因素[5]。THOEFNER等[6]利用过量饲喂低聚果糖建立了奶牛急性蹄叶炎的病理模型，齐长明等[7]通过给奶牛皮下注射二磷酸组织胺后，牛蹄部出现了蹄叶炎的临床症状。因此，奶牛发生蹄叶炎时，机体的碳水化合物代谢[8]和脂质代谢[9]发生了紊乱，但其他物质的代谢是否发生了变化？这些变化在蹄叶炎的发生发展过程中是什么角色，尚无人知晓。因此，对奶牛蹄叶炎发病时体内整体代谢变化的全面了解，将有助于揭示其发病机制。【本研究切入点】代谢组学是研究生物在内外因素影响下，机体代谢轮廓变化规律的学科[10]，通过代谢组学的方法，对患病和健康动物进行代谢物的全谱检测，对比分析代谢轮廓的改变，有助于认识疾病过程中体内物质代谢途径的改变[11]。因此，代谢组学已被广泛地应用于疾病的诊断[12]和生物标记物的筛选[13-14]。目前，代谢组学常用的技术平台包括核磁共振波谱法（NMR），高效液相色谱—质谱法（HPLC-MS），傅里叶变换红外光谱法（FT-IR）和气相色谱—质谱法（GC-MS）[15]，其中由于GC-MS具有较高的分辨度和灵敏度以及良好的重现性而被广泛使用[16-17]。但是，关于奶牛蹄叶炎的代谢谱研究还尚未见报道。因此，对奶牛蹄叶炎发病时体内代谢变化的全面了解，将有助于了解其发病机制。【拟解决的关键问题】拟采用GC-MS技术，结合多元统计分析方法，分析健康和患蹄叶炎病牛的血浆代谢谱，旨在找出疾病过程中变化显著的内源性代谢物和代谢通路，为进一步阐明该病的发病机理提供理论依据，为早期诊断与防治蹄叶炎提供新的思路。

1 材料与方法

1.1 试验时间、地点

本试验于2014年9月至2015年5月执行，其中实验动物筛选和血样采集在甘肃省奶牛繁育中心完成，GC-MS检测由上海博苑生物科技有限公司完成，样品处理、保存和后期的验证实验在中国农业科学院兰州畜牧与兽药研究所完成。

1.2 试剂与仪器

安捷伦7890B GC/5977A MS联用仪（Agilent，USA），配有非极性的DB-5毛细管柱（30m×250μm I.D.，J&W Scientific，Folsom，CA）；高速低温离心机；快速离心浓缩仪；涡旋混合器。

甲氧胺盐酸吡啶溶液，N，O-双（三甲基硅烷基）三氟乙酰胺（BSTFA），三甲基氯硅烷（TMCS）均为Sigma色谱纯。冷甲醇，L-2-氯-苯丙氨酸和正己烷等试剂均为国产分析纯。牛组织胺（Histamine，His）ELISA试剂盒，购自上海桥杜生物技术公司，内毒素检测使用鲎试剂的终点显色法（Chromogenic Endpoint Tachypleus Amebocyte Lysate，CE-

TAL），购自厦门市鲎试剂实验厂有限公司。血浆生化指标的测定[9]使用全自动生化分析仪（深圳迈瑞公司BS-420），抗氧化指标测定所使用的试剂盒购自南京建成生物研究所。

1.3 动物选择和分组

在甘肃省某奶牛场选择健康对照组C和蹄叶炎患病组S奶牛各10头，蹄叶炎奶牛选取标准是通过活动量检测发现近两天内活动量有显著降低的奶牛，再根据有跛行、弓背、蹄部触诊敏感和患蹄肿胀等临床症状，排除腐蹄病、蹄毛疣和趾间增生等病例，经兽医诊断为蹄叶炎且无其他疾病临床症状，近期无用药记录，经组织胺和内毒素检测[18-20]的泌乳牛10头作为患病组S。对照健康奶牛的选取标准是：肢蹄健康，无其他疾病临床症状，近期无用药记录的泌乳奶牛10头作为对照组C。并统计两组奶牛的年龄、胎次、产奶量和体况评分（body condition score，BCS），各组奶牛基本情况见表1。常规饲养管理，两组奶牛均饲喂相同的日粮、自由采食和饮水，该奶牛场生产用日粮配方为：精料10kg、青贮15kg、干草4kg、脂肪350g，其日粮营养成分含量见文献[9]。

表1 试验用奶牛的临床资料和理化指标（均值 ± 标准差）

Table 1 Clinical data and biochemical parameters in experimental cows（mean ± SD）

组别 Group	年龄 Age	泌乳天数 Day in milk（d）	产奶量 Milk yield （$kg \cdot d^{-1}$）	体况评分 Body condition score	组织胺 Histamine （$\mu g \cdot L^{-1}$）	内毒素 Endotoxin （$Eu \cdot mL^{-1}$）
患病Sick	4.5 ± 1.5	121 ± 30	25 ± 4.7	3.42 ± 0.53	36.7 ± 6.3a	0.53 ± 0.12a
健康Health	3.9 ± 1.3	115 ± 26	28 ± 5.3	3.35 ± 0.46	17.5 ± 4.1b	0.23 ± 0.04b

同行数据后所标字母相异表示差异显著（P<0.05），所标字母相同表示差异不显著（P>0.05）

Different letters in the same row means significant difference between the treatments（P<0.05），the same letter in the same row means not significant difference between treatments（P>0.05）

1.4 样品采集

在清晨饲喂前采用含有EDTA K_2抗凝剂的真空采血管尾静脉采集全血8mL，3 000×g离心15min，将血浆分装到EP管内，-80℃冻存备用。

1.5 血浆样品的衍生化

血浆样品在室温下解冻血浆样品后取50μL加入150μL冷甲醇和10μL内标（L-2-氯-苯丙氨酸，0.3mg·mL^{-1}，甲醇配制）涡旋2min，然后超声提取10min，再低温离心10min（14 000r/min，4℃）。取150μL上清液至玻璃衍生瓶中，快速蒸干后加入80μL甲氧胺盐酸吡啶溶液（15mg·mL^{-1}），涡旋震荡2min后，于震荡培养箱中37℃肟化反应90min，取出后再加入80μL BSTFA（含1%TMCS）衍生试剂和20μL正己烷，涡旋震荡2min，于70℃反应60min。室温放置30min后，上机GC/MS分析小分子代谢物。在测定过程中，插入质控样本（QC），质控样本制备方法为取小部分所有待测样本的混合物，对测定过程进行监督。

1.6 测定条件

气相条件：载气为高纯氦气；载气流速为1.0mL·min^{-1}；程序升温为0～2min至80℃；2～12min至80～180℃；12～24min至180～240℃；24～26min至240～280℃；26～35min至280℃。质谱条件：EI源温度为220℃；电压为-70V；质量扫描范围为30～600（m/z）；采集速度为20谱/秒。

1.7 代谢物鉴定

使用GC-MS联用仪对血浆样品进行检测分析，所产生的原始数据文件转换至NetCDF格式，之后采用Leco TOF软件，进行数据归一化，包括降噪，保留时间校正，峰对齐和反卷积等数据标准化分析[21]。根据GC-MS总离子流图中各峰的保留时间选取共有峰（即各图中共有的色谱峰），每个化合物的特征离子片段谱的碎片质荷比和丰度与美国国家标准与技术局化学数据库（national institute of standards and technology，NIST）、Feihn代谢组学数据库的标准离子片段谱库比对，以匹配度超过70%的检测物被认为是与之匹配的代谢物。

1.8 多元统计分析

将归一化后的数据矩阵导入SIMCA-P+11.5软件包（Umetrics，Umea，Sweden），先采用无监督的主成分分析（principal component analysis，PCA）观察各样本间的总体分布，然后用有监督的偏最小二乘判别分析（partial least squares-discriminant analysis PLS-DA）区分各组间代谢轮廓的总体差异，找到组间的差异代谢物。为防止模型过拟合，采用七次循环交互验证和200次响应排序检验的方法来考察PLS-DA模型的质量。通过PLS-DA分析，结合t-检验，变量权重值（variable important in projection，VIP）大于1且P<0.05的变量被认为是差异变量。

1.9 代谢通路富集分析

使用MBRole（http://csbg.cnb.csic.es/mbrole/）并结合KEGG（Kyoto Encyclopedia of Genes and Genomes，http://www.kegg.jp/）进行代谢通路富集分析，鉴别出两组间差异显著的代谢通路，P<0.05说明差异显著。

1.10 验证试验

为进一步验证GC-MS的结果，对患病组和健康组奶牛的血浆进行了相关生化指标和抗氧化能力检测，主要检测了脂质代谢的相关指标TG、TC、HDL-C和LDL-C与抗氧化能力的相关指标T-AOC、SOD、MDA和GSH-Px。

2 结果

2.1 代谢物总离子图分析与代谢物鉴定

对两组血浆样品的总离子流图（TIC图）进行可视化检查（图1），发现所有样品的仪器分析信号强、峰容量大且保留时间重现性好。通过将检测结果与NIST数据库标准质谱图对比，共鉴定出242个代谢物。为了进一步表征两组间代谢差异的特征，采用多元统计和单维统计相结合的方法获得组学特征。

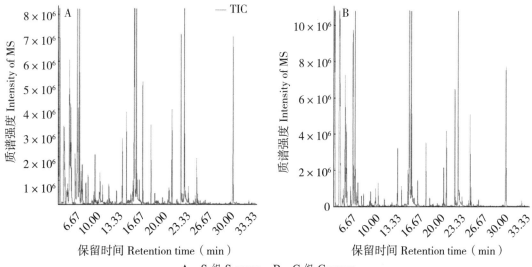

A：S 组 S group；B：C 组 C group

图1 基于GC-MS分析获得的总离子流
Fig. 1 A total ion chromatograms（TICs）of the plasma samples by GC-MS analysis

2.2 多元统计分析

2.2.1 PCA分析

PCA分析结果见图2，图中每个点代表一个样本，所有样本均在95%的置信区间内。PCA分析显示其解释率较高/强（$R^2X=0.559$），证实该分析模型可靠。由图可见，S组主要分布在PC1的左侧，C组主要分布在PC1的右侧，在PC2上的分布虽有一定的重合，但大部分样品分别聚类在PC2的上下两侧，说明S组和C组有一定的组间差异性，并且QC样本聚成一类，说明整个分析方法（前处理方法和仪器分析系统）稳健可靠，各试验组间的差异性是真实存在的。

PCA分析显示：其解释率较高/强（$R^2X=0.559$），证实该分析模型可靠PCA。得分图中（图2），每一个点代表一个样本，所有样品基本在95%的置信区间内。从图中可以看出，S组主要分布在PC1的左侧，C组主要分布在PC1的右侧，在PC2上的分布虽有一定的重合，但大多数样品分别聚类在PC2的上下两侧，说明S组和C组有一定的组间差异性。并且QC样本聚成一类，说明整个分析方法（前处理方法和仪器分析系统）稳健可靠，各试验组间的差异性是真实存在的。

2.2.2 PLS-DA分析

为了消除与分类不相关的噪声信息，在PCA模型的基础上，建立了PLS-DA模型（$R^2X=0.873$，$R^2Y=0.952$，$Q^2=0.766$），进一步分析蹄叶炎与健康对奶牛在血浆代谢水平的变化血液代谢模式的影响。其中Q^2表示模型的预测率，$Q^2>0.5$表示模型具有较好的判别分析能力。PLS-DA模型得分图（图3）显示，两组样品之间有一定的差异，其中S组主要分布在PC1的右侧，C组主要分布在PC1的左侧，说明两组样本在主成分坐标轴上分离较好。

采用200次响应排序的方法对模型的稳健性进行考察，结果见图4，参数为：$R^2=0.673$，$Q2=-0.201$，说明此模型是稳健可靠。

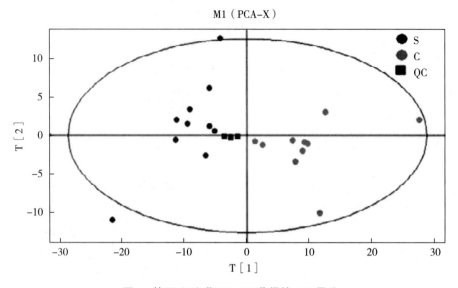

图2 基于2组血浆GC-MS获得的PCA得分
Fig. 2 PCA score scatter plots of plasma samples

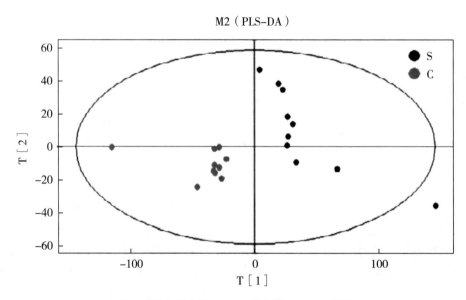

图3 基于2组血浆GC-MS获得的PLS-DA得分
Fig. 3 PLS-DA score scatter plots of plasma samples

2.3 差异代谢物鉴定及相关代谢通路分析

通过PLS-DA分析后,并根据VIP>1及t检验的P<0.05的相对变量标准分析结果,将代谢物的峰面积进行归一化处理,经过相对定量,得到各个结果在患病组和健康组奶牛血浆中共筛选到37个差异代谢物(表2)。其中有3种代谢物上调,分别是氨基氧乙酸、乳糖和油酸;有34种代谢物下调,代谢物在两组间的变化差异倍数(fold change,FC),其中FC(S/C)为代谢物在患病组和健康组的比值。包括氨基酸、饱和脂肪酸以及与之代谢相关的化合物。

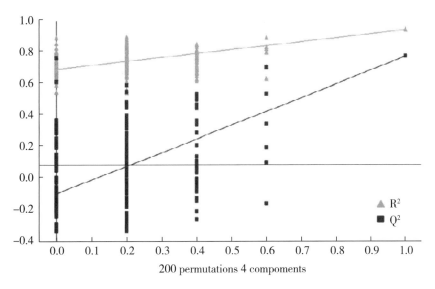

图4 使用200次响应法对PLS-DA进行验证的点
Fig. 4 Validation plots of PLS-DA with 200 permutation and 4 components

通过MBRole和KEGG数据库对差异代谢物进行代谢通路富集分析（图5），两组奶牛血浆中脂肪酸代谢（饱和脂肪酸和不饱和脂肪酸的生物合成）、氨基酸代谢存在极显著差异（$P<0.01$），嘌呤代谢、甲烷代谢和氰基氨基酸代谢等相关代谢途径也存在显著差异（$P<0.05$）。且两组血浆中棕榈酸和硬脂酸相对含量有所差异（图6）。

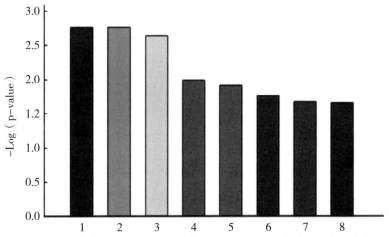

1. 脂肪酸生物合成通路，2. 甘氨酸/丝氨酸/苏氨酸代谢通路，3. 不饱和脂肪酸生物合成通路，4. 嘌呤代谢通路，5. 甲烷代谢通路，6. 苯丙氨酸代谢通路，7. 两组奶牛血浆代谢总体通路，8. 含腈氨基酸代谢通路

1. Fatty acid biosynthesis, 2. Gly, Ser, Thr metabolism, 3. Biosynthesis of unsaturated fatty acids, 4. Purine metabolism, 5. Methane metabolism, 6. Cyanoamino acid metabolism, 7. Total metabolic pathway, 8. Phenylalanine metabolism

图5 代谢通路差异
Fig. 5 Pathway enrichment analysis

表2 S组和C组间的差异代谢物

Table 2　Identification of significantly different metabolites

代谢物 Metabolite name	RT (min)	VIP value	FC (S/C)	P value
氨基氧乙酸 Aminooxyacetic acid	15.68	9.45	1.90	0.001 4
乳糖 Lactose	20.44	6.03	1.99	0.044 4
柠檬酸 Citric acid	12.93	4.07	0.52	0.000 3
尿素 Urea	7.68	3.77	0.58	0.002 2
磷酸盐 Phosphate	26.60	3.69	0.72	0.002 0
缬氨酸 Valine	20.79	3.26	0.43	<0.001
油酸 Oleic acid	15.80	2.46	4.85	0.035 4
二十四烷 Tetracosane	10.45	2.42	0.18	<0.001
异亮氨酸 Isoleucine	4.13	2.34	0.43	<0.001
棕榈酸 Palmitic acid	10.26	2.32	0.62	0.034 5
硬脂酸 Stearic acid	24.25	2.31	0.59	0.014 5
脯氨酸 Proline	5.95	2.13	0.42	<0.001
十六烷 Hexadecane	7.12	2.04	0.21	<0.001
甘氨酸 Glycine	5.91	1.97	0.67	0.014 2
棕榈油酸甲酯 Methyl palmitoleate	10.95	1.97	0.19	<0.001
苏氨酸 L-allothreonine	11.77	LSI	0.61	0.004 8
亚油酸甲酯 Linoleic acid methyl ester	19.06	1.77	0.39	<0.001
Bis (2-hydroxypropyl) amine	12.79	1.74	0.38	0.001 1
氧代脯氨酸 Oxoproline	22.68	1.60	0.61	0.001 4
氨基丙二酸 Aminomalonic acid	9.88	1.54	0.46	<0.001
5-dihydrocortisone	28.07	1.52	0.69	0.041 5
丙氨酸 L-dopa	26.55	1.47	0.41	0.000 1
9-苯基苯甲酰胺 9-fluorenone	9.15	1.46	0.18	<0.001
尿酸 Uric acid	25.52	1.40	0.49	0.002 2
马尿酸 Hippuric acid	26.79	1.40	0.57	0.014 2
3-磷酸甘油 D-(glycerol-phosphate)	26.03	1.34	0.55	<0.001

(continued)

代谢物 Metabolite name	RT(min)	VIP value	FC(S/C)	P value
苯乙醛 Phenylacetaldehyde	4.26	1.20	0.29	0.000 4
氢化肉桂酸 Hydrocinnamic acid	9.41	1.18	0.27	<0.001
吲哚乳酸 Indolelactate	9.62	1.18	0.35	0.000 1
甲氧基酚 Guaiacol	19.42	1.16	0.16	<0.001
丙二酰胺 Malonamide	20.42	1.09	0.72	0.004 4
二磷酸果糖 D-fructose, 6-bisphosphate	22.34	1.07	0.34	0.001 5
丝氨酸 Serine	24.17	1.06	0.69	0.028 8
芴 Fluorene	21.23	1.06	0.17	<0.001
甲基磷酸盐 Methyl phosphate	12.00	1.05	0.24	<0.001
苯乙酰甘氨酸 Phenaceturic acid	5.49	1.03	0.48	0.002 0
Prunin degr. prod	12.06	1.03	0.20	<0.001

RT=保留时间，FC=差异倍数，S组/C组。如果FC>1，代表该物质含量S组高于C组，P值由T-检验获得

RT=retention time. FC=fold change, mean value of peak area obtained from S group/ mean value of peak area obtained from C group. If the FC value greater than 1, means that metabolite are more in S group than in C group. P values were obtained by T-test

2.4 验证试验

奶牛血浆抗氧化能力指标检测结果见表3，患病组奶牛血浆T-AOC和GSH-Px含量均低于健康组，差异显著（P<0.05），MDA显著高于健康对照组（P<0.05），但两组间SOD差异不显著。奶牛血浆生化相关指标检测结果见表4，蹄叶炎奶牛中的血浆含量TC、TG、LDL-C均显著高于健康组奶牛，差异显著（P<0.05）。由此说明奶牛患病后，机体抗氧化能力和脂质代谢发生紊乱，这与代谢组学结果相一致。

表3 两组奶牛血浆抗氧化指标（均值 ± 标准差）
Table 3 Antioxidant ability of plasma in dairy cows（Mean ± SD）

组别 Group	总抗氧化能力 T-AOC（μmol·L^{-1}）	超氧化物歧化酶 SOD（U·mL^{-1}）	丙二醛 MDA（nmol·mL^{-1}）	谷胱甘肽过氧化物酶 GSH-Px（U·mL^{-1}）
患病 Sick	5.82 ± 0.2a	58 ± 7.2a	3.4 ± 0.5a	27 ± 3.4a
健康 Health	7.2 ± 1.45b	64 ± 4.6a	2.2 ± 0.1b	41 ± 4.2b

同列不同小写字母表示差异显著（P<0.05），同列相同小写字母表示差异不显著（P>0.05）

Different letters in the same column mean significant difference（P<0.05）and same letters mean no significant difference（P>0.05）

表4 两组奶牛血浆中脂质代谢相关各指标（均值 ± 标准差）

Table 4 Lipid metabolism related parameters in plasma（Mean ± SD）

组别 Group	甘油三脂 TG（mmol·L^{-1}）	胆固醇 TC（mmol·L^{-1}）	高密度脂蛋白 HDL-C（mmol·L^{-1}）	低密度脂蛋白 LDL-C（mmol·L^{-1}）
患病Sick	0.7 ± 0.494A	5.8 ± 1.749A	1.1 ± 0.215A	1.5 ± 0.594A
健康Health	0.4 ± 0.200B	4.2 ± 1.590B	2.3 ± 0.526B	0.9 ± 0.394B

同列不同大写字母表示差异显著（P<0.05）

Different letters in the same column mean significant difference（P<0.05）

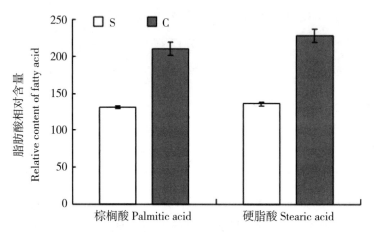

图6 两组奶牛血浆中棕榈酸和硬脂酸的相对含量

Fig. 6 The contents of palmitic acid and stearic acid in plasma from laminitis and healthy cows

3 讨论

3.1 奶牛血浆代谢谱分析

利用代谢组学筛选疾病发生发展过程中的生物标记物，为疾病的早期诊断和治疗提供理论依据，在人类和动物疾病的研究中已有可喜进展[11, 22-23]。但是采用代谢组学方法分析奶牛蹄叶炎的血浆代谢变化，这在国内外尚未见相关报道。目前，普遍认为奶牛蹄叶炎是全身代谢紊乱的局部体现，但具体哪些代谢通路发生了变化还不清楚，本研究利用GC-MS技术发现蹄叶炎患病牛与健康奶牛血浆中存在37种差异代谢物，主要涉及脂肪酸生物合成通路，甘氨酸/丝氨酸/苏氨酸代谢通路和不饱和脂肪酸生物合成通路，另外，嘌呤代谢通路，甲烷代谢通路，苯丙氨酸代谢通路，含腈氨基酸代谢通路也发生了显著变化。本研究全面分析了蹄叶炎患病牛与健康奶牛血浆代谢谱的变化，为进一步解释蹄叶炎的发生提供了一定的理论依据。

目前普遍认为奶牛过量食入精料后，引起亚急性瘤胃酸中毒[24]，从而导致蹄叶炎，并且有很多研究通过饲喂过量碳水化合物，成功建立了蹄叶炎病理模型[6]，但本试验通过代谢通路富集分析，并未发现血浆中与能量代谢相关的通路存在显著差异，只是与能量代谢相关的部分代谢物有所变化，推测可能是当奶牛发生瘤胃酸中毒时，瘤胃微生物异常发酵，释放出内毒素和组胺，经血液循环运送到蹄部，作用于蹄部微循环，改变了血管通透性，造成蹄部的微循环障碍，从而引起局部的能量代谢变化，而并未表现出全身的能量代谢障碍。这与MEDINA[8]的研究结果一致，他发现马发生蹄叶炎时，在蹄部间质组织中能量代谢发生改变，而在血浆中并未表现出能量代谢障碍，这可能与血液的自我稳态调控有关。

3.2 脂质代谢异常

本研究结果发现，蹄叶炎患病组和健康组奶牛相比，血浆中脂质代谢通路发生了显著变化（图3），并经过相关的试验证明，发现蹄叶炎患病牛与健康奶牛血浆中脂质代谢的相关指标有显著差异（表4），说明奶牛发生蹄叶炎时脂质代谢出现紊乱，这与笔者前期的蛋白质组学研究结果一致[25]。显示，蹄叶炎患病牛血浆中棕榈酸和硬脂酸含量低于健康奶牛，说明奶牛发生蹄叶炎时脂肪酸合成受阻（图6）。已知脂肪酸的生物合成起始于乙酰-CoA转换成丙二酸单酰-CoA，这步反应是在乙酰-CoA羧化酶的作用下实现的。乙酰-CoA羧化酶是脂肪酸生物合成的关键酶，生物素是其辅基。因此生物素含量下降时，会通过抑制乙酰-CoA羧化酶的活性而影响脂肪酸的合成。KHALED[26]等发现，跛行牛和健康牛血清中生物素的含量与机体抗氧化能力显著负相关，当机体抗氧化能力下降时，生物素含量随之下降。本研究中进行的试验也表明蹄叶炎患病牛的抗氧化能力降低，因此，生物素含量会降低，抑制乙酰-CoA羧化酶的活性，使脂肪酸生物合成受阻，导致棕榈酸和硬脂酸含量下降。另有研究表明，奶牛发生蹄叶炎时血浆胆固醇含量升高[9]。众所周知，有机体生物合成胆固醇的原料为乙酰-CoA，奶牛发生蹄叶炎时，胆固醇的大量合成也会影响脂肪酸的合成。

本研究中脂肪酸合成通路和不饱和脂肪酸合成通路均异常，因此，脂质代谢通路可作为一条新的线索来进一步明确蹄叶炎的发病机制。

3.3 氨基酸代谢异常

本研究结果显示蹄叶炎患病牛血浆中许多氨基酸含量低于健康组奶牛，推测可能是丙氨酸、苏氨酸、丝氨酸、甘氨酸碳骨架分解后会形成乙酰-CoA，以调节体内的脂肪酸合成异常。甘氨酸和丝氨酸在动物的生命活动中有重要作用，为核酸和脂质的合成提供前体物质[27]，而且在炎症[28]，肿瘤[29]，癌症[30]的发生过程中也有一定的作用，说明甘氨酸和丝氨酸在维持动物健康方面有重要作用。AMELIO[31]研究发现，丝氨酸的合成还会直接影响细胞的抗氧化能力。蹄叶炎发生时，奶牛体内丝氨酸含量下降，细胞抗氧化能力下降，细胞功能受损，更容易受到组胺和内毒素的侵害。许多氨基酸如：甘氨酸、苏氨酸、丝氨酸等还可作为一碳单位的来源，从而参与嘌呤与嘧啶的生物合成，本研究结果中蹄叶炎患病牛嘌呤代谢异常可能是由于氨基酸含量下降引起的。另外，经KEGG数据库查阅发现，甲烷代谢通路（ko00680，KEGG数据库登录号）中甘氨酸和丝氨酸可通过四氢叶酸转移一碳单位参与其中，因此在本试验中甲烷代谢通路在蹄叶炎奶牛中也表现出异常。

综上所述，氨基酸代谢在蹄叶炎的发生过程中至关重要，且甘氨酸/苏氨酸/丝氨酸代谢通路异常会影响其他几条通路，今后应给予进一步研究，为揭示蹄叶炎的机理和防治提供新的线索。

4 结论

本研究基于气相色谱/质谱联用技术的代谢组学研究方法，研究了奶牛发生蹄叶炎时血浆中小分子代谢物的改变，共有37种差异代谢物被鉴别出，其中3种物质上调，34种物质下调，主要涉及脂肪酸合成通路和甘氨酸/丝氨酸/苏氨酸代谢通路异常等。为进一步研究蹄叶炎的机理和防治提供新的线索。

（责任编辑　林鉴非）
（发表于《中国农业科学》，院选中文，4 255-4 264.）

牛源金黄色葡萄球菌耐药性与相关耐药基因和菌株毒力基因的相关性研究

杨 峰，王旭荣，李新圃，罗金印，刘龙海，张 哲，王 玲，张世栋，李宏胜*

（中国农业科学院兰州畜牧与兽药研究所/农业部新兽药工程重点实验室/
甘肃省中兽药工程技术研究中心，兰州 730050）

摘 要：分别采用纸片扩散法和PCR检测了牛源金黄色葡萄球菌对红霉素和四环素的耐药性与相关耐药基因和毒力基因，进而分析了耐药性与相关耐药基因和毒力基因之间的相关性。结果表明，37株金黄色葡萄球菌菌株中，红霉素耐药菌株占27%，而含有红霉素耐药基因ermB或ermC的菌株占19%，其中30%的红霉素耐药表型菌株含有ermB或ermC，所有菌株均未检测到emA；四环素耐药菌株占22%，而含有四环素耐药基因tetK的菌株占27%，其中75%的四环素表型耐药菌株含有tetK，所有菌株均未检测到tetM。金黄色葡萄球菌毒力基因lukED、hlb、hla、hld、ecfin、tst和lukPV的检出率分别为89%、73%、70%、70%、14%、5%和3%，未检测到eta、etb和lukM基因。统计学分析结果表明，金黄色葡萄球菌的红霉素耐药表型与相关耐药基因之间未发现相关性，而与毒力基因tst和lukED的相关性较强；金黄色葡萄球菌四环素耐药表型与耐药基因tetK存在相关性，而未发现四环素耐药表型与菌株毒力基因之间存在相关性。结果提示，金黄色葡萄球菌对红霉素和四环素的耐药率较高；耐药基因并不能作为菌株表型耐药的诊断指标，但其潜在的危险应当引起关注；此外，金黄色葡萄球菌菌株中高频率的毒力基因lukED、hla、hlb和hld编码的毒力因子白细胞毒素和溶血素是引发奶牛乳腺炎的主要因素。

关键词：金黄色葡萄球菌；耐药性；耐药基因；毒力基因；毒力因子

基金项目：甘肃省青年科技基金计划项目（145RJYA311）；中央级公益性科研院所基本科研业务费专项资金项目（1610322015-007）；"十二五"国家科技支撑计划项目（2012BAD12B03）
作者简介：杨峰（1985— ），男，陕西榆林人，助研，硕士。* 通信作者，Tel：0931-2115262，E-mail：lihsheng@sina.com
DOI：10.16656/j.issn.1673-4696.2016.02.020

Study on Relationship Between Resistance and Resistance-and Virulence-related Genes in *Staphylococcus aureus* from Bovine Mastitis Cases

YANG Feng, WANG Xurong, LI Xinpu, LUO Jinyin, LIU Longhai, ZHANG Zhe, WANG Ling, ZHANG Shidong, LI Hongsheng

(Key Laboratory of Veterinary Pharmaceutical Development, Ministry of Agriculture/Engineering & Technology Research Center of Traditional Chinese Veterinary Medicine of Gansu Province/Lanzhou Institute of Husbandry and Pharmaceutical Sciences, Chinese Academy of Agricultural Sciences, Lanzhou 730050, China)

Abstract: The erythromycin and tetracycline resistance were determined by disk diffusion method, andresis tance-and virulence-related genes were detected by PCR to analyze the relationship between the resistance and resistance-and virulence-related genes in *Staphylococcus aureus* isolated from bovine mastitis cases. The results showed that, in 37 *S. aureus* strains, 27% were resistant to erythromycin. ermB or ermC were detected in 19% isolates and found in 30% of erythromycin-resistant isolates. None isolate carried ermA was detected. Meanwhile, 22% of S. aureus isolates were resistant to tetracycline. And 27% isolates carried tetK, which was detected in 75% tetracycline-resistant strains. tetM was not found in any isolates. Virulence genes lukED, hlb, hla, hld, edin, tst and lukPV were detected in 89%, 73%, 70%, 70%, 14%, 5% and 3% of *S. aureus* isolates, respectively. eta, etb and lukM genes were not found in any isolates. The erythromycin resistance and virulence genes tst and lukED showed high relationship. In addition, the relationship between resistance and resistance gene tetK was detected. These results indicated that *S. aureus* exhibited high resistance rates to erythromycin and tetracycline. Although resistance genes can not be used as a diagnostic indicator for phenotypic antimicrobial susceptibility patterns, their potential threat should draw public attention. In addition, leukotoxins and hemolysins encoded by frequent virulence genes lukED, hla, hlb and hld played a crucial role in pathogenesis of bovine mastitis.

Key words: *Staphylococcus aureus*; Resistance; Resistance gene; Virulence gene; Virulence factor

金黄色葡萄球菌是引发奶牛乳腺炎常见的病原菌,给奶牛养殖业造成了巨大的经济损失[1-2]。抗生素是治疗葡萄球菌性奶牛乳腺炎的首选途径[3]。其中,红霉素和四环素经常被用于治疗葡萄球菌性奶牛乳腺炎,然而,由于生产中不合理的滥用,导致金黄色葡萄球菌耐药性菌株逐年增多[4],致使抗生素疗法对葡萄球菌性奶牛乳腺炎的治愈率降低[5]。

金黄色葡萄球菌对抗生素的反应主要依赖于菌株自身携带的各种耐药相关基因,其致病性主要与其毒力相关基因有关[6]。红霉素耐药基因ermA、ermB和ermC可编码N6-甲基转移酶，通过二甲基化23S rRNA上的腺嘌呤，降低对抗生素的亲和力。四环素耐药基因fetK和fetM分别编码四环素外排蛋白和核糖体保护蛋白，通过激活外排泵和保护核糖体从而达到耐药的效果。目前，已报道的金黄色葡萄球菌毒力相关基因有40多种[7]，它们编码的表面蛋白、细胞蛋白、蛋白酶和毒素等毒力因子，可协助细菌快速入侵宿主体内，并在宿主体内长时间存活[8]。

近年来，金黄色葡萄球菌耐药性的分子机制研究日益受到重视，然而有关国内西北地区牛源金黄色葡萄球菌耐药性与其基因特征的研究报道甚少。本研究以中国西北地区分离的牛源金黄色葡萄球菌为研究对象，研究其对红霉素和四环素的耐药性、检测相关耐药基因和毒力基因，进一步分析耐药表型与相关耐药基因、耐药表型与菌株毒力基因之间的相关性，旨在为阐明国内西北地区金黄色葡萄球菌性奶牛乳腺炎的发病机理奠定基础，为采取合理的防治策略提供科学依据。

1 材料与方法

1.1 菌株

37株金黄色葡萄球菌菌株于2015年从甘肃、宁夏采集的奶牛乳腺炎乳汁样品中分离获得，由中国农业科学院兰州畜牧与兽药研究所微生物课题组分离鉴定和保存。

1.2 主要试剂

药敏纸片为Oxoid公司产品。Premix Ex Taq DNA聚合酶、DNA Mrker均为宝生物工程（大连）有限公司产品。细菌总DNA提取试剂盒为Omega公司产品。

1.3 引物的设计与合成

红霉素耐药相关的基因引物参照Sutclife等[9]的方法设计，四环素耐药相关的基因引物参照Strom-menger等[10]的方法设计，毒力相关基因引物参照Jar-raud等[11]的方法设计，具体引物序列见表1。上述引物均由北京六合华大基因科技股份有限公司合成。

1.4 药物敏感性检测

按美国临床实验标准委员会标准[12]，采用纸片扩散法进行药敏试验。药敏纸片包括红霉素和四环素，购自英国Oxoid公司。

1.5 细菌总DNA的提取

将本研究中所用的37株金黄色葡萄球菌分离株在营养肉汤中复壮，分别取对数生长期菌液1.8mL提取细菌总DNA，操作严格按DNA提取试剂盒说明书进行，DNA产物在10g/L琼脂糖凝胶上电泳。

1.6 PCR扩增

按Ex Taq DNA聚合酶使用说明进行各耐药基因和毒力相关基因的PCR扩增，反应体系为25μL：Premix Ex Taq 12.5μL，上、下游引物各0.5μL，DNA模板2μL，ddH$_2$O 9.5μL。反应结束后，取PCR产物5μL在10g/L琼脂糖凝胶上电泳。

1.7 统计分析

使用SPSS 17.0中Fisher's exact test进行相关性分析。

表1 金黄色葡萄球菌耐药基因和毒力基因的引物序列

Table 1 Target genes and oligonucleotide primers used to amplify resistance-and virulence genes of *S. aureus* isolates

基因名称 Gene names	引物序列(5'→3') Forward and reverse primer seaquences	片段大小/bp Amplimer size	退火温度/℃ Annealing temperature
ermA	TCTAAAAAGCATGTAAAAGAA CTTCGATAGTTTATTAATATTAGT	645	45
ermB	GAAAAGGTACTCAACCAAATA AGTAACGGTACTTAAATTGTTTAC	639	45
ermC	TCAAAACATAATATAGATAAA GCTAATATTGTTTAAATCGTCAAT	642	45
tetK	GTAGCGACAATAGGTAATAGT GTAGTGACAATAAACCTCCTA	360	48
tetM	AGTGGAGCGATTACAGAA CATATGTCCTGGCGTGTCTA	158	48
tst	TTCACTATTTGTAAAAGTGTCAGACCCACT TACTAATGAATTTTTTTATCGTAAGCCCTT	180	53
eta	ACTGTAGGAGCTAGTGCATTTGT TGGATACTTTTGTCTATCTTTTTCATCAAC	190	53
etb	CAGATAAAGAGCTTTATACACACATTAC GTGAACTTATCTTTCTATTGAAAAACACTC	612	55
lukPV	TCATTAGGTAAAATGTCTGGACATGATCCA GCATCAASTGTATTGGATAGCAAAAGC	433	55
lukED	TGAAAAAGGTTCAAAGTTGATACGAG TGTATTCGATAGCAAAAGCAGTGCA	269	54
lukM	TGGATGTTACCTATGCAACCTAC GTTCGTTTCCATATAATGAATCACTAC	780	55
hla	CTGATTACTATCCAAGAAATTCGATTG CTTTCCAGCCTACTTTTTTATCAGT	209	54
hlb	GTGCACTTACTGACAATAGTGC GTTGATGAGTAGCTACCTTCAGT	309	54
hld	AGAATTTTTATCTTAATTAAGGAAGGAGTG TTAGTGAATTTGTTCACTGTGTCGA	111	54
edin	GAAGTATCTAATACTTCTTTAGCAGC TCATTTGACAATTCTACACTTCCAAC	625	55

2 结果

2.1 金黄色葡萄球菌药物敏感性检测

在37株金黄色葡萄球菌分离株中，对红霉素和四环素耐药的菌株分别有10株（27%）和8株（22%），其中1株（3%）对红霉素和四环素同时耐药。

2.2 金黄色葡萄球菌红霉素耐药性与相关耐药基因之间的相关性

在37株金黄色葡萄球菌中，含有红霉素耐药基因的菌株共7株（19%），其中3株含有ermB，4株含有ermC，未检测到耐药基因ermA。红霉素表型耐药菌株中，1株含有ermB，2株含有ermC；而红霉素敏感表型菌株中，分别有2株含有ermB和ermC。红霉素耐药性与相关耐药基因之间未发现相关性（表2）。

表 2 金黄色葡萄球菌红霉素耐药性和耐药基因的相关性

Table 2 Relationship between erythromycin resistance and related genes in *S. aureus* isolates

耐药基因 Resistance genes	敏感性 Antibiotic susceptibility		P值 P value
	敏感表型（27）/株 Sensitive	耐药表型（10）/株 Resistant	
ermA	0	0	-
ermB	2	1	1
ermC	2	2	0.291

-表示无法计算，因为耐药基因是一个常量

-i ndi cat e P value cannot be calculated（at least one variable is constant）

2.3 金黄色葡萄球菌四环素耐药表型与相关耐药基因之间的相关性

在37株金黄色葡萄球菌中，含有四环素耐药基因tetK的菌株为10株（2%），其中6株为四环素耐药菌株，4株为四环素敏感菌株。未检测到耐药基因tetM。四环素耐药性与耐药基因tetK存在相关性（$P<0.1$），而与耐药基因tetM未发现相关性（表3）。

表 3 金黄色葡萄球菌四环素耐药性和耐药基因的相关性

Table 3 Relationship between tetracycline resistance and related genes in *S. aureus* isolates

耐药基因 Resistance genes	敏感性 Antibiotic susceptibility		P值 P value
	敏感表型（29）/株 Sensitive	耐药表型（8）/株 Resistant	
tetK	4	6	0.046*
tetM	0	0	-

2.4 金黄色葡萄球菌毒力基因的检测分析

如表4所示，在37株金黄色葡萄球菌中，含有毒力基因lukED、hlb、hla、hld、edin、tstlukPV的菌株分别为33株（89%）、27株（73%）、26株（70%）、26株（70%）、5株（14%）、2株（5%）、1株（3%）；所有菌株均未检测到毒力基因eta，etb和lukM。

2.5 金黄色葡萄球菌红霉素耐药性与毒力基因之间的相关性分析

在10株红霉素耐药表型菌株中，含有毒力基因lukED、HLb、HLd、HLa、tst和LUKpv的菌株分别占70%、70%、70%、60%、20%和10%；27株红霉素敏感表型菌株中含有毒力基因lukED、hla、hlb、hld和edin的菌株分别占96%、74%、74%、70%和19%。红霉素耐药性与毒力基因tst和lukED有较强的相关性（$P<0.1$），未发现与其他毒力基因之间的相关性（表4）。

2.6 金黄色葡萄球菌四环素耐药性与毒力基因之间的相关性分析

在8株四环素耐药表型菌株中，含有毒力基因lukED、hlb、hld和hla的菌株分别占88%、88%、75%和75%；29株四环素敏感菌表型菌株中，含有毒力基因lukED、hlb、hla、hld、tst和lukPV的菌株分别占90%、72%、69%、66%、7%和3%。未发现四环素耐药性与毒力基因之间存在相关性（表4）。

3 讨论

大量耐药性菌株的出现成为金黄色葡萄球菌性奶牛乳腺炎防控的瓶颈。药物敏感性检测不仅可以呈现金黄色葡萄球菌的耐药现状，同时也为兽医工作者选择适宜的治疗药物提供了依据[13-14]。本试验中，金黄色葡萄球菌对红霉素和四环素的耐药率较高，与国内外其他学者的报道相近[15-16]。该结果可能与红霉素和四环素长期、高频、不规范地用于治疗奶牛乳腺炎密切相关，此外，未经诊断而盲目地使用不适当的抗生素也会导致大量耐药菌株的出现[17]。

金黄色葡萄球菌常见耐药机制包括药物捕获、药物靶标的修饰、酶的失活和外排泵，均依赖于耐药基因编码的相关蛋白[18-19]。本研究中，四分之一左右的菌株分别含有红霉素和四环素耐药基因，只有四环素耐药性与耐药基因tetK存在关联。有些耐药菌株未检测到相关耐药基因，而有些敏感菌株却含有相关的耐药基因，这与Frey等[20]和Gao等[21]的报道一致。菌株表型耐药但未检测到相关耐药基因，可能与其他耐药机制如生物膜的生成有关[22]；表型敏感却含有相关耐药基因，可能是由于该耐药基因未被激活，从而不能编码实现耐药的相关蛋白[23]。

表4 金黄色葡萄球菌耐药性和毒力基因的相关性

Table 4 Relationships between drug resistance and virulence genes in *S. aureus* isolates

毒力基因 Virulence genes	红霉素Erythromycin		P值 P value	四环素Tetracycline		P值 P value
	敏感表型（27）/株 Sensitive	耐药表型（10）/株 Resistant		敏感表型（29）/株 Sensitive	耐药表型（8）/株 Resistant	
tst	0(0)	2(20)	0.068*	2(7)	0(0)	1
eta	0(0)	0(0)	−	0(0)	0(0)	−

(continued)

毒力基因 Virulence genes	红霉素 Erythromycin		P值 P value	四环素 Tetracycline		P值 P value
	敏感表型(27)/株 Sensitive	耐药表型(10)/株 Resistant		敏感表型(29)/株 Sensitive	耐药表型(8)/株 Resistant	
etb	0(0)	0(0)	—	0(0)	0(0)	—
lukPV	0(0)	1(10)	0.270	1(3)	0(0)	1
lukED	26(96)	7(70)	0.052*	26(90)	7(88)	0.541
lukM	0(0)	0(0)	—	0(0)	0(0)	—
hla	20(74)	6(60)	0.442	20(69)	6(75)	1
hlb	20(74)	7(70)	1	21(72)	6(75)	1
hld	19(70)	7(70)	1	19(66)	7(88)	0.391
edin	5(19)	0(0)	1	0(0)	0(0)	—

在37株金黄色葡萄球菌中，含有毒力相关基因lukED、hla、hlb和hld的菌株均达70%以上，这与国内东部地区和广西的报道数据相近[24-25]。likED编码的双组分白细胞毒素可杀死宿主防御细菌感染的先天免疫细胞[26]。hla、hlb和hld编码的溶血素可以协助金黄色葡萄球菌快速入侵宿主细胞并引发炎症[27]。故此推断，白细胞毒素和溶血素在金黄色葡萄球菌引发奶牛乳腺炎的过程中起着重要的作用。红霉素耐药性与毒力基因tst和likED存在较强的关联，表明红霉素耐药菌株较敏感菌株含有更多的毒力基因tst和lukED。

综上所述，金黄色葡萄球菌对常用抗生素红霉素和四环素的耐药率较高，建议临床上给药前先做诊断，以便选择合适的抗菌药物。耐药基因并不能作为菌株表型耐药的诊断指标，但其潜在的危险应当引起关注。虽然未能检测所有的毒力相关基因，但高频率的lukED、hla、hlb和hld暗示着其在金黄色葡萄球菌致病性中的重要作用。

（责任编辑　胡弘博）

（发表于《中国兽医科学》，院选中文，247-252.）

SAA与HP在奶牛活体和离体炎性子宫内膜上皮细胞中表达的研究

张世栋，董书伟，王东升，闫宝琪，杨 峰，严作廷*，杨志强*

（中国农业科学院兰州畜牧与兽药研究所/农业部兽用药物创制重点实验室
甘肃省中兽药工程技术研究中心，兰州 730050）

摘 要：为研究奶牛子宫内膜炎时急性期蛋白SAA和HP的分子表达规律，采集奶牛子宫组织，通过病理学研究鉴定出不同炎症强度的子宫组织样本；同时体外培养奶牛子宫内膜上皮细胞，以致炎因子细菌脂多糖（LPS）诱导建立了细胞炎症模型。应用RT-qPCR检测子宫组织和细胞炎症模型中SAA和HP基因的表达变化，以Western-blot检测子宫组织中SAA蛋白的表达变化。结果显示，在子宫组织中，SAA基因与蛋白的表达水平均与子宫内膜炎强度呈正相关，然而HP基因表达水平与组织炎症强度无显著相关性。此外，在子宫内膜上皮细胞体外炎症模型中，SAA基因表达水平与LPS剂量呈正相关，而HP基因表达水平呈负相关。结果表明，子宫组织中和体外培养的奶牛子宫内膜上皮细胞中都能表达SAA和HP。作为急性期反应的生物标记，与SAA相比，HP可能对炎症更敏感，其基因表达更早更迅速。

关键词：子宫内膜炎；炎性子宫内膜上皮细胞；细胞炎症模型；急性期蛋白

Study on expression of SAA and HP of in vivo and in vitro dairy cow inflammatory endometrial epithelial cells

ZHANG Shidong, DONG Shuwei, WANG Dongsheng,
YAN Baoqi, YANG Feng, YAN Zuoting, YANG Zhiqiang

(Key Laboratory of Veterinary Pharmaceutics Discovery, Ministry of Agriculture/Engineering
& Technology Research Center of Traditional Chinese Veterinary Medicine of Gansu Province/

Lanzhou Institute of Husbandry and Pharmaceutical Science, Chinese Academy of Agricultural Sciences, Lanzhou 730050, China)

Abstract: To study molecular expression patterns of acute phase proteins serum amyloid A (SAA) and haptoglobin (HP) in dairy cow's endometritis, uterine tissues were collected, and different inflammation magnitude were also identified by pathological research. At the same time, endometrial epithelial cells (EECs) were cultured in vitro, and a cell inflammation model was established through inducing by proinflammation factor LPS. Gene expression of SAA and HP was determined by RT-qPCR in both of uterine tissues and cell inflammation model, and protein expression of SAA in uterine tissues was also detected by Western-blot. The results indicated that the levels of genes' expression and protein expression of SAA were positively related to inflammation magnitude in uterine tissue, and gene expression of HP, however, was not significantly related to inflammation magnitude of tissue. Additionally, correlation Between gene expression of SAA in cell inflammation model and LPS dosage was still positive, but that of HP was negative. The results showed that genes expression of SAA and HP could take place in both uterine tissue and cultured EECs in vitro. As biomarkers of acute phase response, HP may be more sensitive to inflammation, and its gene expression was also earlier and faster compared to SAA.

Key words: Endometritis; Inflammatory endometrial epithelial cells; Cell inflammation model; Acute phase proteins

子宫内膜炎是奶牛产后多发临床病[1]。国外的临床研究报道，超过40%的产后牛在2周内患有子宫炎，20%左右的牛在3周内患有子宫内膜炎[2-3]。与国外的放牧技术相比，我国普遍采用的集中圈养环境使得奶牛子宫内膜炎患病率实际比国外更高。子宫内膜炎往往导致母牛子宫组织的炎性损伤、胚胎死亡、早产、流产、推迟卵巢周期、延长黄体期、以及降低妊娠率等不良后果[4-6]。所以，子宫内膜炎严重损害奶牛的繁殖性能[7]，阻碍了奶牛养殖业的健康发展[8]，其诊断与治疗相关费用也增加了奶牛养殖业额外的经济负担[9]。

导致奶牛子宫内膜炎的因素很多，比如难产、酮病、分娩等[10]，但细菌感染被认为是主要原因，且最常见的病原菌主要是大肠杆菌和化脓性链球菌[11-13]。尽管不同种属的细菌对子宫疾病的影响没有被完全地揭示[14]，但是细菌感染引发的不受调控的炎症反应会导致子宫炎[15]。由于炎性因子的刺激作用，在炎症的不同阶段均有急性相蛋白（acute phase proteins，APPs）产生，在众多的APPs中，血清淀粉样蛋白（serum amyloid，ASAA）和触珠蛋白（haptoglobin，HP）是最主要的成员[16]。SAA对机体感染革兰阴性和阳性细菌后都具有明显的免疫调节作用[17]。HP在机体中不仅能调节白细胞先天性免疫反应，还具有抗炎、直接抑菌作用和分子伴侣活性[18]。通常状况下，SAA和HP的表达直接反映机体身体部位的病理状态，在机体受到感染和出现炎症时其表达量会增强[19]。因此，APPs与奶牛子宫内膜炎的相关性一直是人们关注的研究热点之一。据报道，非病理状态时，牛的所有组织中几乎都有低水平的SAA表达[20]，而患有子宫内膜炎时牛血清SAA与HP表达水平均显著升高[21]。

本研究采集了临床子宫内膜炎患牛的子宫组织，并进行了急性期蛋白SAA、HP的基因表达检测。同时，在体外培养了奶牛子宫内膜上皮细胞（endome-trial epithelial cells，EECs），以细菌内毒素LPS诱导细胞炎症反应，并检测了SAA和HP基因在体外炎性子宫内膜上皮细胞中的表达变化，为进一步揭示急性相蛋白与奶牛子宫内膜炎的相关性奠定试验基础。

1 材料与方法

1.1 主要试剂和仪器

脂多糖（LPS，E. coliO111：B4）、噻唑兰（MTT）均购自Sigma（上海）公司。胎牛血清、DMEM/F12培养基、胰蛋白酶均是美国Gibco公司产品。牛TNF-α和IL-1β的ELISA检测试剂盒均购自上海蓝基生物科技有限公司。TRIzol购自美国Invitrogen公司。RT-PCR试剂盒、SYBR Green PCR Master Mix试剂均为宝生物工程（大连）有限公司产品。引物由北京六合华大基因科技股份有限公司合成。脱脂奶粉为BD公司产品。硝酸纤维素膜（NC膜）为美国Millipore公司产品。组织总蛋白提取试剂盒购自Best-Bio公司（中国上海）。小鼠抗牛beta-actin单克隆抗体（一抗）购自美国Abcam（ab8226）。SAA山羊多克隆抗体（一抗）购自美国Santa Cruz公司。山羊抗小鼠IgG-HRP（二抗）购自西安壮志生物科技有限公司。家兔抗山羊IgG-HRP（二抗）购自武汉三鹰生物技术有限公司。DMI6000B倒置显微镜为德国Leica公司产品。Spectra Max M2/M2e多功能酶标仪为美国Molecular Devices公司产品。CFX96型定量PCR仪、Trans-Blot SD半干分子转印槽均为美国Bio-Rad公司产品。HeaL Force HF90型CO_2培养箱为中国上海力康有限公司产品。超微量分光光度计Nanodrop2000为Thermo Scientific公司产品。

1.2 动物组织采集

奶牛子宫组织采集于宁夏某屠宰场。将采集到的子宫组织沿基质层横向切开，取子宫内膜层所在组织块速冻于液氮中；同样的另一块组织固定于100mL/L福尔马林溶液中。

1.3 组织病理学观察

取出经福尔马林固定好的组织块，适当修整后进行常规脱水、浸蜡、包埋、切片，然后常规苏木精—伊红（HE）染色后，树胶封片，于光学显微镜下观察并拍照记录。经病理学观察，将病程相似的组织归类用于研究。其中，正常子宫选3份，轻度炎症和重度炎症子宫各选3份。

1.4 子宫内膜上皮细胞的分离与培养

轻轻刮取子宫内膜表面组织，经缓冲液漂洗两次后，将组织转移至20mL胶原酶溶液中（250U/mL），37℃消化1~1.5h，然后将组织转移至直径60mm培养皿中，加入含500mL/L FBS的DMEM/F12培养液，静置于37℃、50mL/L CO_2饱和湿度的培养箱中培养。细胞成功存活后，主要为成纤维细胞与上皮样细胞混合生长。利用两者对胰酶敏感性的差异，以差时消化的方式经2次消化传代可得到纯度较高的上皮细胞。详细的上皮细胞分离培养与鉴定操作参照文献[22]进行。

1.5 LPS对奶牛子宫内膜上皮细胞活力的影响

取纯化培养的奶牛子宫内膜上皮细胞，经胰酶消化后制成$4×10^4$/mL的单细胞悬液，接种于96孔板，每孔100μL，置于CO_2培养箱中培养24h后，吸弃细胞代谢液，加入含有不同浓度LPS的新鲜培养液，作用48h后，用MTT法检测不同浓度LPS对细胞活力的影响。

1.6 体外细胞炎症模型的建立与评价

根据LPS对奶牛子宫内膜上皮细胞活力和增殖的影响，分别以不同剂量的LPS作用于细胞诱导炎症反应，并检测细胞代谢液中IL-1β和TNF-α的含量，以评价子宫内膜上皮细胞炎症模型。取子宫内膜上皮细胞制成$1×10^5$/mL的单细胞悬液，接种于12孔板，每孔0.5mL。培养液为DMEM/F12，含50mL/L胎牛血清、10ng/mL表皮生长因子、5μg/mL牛胰岛素。细胞预培养24h后，吸弃代谢液，分别加入含0、5、10μg/mL LPS的完全培养液，并继续培养细胞。分别在第24和48小时吸取上清液，1 000r/min离心弃沉淀，然后按照ELISA试剂盒说明书操作，检测各组上清液中IL-1β和TNF-α的含量。

1.7 基因表达水平的实时荧光定量PCR检测

应用TRIzol试剂法提取组织和细胞总RNA。将液氮中冻存的组织块取出置于冰块上缓慢融化，然后切取适量组织块并转移至无RNA酶的玻璃匀浆器中，并加入1mL TRIzol试剂，然后进行手动研磨。组织研磨液按照TRIzol试剂操作说明进行总RNA提取。细胞总RNA提取时，首先弃去代谢液，用PBS漂洗细胞2次后，加入TRIzol试剂，室温放置5min后，并用移液枪缓慢地反复吹吸液体使细胞裂解，裂解液亦按照TRIzol试剂操作说明进行总RNA提取。提取总RNA后，立即进行甲醛变性琼脂糖凝胶电泳检测总RNA的提取效果，结果显示总RNA结构完整、无降解。用超微量分光光度计Nanodrop2000检测总RNA含量，并按试剂盒说明书，取500ng总RNA并将mRNA反转录为cDNA，得到的cDNA用于RT-PCR检测SAA和HP基因的相对表达水平。PCR引物信息见表1。PCR反应条件是95℃预变性3min；95℃变性15s，60.5℃或61.0℃退火20s，72℃延伸20s，循环39次。mRNA的相对表达量采用$2^{-\Delta\Delta Ct}$方法计算，并以beta-actin（ACT）为内参基因。

表1 基因扩增引物信息

Table 1 Information of primers used in PCR

基因登录号 Accession No.	基因名称 Gene names	产物长度/bp Product size	退火温度/℃ Annealing temperature	正向序列 Forword	反向序列 Reverse
gb|NM_173979.3	ACT	143	60.5	ATCGGCAATGAG-CGGTTCC	GTGTTGGCGTAGAG-GTCCTTG
gb|AF540564.1	SM	169	60.5	TATGACCAGG-GACCAGGTAC	CTCTGCCCTTAGG-GACTCACAG
gb|NM_001040470.2	HP	128	61.0	GACCAGGTGCA-GAGGATCA	GTGAGGAGCCATC-GTTCATT

1.8 SAA蛋白表达产物的Western-blot检测

取液氮中冻存的组织块，在冰块上将其缓慢融化后，切取适量组织，按照总蛋白提取试剂盒操作说明，用玻璃匀浆器研磨组织。离心匀浆液获得蛋白上清溶液，用BCA法测定各组蛋白质浓度，各组取20μg蛋白样品与SDS蛋白上样缓冲液混匀，沸水中变性3~5min后，在10%分离胶、5%浓缩胶中进行SDS-PAGE电泳，100V 2.5h。电泳结束后，将胶片放入半干转印槽，12V 1h将蛋白转印至NC膜上，用50g/L的脱脂奶粉溶液封闭转印膜1h后，加入一抗（50g/L脱脂奶粉溶液配制），4℃过夜孵育，TBST洗膜4次，每次10min，加入HRP标记的二抗（50g/L脱脂奶粉溶液配制），室温缓慢摇动孵育2h，TBST洗涤4次后，往NC膜上滴加ECL发光液，立即转移至暗室，胶片曝光、显影、定影后显示目标蛋白的表达水平。

1.9 统计分析

实验重复3次，以$\bar{x}\pm s$表示，组间采用单因素方差（One-Way ANOVA）分析，$P<0.05$为差异具有统计学意义。

2 结果

2.1 子宫的组织病理学观察

图1A是正常状态下的子宫组织结构，可看到黏膜上皮细胞呈单层柱状排列，子宫内膜腺内可见到较多的分泌物，固有层可见到少量淋巴细胞及少量红细胞。子宫患有较轻炎症时（见图1B），固有层成纤维细胞增生，淋巴细胞、浆细胞浸润，子宫内膜腺萎缩，腺上皮细胞变性、坏死，部分脱落于管腔；固有层结缔组织大量增生，部分成纤维细胞坏死，大量淋巴细胞、浆细胞浸润，可见少量巨噬细胞与中性粒细胞。当子宫炎症较重时（图1C），组织病理学显示，固有层成纤维细胞增生，部分区域坏死，深染伊红，呈均质玻璃样，有较多的红细胞；固有层亦可见较多的淋巴细胞、浆细胞和少量的中性粒细胞；部分子宫内膜腺萎缩，腺上皮细胞变性、坏死、脱落。

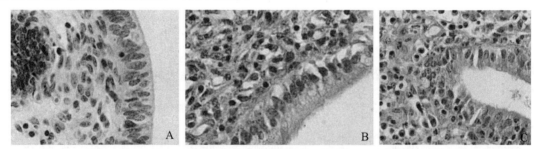

A：正常子宫内膜结构；B：轻度炎症时的子宫内膜组织结构；C：重度炎症时子宫内膜组织结构
A: Normal uterus; B: Uterus with mild inflammation; C: Uterus with severe inflammation

图1 奶牛子宫内膜组织的病理学观察400×
Fig. 1 Pathology observation on dairy cow's uterine tissues

2.2 奶牛子宫内膜上皮细胞的体外培养

从图2中可以看出，纯化后的奶牛子宫内膜上皮细胞生长状态良好，形态呈铺路石状，表现出典型的上皮细胞特征，并与原代细胞形态保持一致。传代培养的第2~8代细胞均表现出稳定一致的生长状态。结果表明，分离的原代上皮细胞能够满足体外细胞学试验要求。

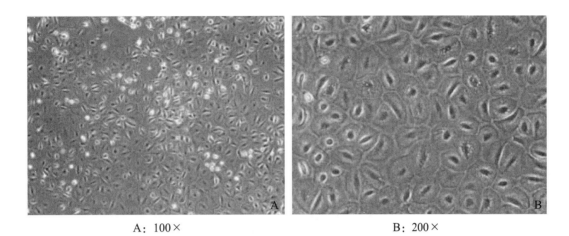

A：100× B：200×

图2 体外培养的奶牛子宫内膜上皮细胞
Fig. 2 Dairy cow's endometrial epithelial cells cultured in vitro

2.3 LPS对子宫内膜上皮细胞炎性因子分泌的影响

图3显示不同剂量LPS诱导子宫内膜上皮细胞一定时间后，细胞分泌产生的炎性细胞因子水平。从图3可见，LPS可诱导细胞炎性因子TNF-α的分泌量显著增高，且在第24和第48小时的分泌水平无显著差异。同样剂量的LPS亦可诱导细胞IL-1β分泌水平显著增高，且第48小时的分泌水平显著高于第24小时的水平。结果显示，10μg/mL以内的LPS诱导的细胞炎性反应与其剂量具有一定正相关性。

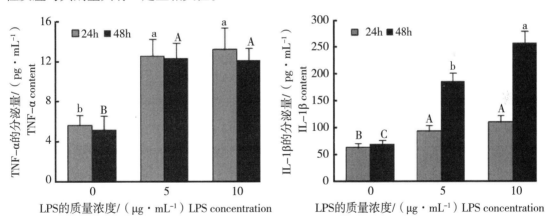

图3 LPS诱导子宫内膜上皮细胞分泌炎性细胞因子的变化
Fig. 3 Changes of inflammatory cytokines secreted from EECs induced by LPS

标注不同字母表示组间差异显著（P<0.05），标注相同字母表示组间差异不显著（P>0.05）。下图同
The difference between means with the different letters is significant（P<0.05），and the difference between means with the same letter is not significant（P>0.05）。The same as follows

2.4 LPS对子宫内膜上皮细胞中SAA与HP基因表达的影响

图4显示，随着LPS剂量的增大，SAA基因的表达水平显著高于对照组，10μg/mLLPS诱

导的SAA表达水平显著高于1和5μg/mL组的表达水平。然而，在1μg/mL的LPS诱导下，HP基因的表达水平最高，且LPS剂量越大，HP基因的表达水平越低。

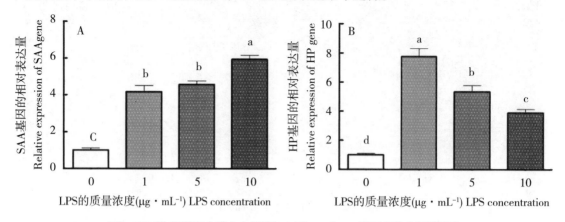

图4　LPS作用子宫内膜上皮细胞24 h后SAA与HP基因表达水平的变化
Fig. 4　Changes of gene expression level of SAA and HP in EECs induced by LPS

2.5　子宫组织中 SAA 与 HP 基因表达水平的实时荧光定量 PCR 检测

由图5可以看出，SAA基因在正常子宫（A）组织中表达水平很低，而在炎症子宫（B、C）组织中表达水平显著升高，且炎症越严重，表达水平越高。然而，HP基因在各组中的表达水平无显著差异（图5）。

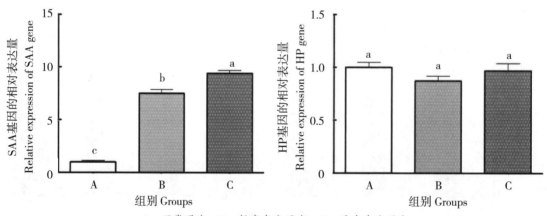

A：正常子宫；B：轻度炎症子宫；C：重度炎症子宫
A：Normal uterus；B：Uterus with mild inflammation；C：Uterus with severe inflammation

图5　子宫组织中SAA与HP基因的表达
Fig. 5　Gene expression of SAA and HP in uterine tissues

2.6　子宫组织中 SAA 蛋白的表达

Western-blot检测结果显示，正常子宫组织中SAA蛋白表达量很低（图6A），子宫炎症时其表达量显著提高，且炎症越严重SAA蛋白的表达量越高（图6B，C）。

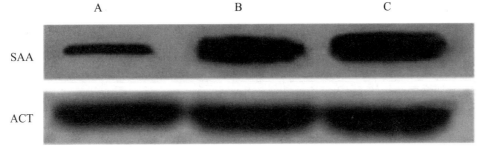

A：正常子宫组织；B：轻度炎症子宫组织；C：重度炎症子宫组织
A：Normal uterus；B：Uterus with mild inflammation；C：Uterus with severe inflammation

图6　子宫组织中SAA蛋白的表达
Fig. 6　Proteins expression of SAA in uterine tissues

3　讨论

　　动物子宫内膜在促进受精、受精卵着床、妊娠、以及分娩等正常的繁殖活动中具有重要作用[9]，而子宫组织中的炎症反应常常会损害子宫内膜对胚胎的容受性[23]。哺乳动物的子宫内膜是由柱状排列的单层上皮细胞组成的黏膜组织，也是子宫预防病原感染的第一道组织防线[1, 24]。当病原菌入侵时，子宫内膜上皮细胞是最先接触、识别和做出免疫应答的细胞，并作为起始防御细胞进一步介导子宫黏膜免疫系统开启先天免疫功能[9]。组织病理学研究表明，牛子宫内膜炎主要是发生在子宫内膜细胞的炎症过程[2]。所以，子宫内膜上皮细胞是子宫内膜功能研究的基本单元，也是子宫内膜炎机制研究的根本对象。

　　SAA和HP主要由肝产生，随着体液循环至机体各个部位。近几年的研究表明，在反刍动物中除肝以外的许多组织中也都有SAA和HP的表达，比如在奶牛乳腺组织、脂肪组织、前胃组织等，均有其基因转录本表达[25]。在本研究中，结合病理学研究（图1）结果显示，临床子宫内膜炎患牛子宫组织中SAA基因和蛋白表达都与炎症程度呈正相关，炎症越重，其表达量越高（图5、图6）；而HP在临床子宫内膜炎患牛子宫组织中的表达量却没有显著变化（图5）。然而，由于没有得到合适的牛HP蛋白抗体，所以不能直接检测到子宫组织中HP蛋白的表达量。

　　虽然SAA和HP在各种组织中被广泛表达，但是在体外培养的牛子宫内膜细胞中是否也存在表达仍然没有被证实[1]，而学者们也只是猜测子宫内膜细胞能表达SAA[26]。本研究通过体外培养奶牛子宫内膜上皮细胞（图2），并以致炎因子LPS诱导了细胞炎性反应（图3）。因为炎性细胞因子TNF-α和IL-1β的水平直接反映了炎症反应的强度，所以不同剂量LPS诱导的子宫内膜上皮细胞体外炎症模型亦可模拟不同病程的子宫内膜炎。本研究中对体外炎症细胞中SAA和HP基因表达的检测结果显示，SAA表达水平与上皮细胞炎症反应强度呈正相关（图4A），而HP表达水平与炎症强度呈负相关，在炎症强度最低时HP基因的表达水平最高（图4B），这说明HP基因表达与炎症早期相关性更大。作为急性期反应的分子标记，与SAA相比，可能HP对炎症更敏感，其基因表达更早更迅速；而炎症强度加大时，子宫内膜上皮细胞亦遭受到更大的炎性损伤，反而限制了HP基因的表达。此外，笔者认为，子宫内膜炎发生时，HP基因的高水平表达主要发生在子宫内膜上皮细胞中。因为在临床采集的子宫组织中，炎症强度实际远远高于体外细胞炎症模型，上皮细胞亦受到很强的炎性损伤，所以组织中HP基因的表达反而被限制在较低水平，而HP基因的表达可能主要体现了基质细胞中的表达水平，且子宫内膜炎并不能显著影响基质细胞中HP基因的表达水平（见图5B）。

总之，体外培养的奶牛子宫内膜上皮细胞中能表达急性期蛋白SAA和HP。体内组织和体外细胞中基因表达规律表明，HP可能是比SAA更敏感、表达更早更迅速的急性期蛋白，而且子宫中的HP更多是在子宫内膜上皮细胞中表达，而不是在基质细胞中。

（编辑　胡志敏）
（发表于《中国兽医科学》，院选中文，921–927.）

金黄地鼠动脉粥样硬化模型的病理观察和生化指标分析

马 宁，杨亚军，刘希望，秦 哲，孔晓军，李世宏，李剑勇*

（中国农业科学院兰州畜牧与兽药研究所/农业部兽用药物创制重点实验室/
甘肃省新药工程重点实验室，兰州 730050）

摘 要：探讨建立金黄地鼠动脉粥样硬化（AS）模型的方法，进而考察该模型病理和生化指标的变化，为该疾病的研究提供相应的试验材料。采用饲喂金黄地鼠纯化高脂饲料的方法建立AS疾病模型，通过主动脉弓的常规病理检查确定金黄地鼠疾病发展情况，并利用主成分分析（PCA）和聚类分析方法对血清生化指标进行分析。结果表明，模型组主动脉弓管壁明显增厚，管腔狭窄，内膜及中膜界限不清，并有大量的脂质积累，平滑肌细胞发生变性、坏死并呈灶性聚集。主成分分析结果将19项生化指标归为五类，分别代表血脂血糖、肾代谢、肝功、胆功及肌肉代谢相关指征。聚类分析结果与样品采集结果一致。生化指标统计结果表明，AS金黄地鼠脂质代谢紊乱，肝功和肾功相关指标差异显著。表明，通过饲喂纯化高脂饲料12周的方法能够建立金黄地鼠AS模型。血清生化指标分析结果表明，高脂饲料可导致金黄地鼠脂质代谢紊乱并可能引起肝肾功能的损伤。

关键词：动脉粥样硬化；金黄地鼠；主成分分析；聚类分析

Pathology and biochemical index analysis of atherosclerosis disease model in hamster

MA Ning, YANG Yajun, LIU Xiwang, QIN Zhe,
KONG Xiaojun, LI Shihong, LI Jianyong

(Key Laboratory of New Animal Drug Project of Gansu Province/Key Laboratory of Veterinary Pharmaceutical Development, Ministry of Agriculture/Lanzhou Institute of Husbandry and Pharmaceutical Science of Chinese Academy Agriculture Sciences, Lanzhou 730050, China)

基金项目：国家自然科学基金项目（31572573）
作者简介：马宁（1990— ），男，河北保定人，博士生，研究方向为兽医药理与毒理学。通信作者，研究员，研究方向为新兽药的研究与开发，Tel：0931-2115290，E-mail：lijy1971@163.com
DOI：10.16656/j.issn.1673-4696.2016.09.017

Abstract: In order to provide experimental material for atherosclerosis (AS) research, the method for establishing atherosclerosis (AS) disease model in hamster was investigated, and then its changes in biochemical indexes and pathology were analyzed. The method of feeding hamster with Western Diet was used to establish AS disease model. The development of AS in aortic arch was evaluated by pathological examination. Principal component analysis (PCA) and cluster analysis were employed in biochemical indexes analysis. The pathological results indicated that blood vessel wall and lumina in model group became thicker and narrower than that in healthy control group. There was no clear distinction between tunica intima and tunica media, and serious accumulation of fat was occurred in model group. Meanwhile, focal aggregation, degeneration and necrosis were observed in smooth muscle cells. Nineteen biochemical indexes were divided into five categories by PCA, which represented blood glucose and lipid, renal metabolism, liver function, gallbladder function and muscle metabolism. The results of cluster analysis were in correspondences with the sample collection. The statistical analysis of biochemical indexes showed that hamster with AS suffered from disorder of lipid metabolism. The indexes which were related with kidney and liver possessed significant differences from control group. In conclusion, the AS disease model in hamster could be successfully established by feeding with Western Diet for 12 weeks. The analysis results of biochemical indexes suggest that Western Diet causes disorder of lipid metabolism and induces damage to liver and kidney.

Key words: Atherosclerosis; Hamster; PCA; Cluster analysis

动脉粥样硬化（Atherosclerosis，AS）是由于脂质代谢紊乱所导致的严重危害人类健康的心血管疾病[1]。AS病理过程与多种因素相关，如高血脂、慢性炎症、血栓等[2]，其病理变化主要体现在动脉及其分支发生脂质沉淀，形成动脉粥样硬化斑块，从而导致动脉内膜增厚，管腔狭窄。自从发现叙利亚金黄地鼠的脂蛋白代谢与人类有许多共同之处后，国内外越来越多的血脂研究将其作为实验动物[3-4]。病理学研究能够展示疾病的发展规律以及发展过程中相应组织器官的病理变化，在医学研究中发挥着不可代替的作用[5-6]。血液生化参数作为疾病发生机理的重要指标和参数，其相关分析有助于识别靶器官的毒性和监测动物的健康状况，同时也可以为疾病的早期诊断提供预警信号[7-9]。本试验以纯化高脂饲料诱导AS金黄地鼠模型，观察主动脉的病理变化，并利用主成分分析和聚类分析对生化指标进行研究，进而探讨AS金黄地鼠模型的组织病理和血液生化特征。本试验可为动脉粥样硬化疾病的相关研究提供试验材料，并为其发病机制提供一定的理论基础。

1 材料与方法

1.1 实验动物

雄性叙利亚金黄地鼠21只，体重100～110g，购自北京维通利华实验动物技术有限公司。随机分2组，对照组11只，模型组10只。

1.2 AS模型的建立

对照组地鼠饲喂标准饲料，模型组地鼠饲喂高脂饲料各12周，诱导AS模型。啮齿类高纯度饲料（Western Diet）购自加拿大Research Diet公司。高脂饲料主要成分为17%蛋白质、43%碳水化合物和40%油脂（kcal%）。普通饲料购自北京维通利华实验动物技术有限公司，主要成分为24.4%蛋白质，63.3%碳水化合物和12.3%油脂（kcal%）。

1.3 主动脉弓与血液采集

饲喂金黄地鼠高脂饲料12周后，禁食10～12h，以1%的戊巴比妥钠腹腔注射麻醉各组地鼠。心脏取血，收集血液标本，4℃ 1 000×g离心15min，制备血清用于生化指标分析。完整采集主动脉弓，石蜡包埋、切片，HE染色，光镜观察。

1.4 生化指标检测

利用德国Erba公司生产的XL—640全自动生化分析仪对相关生化指标进行检测，其中包括高密度脂蛋白（HDL）、低密度脂蛋白（LDL）、尿酸（UA）、肌酐（CREA）、谷氨酰转移酶（GGT）、乳酸脱氢酶（LDH）、钙（Ca）、总胆红素（TBIL）、间接胆红素（DBIL）、丙氨酸氨基转移酶（ALT）、碱性磷酸酶（ALP）、天门冬氨酸氨基转移酶（AST）、总蛋白（TP）、白蛋白（ALB）、尿素氮（BUN）、血糖（GLU）、总胆固醇（TCH）、甘油三酯（TG）和肌酸激酶（CK）。所用生化试剂盒购自宁波美康生物科技股份有限公司。

1.5 数据分析

采用SPSS 13.0统计软件进行数据分析，主成分分析包括样品相关矩阵计算，各主成分的特征值、贡献率和积累贡献率的计算；聚类分析采用聚类树状图分析。组间差异采用非配对t检验。

2 结果

2.1 AS 组织病理

金黄地鼠主动脉弓组织病理检查见图1和图2。低倍镜下，相比于对照组，模型组血管壁明显增厚，造成动脉管腔狭窄。高倍镜下，对照组金黄地鼠主动脉壁各层结构正常，内皮完整、光滑，中层平滑肌细胞排列整齐。模型组主动脉弓内膜不完整，突出于管腔形成纤维帽，平滑肌细胞发生变性、坏死并呈灶性聚集，内膜、中膜界限不清且明显不规则增厚，细胞发生脂质沉淀并伴有大量的泡沫细胞生成。

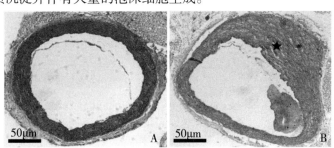

图1 低倍镜下对照组（A）和模型组（B）的主动脉弓病理变化HE50×

Fig. 1 Observation of pathological changes of aortic arch at low magnification on the control group（A）and the model group（B）

★：相比于对照组，模型组主动脉弓的血管壁显著增厚，血管管腔狭窄

★：When compared with controlgroup, blood vesselwalland lumina ofaortic arch in modelgroup became thicker and narrower

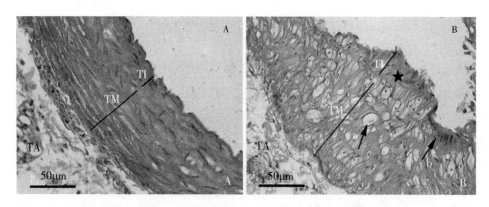

图2 高倍镜下对照组（A）和模型组（B）的主动脉弓病理变化HE400×
Fig. 2 Observation of pathological changes of aortic arch at high magnification on the control group (A) and the model group (B)

TA：外模；TM：中膜；TI：内膜；对照组内膜光滑，血管壁结构完整。模型组中，内膜突出于血管壁形成纤维帽（★），内膜与中膜的界限不清楚，大量泡沫细胞生成并发生严重的脂质沉淀（箭头a），管壁增厚，平滑肌细胞发生变性、坏死并呈灶性聚集（箭头b）

TA: Tunic adventitia; TM: Tunic media; TI: Tunic intima. The structure of blood vessel in control group was integrated and the TI was smooth. In model group, fibrous cap was formed in TI which out into blood vessel lumina (★). No clear distinction between TI and TM was observed in model group. Lots of foam cells were generated and serious accumulation of fat was occurred (arrow a), which thicken the blood vessel wall. Degeneration, focal aggregation (arrow b), and necrosis in smooth muscle cells were observed

2.2 血清生化指标的主成分分析与聚类分析

根据方差累计贡献率达到75%的原则，选择了5个主成分，其特征值分别为$\lambda_1=8.766$，$\lambda_2=2.026$，$\lambda_3=1.771$，$\lambda_4=1.328$，$\lambda_5=1.164$，前5个的特征值均大于1。从主成分的贡献率来看，第一至第五主成分的贡献率分别为46.137%，10.663%，9.321%，6.991%和6.128%，前5个主成分的累计贡献率达到79.24%，即所有变量中79.24%的信息能够被解释（表1）。因此，结合碎石图确定主成分的个数为5个（图3）。

表1 AS金黄地鼠血生化参数矩阵特征值、贡献率及累积贡献率
Table 1 The eigenvalue, contribution rate and cumulative contribution rate of biochemical indexes matrix in AS hamster

项目 Items	Z1	Z2	Z3	Z4	Z5
矩阵特征值Eigenvalue	8.766	2.026	1.771	1.328	1.164
贡献率Contribution rate	0.461 3	0.106 6	0.932 1	0.699 1	0.612 8
累积贡献率Cumulative contribution rate	0.461 3	0.568 0	0.661 2	0.731 1	0.792 4

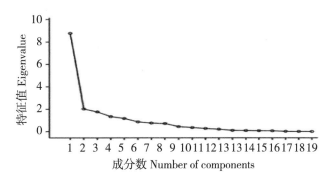

图3　AS金黄地鼠血生化指标特征值分布的碎石图
Fig. 3　The scree plot of biochemical index eigenvalue in AS hamster

5个主成分的表达式为：$Z1=0.800\times TG+0.912\times HDL+TCH\times0.894+0.794\times LDL+0.801\times GLU\cdots$；$Z2=0.537\times UA+0.019\times BUN\cdots$；$Z3=0.268\times TP+0.281\times ALT+0.271\times AST-0.246\times ALP+0.222\times LDH\cdots$；$Z4=0.541\times TBIL+0.372\times DBIL\cdots$；$Z5=0.475\times CK+0.445\times CREA\cdots$；在第一主成分Z1中，主要反映了与血脂和血糖相关指征；在第二主成分Z2中，主要反映了肾代谢相关的指征；第三主成分Z3主要体现了肝功相关的指征；Z4则主要代表与胆功相关的指征；第五主成分Z5则体现了肌肉代谢相关指征。

根据筛选出来的5个主成分对，对21个样本进行系统聚类分析，21个样本中1~11为对照组样本，12~21为模型组样本。系统聚类分析方法相关参数如下：数据的标准化处理采用Squared Euclidean Distance准则，Z Scores算法，聚类方法为Between-groups Linkage，以树状图显示最终结果（图4）。从树状图可以直观的看出，所有的观测血清样本被分为两大类，1~11为一类样本，12~21为一类样本。采用主成分分析法，对血清样本生化指标聚类分析的结果，与样本采集的分类结果保持一致。

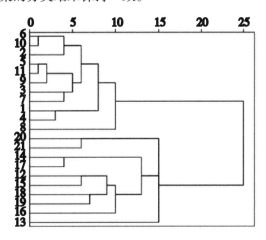

图4　聚类分析结果树状图
Fig. 4　The dendrogram of cluster result

样本1~11来自对照组，12~21号来自模型组。21个样本被分为两大类，且聚类分析的结果与样本采集结果保持一致

Samples of 1 to 11 were from control group, and 12 to 21 were from model group. 21 samples were divided into two categories, and the results of cluster analysis were in correspondences with the sample collection

2.3 血清生化指标的非配对 t 检验

通过与对照组的血生化指标进行比较分析发现，在模型组中，TG、HDL、LDL、TCH、TBIL、DBIL、TP、ALB、ALT、AST、LDH、UA、GLU指标显著升高（P<0.05或P<0.01）；BUN、CREA指标显著降低（P<0.01），详细结果见表2。

表 2 血清生化指标的统计学分析
Table 2 The statistical analysis of serum biochemical indexes

变量 Variables	单位 Units	对照 Control (n=11)	模型 Model (n=10)
TG	mmol·L^{-1}	1.41 ± 0.35	2.71 ± 0.67^A
HDL	mmol·L^{-1}	0.99 ± 0.22	1.95 ± 0.06^A
LDL	mmol·L^{-1}	0.32 ± 0.09	0.61 ± 0.12^A
TCH	mmol·L^{-1}	1.72 ± 0.37	3.48 ± 0.52^A
TBIL	mmol·L^{-1}	1.07 ± 0.24	1.88 ± 0.71^A
DBIL	mmol·L^{-1}	1.04 ± 0.22	2.1 ± 0.83^A
TP	g·L^{-1}	38.45 ± 5.04	49.86 ± 8.08^A
ALB	g·L^{-1}	18.21 ± 3.48	28.13 ± 4.26^A
ALT	U·L^{-1}	48.34 ± 8.25	70.06 ± 15.42^A
AST	U·L^{-1}	70.1 ± 15.1	110.3 ± 33.1^A
ALP	U·L^{-1}	61.3 ± 10.2	65.1 ± 12.0
GGT	U·L^{-1}	4.75 ± 1.38	3.9 ± 0.76
LDH	U·L^{-1}	143.1 ± 39.8	225.1 ± 59.3^A
CK	U·L^{-1}	2 348 ± 444	2 341 ± 325
BUN	mmol·L^{-1}	7.33 ± 1.11	5.47 ± 0.76^A
CREA	umol·L^{-1}	21.12 ± 2.84	19.25 ± 4.42^A
UA	umol·L^{-1}	0.05 ± 0.02	0.08 ± 0.03^a
GLU	mmol·L^{-1}	5.88 ± 1.23	10.19 ± 1.95^A
Ca	mmol·L^{-1}	2.46 ± 0.18	2.57 ± 0.20

与对照组比较，a：P<0.05，A：P<0.01
Compared the control group, a：P<0.05, A：P<0.01

3 讨论

AS是一种与脂质代谢密切相关的慢性、多发性和常见性心血管疾病，严重危害人类健康[10]。建立一种与人类脂质代谢相关的AS疾病动物模型，具有十分重要的意义。金黄地鼠对纯化饲料中的饱和脂肪酸高度敏感，能够有效增加血液中胆固醇的含量，从而导致动脉血管中胆固醇的积累致使发生相应的病理变化[11]。金黄地鼠肝胆固醇合成速度、载脂蛋白、胆固醇酯转运蛋白以及受体依赖型LDL的运输均与人类相似[12-13]。因此，利用金黄地鼠建立的高血脂和AS疾病模型在血脂相关研究中被普遍使用，并逐步得到认可[14-16]。本试验中通过纯化高脂饲料成功建立了AS实验模型。组织病理结果表明，脂质在主动脉弓血管壁中发生严重积累，造成管壁增厚，管腔狭小。该结果证实，本方法可以在较短时间内造成金黄地鼠发生动脉粥样硬化病变，并且方法简单，成本低，可为动脉粥样硬化疾病的研究提供相应的试验材料。

主成分分析通过降维方法，获得尽可能少的新的综合变量。在反映或者解释原有变量大部分信息的基础上，综合变量能够用来发掘研究对象和变量之间的关系[17]。聚类分析通过数学统计方法对样本进行分类，从而挖掘样本之间的联系，进而展示大量样本的内部结构[18]。主成分分析和聚类分析可以有效发掘数据中所隐藏的信息，使数据分析更具客观性，从而为进一步研究提供依据和方法。经主成分分析，本试验中金黄地鼠19项血清生化指标转化为5项能够解释原来样本基本信息的综合指标，且反映了血脂血糖水平、肾代谢、肝胆代谢及肌肉代谢的情况。以上5种指征均与脂质代谢紊乱有紧密的联系，而AS疾病的发展以脂质代谢障碍为基础[19-20]。因此，上述指征可用于AS金黄地鼠疾病模型早期的预判与诊断。聚类结果与样本采集结果保持一致，能将对照组与模型组分为两类，其结果进一步确认能够通过血清生化指标有效反映出对照组与AS模型组之间的差异。

生化指标能够反映机体的生理状态，在一定程度上能够体现某些组织或者器官的代谢状态。研究表明，脂质代谢紊乱是AS的物质基础。本试验生化分析结果表明，TG、TCH、HDL、LDL及GLU水平明显增高。纯化的高脂饲料含有大量的饱和脂肪酸及高能量物质，经消化吸收可显著提升血脂、血糖水平，干扰机体正常的脂质代谢[21]。相比于对照组，TP、ALB、LDH、ALT和AST在模型组中显著增高，说明肝功能和代谢发生显著变化。血脂水平过高与肝的脂肪变性有着紧密联系。高脂饲料能够引起肝细胞合成大量的脂类物质，造成脂肪在肝细胞中的积累，继而引起肝细胞的损伤，导致相关指标差异的产生[22]。升高的TBIL和DBIL在一定的程度上也能反映肝的损伤。受损的肝细胞不能使胆红素转化为胆汁以及自身代谢功能的降低可能是TBIL和DBIL升高的主要原因[23]。在模型组中，BUN水平下降，UA水平的增高，则主要可能是由于饲料成分的改变导致。在高脂饲料中，蛋白质含量明显降低，蛋白质的缺失导致肾含氮物质排泄减少，引起BUN的降低。高脂饲料中的油脂成分能够抑制尿酸的排泄，引起UA水平升高[24]。CREA是肌肉代谢的小分子产物，主要通过肾排出体外。肾代谢异常可能导致CREA的增高，其结果预示着高脂饲料可能引起肾损伤。

综上所述，本试验中，通过饲喂纯化高脂质饲料的方法成功建立了AS金黄地鼠模型。组织病理结果显示，主动脉弓具有AS的典型病理变化。血液生化指标经主成分和聚类分析能够有效挖掘隐藏信息，反应模型组和对照组之间的差异。生化指标的统计学分析表明，在纯化高脂饲料诱导的金黄地鼠AS模型中，血脂、血糖等水平显著增高，说明脂质代谢紊乱，同时可能伴有肝和肾的损伤及功能的改变。

（责任编辑　何知宇）

（发表于《中国兽医科学》，院选中文，1 177–1 182.）

青蒿素的来源及其抗鸡球虫作用机制研究进展

黄鑫，郭文柱，高旭东，郝宝成，梁剑平*

（中国农业科学院兰州畜牧与兽药研究所/农业部兽用药物创制重点实验室/
甘肃省新兽药工程重点实验室，兰州　730050）

　　青蒿素（Arteminsinin，ART）为菊科青蒿属植物黄花蒿的有效药用成分之一，分子式为$C_{15}H_{22}O$，分子结构属于倍半萜内酯类过氧化物[1-2]。1971年，我国科学家屠呦呦首次从植物黄花蒿中利用乙醚提取获得青蒿素，并经过大量的临床实验证明其具有良好的抗疟活性，而且与原有抗疟药物无交叉耐药性。青蒿素作为治疗疟疾的特效药，拯救了全球尤其是发展中国家数百万人的生命[3]。作为在青蒿素的发现中做出巨大贡献者，屠呦呦获得了2015年诺贝尔生理学或医学奖。

　　根据文献报道，青蒿素抗疟作用是因结构中过氧基团发挥生物活性作用[4]。除优良的抗疟作用外，青蒿素还具有调节免疫、抗血吸虫病、祛痰、镇咳、平喘、抑制病毒、抗菌等作用[5-7]；近年来，也有关于青蒿素具有抗鸡球虫病作用的报道，药理作用良好。但其抗球虫作用机制复杂，并且对不同球虫或同种球虫不同发育阶段的作用不尽相同。Mo等的研究结果显示，青蒿素能够引起艾美耳球虫（E. tenella）第二代裂殖子的线粒体膜电位降低47.42%（$P<0.01$），这表明青蒿素不仅可以促进裂殖子的凋亡，还能够促进被感染盲肠组织的凋亡[8]。青蒿素自20世纪60年代发现其优良的抗疟活性后，国内外学者对其进行了大量的研究，但大多数研究都只证明其具有广泛的药理作用，而对于其发挥部分药理作用的机制并不十分明确。

　　青蒿素作为新型抗疟药物已经广泛用于人类疾病治疗；同样地，因其具有良好的抗球虫作用，也可广泛应用于各种球虫病的预防和治疗，但青蒿素抗球虫的药用机理尚不明确，来源受各种因素限制，影响了其在兽用抗球虫病方面的开发和应用。因此，本文综述了近年来对青蒿素来源及抗球虫作用及药理机制的研究，为其在抗球虫病的临床应用相关的信息。

1　青蒿素的来源

　　青蒿素是我国发现的抗疟特效药，但其只能通过从青蒿植物中提取获得，来源比较单一[9]。虽然天然青蒿分布广泛，但其资源问题仍未解决，主要原因有：第一，天然青蒿植物中青蒿素含量较低；第二，青蒿植物合成青蒿素的高峰期是在植物形成花蕾的时候，其主要存在于青蒿植物的叶片中，所以为了在开花期之前将其采收，使天然青蒿资源日益匮乏；第三，一般青蒿植株具有自交不亲和性，所以人工栽培很难获得能够稳定遗传的高产

基金项目：中国农业科学院科技创新工程（CAAS-ASTIP-2014-LZHPS-04）
作者简介：黄鑫（1992—），男，甘肃天水人，硕士研究生，主要从事天然兽药开发研究
＊通信作者：E-mail：liangjp100@sina.com

植株。因此，科学家们一直致力于青蒿素的生物合成和化学合成，并在各个领域均取得了一定的突破和进展，但距离其商品上市还需要一段时间。

1.1 植物中提取

青蒿素首先在中药黄花蒿中提取获得，其经典而独特提取方法为乙醚低温提取方法[10]。目前，从植物中提取青蒿素的提取工艺已经非常成熟，并广泛应用于其工业化的生产制备。其中，常用的从植物中提取青蒿素的方法有：Ⅰ.采索氏提取法或者室温提取等传统方法，王轶等以干燥的青蒿叶粉末为原料，其最佳的提取条件为：溶剂量60mL/g，原料粒度0.25mm，在提取温度50℃下800r/min搅拌2h，青蒿素的提取效率最高，提取率可达78.19%[11]。Ⅱ.超声波萃取法，该方法是指利用超声波的空化作用和次级效应（如机械振动、扩散、乳化、击碎、化学效应等），加速萃取有效成分的扩散和释放并使萃取溶剂能充分混合，从而提高萃取效率的一种现代高新技术手段。李再新等选用超声波法黄花蒿植物中提取青蒿素，他采用强度为80W超声波超声20min，在提取温度为50℃条件下800r/min搅拌提取2h，提取效率最高[12]。Ⅲ.微波萃取法，该方法是利用电磁场的作用使食品或中药等固体或半固体物质中的某些有效成分与基体有效分离，并能保持分析对象的化合物状态不改变的一种分离方法。李志英等研究显示：在温度40℃的条件下料液比1∶40，微波功率为500W提取120s时其提取效率最佳[13]。Ⅳ.超临界CO_2萃取法，即超临界的CO_2对某些特殊天然产物具有特殊溶解性，利用压力和温度等对超临界CO_2溶解能力的影响而使有效成分溶解分离的方法。胡淼等研究显示，最佳工艺条件是在萃取温度为50℃压力20MPa，CO_2质量流量每千克原料1kg/h，流速压力14MPa时，青蒿素的萃取效率最高，可达95%以上[14]。

目前，青蒿素主要是从野生黄花蒿和人工栽培的黄花蒿中提取，虽然其提取工艺比较成熟，但性质极不稳定，提取率较低、成本过高。而且野生黄花蒿中青蒿素的含量会受气候环境的影响与季节的限制，加之采收过度，使天然药物资源日趋匮乏；人工种植占地面积大，不容易成活，耗力费时，经过人工培育的植株易变异，产量不稳定，因此迫切需要开拓青蒿素的新药源[15]。

1.2 化学合成

在过去30年中，青蒿素的全合成方法层出不穷[16-17]。1983年，Schmid等首次采用通过化学方法全合成青蒿素，即以单萜（-）-异胡薄荷醇和（+）-香茅醛为原料，经过10步反应，最终获得纯品青蒿素[18]。1986年，中国科学院上海有机研究所经过长期探索，终于解决了架设过氧桥的难题，完成了香茅醛化学全合成青蒿素的研究[19]。2010年，Yadav等报道了以香茅醛为原料，经过脯氨酸衍生物和3,4-二羟基苯甲酸乙酯共催化反应、分子内羟醛缩合、甲基格式试剂加成、$SnCl_4$催化、立体选择性氢化反应，最后经过氧化反应获得青蒿素，共经过12步，总产率仅为5%[20]。2012年，Zhu等设计了一条简洁、快速的青蒿素全合成路线，他采用环己烯酮为合成原料，通过串联反应的应用，避免了使用保护基，共经过5步反应获得青蒿素，总产率达10%以上[21]。同年，Zhang等研发出一种简单而有效的青蒿素的化学半合成法，他以青蒿酸为原料，还原得到双氢青蒿素后，再经过特定的催化剂，将二氢青蒿素氧化为过氧化二氢青蒿酸，最后经过氧化重排反应获得青蒿素，总产率达到60%[22]。该方法合成路线短，条件温和，产率高，可以应用于青蒿素的工业大规模生产中。

1.3 生物合成

青蒿素生物合成途径包括3种：Ⅰ.添加生物合成的前体物质，提高青蒿素的产量；Ⅱ.青蒿素合成过程中关键酶基因的调控及其酶活性的控制；Ⅲ.利用基因工程技术进行生产。Farhi等将合成青蒿素的关键酶控制基因整合于烟草植物的线粒体中，在保证前体物质充足的情况下，使其由前体物质甲羟戊酸转变为青蒿素，通过烟草生产青蒿素[23]。Xu等在研究中发现，刺激转基因青蒿植物中丙二烯氧化环化酶的过量表达，能够提高青蒿素在植物体内的产量，这表明，丙二烯氧化环化酶是青蒿素合成过程中的关键酶[24]。Zhang等发现在青蒿素的生物合成过程中，一种亮氨酸转录子AabZIP1的表达能够诱导青蒿素合成的关键基因紫穗槐-4,11-二烯合成酶的启动子区域（ADS）基因与CYP71AV1基因的高表达，而且其出现的植株部位与青蒿素含量具有相关性，即AabZIP1表达量高的植物组织中青蒿素的含量也相应的高，反之亦然[25]；这一发现对青蒿素的生物合成和遗传育种具有非常重要的意义。

2 青蒿素对球虫病的防治

2.1 鸡球虫病的发病特点与防治现状

鸡球虫病是由顶复器门艾美耳科艾美耳属的各种艾美耳球虫寄生于鸡的肠道内，并引起一系列临床症状的寄生虫疾病。该病主要感染3~6周龄的雏鸡，感染后引起雏鸡的腹泻和死亡，严重时死亡率可达80%，球虫病通常对集约化养鸡场危害严重。在鸡的球虫病病原中，柔嫩艾美耳球虫是致病能力最强的一种病原，严重影响养禽业的发展[26-28]。

目前，我国对于鸡球虫病的防治主要是以抗球虫药物预防为主，但现有的大多数球虫均已对抗球虫药物产生交叉性或者多重性耐药，导致鸡球虫病频发。据估计每年因生产损失，再结合该病的预防和治疗费用将超过30亿美元[29]。为了防治球虫病的暴发和流行，一方面养殖户应该改善现有的饲养管理模式，另一方面应该积极寻找天然代替药物、加强球虫的耐药机制以及现有抗球虫药增敏机理的研究，将有利于球虫病的有效防治。

2.2 青蒿素对鸡球虫病的防治

近年来的研究表明，抗疟疾药物青蒿素可以抑制球虫肌内质网Ca^{2+}-ATP酶（SER-CA）的表达，从而影响卵囊壁形成，使感染鸡的卵囊排放量减少，同时减少卵囊排出的数量，具有一定的抗球虫效果[30]。Thφfner利用二氯甲烷提取的青蒿提取物饲喂第5d感染的动物和第7d感染的鸡，经粪便检查发现其对球虫的抑制作用没有明显差异，而第5d感染治疗组与第5d感染不治疗组之间球虫的感染率差异极其显著，这表明青蒿提取物具有明显的抗球虫作用，在抗球虫作用上感染越严重的病鸡其所需青蒿提取物治疗的剂量越大[31]。Ah-madreza比较了采用石油醚（PE）、96℃乙醇（E）和水（W）来萃取制备的3种类型的青蒿提取物对鸡球虫卵囊孢子化的抑制程度，实验结果证明了不同萃取液制备的青蒿提取物不一定均具有抗球虫活性，其中用石油醚和96℃乙醇萃取的青蒿提取物抗球虫活性较高，而且其抗球虫作用可能是提取物刺激或抑制球虫卵囊孢子的形成而发挥的[32]。此外，宋萍等对青蒿素给药剂量研究结果表明，与对照组相比，各治疗组的相对增重率、成活率和抗球虫指数（ACI）均有明显改善。卵囊值与盲肠病变值也呈下降趋势，其中给药剂量最高组（40mg/kg）ACI可达147，即青蒿素抗球虫作用与剂量呈正相关[33]。上述研究中，青蒿素主要用于临床药物疗效评价，获得的抗球虫指数不一致，但整体看来，单味药青蒿的抗球虫效果中等，与其他中草药组成复方或与西药配伍可以提高药物疗效作用，并且减少抗生素药类的用量。

3 青蒿素治疗鸡球虫病的作用机制

随着人们对抗生素滥用、药物残留和食品安全等问题的日益关注，欧美发达国家开始禁止在饲料中添加任何抗生素，全球养殖业正走向绿色安全环保的养殖模式[34-35]。因此，除了改善饲养管理模式，积极寻找天然产物中的替代药物，加强对球虫耐药机理的研究迫在眉睫。近30年来，我国传统中药黄花蒿中的有效成分青蒿素，被世界公认为是最有效的抗疟药物之一[36]，近年来发现，青蒿素还具有良好的防治鸡球虫病的作用。

为鉴定青蒿素药物的作用靶点，Amy等建立了完整的疟原虫细胞模型，在已知量药物作用下分析青蒿素化合物对凋亡细胞线粒体和细胞血红素的作用机理，选用电子传递呼吸链缺失却能正常存活的HeLaρ0细胞作为模式细胞，研究发现青蒿素化学结构中内过氧基团在线粒体中形成活性氧（ROS），破坏了线粒体中的电子呼吸链结构，使线粒体功能障碍，最终导致细胞凋亡；此外还发现青蒿素化合物中的内过氧基团能诱导细胞内游离或结合的血红素而发生细胞毒性作用[37]。王娟等在实验中建立了疟原虫模型，观察发现青蒿素能直接作用于疟原虫细胞的线粒体使其肿胀，从而导致线粒体功能障碍，最终导致原虫细胞凋亡；此外，线粒体抑制剂对线粒体膜通透性的抑制作用，可降低青蒿素的抗疟作用[38]。Mo等采用荧光探针技术和流式细胞仪检测技术观察青蒿素对柔嫩艾美耳球虫的第二代裂殖子线粒体膜电位的影响，观察发现青蒿素能明显降低其二代裂殖子中线粒体的膜电位，结果证明青蒿素能使第二代裂殖子大量凋亡，并且推测第二代裂殖子的凋亡可能是由于线粒体膜电位降低、功能受阻，从而导致裂殖子细胞凋亡[8]。这一结果证明青蒿素不仅能激活宿主细胞的凋亡途径，而且还能直接作用于球虫体细胞的线粒体，诱导宿主细胞和球虫体细胞的共同凋亡。以上报道表明线粒体可能是青蒿素类药物的主要作用靶点之一。线粒体在多细胞真核生物的细胞凋亡过程中具有重要的作用，线粒体膜通透性改变是细胞凋亡前的重要特征[39]。现已知球虫发育的每个阶段都伴随着大量的线粒体存在，在球虫无性增殖阶段尤为明显，而且线粒体的功能也更加活跃。这也说明了青蒿素类药物能有效的抑制球虫的生长发育，从而发挥良好的抗球虫作用。

4 小结与展望

鸡球虫病是一种全球性的原虫病，因其发病广、传播快、危害严重且防治困难，成为养鸡业最严重的疾病，以集约化养鸡业尤为严重，造成惨重的经济损失。青蒿素作为天然源抗球虫药具有药用效果广泛、吸收代谢快、毒副作用小等特点，越来越受到养鸡业的重视。但由于青蒿素是从天然青蒿植物中提取的，所以其来源受到了限制，加之其结构不稳定、提取工艺不太完善，所以开发青蒿素新的来源十分必要。目前，生物合成法被认为是最有潜力的来源之一，利用基因工程技术获得青蒿植物中合成青蒿素的相关酶类的基因序列，从而将其组装后导入工程菌中进行发酵工程培养；因生物合成法具有成本低、产率高、易控制、对环境的损害小等特点，越来越受科学家们关注，其也必将成为天然产物解决来源的重要途径之一。

目前，青蒿素抗疟原虫的作用机理已经较明确[39]，但其抗球虫作用机理的报道较少。莫平华等的研究发现青蒿素能直接作用于柔嫩艾美耳球虫第二代裂殖子的线粒体中，降低线粒体的膜电位，从而使其细胞发生凋亡[8]；这表明青蒿素是通过诱导裂殖子中线粒体的功能发生障碍，最终导致球虫体细胞凋亡而发挥药效的。这一原理与青蒿素抗疟原虫的机理相似[40]，这提示线粒体可能是青蒿素类药物的作用靶点。青蒿素的抗疟机制"铁参与理论""血红素参与理论"和"线粒体参与理论"等的提出[39]，将有助于青蒿素抗球虫机制的研究。虽然已有文献对青蒿素抗球虫的机制做了研究，但内容不够全面，青蒿素在球虫3个发育阶段中的作用并不清楚，青蒿素抗球虫的作用靶点也并未证实。因此，青蒿素抗球虫的机制研究将是下一步研究的重点，其有助于更好地揭示青蒿素抗球虫的作用过程；也是制定青蒿素临床安全给药方案，避免耐青蒿素球虫毒株出现的理论指导。

（本文编辑：赵晓岩）
（发表于《中国预防兽医学报》，院选中文，503-506.）

金黄色葡萄球菌中耐甲氧西林抗性基因mecC的研究进展

陈 鑫[1,2]，蒲万霞[1*]，吴 润[2]，梁红雁[1,2]，李昱辉[1,2]

（1.中国农业科学院兰州畜牧与兽药研究所/农业部兽用药物创制重点实验室/甘肃省新兽药工程重点实验室，兰州 730050；2.甘肃农业大学动物医学院，兰州 730070）

近几年耐甲氧西林金黄色葡萄球菌（Methicillin-resistant Staphylococcus aureus，MRSA）的mecA同源基因mecAL-GA251在欧洲各地陆续出现报道[1]，在2012年将该同源基因正式命名为mecC。作为一种新的耐药基因它与mecA基因在DNA水平上有69%的同源性，其编码的PBP2a在氨基酸水平上有63%的同源性[2]，因此在临床上易将MRSA感染误诊为甲氧西林敏感的金黄色葡萄球菌（Methicillin-sensitive Staphylococcus aureus，MSSA）感染，从而给临床治疗MRSA感染带来了潜在的问题和困难。目前，mecC MRSA已在13个欧洲国家发现并报道，同时也已发现其宿主有14个不同的物种[3]。mecC MRSA的出现，将成为当下研究人畜共患病的又一极其重要的主题。

1 耐药基因MECC的发现及对MECC的基因分析

2007年对英格兰西南部的散装奶粉样品进行了奶牛乳房炎流行病学调查，发现了一些表型为耐苯唑西林和头孢西丁，但在验证mecA基因和蛋白PBP2a时结果却为阴性的金黄色葡萄球菌（S.aureus）菌株（LGA251）[4]。对菌株进行了全基因组测序，其测序结果已录入NCBI数据库（GenBank：FR821779.1），对结果进行分析后发现，该菌株携带有一种新的同源mecA耐药基因，最初人们将该同源基因称为mecALGA251，并于2012年正式更名为mecC[5]。该基因与mecA基因在DNA水平上只存在有69%的同源性，所编码的PBP2a在氨基酸水平上也只有63%的同源性，正因为如此，在利用PCR检测mecA基因和进行蛋白PBP2a凝集试验时则会出现阴性结果，误判为MSSA。回顾之前在英国[6]和丹麦[7]所发现的65株mecALGA251基因阳性菌株，这些菌株来源虽然以患病奶牛的奶样为主，但也包括有1975年丹麦在人体血液所分离得到的（S.aureus）菌株，这说明mecALGA251 MRSA可能已经感染人类超过38年之上，而近几年人们对它才有所进一步的发现和了解。

现已研究发现MSSA转变为MRSA是通过一种具有移动能力的基因元件SCCmec的转移来实现的，通过其自身位点特异性重组酶ccrA和ccrB精确的切除和整合作用，将SCCmec整合到葡萄球菌属的染色体上，使其携带mecA基因，从而获得对β-内酰胺类抗生素的耐药性。而耐药基因mecALGA251同样位于基因元件SCCmec上，其水平转移机制与mecA基因相

基金项目：中央级公益性科研所基本科研业务费专项资金项目（161032201100）
作者简介：陈鑫（1989—），男，江苏泰兴人，硕士研究生，主要从事兽医病原微生物耐药性研究
*通信作者：E-mail：puwanxia@caas.cn

同,同样通过可移动基因元件SCCmec的水平转移来实现,LGA251型SCCmec的ccr复合体为8型,由ccrA1和ccrB3组合而成,mec复合体为E类(即位于mecI-mecR1-mecC1-blaZ上),因此将LGA251型SCCmec确定为SCCmecⅪ型(图1)。

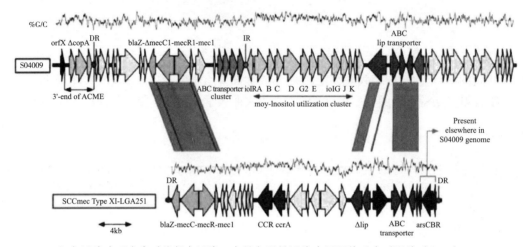

红色区域为两个序列的保守区域,亮绿和深绿区域为同源编码序列区域(CDS)。
其他CDS以其他亮色显示,蓝色和红色点分别表示同向重复(DR)和反向重复(IR)

图1 S.xylosus S04009 orfX区域(EMBL登录号HE993884)与LGA251 SCCmecⅪ型orfX区域的比较(EMBL登录号FR821779)[8]

2 MECC 编码的 PBP2A 功能分析

Kim等对mecC编码的PBP2a蛋白进行了功能研究,由于mecC在编码PBP2a时会受到mecI/mecR的调控,因此其最低抑菌浓度(MIC)值很可能是由于诱导产生的PBP2a低水平表达所致,为了解决这个问题采取了构建携带有mecC基因的质粒并在MSSA中表达的试验方法进行研究,并对其进行了各种生物化学测试、表型抗性的表达和微生物检测。经过研究发现,携带有mecC基因的S.aureus菌株具有较高的MIC值,值得注意的是mecC所编码的PBP2a LGA251相比mecA编码的PBP2a具有对药物的偏好性,具体为mecC MRSA相比MSSA对苯唑西林具有较高的MIC,同时PBP2a LGA251相对头孢西丁而言对苯唑西林则具有较高的亲和力,因此表现为对头孢西丁的高度耐药、对苯唑西林处于中介耐药水平。MRSA的mecA基因所编码的PBP2a则不存在这种现象,但PBP2a LGA251与PBP2a对头孢菌素、头孢西丁的亲和力则几乎是相同的,这一现象表明mecC MRSA最有可能首先接触到的药物是头孢菌素类抗生素。两种蛋白在耐热性上也表现出了差异,在37℃时,相比mecAPBP2a mecC编码的PBP2a表现的更加不稳定,有意思的一点是由于可能缺少mecI/mecR的调控,携带有mecC质粒的MSSA表现出了比原菌株LGA251更高的耐药性,最后值得注意的是无论是PBP2a还是PBP2a LGA251都不具有PBP2所拥有的转谷氨酰胺酶活性[9]。

3 MECC MRSA 的检测

无论是mecC MRSA还是常规MRSA它们都具有甲氧西林抗性,因此需要对其进行准确的区别。目前,市面上检测mecA基因、PBP2a的试剂盒在检测mecC MRSA时还存在缺陷,急需对其进行弥补，而实验室诊断则可利用药敏试验和PCR方法进行。

3.1 药敏试验

一般而言，常规的MRSA无论对苯唑西林还是头孢西丁均表现出明显的耐药性，但大部分mecC MRSA的表型却表现为对头孢西丁的显著耐药且对苯唑西林敏感，导致在进行药敏试验时极易误判为MSSA。由于PBP2a LGA251对药物存在偏好性，因此当进行药敏试验时，头孢西丁作为试验药物比苯唑西林更为可靠，对mecC MRSA的检测也更为准确[10]。研究表明当使用头孢西丁进行Vitek 2系统检测mecC MRSA时，Vitek 2系统具有88.7%的灵敏度和99.5%识别特异性，但是其结果还需要进一步用PCR对mecC基因进行验证[11]。

3.2 mecC 的 PCR 检测

根据上述研究结果我们认为对于携带有mecC的MRSA需要用更加可靠的方法进行检测，其中利用PCR方法对mecC基因进行验证是检测mecC MRSA的金标准。Stegger等建立了金黄色葡萄球菌mecA基因或同源基因mecALGA251的多重PCR快速检测方法，该方法所设计的mecA、mecC引物具有显著的特异性，为了证明其PCR扩增产物的保守性，他们选择了79株mecC MRSA进行了验证。该方法利用多重PCR方法来区分mecA和mecC的同时还可进行spa分型和PVL毒力基因的检测，该方法的建立为mecC MRSA的检测提供了有效的方法[12]。

综上所述，由于mecC MRSA单一方法的检测有着潜在的不准确性，因此需要将蛋白水平与分子水平检测两者互相结合以确保实验室诊断的准确性，两者均为检测mecC MRSA不可缺少的方法。

4 MECC MRSA 在人和动物之间的流行情况

目前，mecC MRSA已在13个欧洲国家发现并报道，同时也已发现其宿主有14个不同的物种。在丹麦最早mecC MRSA的感染率占整个MRSA的1.9%，而2010年起有明显增加，截至2011年达到了2.8%[13]。在德国，从人体分离得到的3207株MRSA菌株中只发现两例mecC MRSA，感染率仅为0.06%[14]。在英国的一项调查中，对2011年至2012年间临床分离得到的335株MRSA菌株进行检测时发现其mecC MRSA的感染率为0.45%[6]。在瑞士西部地区，对2005年至2011年间所分离得到的565株金黄色葡萄球菌菌株进行了检测鉴定，未发现mecC MRSA[15]，但是最近有报道指出已在该地区刺猬身上分离得到CC130谱系的mecC MRSA[16]。本实验室也对我国6个地区（四川、内蒙古、上海、甘肃、贵州和华南地区）共206株金黄色葡萄球菌菌株进行了分离鉴定，并未发现mecC MRSA。根据上述感染率显示其流行存在地区性的差异，但是值得注意的是丹麦mecC MRSA有明显的增长趋势。

在对已知的mecC MRSA菌株进行多位点序列分型（MLST）时发现，mecC MRSA菌株绝大多数属于两个谱系（即CC130和ST425），而在这两个谱系中CC130占主导地位。在mecC MRSA从人类身体中分离得到并发现之前，其CC130[17]和ST425一直被认为是流行于动物之间的谱系[18]，这说明mecC MRSA很有可能是首先流行于动物之间后才传播给人类。目前在欧洲已知的mecC MRSA宿主主要包括有家畜[19]、野生动物[20]和宠物[21]等，其感染物种非常广泛[22]，因此它将成为今后研究人畜共患病原菌的重要对象之一。

5 结语

虽然mecC MRSA的来源尚不清楚，但作为人畜共患病原菌，它对人类的健康和临床疾病治疗带来潜在的威胁和困难，值得注意的是针对其可以跨物种传播的特性将它归类为家畜相关MRSA（LA-MRSA），也正因为如此它极易与人类产生接触和发生感染。虽然迄今只有欧洲有过报道，还属于罕见菌株，但它已成为一个潜在的诊断问题并且亟待解决，目前对于mecC基因的研究还不是很完善，未来还需要我们做进一步更深入的研究。

（本文编辑：赵晓岩）

（发表于《中国预防兽医学报》，院选中文，672–674.）

2种甾体抗原对甘肃高山细毛羊繁殖率的影响

冯瑞林[1,2]，郭　健[1,2]，裴　杰[1,2]，刘建斌[1,2]，岳耀敬[1,2]，
郭婷婷[1,2]，牛春娥[1,2]，孙晓萍[1,2]，杨博辉[1,2]

（1.中国农业科学院兰州畜牧与兽药研究所，兰州　730050，
2.中国农业科学院羊育种工程技术研究中心，兰州　730050）

摘　要：[目的]观察2种甾体抗原对甘肃高山细毛羊排卵数的作用。[方法]在甘肃省张掖市肃南县和武威市天祝县，用2种甾体抗原免疫甘肃高山细毛羊，研究2种甾体抗原对甘肃高山细毛羊繁殖率的影响。[结果]肃南县试验点，试验组甘肃高山细毛羊的排卵率为168.00%，对照组排卵率为108.62%，试验组比对照组排卵率提高59.38%，差异极显著（$P<0.01$）。天祝县试验点，试验Ⅰ组排卵率为158.97%，对照组排卵率为103.92%，试验Ⅰ组比对照组排卵率提高55.05%，差异极显著（$P<0.01$）。试验Ⅱ组排卵率为137.92%，试验Ⅱ组比对照组排卵率提高34.00%，差异极显著（$P<0.01$）。[结论]2种甾体抗原均可以提高甘肃高山细毛羊的排卵率。

关键词：甾体抗原；睾酮抗原；雄烯二酮抗原；甘肃高山细毛羊；排卵率

Effects of Two Kinds of Steroid Antigens on the Reproduction Rate of Alpine Fine-wool Sheep in Gansu Province

FENG Ruilin[1,2], GUO Jian[1,2], PEI Jie[1,2], et al

(1. Lanzhou Institute of Husbandry and Pharmaceutical Sciences of CAAS, Lanzhou, Gansu 730050; 2. Sheep Breeding Engineering Technology Center of CAAS, Lanzhou, Gansu 730050)

Abstract: [Objective] To observe the effects of two kinds of steroid antigens on the reproduction rate of alpine fine-wool sheep in Gansu Province. [Method] In Sunan County of Zhangye City in Gansu Province and Tianzhu County in Wuwei City, effects of two kinds of steroid antigens on the reproduction rate of alpine fine-wool sheep in Gansu Province were researched. [Result] Ovulation rate of alpine fine-wool sheep in Sunan County was 168.00% in test group; ovulation rate was 108.62% in control group. Thus, the ovulation rate in test group enhanced by 59.38% in control group, showing extremely significant differences ($P<0.01$). Ovulation rate in test group Ⅰ in Tianzhu County was 158.97%; the ovulation rate was 103.92%

in control group. Thus, the ovulation rate in test group Ⅰ enhanced by 55.05% compared with control group, showing extremely significant differences (P<0.01). Ovulation rate of test group Ⅱ was 137.92%, which enhanced by 34.00% compared with test group Ⅱ and showed extremely significant differences (P<0.01). [Conclusion] The two steroid antigens can both enhance the ovulation rate of alpine fine-wool sheep in Gansu Province.

Key words: Steriod antigen; Testosterone antigen; Androstenedione antigen; Alpine fine-wool sheep in Gansu Province; Ovulation rate

（发表于《安徽农业科学》，中文核心，127–128，190.）

薄层色谱法对宫衣净酊质量标准的研究

朱永刚,王 磊,王旭荣,崔东安,张景艳,张 凯,张 康,李建喜,杨志强[*]

(中国农业科学院兰州畜牧与兽药研究所/甘肃省中兽药工程技术研究中心,兰州 730050)

摘 要:[目的]建立治疗奶牛胎衣不下的中兽药宫衣净酊的质量标准。[方法]采用薄层色谱(TLC)法对该酊剂中的中药材进行定性鉴别,确定该酊剂中葛根、红花、急性子和川牛膝的薄层鉴别方法。[结果]在选定的色谱条件下,色谱斑点分离较好,该鉴别方法能够有效鉴别葛根、红花和急性子。[结论]该试验中所建立的TLC方法简单、易操作、专一性强、重复性好,可为宫衣酊剂的质量控制提供依据。

关键词:宫衣净酊;薄层色谱法;质量标准

Quality Standard of Gongyijing Tincture by Thin Layer Chromatography

ZHU Yonggang, WANG Lei, WANG Xurong, YANG Zhiqiang[*], et al

(Lanzhou Institute of Husbandry and Pharmaceutical Science of CAAS, Engineering & Technology Research Center of Traditional Chinese Veterinary Medicine of Gansu Province, Lanzhou, Gansu 730050)

Abstract: [Objective] To establish quality standards of Gongyijing tincture to cure retained fetal membranes. [Method] Thin layer chromatography (TLC) method was used for qualitative identification of the traditional Chinese medicine in Gongyijing tincture. Puerarin, safflower and garden balsam seed and Radix Cyathulae were identified by TLC in Gongyijing tincture. [Result] Under the selected chromatographic conditions, chromatographic spots were better separated. This identification method could effectively identify puerarin, safflower, and garden balsam seed. [Conclusion] The established TLC method was simple, easy to operate, has strong specificity and good repeatability, which provides references for the quality control of Gongyijing tincture.

Key words: Gongyijing tincture; Thin layer chromatography (TLC); Quality standard

(发表于《安徽农业科学》,中文核心,114-116.)

绵山羊双羔素提高细毛羊繁殖率的研究

冯瑞林[1,2]，郭 健[1,2]，裴 杰[1,2]，刘建斌[1,2]，岳耀敬[1,2]，
郭婷婷[1,2]，牛春娥[1,2]，孙晓萍[1,2]，杨博辉[1,2]

（1. 中国农业科学院兰州畜牧与兽药研究所，兰州 730050；
2. 中国农业科学院羊育种工程技术研究中心，兰州 730050）

摘 要：[目的]提高细毛羊的繁殖潜能，增加农牧区养殖者的经济效益。[方法]应用绵山羊双羔素（睾酮-3-羧甲基肟·牛血清白蛋白），在新疆巩乃斯种羊场、甘肃天祝种羊场、青海三角城种羊场和黑龙江齐齐哈尔种羊场开展免疫试验。[结果]巩乃斯种羊场试验组产羔率为143.68%，对照组产羔率为125.10%，试验组产羔率比对照组提高了18.58%；天祝种羊场试验组产羔率为129.34%，对照组产羔率为102.71%，试验组产羔率比对照组提高了26.63%；三角城种羊场试验组产羔率为114.80%，对照组产羔率为104.84%，试验组产羔率比对照组提高了9.96%；齐齐哈尔种羊场试验组产羔率为150.62%，对照组产羔率为119.67%，试验组产羔率比对照组提高了30.95%。[结论]只要加强饲养管理，绵山羊双羔素可以提高细毛羊的繁殖率。

关键词：绵山羊双羔素；细毛羊；繁殖率；双羔率；产羔率

Study on Improvement of Reproductive Rate of Fine-wool Sheep by Fecundin

FENG Ruilin[1,2], GUO Jian[1,2], PEI Jie[1,2], et al

(1. Lanzhou Institute of Animal Husbandry and Veterinary Medicine, Chinese Academy of Agricultural Sciences, Lanzhou, Gansu 730050; 2. Sheep Breeding Engineering Technology Research Center, Chinese Academy of Agricultural Sciences, Lanzhou, Gansu 730050)

Abstract: [Objective] The aims were to excavate reproductive potential of fine-wool sheep, to improve the economic benefits of farmers in agricultural and pastoral areas. [Method] The steroidal antigen (testosterone-3-ethyloic-oxime·BSA) was used as fecundin to immunize fine-wool sheep for inoculation test in Gongnaisi Sheep Breeding Farm of Xinjiang Province, Tianzhu Sheep Breeding Farm of Gansu Province, Sanjiaocheng Sheep Breeding Farm of Qinghai Province and Qiqihaer Sheep Breeding Farm of Heilongjiang Province. [Result] In Gongnaisi Sheep Breeding

Farm, the lambing rate was 143.68% in experimental group and that was 125.10% in control group, the lambing rate was improved by 18.58%. In Tianzhu Sheep Breeding Farm, the lambing rate was 129.34% in experimental group and that was 102.71% in control group, the lambing rate was improved by 26.63%. In Sanjiaocheng Sheep Breeding Farm, the lambing rate was 114.80% in experimental group and that was 104.84% in control group, the lambing rate was improved by 9.96%. In Qiqihaer Sheep Breeding Farm, the lambing rate was 150.62% in experimental group and, that was 119.67% in control group, the lambing rate was improved by 30.95%. [Conclusion] The reproductive potential of fine-wool sheep can he greatly improved through application of fecundin, provided feeding and management were strengthened.

Key words: Fecundin; Fine-wool sheep; Reproductive rate; Double rate of lamb; Lambing rate

(发表于《安徽农业科学》，中文核心.)

小鼠口服五氯柳胺的急性毒性研究

张吉丽，李 冰，司鸿飞，周绪正，程富胜，张继瑜[*]

（中国农业科学院兰州畜牧与兽药研究所，农业部兽用药物创制重点实验室，甘肃省新兽药工程重点实验室，兰州 730050）

摘 要：[目的]为初步评价五氯柳胺的安全性，对其进行小鼠的急性毒性试验。[方法]预试验采用递增法确定给药的剂量范围和正式试验分组。正式试验采用简化寇氏法，灌胃给药，观察7d，观察给药后小鼠的体征变化，统计死亡率和死亡时间，计算LD_{50}值及其95%置信区间。[结果]小鼠口服五氯柳胺的LD_{50}为1.130g/kg，95%置信区间为0.940~1.359g/kg。[结论]按照化学物的急性毒性剂量分级，五氯柳胺为低毒化学物。

关键词：五氯柳胺；小鼠；急性毒性试验；LD_{50}

Study on the Acute Toxicity of Oxyclozanide in Mice by Oral Administration

ZHANG Jili, LI Bing, SI Hongfei, ZHANG Jiyu[*], et al

(Lanzhou Institute of Husbandry and Pharmaceutical Sciences of CAAS, Key Laboratory of Veterinary Pharmaceutical Development of Ministry of Agriculture, Key Laboratory of New Animal Drug Project of Gansu Province, Lanzhou, Gansu 730050)

Abstract: [Objective] In order to evaluate the safety of Oxyclozanide, the acute toxicity test in mice was conducted. [Method] The incremental method was used to determine the groups and dosage of the formal experiment in preliminary experiment. According to the simplified Karber's method in the formal experiment, after giving medicine lavage and observing for 7 days, the change of sign was observed after administration in mice, the mortality and death time was recorded, the LD_{50} and 95% confidence interval was calculated. [Result] The LD_{50} of Oxyclozanide in mice was 1.130 g/kg and 95% confidence interval was 0.940 ~ 1.359 g/kg. [Conclusion] According to the classification of acute toxicity dosages of chemicals, Oxyclozanide is the low toxic chemical.

Key words: Oxyclozanide; Mice; Acute toxicity test; LD_{50}

（发表于《安徽农业科学》，中文核心，148-149.）

绵山羊双羔素提高黑山羊繁殖率的研究

冯瑞林[1,2]，郭 健[1,2]，裴 杰[1,2]，刘建斌[1,2]，岳耀敬[1,2]，
郭婷婷[1,2]，牛春娥[1,2]，孙晓萍[1,2]，杨博辉[1,2]

（1.中国农业科学院兰州畜牧与兽药研究所，兰州 730050；
2.中国农业科学院羊育种工程技术研究中心，兰州 730050）

摘 要：[目的]提高南方黑山羊的繁殖率，增加养殖户的经济效益。[方法]应用绵山羊双羔素（睾酮-3-羧甲基肟·牛血清白蛋白），免疫贵州黑山羊80只和云岭黑山羊275只，开展提高黑山羊繁殖率的试验。[结果]贵州黑山羊试验点试验组产羔率为166.15%，对照组产羔率为143.75%，试验组产羔率比对照组提高了22.40%；云岭黑山羊试验点试验组产羔率为150.98%，对照组产羔率为107.35%，试验组比对照组提高了43.63%。[结论]只要加强南方黑山羊的饲养管理，使用绵山羊双羔素就可以提高繁殖率。

关键词：绵山羊双羔素；贵州黑山羊；云岭黑山羊；繁殖率；双羔率；产羔率

Study on Improvement of Black Goat Lambing Rate by Fecundin

FENG Ruilin[1,2], GUO Jian[1,2], PEI Jie[1,2], et al

(1. Lanzhou Institute of Animal Husbandry and Veterinary Medicine, CAAS, Lanzhou, Gansu 730050; 2. Sheep Breeding Engineering Technology Research Center, CAAS, Lanzhou, Gansu 730050)

Abstract: [Objective] The study aimed of improving reproductive rate of southern black goat, and increasing economic benefits for farmers. [Method] With fecundin (testosterone-3-ethyloic-oxime·BSA) as an immunogen, 80 Guizhou black goats and 275 Yunling black goats were selected to carry out experiment for improving black goat reproductive rate. [Result] In Guizhou black goat test point, lambing rate of experimental group was 166.15% and lambing rate of control group was 143.75%, the lambing rate was improved by 22.40%; in Yunling black goat test area, lambing rate of experimental group was 150.98% and lambing rate of control group was 107.35%, the lambing rate was improved by 43.63%. [Conclusion] The results suggested that the fecundin can improve reproductive rate of southern black goat, as long as the feeding management is enhanced.

Key words: Fecundin; Guizhou black goat; Yunling black goat; Reproductive rate; Double rate of lamb; Lambing rate

（发表于《安徽农业科学》，中文核心，122-123，144.）

紫花苜蓿草地土壤碳密度年际变化研究

田福平[1,2]，师尚礼[1]

（1.甘肃农业大学草业学院/草业生态系统教育部重点实验室/甘肃省草业工程实验室/中—美草地畜牧业可持续发展研究中心，兰州 730070；2.中国农业科学院兰州畜牧与兽药研究所/农业部兰州黄土高原生态环境重点野外科学观测试验站，兰州 730050）

摘 要：为了阐明紫花苜蓿草地土壤碳密度的年际变化规律，以长期定位试验的紫花苜蓿（*Medicago sativa*）草地为研究对象，通过大田测定和实验室分析，研究了不同生长年限紫花苜蓿草地土壤碳密度。结果表明：随着紫花苜蓿生长年限的增加，土壤有机碳含量增加，土壤碳密度提高。不同生长年限的苜蓿草地土壤有机碳含量、土壤有机碳密度在0～100cm土层的分布规律为随土层深度的增加而降低。不同年限的苜蓿草地土壤有机碳含量均为0～10cm表层最高，其土壤有机碳含量大小顺序为：1年龄（6.39g/kg）<2年龄（6.75g/kg）<3年龄（7.52g/kg）<4年龄（8.76g/kg）<5年龄（9.16g/kg）。不同年限的紫花苜蓿草地土壤有机碳密度差异显著（$P<0.05$），随着生长年限的增加，土壤有机碳密度显著提高，其中100cm的土壤有机碳密度平均值5年龄最高（8.38kg/m^2），比1年龄、2年龄、3年龄和4年龄分别增加了91.3%、56.1%、32.4%和8.1%。

关键词：紫花苜蓿；年际变化；土壤有机碳；土壤碳密度

（发表于《草原与草坪》，中文核心，8–13.）

中兽药治疗奶牛胎衣不下的系统评价

崔东安,王胜义,王 磊,王 慧,妥 鑫,黄美洲,刘永明*

(中国农业科学院兰州畜牧与兽药研究所/农业部兽用药物创制重点实验室/
甘肃省中兽药工程技术研究中心,兰州 730050)

摘 要:为评价中兽药治疗奶牛胎衣不下的临床疗效和安全性,通过检索电子资料库,包括中文生物医学数据库及外文文献检索PubMed、MEDILNE及Cochrane图书馆(建库至2013年),纳入比较中兽药治疗奶牛胎衣不下随机或半随机对照试验,进行异质性检验和Meta分析分析。共纳入7个随机或半随机对照试验,质量评价结果表明,所纳入文献的方法学质量较低。纳入的7项研究显示中兽药疗法能够有效促进滞留胎衣的完全自主排出(OR=8.08,95% CI3.65~17.60,Z=5.19,$P<0.00001$);2项研究显示,中兽药疗法可显著缩短患胎衣不下奶牛产后至首次发情(配种)时间(Z=2.95,P=0.003);2项研究显示中兽药治疗胎衣不下奶牛可明显提高受试奶牛在第一情期的受胎率(Z=2.20,P=0.03);4项研究显示中兽药治疗胎衣不下奶牛可显著提高受试奶牛产后3个情期内受胎率(Z=3.06,P=0.002);仅有1篇文献报道了不良反应。根据现有证据,中兽药疗法在促进滞留胎衣的完全自主排出和改善繁殖性能方面优于西药/综合疗法的治疗效果。但由于纳入研究数量少且质量不高,因此其论证强度较弱,尚需更多高质量研究以增加论证强度。

关键词:中兽药;系统评价;胎衣不下;奶牛

(发表于《畜牧与兽医》,中文核心,98-104.)

复方板黄口服液制备工艺研究

李　冰，牛建荣，周绪正，李剑勇，魏小娟，杨亚军，刘希望，张继瑜[*]

（中国农业科学院兰州畜牧与兽药研究所/农业部兽用药物创制重点实验室/
甘肃省新兽药工程重点实验室，兰州　730050）

摘　要：为确定复方板黄口服液的最佳制备工艺，采用正交试验等方法，对复方板黄口服液制备工艺各条件进行优选。确定的最佳工艺为药材加纯净水浸渍12h，煎煮2次，每次1h，滤液80℃浓缩至药液相对密度为1.18～1.25g生药/mL，醇沉2次（含醇量分别为600mL/L和750mL/L），静置24h，回收乙醇，加纯净水定容。该制备工艺简便易行，质量稳定，可作为复方板黄口服液的最佳制备工艺。

关键词：复方板黄口服液；制备工艺；正交试验

（发表于《动物医学进展》，中文核心，15–18.）

金黄色葡萄球菌生物被膜研究知识图谱分析

杨　峰，王旭荣，王　玲，李新圃，罗金印，张世栋，李宏胜*

（中国农业科学院兰州畜牧与兽药研究所/农业部兽用药物创制重点实验室/
甘肃省新兽药工程重点实验室，兰州　730050）

摘　要：为了展示金黄色葡萄球菌生物被膜领域的整体研究状况和研究热点，基于 CitespaceⅢ研究了1995—2015年Web of Science™核心合集数据库所收录的有关金黄色葡萄球菌生物被膜的3 582条科技文献，分析了年度发文量，绘制了国家、机构、作者、被引期刊、被引文献及关键词的知识图谱。金黄色葡萄球菌生物被膜研究领域的发文量从1995—2014年持续上升，2015年有延续的趋势；研究金黄色葡萄球菌生物被膜的国家和机构及刊登该领域文献的期刊主要分布在北美洲和欧洲，大学是该研究领域的主要研究机构；金黄色葡萄球菌生物被膜的研究热点集中于生物被膜的形成机制和抑制药物的研发两个方面，未来一段时间的研究方向将侧重于阻碍生物被膜形成和促进已形成生物被膜瓦解的药物研发。

关键词：金黄色葡萄球菌；生物被膜；知识图谱；Web of Science™核心合集；CitespaceⅢ

（发表于《动物医学进展》，中文核心，24-31.）

金黄色葡萄球菌性奶牛乳房炎研究的知识图谱分析

杨 峰，王 玲，王旭荣，李新圃，罗金印，张世栋，李宏胜*

（中国农业科学院兰州畜牧与兽药研究所/农业部兽用药物创制重点实验室/
甘肃省新兽药工程重点实验室，兰州 730050）

摘 要：基于CitespaceⅢ研究了1995—2015年Web of Science™核心合集数据库所收录的金黄色葡萄球菌性奶牛乳房炎相关的1 797条科技文献，分析了年度发文量，绘制了国家、机构、作者、被引期刊、被引文献及关键词的知识图谱。全世界金黄色葡萄球菌性奶牛乳房炎研究领域的发文量从1995—2012年持续上升，2013年发文量开始呈下滑的趋势；研究金黄色葡萄球菌性奶牛乳房炎的国家和机构主要分布在美洲和欧洲，大学是该研究领域的主要研究机构；金黄色葡萄球菌性奶牛乳房炎的研究热点集中于病原菌的分离鉴定、药物敏感性检测和致病机理探究，该领域未来一段时间的研究方向将侧重于金黄色葡萄球菌生物被膜相关的研究。

关键词：金黄色葡萄球菌；奶牛乳房炎；知识图谱；Web of Science™核心合集；CitespaceⅢ

（发表于《动物医学进展》，中文核心，38-43.）

甘肃省某奶牛场肠杆菌性乳房炎主要病原菌的分离鉴定与耐药性分析

王海瑞，王旭荣，张景艳，王 磊，曹明泽，李建喜*，王学智*

（中国农业科学院兰州畜牧与兽药研究所/
甘肃省新兽药工程重点实验室，兰州 730050）

摘 要：为查明甘肃省某奶牛场发生临床型奶牛乳房炎的主要病原菌并选择治疗用敏感药物，随机无菌采集了6份临床型病牛乳样和1份大罐乳样，采用划线分离、菌落形态观察、革兰染色镜检、生化鉴定及16S rDNA序列分析等方法进行细菌分离鉴定，同时采用K-B法测定主要分离菌株对抗菌药物的敏感性。结果表明，从临床型乳房炎乳样中分离鉴定出大肠埃希菌、芽孢杆菌、凝固酶阴性葡萄球菌（腐生葡萄球菌）3种细菌，其中大肠埃希菌的分离率最高，为66.67%；从大罐奶样中也分离出大肠埃希菌。药敏试验结果表明，大肠埃希菌对氨苄青霉素、阿莫西林、头孢噻吩、杆菌肽、红霉素均存在不同程度的耐药，耐药率为29%～100%；对诺氟沙星、恩诺沙星、氟苯尼考、头孢哌酮等药物敏感。说明此次牛场临床型乳房炎的主要病原菌是大肠埃希菌、葡萄球菌和芽孢杆菌，而且分离的大肠埃希菌具有多重耐药性。根据检测分析结果，给奶牛场提出针对性的防控措施，临床型乳房炎的发病率下降，病情得以控制。

关键词：奶牛；肠杆菌；乳房炎；病原菌；药敏试验；耐药

（发表于《动物医学进展》，中文核心，63-68.）

HPLC测定宫衣净酊中葛根素含量

朱永刚，崔东安，王 磊，王旭荣，张景艳，林 杰，杨志强，李建喜[*]

（中国农业科学院兰州畜牧与兽药研究所/
甘肃省中兽药工程技术研究中心，兰州 730050）

摘 要：为了建立宫衣净酊中葛根素的高效液相色谱（HPLC）测定方法，应用色谱柱为安捷伦Eclipse XDB-C18（4.6mm×250mm，5μm），流动相为甲醇-水（25∶75），流速为1mL/min，检测波长为250nm，柱温为30℃。结果表明，在此色谱条件下，葛根素的线性范围为18.25~584.00μg/mL（r=0.999 9，n=6），平均回收率为108.83%（RSD=1.03%）。说明本方法简便、准确、重现性好，可以作为宫衣净酊的质量控制方法。

关键词：高效液相色谱；宫衣净酊；葛根素；含量测定

（发表于《动物医学进展》，中文核心，74-76.）

Wnt10b、β-catenin、FGF18基因在甘肃高山细毛羊胎儿皮肤毛囊中的表达规律研究

赵 帅[1]，岳耀敬[1,2]，郭婷婷[2]，吴瑜瑜[2]，刘建斌[2]，韩吉龙[2]，郭 健[2]，牛春娥[2]，孙晓萍[2]，冯瑞林[2]，王天翔[3]，李桂英[3]，李范文[3]，史兆国[1]，杨博辉[1,2]

（1.甘肃农业大学动物科学技术学院，兰州 730070；2.中国农业科学院兰州畜牧与兽药研究所，兰州 730050；3.甘肃省绵羊繁育技术推广站，张掖 734031）

摘 要：以81~147d胎龄的'甘肃高山细毛羊'胚胎体侧部皮肤作为研究材料，通过冰冻切片、HE染色、荧光定量PCR、免疫组化技术研究Wnt10b、β-catenin、FGF18基因在次级毛囊形态发生过程中的表达规律. 结果表明：从次级毛囊形态发生诱导期到器官形成阶段（胎龄87~108d）Wnt10b、FGF18基因表达量逐渐升高，在胎龄第108天Wnt10b、FGF18基因相对表达量分别达到（43.652±0.425）（13.67±0.207），在细胞分化期表达量逐渐下降；而β-catenin基因的表达量始终维持较低水平，相对表达量仅为（0.58±015）；Wnt10b基因分别与β-catenin、FGF18基因表达水平呈显著正相关（$r=0.85$，$P<0.01$；$r=0.58$，$P<0.05$）；β-catenin基因与FGF18基因表达水平呈正相关（$r=0.43$，$P>0.05$）. 免疫组化结果显示在次级毛囊形态发生过程中Wnt10b、β-catenin基因主要在表皮、基板、毛芽、毛钉和成熟毛囊的内根鞘、毛干部表达；而FGF18不仅在表皮、基板、毛芽、毛钉和成熟毛囊的内根鞘、毛干部表达，而且在真皮中表达. 本研究初步表明，Wnt10b、β-catenin、FGF18基因通过Wnt/β-catenin基因经典信号通路参与调控'甘肃高山细毛羊'次级毛囊形态发生和再分化过程.

关键词：绵羊；次级毛囊；Wnt10b；β-catenin；FGF18

Expression patterns of Wnt10b, β-catenin and FGF18 genes in fetal hair follicles of Gansu alpine fine-wool sheep

ZHAO Shuai[1], YUE Yaojing[1,2], GUO Tingting[2], WU Yuyu[2], LIU Jianbin[2],
HAN Jilong[1], GUO Jian[2], NIU Chune[2], SUN Xiaoping[2], FENG Ruilin[2],
WANG Tianxiang[3], LI Guiying[3], LI Fanwen[3], SHI Zhaoguo[1], YANG Bohui[1,2]

(1. College of Animal Science and Technology, Gansu Agricultural University, Lanzhou 730070, China; 2. Lanzhou Institute of Husbandry and Pharmaceutical Sciences, Chinese Acad-

emy of Agricultural Sciences, Lanzhou 730050, China; 3. Gansu Provincial Sheep Breeding Technology Extension Station, Zhangye 734031, China)

Abstract: The expression patterns of Wnt10b, β-catenin and FGF18 genes during the morphogenesis of secondary follicles (SF) in fetal hair follicles of Gansu alpine fine-wool sheep were analyzed in the present study. The midside skins of the embryos of Gansu alpine fine-wool sheep with a gestational age (GA) of 81-147 d were used as the study objects. The expression patterns of Wnt10b, β-catenin and FGF18 genes during the morphogenesis of SF were studied through frozen sections, hematoxylin and eosin (H.E) staining, real-time quantitative polymerase chain reaction (PCR) and immunohistochemical techniques. The results showed that the morphogenesis of the secondary-original follicles (SO) of the Gansu alpine fine-wool sheep underwent 3 periods and 8 steps before the follicles matured. The original SO could further differentiate into secondary-derived follicles (SD). The gene expression levels of Wnt10b, β-catenin and FGF18 gradually increased from the induction phase of the SF morphogenesis to the organogenesis period (GA 87—108 d). At GA 108 d, the relative expression levels of Wnt10b and FGF18 reached (43.6521 ± 0.425) and (13.67 ± 0.207), respectively, and then gradually decreased during the cell differentiation period. However, the gene expression of β-catenin remained at a relatively low level, while its relative expression level was only (0.58 ± 015). The correlations among Wnt10b, β-catenin and FGF18 expressions showed that FGF18 and Wnt10b, beta catenin expression quantity were positively correlated (r=0.58, $P<0.05$ and r=0.05, $P>0.05$). The immunohistochemical results showed that Wnt10b and β-catenin were expressed primarily in the epidermis, the hair placode, the hair buds, the hair nails, the inner root sheaths and the hair shafts of the mature follicles. FGF18 was not only expressed in the epidermis, the hair placode, the hair buds, the hair nails, the inner root sheaths and the hair shafts of the mature follicles, but also expressed in the dermis. So, we proposed that Wnt10b, β-catenin and FGF18 genes may be involved in regulating of SF morphogenesis through the canonical Wnt/β-catenin signaling pathways.

Key words: Sheep; Secondary follicle morphogenesis; Wnt10b; β-catenin; FGF18

（发表于《甘肃农业大学学报》，中文核心，6-14.）

不同栽培模式下甜菜和籽立苋对次生盐渍化土壤的抑盐效应

代立兰[1]，张怀山[2]，夏曾润[3]，王 平[1]，王国宇[1]，杨世柱[2]

（1.甘肃省兰州市农业科技研究推广中心，兰州 730000；2.中国农业科学院兰州畜牧与兽药研究所/农业部兰州黄土高原生态环境重点野外科学观测试验站，兰州 730050；3.兰州大学草地农业科技学院/草地农业生态系统国家重点实验室，兰州 730020）

摘 要：针对甘肃省秦王川引大灌区的土壤次生盐渍化愈发严重的农业和生态问题，以露地平播为对照（CK），研究了垄沟露地（R）、垄沟覆膜（R+P）两种栽培模式下耐盐作物甜菜和籽粒苋对该地区次生盐渍化土壤的pH值、盐分离子Na^+、Ca^{2+}、Cl^-、SO_4^{2-}）含量和电导率（EC）及植物电导率、成苗率和产量的影响。结果表明，种植耐盐作物对次生盐渍化土壤的脱盐改良效应十分明显，配合R、R+P的栽培模式效果更加显著；R+P模式下耕种一季作物后，与CK相比，种植甜菜和籽粒苋的土壤盐分离子浓度分别下降4.8%～7.4%和4.7%～6.5%，土壤电导率分别下降16.7%～28.6%和12.8%～30.9%，甜菜和籽粒苋叶片电导率分别下降22.9%和14.9%；R+P模式下籽粒苋成苗率显著提高，但栽培模式对两种作物的产量无显著影响。

关键词：栽培模式；甜菜；籽粒苋；次生盐渍化土壤；抑盐效应

Effect of Beta vulgaris and Amaranthus Hypochondriacus on Secondary Salinization Soil under Different Cultivation Patterns

DAI Lilan[1], ZHANG Huaishan[2], XIA Zengrun[3],
WANG Ping[1], WANG Guoyu[1], YANG Shizhu[2]

(1. Lanzhou Agriculture Science Research Center, Lanzhou, Gansu 730000, China; 2. Lanzhou Institute of Animal Sciences and Veterinary Pharmaceutics, Chinese Academy of Agriculture Science/Lanzhou Scientific Observation and Experiment Field Station of Ministry of Agriculture for Ecological System in Loess Plateau Areas, Lanzhou, Gansu 730050, China; 3. College of Pastoral Agriculture Science and Technology, Lanzhou University/State Key Laboratory of Grassland Agro-ecosystems, Lanzhou, Gansu 730020, China)

Abstract: To address the agricultural and ecological problems that are becoming more severe, secondary salinization of soil in Qinwangchuan irrigated area, Gansu Province, was selected to investigate the effects of B. vulgaris and A. hypochondriacus on pH value, salt ion (Na^+、Ca^{2+}、Cl^-、SO_4^{2-}) contents, electronic conductivity (EC) of soil, seedling rate, EC and yield of crops. In this research, three cultivation patterns were used, including flat sowing without mulch (CK), ridge sowing without mulch (R) and ridge sowing with plastic mulch (R+P). The result indicated that significant effects of salt-tolerant crop on desalination improvement of soil were discovered, especially under the R+P and R cultivation patterns. Under R+P pattern, after harvesting B. vulgaris and A. hypochondriacus, soil salt ion concentration was decreased by 4.8%~7.4%, and 4.7%~6.5%, respectively, and EC of soil was significantly decreased by 16.7%~28.6% and 12.8%~30.9% from CK, respectively. Under R+P cultivation pattern, the leaf EC of B. vulgaris and A. hypochondriacus became significantly decreased by 22.9% and 14.9%, respectively. The successful seedling rate was significantly enhanced for A. hypochondriacus. However, cultivation patterns were found to have no influence on the yield of both crops.

Key words: Cultivation patterns; Beta vulgaris; Amaranthus hypochondriacus; Secondary salinization soil; Effect of salt restraint

（发表于《干旱地区农业研究》，中文核心，257-263.）

银翘蓝芩口服液中绿原酸含量测定方法的建立

许春燕[1,2]，刘希望[2]，杨亚军[2]，孔晓军[2]，
李世宏[2]，秦　哲[2]，杨孝朴[1]，李剑勇[2*]

（1.甘肃农业大学动物医学院，兰州　730070；2.中国农业科学院兰州畜牧与兽药研究所/农业部兽用药物创制重点实验室/甘肃省新兽药工程重点实验室，兰州　730050）

摘　要：为建立银翘蓝芩口服液中绿原酸含量的高效液相色谱测定方法，选用Scienhome Kromasil ODS-1 C18（250mm×4.6mm，5μm）色谱柱，流动相为甲醇-0.1%磷酸水（体积比为20∶80），检测波长324nm，柱温30℃，流速为1mL/min进行测定。结果显示，绿原酸在该色谱条件下，系统适应性良好，在质量浓度为19.32~96.60μg/mL时线性关系良好，回归方程为Y=29 059X-117 695，R^2=0.999；加样平均回收率为94.57%，RSD为1.93%；对3批样品的绿原酸含量进行测定，RSD为1.76%。表明建立的方法可用于银翘蓝芩口服液中绿原酸含量的测定。

关键词：银翘蓝芩口服液；绿原酸；高效液相色谱法

Establishment of Determination Method of Chlorogenic Acid Content in Yinqiaolanqin Oral Liquid by High Performance Liquid Chromatography

XU hunyan[1,2], LIU Xiwang[2], YANG Yajun[2], KONG Xiaojun[2],
LI Shihong[2], QIN Zhe[2], YANG Xiaopu[1], LI Jianyong[2*]

（1. College of Veterinary Medicine, Gansu Agricultural University, Lanzhou 730070, China;
2. Lanzhou Institute of Husbandry and Pharmaceutical Sciences of CAAS/Key Laboratory of Veterinary Pharmaceutical Development, Ministry of Agriculture/Key Laboratory of New Animal Drug Project of Gansu Province, Lanzhou 730050, China）

Abstract: In order to establish the determination method of chlorogenic acid content in Yinqiaolanqin oral liquid by high performance liquid chromatography（HPLC）, Scienhome Kromasil ODS-1 C18 column（250 mm×4.6 mm, 5 μm）was used. The HPLC system used a mobile phase composed of Methanol-0.1% phosphoric acid water（20∶80, V/V）at a flow rate of 1 mL/min, the ultraviolet detector was set at 324 nm with column temperature of 30℃. The HPLC system suitability of chlorogenic acid was good. A good linear calibration of chlorogenic acid was observed within the concentration rang of 19.32－96.60 μg/mL (R2=0.999), and the average recovery rate was 94.57% with relative standard deviation (RSD) of 1.93%, the regression equation was Y=29 059X－117 695. The RSD of chlorogenic acid content in three lots of Yinqiaolanqin oral liquid was 1.76% indicating the method could be employed to determine the content of chlorogenic acid in Yinqiaolanqin oral liquid.

Key words: Yinqiaolanqin oral liquid; Chlorogenic acid; High performance liquid chromatography (HPLC)

（发表于《河南农业科学》，中文核心，126-129.）

基于Web of Science的"阿维菌素类药物"的文献计量研究

文 豪,周绪正,李 冰,牛建荣,魏小娟,张继瑜

(中国农业科学院兰州畜牧与兽药研究所/农业部兽用药物创制重点实验室,兰州 730050)

A bibliometric Study of the Avermectin Class of Drugs Based on the Web of Science

WEN Hao, ZHOU Xuzheng, LI Bing,
NIU Jianrong, WEI Xiaojuan, ZHANG Jiyu

(Lanzhou Institute of Husbandry and Pharmaceutical Sciences, Chinese Academy of Agricultural Sciences/Key Laboratory of New Animal Drug Discovery, Ministry of Agriculture, Lanzhou 730050, China)

Key words: Avermectin; Web of Science; Bibliometrics; Research status; Paper; Analysis
DOI: 10.13881/j.cnki.hljxmsy.2016.0052

(发表于《黑龙江畜牧兽医》,中文核心,165-168.)

巴尔吡尔对畜禽常见病原菌体外抑菌效果研究

张 哲，李新圃，杨 峰，罗金印，刘龙海，李宏胜

（中国农业科学院兰州畜牧与兽药研究所/农业部兽用药物创制重点实验室/
甘肃省新兽药工程重点实验室，兰州 730050）

（发表于《黑龙江畜牧兽医》，中文核心，194–196.）

抗球虫药青蒿散中青蒿素UPLC检测方法的建立

黄 鑫[1]，郭文柱[1]，郝宝成[1]，高旭东[1]，梁剑平[1]，于静怡[2]，杜宪文[3]

（1.中国农业科学院兰州畜牧与兽药研究所/农业部兽用药物创制重点实验室/甘肃省新兽药工程重点实验室，兰州 730050；2.克什克腾旗水利局，赤峰 025300；3.克什克腾旗农牧业局，赤峰 025300）

关键词： 超高效液相色谱；青蒿素；含量测定；抗球虫药；方法建立

Establishment of a UPLC method for detecting artemisinin in anticoccidial drug Artemisia annua powder

HUANG Xin[1], GUO Wenzhu[1], HAO Baocheng[1],
GAO Xudong[1], LIANG Jianping[1], YU Jingyi[2], DU Xianwen[3]

(1. Key Laboratory of New Animal Drug Project of Gansu Province/Key Laboratory of Veterinary Pharmaceutical Development, Ministry of Agriculture/Lanzhou Institute of Husbandry and Pharmaceutical Sciences of CAAS, Lanzhou 730050, China; 2. Hexigten Banner Water Conservancy Bureau, Chifeng 025300, China; 3. Hexigten Banner Bureau of Agriculture and Animal Husbandry, Chifeng 025300, China)

Key words: Ultra performance liquid chromatography (UPLC); Artemisinin; Content determination; Anticoccidial drug; Method establishment

（发表于《黑龙江畜牧兽医》，中文核心，152-154.）

绵山羊双羔素提高东北细毛羊繁殖率的研究

冯瑞林，郭 健，裴 杰，刘建斌，岳耀敬，郭婷婷，孙晓萍，牛春娥，杨博辉

（中国农业科学院兰州畜牧与兽药研究所，兰州 730050）

关键词：绵山羊双羔素；东北细毛羊；繁殖率；双羔率；产羔率；免疫试验

Study on Using Fecundin to Improve the Reproductive Rate of Chinese Northeast Fine-wool Sheep

FENG Ruilin, GUO Jian, PEI Jie, LIU Jianbin, YUE Yaojing,
GUO Tingting, SUN Xiaoping, NIU Chune, YANG Bohui

(Lanzhou Institute of Husbandry and Pharmaceutical Sciences, Chinese Academy of Agricultural Sciences, Lanzhou 730050, China)

Key words: Fecundin; Chinese northeast fine-wool sheep; Reproductive rate; Twin-lamb rate; Lambing rate; Immune test

DOI: 10.13881/j.cnki.hljxmsy.2016.0029

（发表于《黑龙江畜牧兽医》，中文核心，97-100.）

甘肃省售牛羊肉中雌激素残留状况分析

李维红[1,2,3]，熊　琳[1,2,3]，高雅琴[1,2,3]，杨晓玲[1,2,3]

［1. 中国农业科学院兰州畜牧与兽药研究所，兰州　730050；2. 农业部动物毛皮及制品质量监督检验测试中心（兰州），兰州　730050；3. 农业部畜产品质量安全风险评估实验室（兰州），兰州　730050］

摘　要：以甘肃省兰州市、靖远县、临夏州、张掖地区牛羊肉为试验材料，分析了甘肃省市售牛羊肉中己烯雌酚、雌二醇、戊酸雌二醇和苯甲酸雌二醇等4种雌激素残留状况。结果表明，80份样品中（4个地方牛羊肉各10份）所测定的4种雌激素均未检出，甘肃省市售牛羊肉中无论是天然的雌二醇还是己烯雌酚、戊酸雌二醇和苯甲酸雌二醇3种人工合成雌激素均没有残留。

关键词：牛羊肉；雌激素；残留；甘肃省

Analyses of Hormone Residue in Beef and Mutton Sold in Gansu Province

LI Weihong[1,2,3], XIONG Lin[1,2,3], GAO Yaqin[1,2,3], YANG Xiaoling[1,2,3]

［1. Lanzhou Institute of Husbandry and Pharmaceutical Sciences of CAAS, Lanzhou 730050, China; 2. Quality Supervising, Inspecting and Testing Center for Animal Fiber, Fur, Leather and Products (Lanzhou), Ministry of Agriculture, Lanzhou 730050, China; 3. Quality Safety Risk Assessment of Animal Products (Lanzhou), Ministry of Agriculture, Lanzhou 730050, China］

Abstract: As the test material of beef and mutton of Lanzhou city, Jingyuan county, Linxia prefecture, Zhangye region in Gansu province, the hormone residue of diethylls tilbestrol, estradiol, estradiol valerate and estradiol benzoate in beef and mutton in Gansu province were analyzed. The results showed that 80 samples (10 copies of each of the 4 local beef and mutton) the determination of 4 estrogens were not detected. Gansu province commercial beef and muttonin both natural estradil or diethyl lstilbestrol, estradiol valerate and estradiol benzoate 3 synthetic estrogen were no residue.

Key words: Beef and mutton; Estrogen; Residues; Gansu province

（发表于《湖北农业科学》，中文核心，1 538-1 540.）

基于AHP、负权重和模糊数学的土壤质量评价研究

李润林[1,2]，姚艳敏[3]，董鹏程[1,2]

（1.农业部兰州黄土高原生态环境重点野外科学观测试验站，兰州 730050；2.中国农业科学院兰州畜牧与兽药研究所，兰州 730050；3.中国农业科学院农业资源与农业区划研究所，北京 100081）

摘 要：以北京市房山区为研究区域，选用土壤肥力和土壤环境质量的8个指标，采用层次分析法（AHP）、负权重和模糊数学对北京房山区土壤质量进行综合评价，发现房山区土壤综合指数在0.247~0.996，其中燕山地区土壤综合指数最低（0.247），蒲洼乡的土壤综合指数最高（0.996）。根据耕地地力等级的划分范围，北京市房山区Ⅰ级占9%，Ⅱ级占23%，Ⅲ级占55%，Ⅳ级占14%，说明土壤质量总体处于中等水平。通过对比分析Ⅰ级和Ⅳ级指标含量，发现土壤综合指数高的区域具有土壤肥力指标高、土壤环境质量指标低、没有指标超标的特点，而土壤综合指数低的区域具有土壤肥力指标低、土壤环境质量指标高、至少有一个指标超过临界值的特点。

关键词：土壤质量；负权重；层次分析法（AHP）；模糊数学

The Assessment of Soil Quality Based on AHP, Negative Weights and Fuzzy Math

LI Runlin[1,2], YAO Yanmin[3], DONG Pengcheng[1,2]

(1. The Lanzhou Scientific Observation and Experiment Field Station of Ministry of Agriculture for Ecological System in Loess Plateau Areas, Lanzhou 730050, China; 2. Lanzhou Institute of Husbandry and Pharmaceutical Sciences, Chinese Academy of Agricultural Sciences, Lanzhou 730050, China; 3. Institute of Agricultural Resources and Regional Planning, Chinese Academy of Agricultural Sciences, Beijing 100081, China)

Abstract: The research area located at Beijing Fangshan district, soil fertility and soil environment quality was taken as evaluating indicators which included soil organic matter, alkaline N, available P, available K, Cu, Pb, Cr, Cd. Comprehensive evaluation of soil quality was carried out by analytic hierarchy process, negative weights and fuzzy in Fangshan district of Beijing city, the results showed that the soil quality index was between 0.247 and 0.996, in which soil quality index of Yanshan area was the lowest of 0.247, soil quality index of Puwa town was the higest of 0.996. According to the classification cultivated land fertility range, Ⅰ grade accounted for 9%, Ⅱ grade accounted for 23%, Ⅲ grade 55%, Ⅳ grade 14%. Through comparative analysis it was found that areas of high soil quality index had high soil fertility, low soil environmental quality indicators, no indicators exceeded the threshold. Areas with low soil quality index had low soil fertility, soil environmental quality index was higher, and there was at least a characteristic index exceeded the threshold.

Key words: Soil quality; Negative weights; Analytic hierarchy process（AHP）; Fuzzy math

（发表于《湖北农业科学》，中文核心，4 480-4 483.）

"康毒威"治疗鸡新城疫扭颈的效果

张仁福[1],谢家声[2*]

(1.甘肃省甘谷县畜牧兽医局,甘谷 741200;
2.中国农业科学院兰州畜牧与兽药研究所,兰州 730050)

摘　要:正"康毒威"是中国农业科学院兰州畜牧与兽药研究所研制的纯中药制剂,主要由丹皮、葛根、桂枝、白芍、白芨、五味子、甘草等组成,主要用于缓解非典型性新城疫引起的腹泻、呼吸障碍等症状。使用过程中发现该药对新城疫后期发生的扭颈等典型神经症状也具有良好治疗效果。2006年将其专门用于新城疫扭颈病鸡的治疗,先后在天水市兴旺商品蛋鸡场共收治新城疫扭颈病鸡165只。

(发表于《家禽科学》,中文核心,34-35.)

中药治疗种公鸡冠癣有效

谢家声，王贵波，李锦宇，罗超应

（中国农业科学院兰州畜牧与兽药研究所，兰州 730050）

摘　要：正冠癣又称头癣或黄癣，主要发生于鸡，是由头癣真菌引起的一种传染病，其特征是在鸡头部无羽毛处，特别是鸡冠上形成黄白色、鳞片状的癣痂，重症病例病变也可扩展到有羽毛处。重型品种鸡较易感染，六月龄之内的小鸡很少发生，偶见于蛋鸡和其他禽类，也可感染畜类和人。本病多发于夏、秋多雨潮湿的季节，一般通过皮肤伤口传染或接触传染。

（发表于《家禽科学》，中文核心，35.）

小鼠口服五氯柳胺混悬剂的急性毒性研究

张吉丽,司鸿飞,李 冰,程富胜,周绪正,张继瑜

(中国农业科学院兰州畜牧与兽药研究所/农业部兽用药物创制重点实验室/甘肃省新兽药工程重点实验室,兰州 730050)

摘 要:通过进行小鼠急性毒性试验,对五氯柳胺混悬剂的安全性进行初步评价。预试验采用递增法确定给药的剂量范围、正式试验分组以及组距等。正式试验按照简化寇氏法进行,观察给药后小鼠的体征变化,统计死亡率,记录死亡时间,计算五氯柳胺混悬剂的半致死剂量(LD_{50})以及95%的可信限。结果表明,小鼠口服五氯柳胺混悬剂的LD_{50}值为1.679g/kg,95%的置信区间为1.439~1.947g/kg,按照化学物毒性分级可得:五氯柳胺混悬剂为低毒物质。

关键词:五氯柳胺混悬剂; 小鼠; 急性毒性试验; 半致死剂量(LD_{50})

(发表于《江苏农业科学》,中文核心,340–341.)

藿芪灌注液局部刺激性试验

王东升,张世栋,苗小楼,董书伟,魏立琴,邝晓娇,那立冬,闫宝琪,严作廷

(中国农业科学院兰州畜牧与兽药研究所/农业部兽用药物创制重点实验室/甘肃省中兽药工程技术研究中心/甘肃省新兽药工程重点实验室,兰州 730050)

摘 要:为了评价藿芪灌注液的刺激性,为临床安全用药提供依据,采用眼睛刺激性和阴道黏膜刺激性试验进行局部刺激性研究。眼睛刺激性试验选择家兔5只,以左眼为对照结膜囊内滴入生理盐水0.1mL,右眼结膜囊内滴入藿芪灌注液0.1mL,给药后观察7d,考察藿芪灌注液对家兔眼睛的刺激性。阴道黏膜刺激性试验选择家兔10只,随机分为试验组和对照组,试验组家兔阴道灌注藿芪灌注液2mL,对照组家兔阴道灌注生理盐水2mL,给药24h后处死家兔,对家兔阴道进行病理组织学检查,评价其对家兔阴道黏膜的刺激性。结果表明,藿芪灌注液对家兔眼睛和阴道黏膜无刺激性,说明藿芪灌注液无刺激性。

关键词:藿芪灌注液;刺激性;阴道;眼睛;家兔

(发表于《江苏农业科学》,中文核心,303–305.)

中药治疗奶牛乳房炎的系统评价与Meta分析

杨 健[1,2]，严作廷[1]，王东升[1]，张世栋[1]，杨志强[1]，董书伟[1*]

（1.中国农业科学院兰州畜牧与兽药研究所/农业部兽用药物创制重点实验室/甘肃省新兽药工程重点实验室，兰州 730050；2.北京农学院动物科学技术学院，北京 102206）

摘 要：[目的]利用Meta分析方法对已发表的中药治疗奶牛乳房炎临床试验研究进行系统评价，为防治奶牛乳房炎提供更有价值的参考依据。[方法]通过计算机检索中国知网（CNKI）和万方数据库，检索时限均为从建库起至2015年4月，全面收集中药治疗奶牛乳房炎的临床试验研究。按照Cochrane方法纳入文献，采用RevMan 5.3软件进行Meta分析。[结果]共纳入26篇文献，3 419头患病牛，其中试验组2 481例，对照组938例，试验组均采用中药治疗，对照组采用青链霉素治疗。复方中药治疗牛乳房炎的总治愈率显著高于青链霉素对照组[OR=2.03，95% CI（1.56，2.63），$P<0.05$]，总有效率显著高于青链霉素对照组[OR=2.25，95% CI（1.57，3.22），$P<0.05$]，乳区总治愈率高于青链霉素对照组[OR=1.54，95% CI（1.43，1.66），$P<0.05$]，乳区总有效率极显著高于青链霉素对照组[OR=1.20，95% CI（1.15，1.26），$P<0.01$]。发表偏倚分析结果显示，所纳入文献存在一定的发表偏倚性。[结论]复方中药对奶牛乳房炎的疗效显著优于青链霉素，即奶牛乳房炎的中药防治研究存在重要潜力。

关键词：奶牛；乳房炎；Meta分析；中药；青链霉素

Systematic Evaluation and Meta-analysis on Efficacy of Chinese Herbal Medicine on Cow Mastitis

YANG Jian[1,2], YAN Zuoting[1], WANG Dongsheng[1], ZHANG Shidong[1], YANG Zhiqiang[1], DONG Shuwei[1*]

(1. Lanzhou Institute of Husbandry and Pharmaceutical Sciences, Chinese Academy of Agricultural Sciences/Key Laboratory of Veterinary Pharmaceutical Development, Ministry of Agriculture/Key Laboratory of New Animal Drug Project, Lanzhou 730050, China; 2. College of Animal Science and Technology, Beijing University of Agriculture, Beijing 102206, China)

Abstract: 〔Objective〕The published clinical trial about treatment of cow mastitis with Chinese herbal medicine were systemically evaluated by meta-analysis method, in order to provide valuable reference for treatment of cow mastitis. 〔Method〕The clinical randomised controlled trials (RCTs) about treatment of cow mastitis with Chinese herbal medicine were searched completely from databases of CNKI and WANFANG DATA. According to Cochrane system instruction, meta-analysis was conducted by using RevMan 5.3 software. 〔Result〕A total of 26 RCTs and 3 419 diseased cows were included, among them, 2 481 cows in treatment group treated with Chinese herbal medicine, 938 cows in control group treated with penicillin-stretomycin. In addition, The cure rate of total cow number in Chinese herbal medicine group was significantly higher than that in penicillin-stretomycin group [OR=2.03, 95% CI (1.56, 2.63), P<0.05]. The effective rate of total cows number in Chinese herbal medicine group was significantly higher than that in penicillin-stretomycin group [OR=2.25, 95% CI (1.57, 3.22), P<0.05]. The cure rate of total breast area number in Chinese herbal medicine group was higher than that in penicillin-stretomycin group [OR=1.54, 95% CI (1.43, 1.66), P<0.05]. The effective rate of total breast area number in Chinese herbal medicine group was higher than that in penicillin-stretomycin group [OR=1.20, 95% CI (1.15, 1.26), P<0.01]. The analysis results of publication bias showed that the included literatures had some certain bias. 〔Conclusion〕Chinese herbal medicine has more significant curative effect on cow mastitis than penicillin-streptomycin.

Key words: Dairy cow; Mastitis; Meta-analysis; Chinese herbal medicine; Penicillin-streptomycin

（发表于《南方农业学报》，中文核心，656–663.）

多菌种协同发酵啤酒糟渣和苹果渣生产蛋白饲料的研究

王晓力[1]，王　帆[2]，孙尚琛[2]，王永刚[2]，李　想[2]，王春梅[1]，朱新强[1]，孙启忠[3]

（1.中国农业科学院兰州畜牧与兽药研究所，兰州　730050；2.兰州理工大学生命科学与工程学院，兰州　730050；3.中国农业科学院草原研究所，呼和浩特　010010）

摘　要：以苹果渣和啤酒糟为发酵基质，利用里氏木霉、黑曲霉、产朊假丝酵母为发酵菌种，采用单因素3验和正交试验对固态发酵工艺进行了优化，并对发酵饲料品质进行了评定，利用扫描电子显微镜技术探究了微生物在固态发酵中的作用。结果表明，最佳的基质配比为1∶1，3种微生物配比为1∶1∶1，接种量为8%，含水量、发酵温度和时间分别为60%、35℃和48h，在此条件下，发酵后饲料中粗蛋白含量提高34.52%，酸性和中性纤维素含量分别降低22.93%和9.08%。饲料品质良好，无结块，伴有淡淡的酸香味和苹果醇香味。尼龙袋瘘管羊瘤胃消化试验表明，72h后粗蛋白、酸性和中性纤维素消化率分别为79.78%、51.96%和39.88%。扫描电镜结果显示微生物在发酵过程中破坏了饲料结晶区表面结构，纤维素分子无定形区消失，大量酵母细胞分布在饲料表面，汲取养分生长繁殖成为高蛋白饲料。

关键词：苹果渣；啤酒糟；固态发酵；蛋白饲料；扫描电镜

Study on Production of Protein Feedstuff from Apple Pomace and Brewer's Grains by Cooperation of Different Microbial Strains

WANG Xiaoli，WANG Fan，SUN Shangchen，WANG Yonggang，
LI Xiang，WANG Chunmei，ZHU Xinqiang，SUN Qizhong

Abstract: Optimization the solid-state fermentation process was studied by single factor experiment and orthogonal experiment using apple pomace and brewer's grains as fermentation substrate, and Trichodermareesei, Aspergillus niger and Candida utilis as for fermentation strains. As well as, evaluation of the fermented feedstuff quality and exploration the role of microorganisms in solid-state fermentation using scanning electron microscopy were investigated. The results showed that the proportion of fermentation substrate and three microorganisms were 1∶1 and 1∶1∶1 respectively, inoculum amount was 8%, the water content, fermentation temperature and time was 60%, 35℃ and 48 h. Under optimized conditions, the protein content was increased

34.52%, while the acidic and neutral cellulose content were decreased 22.93% and 9.08% respectively. The feedstuff with a fantastic quality and no agglomerate was accompanied by mild acid and apple aroma. The degradability of crude protein, acidic and neutral cellulose in feedstuff using the nylon-bag method in the sheep rumen were 79.78%, 51.96% and 39.88% respectively that 72 h after. Scanning electron microscopy results indicated that the microbes in the fermentation process destroyed the surface structure of feedstuff, amorphous area of cellulose molecules dispersed, and a large number of yeast cells distributed in the surface of the feedstuff to absorb nutrients growing to feedstuff with high protein content.

Key words: Apple pomace; Brewer's grains; Protein feed; Digestibility; Scanning electron microscopy

（发表于《饲料工业》，中文核心，32-38.）

VITEK2 Compact全自动微生物分析系统对牛乳中葡萄球菌的鉴定效果评价

林 杰,张景艳,王 磊,王海瑞,邹 璐,朱永刚,
边亚彬,李建喜,杨志强,王旭荣*

(中国农业科学院兰州畜牧与兽药研究所/甘肃省中兽药工程技术研究中心 兰州 730050)

摘 要:〔目的〕评价VITEK2 Compact全自动微生物分析系统对牛乳中葡萄球菌属细菌鉴定效果。〔方法〕对从西宁、白银、西安三地乳样中分离的52株葡萄球菌通过VITEK2 Compact全自动微生物分析系统进行鉴定,并与革兰氏染色、触酶试验、血浆凝固酶试验以及16S rRNA基因序列分析结果进行比对;通过对凝固酶阴性葡萄球菌和凝固酶阳性葡萄球菌在不同麦氏浓度的鉴定,评价该系统的稳定性。〔结果〕VITEK2 Compact全自动微生物分析系统对葡萄球菌在"属"水平的鉴定结果符合率达96.15%,与血浆凝固酶结果对比,凝固酶阴性葡萄球菌属细菌鉴定结果符合率为92.86%,对凝固酶阳性葡萄球菌属细菌鉴定结果符合率为50%,但对具体种鉴定时,结果符合率仅为44.23%。在稳定性检测的试验中,鉴定符合率为100%,鉴定结果可靠性不随菌液浓度变化而变化。〔结论〕该系统可用于牛乳中葡萄球菌在"属"水平的鉴定。

关键词: VITEK2 Compact全自动微生物分析系统,葡萄球菌,16S rRNA基因序列分析,准确性,稳定性

Assessment of VITEK2 Compact Automatic Microbial Analysis System for *Staphylococcus* in Raw Milk

LIN Jie, ZHANG JingYan, WANG Lei, WANG HaiRui, ZOU Lu,
ZHU YongGang, BIAN YaBin, LI JianXi, YANG ZhiQiang, WANG XuRong*

(Lanzhou Institute of Husbandry and Pharmaceutical Science of CAAS, Engineering & Technology Research Center of Traditional Chinese Veterinary Medicine of Gansu Province, Lanzhou, Gansu 730050, China)

Abstract: [Objective] In order to evaluate the VITEK2 Compact automatic microbial analysis system in identifying Staphylococcus in raw milk. [Methods] We isolated 52 strains of Staphylococcus from milk samples from Xining, Baiyin and Xi'an, and these strains were identified by VITEK2 Compact. Then, we compared the results obtained from VITEK2 Compact to the results from Gram's staining, catalase test, plasma coagulase test and 16S rRNA gene sequence analysis. Furthermore, the stability of this system was tested by identifying the coagulase negative and positive Staphylococcus at different concentrations. [Results] With 16S rRNA gene sequence analysis as the reference standard, the coincidence rate of this system reached 96.15% for Staphylococcus at "Genus" level. With plasma coagulase test as the reference standard, the coincidence rate of this system was 92.86% for the coagulase negative Staphylococcus while 50% for the coagulase positive Staphylococcus. When it came to "Species" level, the coincidence rate of this system was only 44.23%. In stability test, coincidence rate was 100% and the reliability of appraisal result did not change by bacteria liquid concentration. [Conclusion] In conclusion, VITEK2 Compact automatic microbial analysis system could be used to identify Staphylococcus in milk at "Genus" level.

Key words: VITEK2 Compact automatic microbial analysis system; Staphylococcus; 16S rRNA gene sequence analysis; Accuracy; Stability

（发表于《微生物学通报》，中文核心，2 514–2 520.）

固态发酵豆渣、葡萄渣和苹果渣复合蛋白饲料的研究

朱新强[1]，魏清伟[2]，王永刚[2]，冷非凡[2]，郭陈真[2]，
庄　岩[2]，王春梅[1]，王晓力[1]，孙启忠[3]

（1.中国农业科学院兰州畜牧与兽药研究所，兰州　730050；2.兰州理工大学生命科学与工程学院，兰州　730050；3.中国农业科学院草原研究所，呼和浩特　010010）

摘　要：以豆渣、苹果渣和葡萄渣为发酵原料，利用酿酒酵母菌和植物乳酸杆菌为发酵菌种，对微生物发酵效果进行比较分析。采用单因素和正交试验方法确定最佳发酵条件为葡萄渣、苹果渣和豆渣配比为2∶3∶5；发酵菌剂组合为酿酒酵母，乳酸杆菌和纤维素酶；发酵时间为72h，发酵温度为32.5℃，接种量为5%；通过体外消化试验得到粗蛋白的消化率随时间成梯度变化；通过扫描电子显微镜观察，确定酵母菌体蛋白是饲料总蛋白含量提高的主要原因之一；最后对发酵饲料营养价值评定，发酵饲料pH值为6.2，粗蛋白质含量11.25%，酸性洗涤纤维素含量为28.43%，中性洗涤纤维素含量为24.68%，达到优质蛋白饲料标准。

关键词：苹果渣；葡萄渣；豆渣；固态发酵；蛋白饲料

（发表于《饲料研究》，中文核心，54–59.）

奶牛生乳中洛菲不动杆菌的分离鉴定与耐药性分析

林 杰，王旭荣，王 磊，王海瑞，朱永刚，杨志强，李建喜，张景艳

(中国农业科学院兰州畜牧与兽药研究所/
甘肃省中兽药工程技术研究中心，兰州 730050)

摘 要：对29份外观正常的奶牛生乳乳样进行乳品质分析和细菌学检测，利用16S rDNA序列分析鉴定乳样中分离的菌株，并进行14种药物的敏感性试验。结果表明，10份乳样有细菌检出，其中4份乳样中分离得到洛菲不动杆菌；药物敏感性结果表明，洛菲不动杆菌对复方新诺明具有较强的耐药性；乳品质分析结果表明，该4份乳样体细胞数（SCC）偏高，乳成分和部分理化性质与其他正常乳无明显区别。

关键词：洛菲不动杆菌；鉴定；耐药性；乳品质

(发表于《西北农业学报》，中文核心，811-815.)

不同生长年限紫花苜蓿地下生物量的空间分布格局

周恒[1]，时永杰[1]，胡宇[1]，陈璐[2]，路远[1]，田福平[1,*]

（1.中国农业科学院兰州畜牧与兽药研究所/农业部兰州黄土高原生态环境重点野外科学观测试验站，兰州 730050；2.甘肃农业大学草业学院，兰州 730070）

摘 要：采用分层取样方法对不同生长年限（10a、15a、3a、1a）紫花苜蓿 0~150cm生物量空间分布特征进行的研究结果表明：不同生长年限紫花苜蓿地上生物量变化范围为1 171.67~4 085.33g/m^2，其顺序为3a>10a>1a>15a；地下生物量变化范围为648.62~2 670.09g/m^2，其顺序为10a>15a>3a>1a；地下生物量与地上生物量的比值变化范围为0.23~2.12，其顺序为15a>10a>1a>3a。不同生长年限紫花苜蓿地下生物量均主要分布0~30cm土层，1a占总地下生物量76.6%，3a占45.6%，10a占55.82%，15a占57.22%；而相同土层不同生长年限紫花苜蓿在0~30cm土层均以15a紫花苜蓿地下生物量最高，30~100cm土层则以10a紫花苜蓿最高。

关键词：紫花苜蓿；地下生物量；根系；生长年限

（发表于《中国草地学报》，中文核心，47–51.）

基于Web of Science™的"树突状细胞"研究论文产出分析

边亚彬，张景艳，王 磊，王旭荣，杨志强，王学智，孟嘉仁，李建喜*

（中国农业科学院兰州畜牧与兽药研究所，甘肃省中兽药工程技术研究中心，兰州 730050）

摘 要：作者通过对2006—2015年Web of Science™数据库有关"树突状细胞"研究论文产出统计比较，分析全球范围内有关"树突状细胞"研究概况。根据Web of Science™核心合集数据库，以"树突状细胞"作为标题词，从"树突状细胞"研究领域的研究方向、国家/地区、科研机构、主要期刊分布及论文发表量高的作者及被引次数高的文献这几方面总结了"树突状细胞"的研究热点。结果显示，美国在"树突状细胞"发表的论文数量排在第一位，德国和中国分别排在第二、第三位；期刊《Journal of Immunology》载文量最多；法国国家医学与健康研究院、美国匹兹堡大学和哈佛大学论文发表数量居全球前三，上海交通大学和浙江大学的论文发表数量居国内前两位；被引次数最多的文章来自牛津大学；"树突状细胞"研究方向主要集中于免疫学、细胞生物学和肿瘤学。综上所述，中国有关树突状细胞研究的主要追踪目标是美国，重点研究集中于免疫学、细胞生物学和肿瘤学。

关键词：Web of Science™；树突状细胞；文献计量；论文

Analysis on Outputs of Dendritic Cells Research Papers Based on Web of Science™

BIAN Yabin, ZHANG Jingyan, WANG Lei, WANG Xurong,
YANG Zhiqiang, WANG Xuezhi, MENG Jiaren, LI Jianxi*

(Engineering & Technology Research Center of Traditional Chinese Veterinary Medicine of Gansu Province, Lanzhou Institute of Husbandry and Pharmaceutical Science, CAAS, Lanzhou 730050, China)

Abstract: This paper analysed the outputs of dendritic cells research articles based on Web of Science™ from 2006 to 2015 in order to know more about the overview of the global "dendritic cells" research. According to Web of Science™ core collection, the direction of research, countries/regions, scientific research institutions, key journals and the published articles of authors and top 10 cited articles were analysed using the title of dendritic cells. The United States was the leading country in the field of dendritic cells. In the quantity of paper, USA toped the list, followed by Germany and China. The core journal about dendritic cells research was "Journal of Immunology". Number of papers published in the first three were Inserm, University of Pittsburgh and Harvard University. Number of papers published in the former two were Shanghai Jiao Tong University and Zhejiang University in the domestic. The most citation paper was from University of Oxford. The direction of research about dendritic cells were mainly concentrated in immunology, cell biology and oncology. So the main target of the study of dendritic cells in China was the United States, the key research directions were immunology, cell biology and oncology.

Key words: Web of Science™; Dendritic cells; Bibliometrics; Paper

（发表于《中国畜牧兽医》，中文核心，423-430.）

青海高原牦牛PRDM16基因克隆、生物信息学及组织差异表达分析

赵生军[1,2]，张勇[2]，郭宪[1*]

（1.中国农业科学院兰州畜牧与兽药研究所/甘肃省牦牛繁育工程重点实验室，兰州 730050；2.甘肃农业大学动物医学院，兰州 730070）

摘　要：本研究对青海高原牦牛PRDM16（PR domain containing 16）基因部分编码区进行克隆及生物信息学分析，同时对PRDM16基因在雌、雄牦牛背最长肌肌肉组织中的表达差异进行了分析。选取雌、雄青海高原牦牛各5头，屠宰后采集背最长肌肌肉组织，克隆牦牛PRDM16基因部分CDS区序列，分析其生物信息学特征；应用实时荧光定量PCR技术检测PRDM16基因在雌、雄牦牛肌肉组织中表达水平。结果显示：克隆所得片段序列长323bp，与黄牛同源性为100%，编码99个氨基酸，具有MDSl-FVI1（complex locus protein MDS1）家族蛋白功能，具有CATH蛋白功能活性，包含卷曲螺旋等典型结构域，与人甲基转移酶蛋白结构域PR蛋白1有20%的相似性；PRDM16基因在雌性牦牛肌肉组织中表达水平极显著高于雄性牦牛（$P<0.01$）。本试验结果为进一步研究青海高原牦牛PRDM16基因奠定了基础，为牦牛肉品质分析提供参考。

关键词：青海高原牦牛；PRDM16基因；背最长肌；基因表达

Cloning, Bioinformatics and Tissue Differential Expression Analysis of PRDM16 Gene in Qinghai Plateau Yak

ZHAO Shengjun[1,2], ZHANG Yong[2], GUO Xian[1*]

(1. Key Laboratory of Yak Breeding Engineering of Gansu Province, Lanzhou Institute of Husbandry and Pharmaceutical Sciences, Chinese Academy of Agricultural Sciences, Lanzhou 730050, China; 2. College of Veterinary Medicine, Gansu Agricultural University, Lanzhou 730070, China)

Abstract: In this study, the part coding region of PRDM16 gene of the Qinghai plateau yak was cloned and its bioinformatics and differential expression of PRDM16 gene of muscle tissue were analyzed in different sex individual. The PRDM16 gene part of CDS region was cloned from the longissimus dorsi muscle tissue of Qinghai plateau yak, analyzed its characteristics of bioinformatics and detected gene expression levels of different sex individuals in muscle tissue by Real-time PCR. The results showed that the sequence length of the cloned fragment was 323 bp, the homology between Qinghai plateau yak and cattle was 100%. It encoded 99 amino acids, containing domains of MDSl-EVI1 (complex locus protein MDS1) family proteins function, CATH protein function and typical structure of the curly spiral. PRDM16 protein was 20% of the similarity with people methyltransferase protein domain PR protein 1. PRDM16 gene expression of female yak was significantly higher than male yak's in longissimus dorsi muscle tissue (P<0.01). The test results provided not only research foundation for genetic progress but also technical references for yak meat quality analysis.

Key words: Qinghai plateau yak; PRDM16 gene; Longissimus dorsi; Gene expression

（发表于《中国畜牧兽医》，中文核心，363-370.）

小鼠溃疡性结肠炎模型的建立与评价

曹明泽,王旭荣,王 磊,张景艳,王海瑞,李建喜[*],王学智[*]

(中国农业科学院兰州畜牧与兽药研究所/农业部兽用药物创制重点实验室/
甘肃省新兽药工程重点实验室,兰州 730050)

摘 要:本试验旨在通过筛选葡聚糖硫酸钠盐(DSS)最佳浓度建立小鼠溃疡性结肠炎模型。将30只BALB/c小鼠随机分为对照组、3.5% DSS组、5%DSS组,每组10只。小鼠自由饮水5d,每天记录小鼠体重,观察粪便性状,测便潜血。试验结束后测血清TNF-α、结肠组织髓过氧化物酶(MPO)、丙二醛(MDA)、还原性谷胱甘肽(GSH)活性变化等指标,计算结肠重量/长度比值。与对照组相比,两试验组结肠组织中MPO活性显著升高(P<0.05),MDA含量显著升高(P<0.05),GSH含量显著降低(P<0.05)。结果表明,两种浓度葡聚糖硫酸钠盐均可造模成功,5%DSS组小鼠精神状态很差,表现嗜睡、萎靡状态,而3.5%DSS组小鼠精神状态良好,且组织中MPO、MDA、GSH与对照组相比均差异显著(P<0.05),与5%DSS组相比差异不显著(P>0.05),因此,选用浓度为3.5%DSS造模更合适。

关键词:溃疡性结肠炎;疾病活动指数;TNF-α;髓过氧化物酶;丙二醛;还原型谷胱甘肽

Establishment and Evaluation of Mouse Model of Ulcerative Colitis

CAO Mingze, WANG Xurong, WANG Lei, ZHANG Jingyan,
WANG Hairui, LI Jianxi[*], WANG Xuezhi[*]

(Key Laboratory of New Animal Drug Project of Gansu Province/Key Laboratory of Veterinary Pharmaceutics Discovery, Ministry of Agriculture/Lanzhou Institute of Husbandry and Pharmaceutical Science, Chinese Academy of Agricultural Sciences, Lanzhou 730050, China)

Abstract: The experiment was aimed to establish mouse model of ulcerative colitis (uc) by screening the optimum concentration of dextran sulfate sodium salt (DSS). Thirty BALB/c mice were randomly assigned to control group, 3.5% DSS group and 5% DSS group, ten mice each group. Mice drank water freely for 5 days, the body weights of everyday were recorded, stool was observed and stool occult blood was tested. After the experiment, the changes of TNF-α, MPO, MDA and GSH were tested, and the colon weight/length ratio was calculated. Compared with control group, the activity of MPO and content of MDA in the experiment groups were significantly increased (P<0.05), and content of GSH was significantly decreased (P<0.05). 3.5% and 5% DSS both could successfully establish mouse model of ulcerative colitis. Mice in 5% DSS group had poor mental state, such as lethargy, malaise; Mice in 3.5% DSS group were appropriate, the mice mental was good, MPO, MDA and GSH were significantly different compared with control group (P<0.05), but there were no difference compared with 5% DSS group (P>0.05). So 3.5% DSS was more appropriate than 5% DSS to establish mouse model of ulcerative colitis.

Key words: Ulcerative colitis (UC); DAI; TNF-α; MPO; MDA; GSH

(发表于《中国畜牧兽医》，中文核心，171-175.)

紫菀不同极性段提取物对SD大鼠亚慢性毒性试验研究

彭文静,辛蕊华,任丽花,罗永江,王贵波,
罗超应,谢家声,李锦宇,郑继方[*]

(中国农业科学院兰州畜牧与兽药研究所/甘肃省新兽药工程重点实验室/
甘肃省中兽药工程技术研究中心/农业部兽用药物创制重点实验室,兰州 730050)

摘 要:本试验旨在观察紫菀不同极性段提取物对SD大鼠的肝脏毒性损伤。SD大鼠随机分为空白组、石油醚组、乙酸乙酯组、正丁醇组、母液组、75%乙醇组,每组10只,灌胃给药0.34g生药/kg体重,连续28d,观察对大鼠摄食量、体增重、脏器系数、尿液指标、血常规、血清生化指标、肝脏组织抗氧化酶、病理切片等的影响。试验结果显示各组大鼠体增重、肝脏组织抗氧化酶、尿液指标、脏器系数均无显著变化($P>0.05$);雄鼠乙酸乙酯组、母液组白细胞计数(WBC)显著降低($P<0.05$),正丁醇组极显著降低($P<0.01$),雌鼠石油醚组、乙酸乙酯组WBC极显著降低($P<0.01$),正丁醇组显著降低($P<0.05$);雌鼠乙酸乙酯组乳酸脱氢酶(LDH)显著升高($P<0.05$);病理组织学检查可见石油醚组和乙酸乙酯组大鼠肝脏出现轻微的肝索紊乱、颗粒变性,其余组与空白组相比无明显差异。结果表明在0.34g生药/kg体重给药条件下,紫菀的石油醚和乙酸乙酯提取物对大鼠产生轻微的肝脏毒性。

关键词:紫菀;亚慢性毒性;抗氧化作用

Study on Subchronic Toxicity Test of Different Polarity Section Extracts from *Aster tataricus* L. f. on SD rats

PENG Wenjing, XIN Ruihua, REN Lihua, LUO Yongjiang, WANG Guibo,
LUO Chaoying, XIE Jiasheng, LI Jinyu, ZHENG Jifang[*]

(Key Laboratory of Veterinary Pharmaceutics Development, Ministry of Agriculture/Engineering & Technology Research Center of Traditional Chinese Veterinary Medicine of Gansu Province/ Key Laboratory of New Animal Drug Project of Gansu Province/Lanzhou Institute of Husbandry and Pharmaceutical Sciences of CAAS, Lanzhou 730050, China)

Abstract: The aim of the experiment was to observe the liver toxicity of different polarity section extracts from *Aster tataricus* L. f. on SD rats. The subchronic test was performed via the daily oral administration of *Aster tataricus* L. f. at dose of 0.34 g/kg body weight in SD rats which were divided into six groups (control group, petroleum ether group, ethyl acetate group, N-butyl alcohol group, mother liquid group and 75% ethanol group), observing the effects on food consumption, body weight gain, viscera coefficient, urine examination, blood routine examination, serum biochemistry indices, liver antioxidant function analysis and histopathological observation. There was no significant changes of body weight gain, viscera coefficient, urine examination and liver antioxidant function among six groups ($P>0.05$), but WBC significantly and extremely significantly decreased in male rats of ethyl acetate group, mother liquid group and N-butyl alcohol group ($P<0.05$; $P<0.01$), and significantly and extremely significantly decreased in female rats of N-butyl alcohol group and petroleum ether group, ethyl acetate group ($P<0.05$; $P<0.01$); The LDH significantly increased in female rats of ethyl acetate group ($P<0.05$); Slight congestion and necrosis were showed in liver in petroleum ether group and ethyl acetate group, there was no differences observed in other groups. The extracts of *Aster tataricus* L. f., especially petroleum ether extract and ethyl acetate extract could cause slight liver toxicity under the dose of 0.34 g/kg body weight.

Key words: *Aster tataricus* L. f.; Subchronic toxicity; Antioxidation

（发表于《中国畜牧兽医》，中文核心，147-156.）

GnIH与INH表位多肽疫苗主动免疫对甘肃高山细毛羊生殖激素的影响

张玲玲，杨博辉*，岳耀敬，郭婷婷，刘建斌，袁　超，
牛春娥，冯瑞林，郭　健，孙晓萍，刘善博

（中国农业科学院兰州畜牧与兽药研究所，兰州　730050）

摘　要： 本试验通过合成促性腺激素类的抑制激素——抑制素（inhibit hormone，INH）与促性腺激素抑制激素（gonadotropin inhibiting hormone，GnIH）的INH表位多肽疫苗和GnIH表位多肽疫苗，主动免疫甘肃高山细毛羊并对促卵泡素（follicle stimulating hormone，FSH）和促黄体素（luteinizing hormone，LH）激素水平进行比较分析。试验选用15只处在非发情季节的3.5岁的甘肃高山细毛羊，注射INH表位多肽疫苗与GnIH表位多肽疫苗分别为表位多肽疫苗A组和表位多肽疫苗B组，对照组注射等量生理盐水。采血后进行免疫，每隔7d免疫和采血，同时在发情后每隔15min采集一次血液，分离血清，用ELISA试剂盒检测抗体水平及FSH和LH激素变化。结果表明，表位多肽疫苗A组和表位多肽疫苗B组在初次免疫后都产生了相应的抗体，且表位多肽疫苗B组较表位多肽疫苗A组产生的抗体浓度高，75、90、105min时，表位多肽疫苗B组相对表位多肽疫苗A组显著促进了FSH的分泌水平（$P<0.05$）。在45~105min时，表位多肽疫苗B组相对表位多肽疫苗A组显著促进了LH的分泌水平（$P<0.05$），且在90到105min时表位多肽疫苗A组和表位多肽疫苗B组血清中FSH和LH的水平均显著高于对照组（$P<0.05$）。本试验结果表明，INH表位多肽疫苗和GnIH表位多肽疫苗主动免疫促进了甘肃高山细毛羊的FSH、LH的分泌，GnIH表位多肽疫苗在甘肃高山细毛羊上具有更好的免疫作用，本研究为该表位多肽疫苗在绵羊上的运用奠定了基础。

关键词： 促性腺激素抑制激素；抑制素；主动免疫；甘肃高山细毛羊；生殖激素

Effects of Active Immunization with GnIH and INH Epitope Peptide Vaccine on Reproductive Hormones of Gansu Alpine Fine-wool Sheep

ZHANG Lingling, YANG Bohui*, YUE Yaojing, GUO Tingting,
LIU Jianbin, YUAN Chao, NIU Chune, FENG Ruilin, GUO Jian,
SUN Xiaoping, LIU Shanbo

(Lanzhou Institute of Animal Science and Veterinary Pharmaceutics, Chinese Academy of Agricultural Sciences, Lanzhou 730050, China)

Abstract: This study aimed to detect the reproductive hormones levels of follicle-stimulating hormone (FSH) and luteinizing hormone (LH) in Gansu Alpine Fine-wool sheep by active immunization of two synthesized epitope peptide vaccines A and B which were combined inhibit hormone (INH) and gonadotropin inhibit hormone (GnIH). 15 Gansu Alpine Fine-wool sheep were selected which were 3.5 year old and lived in anoestrus season, 3 groups each had 5 sheep, GnIH epitope peptide vaccine and INH epitope peptide vaccine were named epitope peptide vaccine A and B groups, while blank group was injected with saline. The blood was collected before immunization and immuned, and bloods were collected once every 7 days, meanwhile collected blood every 15 min after estrus. The serum was used to detect antibody, FSH and LH levels by ELISA Kits. The results showed that the epitope peptide vaccine A and B groups produced the corresponding antibodies after the initial immunization, and epitope peptide vaccine B group produced more antibodies than epitope peptide vaccine A group. Epitope peptide vaccine B group promoted the FSH secretion levels was significantly higher than epitope peptide vaccine A group in 75, 90 and 105 min ($P<0.05$); In 45 to 105 minutes, the LH secretion levels of epitope peptide vaccine B group was significantly higher than epitope peptide vaccine A group ($P<0.05$), while the LH and FSH levels in epitope peptide vaccine A and B groups were significantly higher than control group ($P<0.05$). This study indicated that the epitope peptide vaccine A and B groups could promote the FSH and LH secretion levels through active immunity in Gansu Alpine Fine-wool sheep and had a good immune function. This study laid a foundation for the application of epitope peptide vaccine on sheep.

Key words: Gonadotropin inhibit hormone (GnIH); Inhibit hormone (INH); Active immune; Gansu Alpine Fine-wool sheep; Reproductive hormones

(发表于《中国畜牧兽医》，中文核心，1 039-1 044.)

奶牛饲料中酿酒酵母菌的分离鉴定及其发酵液的体外抑菌活性研究

刘龙海[1]，李新圃[1]，杨 峰[1]，罗金印[1]，张 哲[1,2]，李宏胜[1*]

（1. 中国农业科学院兰州畜牧与兽药研究所/农业部兽用药物创制重点实验室/甘肃省新兽药工程重点实验室，兰州 730050；2. 甘肃农业大学动物医学院，兰州 730070）

摘 要：本试验旨在分离鉴定甘肃地区奶牛场饲料中的酿酒酵母菌，进而研究酿酒酵母菌发酵液的体外抑菌活性。通过YPD固体培养基对采集的5份饲草和5份饲料样品中的酵母菌进行分离，分别进行生化鉴定和26S rDNA测序鉴定，进而检测酿酒酵母发酵液的体外抑菌活性。结果表明，从样品中分离出4种共12株酵母菌，分别为3株酿酒酵母菌（Y1、Y7和Y11）、4株汉逊德巴利酵母菌、3株浅黄隐球酵母菌和2株戴尔有孢圆酵母菌；3株酿酒酵母菌的发酵液对奶牛乳房炎金黄色葡萄球菌、大肠杆菌和无乳链球菌的抑菌活性存在差异，其中酵母菌Y7和Y11的发酵液对3种病原菌均具有抑菌活性，而Y1的发酵液只对大肠杆菌具有抑制作用。本试验为进一步筛选性能优异的酿酒酵母菌用于微生态制剂研发奠定了基础。

关键词：酿酒酵母菌；奶牛饲料；分离鉴定；抑菌活性

The Study of Isolation, Identification of *S. cerevisiae* from Dairy Feed and its Antibacterical Activity in the Fermentation Brothe in vitro

LIU Longhai[1], LI Xinpu[1], YANG Feng[1], LUO Jinyin[1], ZHANG Zhe[1,2], LI Hongsheng[1*]

（1. Key Laboratory of New Animal Drug Project of Gansu Province/Key Laboratory of Veterinary Pharmaceutical Discovery, Ministry of Agriculture/Lanzhou Institute of Animal Husbandry and Pharmaceutical Sciences, Chinese Academy of Agricultural Sciences, Lanzhou 730050, China；2. College of Veterinary Medicine, Gansu Agricultural University, Lanzhou 730070, China）

Abstract: The aim of the study was to isolate and identify *S. cerevisiae* in the dairy feed and analyze antibacterial activity of *S. cerevisiae* fermentation brothe in vitro. 5 forage samples and 5 feed samples were collected to isolate yeast by YPD solid medium. Biochemical identification and 26S rDNA sequence were used to identify the isolated strains. The antibacterial activity of *S. cerevisiae* strains was tested by cylinder plate method. A total of 12 yeast belonged to 4 species were isolated and identified, including 3 *S. cerevisiae* (Y1, Y7 and Y11), 4 Debaryomyces hansenii, 3 Cryptococcus flavescens and 2 Torulaspora debrueckii. The antibacterial activities of the 3 *S. cerevisiae* fermentation brothe to Staphylococcus aureus, *E. coli* and Streptococcus agalactiae were different. Y7 and Y11 fermentation brothe showed antibacterial activities to all the tested pathogens, while Y1 inhibit *E. coli* alone. This study laid the foundations for further research of probiotics containing *S. cerevisiae* with excellent performance.

Key words: *S. cerevisiae*; Cows feed; Isolation and identification; Antibacterical activity

(发表于《中国畜牧兽医》，中文核心，1 226-1 231.)

黄花补血草总黄酮亚急性毒性试验

刘 宇[1*]，曾豪杰[2]，尚若锋[1]，郝宝成[1]，杨 珍[1]，
郭文柱[1]，程富胜[1]，王学红[1]，梁剑平[1]

（1.中国农业科学院兰州畜牧与兽药研究所/农业部兽用药物创制重点实验室/甘肃省新兽药工程重点实验室，兰州 730050；2.国土资源部兰州矿产资源监督检测中心，兰州 730050）

摘 要：试验旨在评价黄花补血草总黄酮的安全性，为今后系统研究其药理作用和作为临床安全用药提供试验依据。选取80只1日龄、体重接近的大白鼠，随机分为2大组（Ⅰ、Ⅱ组），其中每大组分成黄花补血草总黄酮高、中、低剂量组和空白对照组，药物组按照15、10、5g/kg的剂量灌胃给药，对照组不给药，在给药10（Ⅰ组）、21d（Ⅱ组）后采血测定试验指标。结果显示，Ⅰ组中、低剂量组的大白鼠体增重与对照组相比差异不显著（P>0.05），而高剂量组体增重显著低于对照组（P<0.05）。Ⅱ组低剂量组的大白鼠体增重与对照组相比差异不显著（P>0.05），而中、高剂量组体增重显著低于对照组（P<0.05）；试验组大白鼠的各脏器系数与对照组差异不显著（P>0.05）；在病理解剖过程中发现，随着剂量组浓度的增大，黄花补血草总黄酮对主要器官有一定的病理性损伤，21d中、高剂量组的肺部组织出现白色肿块，肝脏出现红色出血点。结果表明，黄花补血草总黄酮毒性低，可以安全用于动物，但其剂量不宜过高。

关键词：黄花补血草总黄酮；亚急性毒性试验；大鼠；安全评价

Subacute Toxicity Test of the Total Flavonoids of *Limonium aureum*（L.）Hill.

LIU Yu[1*], ZENG Haojie[2], SHANG Ruofeng[1], HAO Baocheng[1], YANG Zhen[1],
GUO Wenzhu[1], CHENG Fusheng[1], WANG Xuehong[1], LIANGJianping[1]

（1. Key Laboratory of New Animal Drug Project, Gansu Province/Key Laboratory of Veterinary Pharmaceutical Development, Ministry of Agriculture/Lanzhou Institute of Husbandry and Pharmaceutical Sciences of CAAS, Lanzhou 730050, China; 2. Lanzhou Testing and Quality Supervision Center for Geological and Mineral Products, Ministry of Land Resources, Lanzhou 730050, China）

Abstract: To evaluate the safety of the total flavonoids of *Limonium aureum* (L.) Hill. and lay an experimental basis to study the pharmacological effects and clinical safety in the future, 80 Wistar rats with same age and body weight were randomly divided into two groups (group Ⅰ, Ⅱ), and every group was divided into the high dose group (HG), medium dose group (MG), low dose group (LG) and control group (CG), respectively, of which the mice were given to the total flavonoids of *Limonium aureum* (L.) Hill. by gavage at the doses of 15, 10 and 5 g/kg while the mice of CG were not given any medicine. The blood sample were collected and experimental indexes were measured at 10 (group Ⅰ) and 21 d (group Ⅱ). The results showed that comparing with CG, the weight gain of rats in MG and LG of group Ⅰ had no significant difference ($P>0.05$), and that of HG were significantly decreased ($P<0.05$); The weight gain of rats in LG of group Ⅱ had no significant difference ($P>0.05$), and that of MG and HG were significantly decreased ($P<0.05$); Organ indexes of experimental groups had no significant difference with CG ($P>0.05$); With increasing concentration of the total flavonoids of *Limonium aureum* (L.) Hill., there was pathological damage to major organs and white lumps in lung and red bleeder existed in liver of rats in HG when given total flavonoids of *Limonium aureum*. (L.) Hill. for 21 d. The results indicated that total flavonoids of *Limonium. aureum* (L.) Hill. was low toxicity and could be safely used for animals with minding the dose was not too high.

Key words: Total flavonoids of *Limonium aureum* (L.) Hill.; Subacute toxicity test; Rat; Safety evaluation

（发表于《中国畜牧兽医》，中文核心，2 135-2 142.）

抗鸡球虫药常山口服液对小鼠的急性毒性作用评价

王 玲[1]，郭志廷[1*]，张晓松[2]，林春全[3]，罗小琴[4]，杨 峰[1]，杨 珍[1]

（1.中国农业科学院兰州畜牧与兽药研究所/农业部兽用药物创制重点实验室/甘肃省新兽药工程重点实验室，兰州 730050；2.甘肃农业大学动物医学院，兰州 730070；3.兰州理工大学生命科学与工程学院，兰州 730050；4.兰州市动物卫生监督所，兰州 730050）

摘 要：为评价抗鸡球虫病药物常山口服液的安全性，本试验采用改良寇氏法和剂量递增法对小鼠进行急性毒性试验研究，并通过病理组织学观察来明确常山口服液毒性损伤的主要靶器官。取60只昆明系小鼠，雌雄各半，随机分为6组，1个对照组和5个给药组，给药组分别按1.00g/kg、1.40g/kg、1.96g/kg、2.74g/kg和3.84g/kg体重的剂量灌胃给药，连续观察7d，记录小鼠急性毒性反应过程，并运用改良寇氏法公式计算半数致死量（LD_{50}）及LD_{50}的95%可信限。结果显示，小鼠经口灌胃LD_{50}为2.096 1g/kg体重，LD_{50}的95%可信限为1.741 4～2.542 9g/kg体重；剖检可见1.96、2.74和3.84g/kg体重剂量组部分小鼠的肝脏、肾脏出现较严重的水肿、充血和明显的白色坏死灶；病理组织学检查显示小鼠出现急性肝脏损伤，表现为肝细胞退行性改变，病变主要是小叶周边变性、坏死及崩解，而肾脏损伤表现为肾淤血、间质水肿、肾曲小管上皮浊肿和广泛变性，各主要脏器以肝脏损害较为严重，可确定常山口服液损伤的靶器官主要为肝脏。结果表明，常山口服液按常规剂量使用是安全的，临床用于抗球虫病安全性较高，但大剂量、长疗程的用药会出现毒性反应，由于急性毒性试验观察时间较短，故还应结合长期毒性试验的毒性表现及多种检查结果综合分析评价其毒性。

关键词：常山口服液；急性毒性；组织病理；靶器官

Study on Acute Toxicity Test of Anticoccidial Oral Liquid from Dichroa febrifuga in Mice

WANG Ling[1], GUO Zhiting[1*], ZAHNG Xiaosong[2], LIN Chunquan[3],
LUO Xiaoqin[4], YANG Feng[1], YANG Zhen[1]

（1. Key Laboratory of New Animal Drug Project, Gansu Province/Key Laboratory of Veterinary Pharmaceutics Discovery, Ministry of Agriculture/Lanzhou Institute of Animal Science and Veterinary Pharmaceutics, Chinese Academy or Agricultural Sciences, Lanzhou 730050, China;
2. College of Veterinary Medicine, Gansu Agricultural University, Lanzhou 730070, China;
3. School of Life Science and Engineering, Lanzhou University of Technology, Lanzhou

730050, China; 4. Lanzhou Institute of Animal Health Supervision, Lanzhou 730050, China)

Abstract: The purpose of this study was to evaluate drug safety of Dichroa febrifuga oral liquid (DFOL), and the acute toxicity test of DFOL was established in mice based on modified Karber's method and dose increasing method, and histopathological examination had been applied to explore the main target organs of its toxic injury. The dosage was determined by preliminary experiment, a total of 60 Kunming mice were selected and divided randomly into one control group and five experimental groups, DFOL was administered by intragastric gavage at five different doses (1.00, 1.40, 1.96, 2.74, 3.84 g/(kg·BW)), and observed for 7 days continuously, the LD50 value and 95% confidence interval of LD50 were determined by the modified Karbers method. It showed that LD50 was 2.096 1 g/(kg·BW) and its 95% confidence interval was 1.741 4 to 2.542 9 g/(kg·BW). The liver and kidney of some mice in 1.96, 2.74 and 3.84 g/(kg·BW) drug groups appeared severe edema, congestion and white necrotic foci by necropsy. Acute liver injury was found in mice by histopathological examination and the liver cell showed degenerative change, kidney damage showed renal congestion, interstitial edema, renal tubule epithelial cloudy swelling and degeneration. The liver was damaged seriously in the main organ and the main toxic organ of DFOL was liver. The test showed that the toxicity of DFOL was lower and it had high security in a conventional doses for clinical anti-coccidiosis application, however, high dose, long course of medication had toxicity, and the toxicity of DFOL should be analyzed and evaluated combining with long-term toxicity tests and many other inspection results.

Key words: Dichroa febrifuga oral liquid (DFOL); Acute toxicity; Tissue pathology; Target organ

(发表于《中国畜牧兽医》，中文核心，247-252.)

INH表位多肽疫苗抗体间接ELISA测定方法的建立及优化

张玲玲[1]，岳耀敬[1]，冯瑞林[1]，李红峰[2]，郭婷婷[1]，袁　超[1]，牛春娥[1]，刘建斌[1]，孙晓萍[1]，韩吉龙[1]，刘善博[1]，杨博辉[1*]

（1.中国农业科学院兰州畜牧与兽药研究所/中国农业科学院羊育种工程技术研究中心，兰州　730050；2.甘肃省武威市天祝县畜牧技术推广站，天祝　733200）

摘　要：本试验在间接ELISA的基础上建立了一种检测抑制素（inhibit hormone，INH）表位多肽疫苗抗体的方法并对ELISA试剂盒进行优化，为今后测定主动免疫后细毛羊体内INH表位多肽疫苗抗体提供理论参考。在前人研究的基础上，采用间接ELISA法测定血清中INH表位多肽疫苗抗体的含量，通过对不同试验条件的控制摸索出最佳试验条件，即以脱脂奶粉为封闭液，INH及GnIH以表位多肽疫苗抗体稀释度为20 000倍，最佳反应时间60min，最佳显色时间为15min。本试验建立了一种检测细毛羊体内INH及GnIH抗体的方法，为今后对INH表位多肽疫苗抗体的研究提供了参考。

关键词：抑制素；细毛羊；主动免疫；生殖激素

Establishment and Optimization of INH Epitope Peptide Vaccine for Detection of Antibody

ZHANG Lingling[1], YUE Yaojing[1], FENG Ruilin[1], LI Hongfeng[2], GUO Tingting[1], YUAN Chao[1], NIU Chune[1], LIU Jianbin[1], SUN Xiaoping[1], HAN Jilong[1], LIU Shanbo[1], YANG Bohui[1*]

（1. Lanzhou Institute of Animal Science & Veterinary Pharmaceutics, Chinese Academy of Agricultural Sciences, Lanzhou 730050, China; 2. Tianzhu County Animal Husbandry Technology Extending Stations, Tianzhu 733200, China）

Abstract: In this study, an indirect ELISA method was established to detect inhibin hormone (INH) epitope peptide vaccine antibody, it would provide oretical reference for the determination of Fine-wool sheep after active immune body INH epitope peptide vaccine antibody. On the basis of the predecessors, using indirect ELISA method to determinate serum INH epitope peptide antibody levels of sheep, and through control different experimental conditions to look for the best experimental conditions. Through explorating the experimental conditions, finally, the testing experiment conditions were determined, which was blocked solution with skimmed milk powder, INH and GnIH synthetic peptides dilution degrees for 20 000 times, the optimum reaction time was 60 min, the best color action time was 15 min. In this experiment, a kind of method to detection antibody in the body after INH active immune sheep was built, it would provide a reference for future research.

Key words: Inhibin hormone; Fine-wool sheep; Active immunization; Reproductive hormone

(发表于《中国畜牧兽医》，中文核心，2 156–2 163.)

紫菀乙醇提取物对豚鼠离体气管平滑肌收缩功能的影响

彭文静[1]，辛蕊华[1,2]，刘艳[1]，罗永江[1,2]，王贵波[1,2]，
罗超应[1,2*]，谢家声[1,2]，李锦宇[1,2]，郑继方[1,2*]

（1. 中国农业科学院兰州畜牧与兽药研究所，兰州 730050；2. 甘肃省新兽药工程重点实验室/甘肃省中兽药工程技术研究中心/农业部兽用药物创制重点实验室，兰州 730050）

摘 要：本试验旨在研究紫菀乙醇提物对豚鼠离体气管平滑肌收缩功能的影响。采用离体气管恒温灌流的方法，将豚鼠气管制成气管螺旋条，通过BL-420F生物机能实验系统测定其张力的变化，观察紫菀乙醇提物对离体气管平滑肌静息状态下的舒张作用，以及在氯化乙酰胆碱（Ach）、磷酸组胺（His）、$CaCl_2$条件下和无钙下Ach诱导细胞内钙释放和外钙内流所致两种收缩条件下对离体气管平滑肌张力变化的影响。试验结果显示，低浓度紫菀乙醇提取物（0.002~0.008g/mL）对静息状态下豚鼠离体气管平滑肌具有一定的收缩作用，高浓度（0.008~0.196g/mL）时具有舒张作用；低、中、高3个剂量组均可抑制Ach、His和$CaCl_2$引起的气管平滑肌收缩强度，使各致痉剂的量效曲线非平行右移并降低最大效应。紫菀乙醇提取物对His的抑制作用强度最大，其次为Ach，最后是$CaCl_2$，且均呈剂量依赖性。以上结果表明紫菀乙醇提物具有舒张气管平滑肌的作用，其机制可能与抑制豚鼠气管平滑肌M受体、H_1受体和阻断Ca^{2+}通道从而抑制细胞Ca^{2+}内流有关。

关键词：紫菀乙醇提取物；豚鼠；气管平滑肌；张力变化

Effects of Alcohol Extract of *Aster tataricus* L. f. on the Contraction of Guinea Pig Tracheal Smooth Muscle in vitro

PENG Wenjing[1], XIN Ruihua[1,2], LIU Yan[1], LUO Yongjiang[1,2], WANG Guibo[1,2],
LUO Chaoying[1,2], XIE Jiasheng[1,2], LI Jinyu[1,2], ZHENG Jifang[1,2*]

(1. Lanzhou Institute of Husbandry and Pharmaceutual Sciences of CAAS, Lanzhou 730050, China; 2. Key Laboratory of New Animal Drug Project of Gansu Province/Engineering & Technology Research Center of Traditional Chinese Veterinary Medicine of Gansu Province/Key Laboratory of Veterinary Pharmaceutics Development, Ministry of Agriculture, Lanzhou 730050, China)

Abstract: The aim of the experiment was to study the effect of alcohol extract of *Aster tataricus* L. f. on the contraction of isolated guinea pig tracheal smooth muscle. The guinea pig tracheal smooth muscle was prepared by isolated tracheal thermostatic perfusion method. The effects of *Aster tataricus* L. f. on tracheal smooth muscle under the resting state and contraction stimulated by acetylcholine (Ach), histamine (His), $CaCl_2$, Ca^{2+} release and influx in cells without calcium were measured by BL-420F biological function experiment systems. The results showed that *Aster tataricus* L. f. could induce the contraction of isolated trachea at resting state at a low concentrations of 0.002 to 0.008 g/mL and exerted a relaxation on the isolated trachea when at the higher concentrations of 0.008 to 0.196 g/mL. All of the three concentrations of *Aster tataricus* L. f. could inhibit the tension stimulated by Ach, His and $CaCl_2$ which were antispasmodic agents, led to the non-parallelled rightward moving of cumulative concentration-response curve and the decline of maximal responses. The suppressive effect on His was the strongest, followed by Ach and $CaCl_2$, and the suppression could counteract the contraction induced by extracellular Ca^{2+} influx significantly, and they were all expressed concentration-dependent manner. These results indicated the alcohol extract of *Aster tataricus* L. f. could relax tracheal smooth muscle by acting on M-re-ceptor and histamine receptor and blocking Ca^{2+} passage, thus inhibiting extracellular Ca^{2+} influx.

Key words: Alcohol extract of *Aster tataricus* L. f.; Guinea pig; Tracheal smooth muscle; Changes of tension

（发表于《中国畜牧兽医》，中文核心，1 572-1 578.）

藏药雪山杜鹃叶挥发油成分的GC-MS分析

郭 肖，周绪正，朱 阵，文 豪，张吉丽，张继瑜[*]

（中国农业科学院兰州畜牧与兽药研究所，兰州 730050）

摘 要： 目的：通过研究雪山杜鹃叶挥发油的化学成分，为雪山杜鹃的药用及开发利用提供依据。方法：采用水蒸气蒸馏法提取雪山杜鹃叶挥发油，并运用气相色谱—质谱联用技术（GC-MS）对其挥发油进行了系统的分离和鉴定。结果：从雪山杜鹃叶中共鉴定出71种挥发性成分，占挥发油总量的80.51%。雪山杜鹃叶挥发油主要成分为：芳樟醇10.66%、白菖烯4.35%、epimanoyl oxide3.85%、α-松油醇3.46%、α-杜松醇3.27%。结论：本实验采用GC-MS分析方法对雪山杜鹃叶挥发油进行分析，可为合理开发利用雪山杜鹃资源提供科学依据。

关键词： 雪山杜鹃；挥发油；GC-MS

（发表于《中药材》，中文核心，1 319–1 322.）

奶牛乳房炎疫苗研究进展

刘龙海[1]，李新圃[1]，杨 峰[1]，罗金印[1]，王旭荣[1]，张 哲[2]，罗增辉[3]，李宏胜[1*]

（1.中国农业科学院兰州畜牧与兽药研究所/农业部兽用药物创制重点实验室/甘肃省新兽药工程重点实验室，兰州 730050；2.甘肃农业大学动物医学院，兰州 730070；3.莒县畜牧局桑园兽医站 山东莒县 276500）

摘 要：乳房炎是奶牛最常发生的疾病之一，降低了牛奶的产量和质量，严重时会致使奶牛淘汰甚至死亡，给奶牛养殖业造成巨大的经济损失。基于疫苗预防安全、高效、无毒等特点，奶牛乳房炎疫苗受到广大学者的重视。论文对奶牛乳房炎的免疫机制，奶牛乳房炎灭活苗、活苗、亚单位疫苗、载体疫苗和核酸疫苗的研究进展及疫苗研发存在的难点、解决方案和发展前景进行了综述，以期对奶牛乳房炎疫苗研究有一定的借鉴作用。

关键词：奶牛；乳房炎；疫苗；免疫机制

（发表于《动物医学进展》，中文核心，IF：综述.100-105.）

中药对牛免疫调节作用研究进展

刘 艳[1]，罗永江[1]，辛蕊华[1]，彭文静[1]，梁 歌[2]，郑继方[1*]

（1.中国农业科学院兰州畜牧与兽药研究所，兰州 730050；
2.四川省畜牧科学研究院，成都 619066）

摘 要：中药的性味归经不同、炮制方法不同、配伍不同，其药理作用也不同，已有研究表明，多种单味中药和复方中药具有免疫调节作用。以传统中药为原料研制的免疫调节剂，凭借其高安全、低残留、低毒性、不易产生耐药性等独特优势，在动物临床上的应用越来越广泛。论文分别从非特异性免疫、体液免疫、细胞免疫等方面对单味中药及复方中药制剂对牛免疫功能的调节作用的研究进展进行综述，以期为进一步研制效果良好、使用方便、成本低、副作用小的牛中药免疫调节剂及其临床应用提供参考。

关键词：中药；牛；免疫调节

（发表于《动物医学进展》，中文核心，IF：综述.95-98.）

黄芪多糖对"树突状细胞"形态和功能影响研究进展

边亚彬[△]，张景艳[△]，王 磊，王旭荣，张 康，王学智，孟嘉仁，李建喜[*]

（中国农业科学院兰州畜牧与兽药研究所/甘肃省中兽药工程技术研究中心，兰州 730050）

摘 要：树突状细胞（DC）是目前发现功能最强的一类抗原递呈细胞（APC），因其在机体免疫应答及免疫功能发挥中担当重要角色，成为免疫学研究热点。多糖是中药黄芪的重要有效成分，由己糖醛酸、果糖、阿拉伯糖、半乳糖醛酸等组成，具有免疫调节、抗肿瘤、抗应激和抗氧化等作用。黄芪多糖的免疫调节一直是人们研究的重点，研究发现其可通过促进DC成熟而发挥免疫调节作用。论文从DC的形态和功能变化、DC的成熟和DC表面标志CD80、CD86、CD83和相关细胞因子表达等方面，综合分析了黄芪多糖对DC抗原递呈能力的影响。

关键词：黄芪多糖；树突状细胞；细胞因子

（发表于《动物医学进展》，中文核心，IF：综述.86-89.）

奶牛子宫内膜炎病因学研究进展

那立冬[1,2]，王东升[1]，董书伟[1]，张世栋[1]，
闫宝琪[1]，杨洪早[1,2]，桑梦琪[1]，严作廷[1*]

（1.中国农业科学院兰州畜牧与兽药研究所，兰州　730050；
2.甘肃农业大学动物医学院，兰州　730070）

摘　要：奶牛子宫内膜炎是奶牛养殖业最常见的疾病之一，每年因该病给养殖业带来的损失不可估量。引起该病的主要病因除了有细菌、真菌、支原体等一系列病原微生物以外，激素紊乱、遗传、疾病等奶牛自身因素以及营养、环境和不科学的饲养管理因素同样也会导致本病发生。因其病因因素较多，加大了治疗难度。因此，论文以近年来国内外最新的研究资料为基础，从诱发该疾病的病原微生物、奶牛自身因素和不当的人为因素等方面对奶牛子宫内膜炎的病因进行系统的综述，以期为预防和治疗此类疾病提供参考。

关键词：奶牛；子宫内膜炎；病因；病原微生物

（发表于《动物医学进展》，中文核心，IF：综述.103-107.）

仔猪腹泻的病因及中药防治研究进展

杨洪早[1,2]，王东升[1]，董书伟[1]，张世栋[1]，
那立冬[1,2]，闫宝琪[1]，桑梦琪[1]，严作廷[1*]

（1.中国农业科学院兰州畜牧与兽药研究所，兰州 730050；
2.甘肃农业大学动物医学院，兰州 730070）

摘 要：仔猪腹泻是集约化养猪生产条件下的一种典型的多因素性疾病，是目前较为严重的仔猪疾病之一，也是引起仔猪死亡的重要原因。随着化学抗病毒药物的禁用与抗生素耐药性的产生，中药及其复方制剂在动物疾病防控中已成为疫苗的有益补充，开展中药防治仔猪腹泻具有一定的实践意义。论文主要从仔猪腹泻的发病原因与临床特征、中药治疗及日常管理方面进行阐述总结，为诊断与治疗该病提供一定的参考。

关键词：仔猪；腹泻；细菌；病毒；中药

（发表于《动物医学进展》，中文核心，IF：综述.89-93.）

可递送siRNA的非病毒纳米载体的设计

赵晓乐，孔晓军，李剑勇*

中国农业科学院兰州畜牧与兽药研究所/农业部兽用药物创新重点实验室/
甘肃省新兽药工程重点实验室

摘 要：RNA干扰（RNA interference，RNAi）技术可让不同水平的基因沉默，主要是通过双链RNA（dsRNA）在体内诱导靶基因mRNA产生特异性降解而实现的。小干扰RNA（small interfering RNA，siRNA）是RNAi的效应分子，需要借助递送载体进入细胞内发挥治疗作用。纳米载体具有很多优势：增加生物膜通透性、控制药物释放速度、改变体内分布、提高生物利用度等。本文综述了siRNA非病毒递送载体的相关研究，重点阐述了其结构和生物学特征。

关键词：小干扰RNA；纳米载体；非病毒递送载体

The Design of Non-viral Nanometer Vectors for Delivery of Smallinterfering RNA

ZHAO Xiaole, KONG Xiaojun, LI Jianyong

(Key Laboratory of New Animal Drug Project of Gansu Province/Key Laboratory of Veterinary Pharmaceutical Development, Ministry of Agriculture/Lanzhou Institute of Husbandry and Pharmaceutical Science of CAAS, Lanzhou 730050, China)

Abstract: RNA interference (RNAi) is a method that allows double-stranded RNA (dsRNA) to induce the target gene mRNA to degrade in the body and lead to different levels of gene silencing. Small interfering RNA (siRNA) is the effector molecules of RNAi, and needs the delivery vehicle into the cells to play a therapeutic role. Nanoscale carrier can adjust the rate of drug release, increase permeability of biofilm, change distribution in the body, and improve bioavailability. In this manuscript, materials, structure and biological characteristics of existing non-viral nanometer vectors delivery for siRNA are summarized.

Key words: Small interfering RNA; Nanoscale vectors; Non-viral vectors

（发表于《国际药学研究杂志》，中文核心，IF：综述.677-681.）

动物抗寄生虫药物的研究与应用进展

张吉丽,李 冰,张继瑜

(中国农业科学院兰州畜牧与兽药研究所/农业部兽药创制重点实验室,兰州 730050)

关键词:动物;抗原虫药;抗蠕虫药;杀虫药;研究应用;展望

Progress in Research and Application of Animal Antiparasitic Drugs

ZHANG Jili, LI Bing, ZHANG Jiyu

(Key Laboratory of Veterinary Pharmaceutics Discovery, Ministry of Agriculture/Lanzhou Institute of Animal Science and Pharmaceutical Sciences, Chinese Academy of Agricultural Sciences, Lanzhou 730050, China)

Key words: Animal; Antiprotozoal; Anthelmintic; Insecticide; Research and application; Prospect

(发表于《黑龙江畜牧兽医》,中文核心,IF:综述.74-79.)

肝片吸虫病的研究进展

张吉丽，朱 阵，李 冰，周绪正，张继瑜

（中国农业科学院兰州畜牧与兽药研究所/农业部兽药创制重点实验室，兰州 730050）

关键词：肝片吸虫；流行病学特点；临床表现；诊断方法；研究现状；防治措施

Research progress in hepatic fascioliasis

ZHANG Jili, ZHU Zhen, LI Bing, ZHOU Xuzheng, ZHANG Jiyu

(Key Laboratory of Veterinary Pharmaceutics Discovery, Ministry of Agriculture/Lanzhou Institute of Husbandry and Pharmaceutical Sciences of CAAS, Lanzhou 730050, China)

Key words: Fasciola hepatica; Epidemiological characteristic; Clinical manifestation; Diagnostic method; Research status; Prevention and control measure

（发表于《黑龙江畜牧兽医》，中文核心，IF：综述.58–61, 65.）

纤维素酶及其在中药发酵中的运用

苏贵龙,李建喜

(中国农业科学院兰州畜牧与兽药研究所/甘肃省中兽药工程技术研究中心,兰州 730050)

关键词:纤维素;纤维素酶;来源;产纤维素酶微生物;运用;发酵中药;生物转化

The Cellulase and its Application in the Fermentation of Traditional Chinese Medicine

SU Guilong, LI Jianxi

(Technology Research Center of Traditional Chinese Veterinary Medicine of Gansu Province/Lanzhou Institute of Husbandry and Pharmaceutical Sciences of CAAS, Lanzhou 730050, China)

Key words: Cellulose; Cellulase; Source; Microorganisms producing cellulase; Application; Fermented traditional Chinese medicine; Biotransformation

(发表于《黑龙江畜牧兽医》,中文核心,IF:综述.65-67,72.)

饮食与疾病：由牛奶致癌说引发的思考*

罗超应[1]，罗磐真[2]，李锦宇[1]，王贵波[1]，谢家声[1]

摘　要： 从"一杯牛奶强壮一个民族"到"牛奶致癌"，着实让人吃惊。尽管国家权威机构、主流媒体曾多次辟谣，但"牛奶致癌说"在网上却依然甚是流行。从复杂性科学的角度在对其进行认真考察分析后认为，转变科学观念，并以复杂性科学为指导，从整体角度正确理解饮食与疾病"相关性并不等于因果性"的复杂性关系，做到既不要恐慌牛奶的促癌作用，也不能忽视西方的膳食经验教训，保持膳食营养平衡，从而做到最大限度地利用好牛奶这一健康资源，而又不受害于它。

关键词： 牛奶，促癌，相关性，因果性，复杂性科学

Diet and Disease: Thinking from the Rumer of "Milk Promotes Cancer"

LUO Chaoying, LUO Panzhen, LI Jinyu, et al.

(Lanzhou Institute of Husbandry & Pharmaceutics Science, Chinese Academy of Agricultural Sciences/Engineering & Technology Research Center of Traditional Chinese Veterinary Medicine of Gansu Province, Lanzhou 730050, China)

Abstract: The public recognitions change dramatically from "one cup of milk, one strong nation" to "milk promotes cancer", surprisingly, which is still a very popular perception on the Internet, although the national authorities and the mainstream medias of China have refuted the rumor many times. By analyzing the phenomenon from the perspective of complexity science, this paper argued that scientific concept needed to be shifted and be directed by complexity science, and a correct understanding of the complex relationship that correlation does not equal causation should be built. In conclusion, we should neither be panic by the rumor of "milk promotes cancer", nor ignore the lessons of western diet. Meanwhile, we should keep the balanced diet with maximizing the benefit of having milk.

Key Words: Milk; Promot cancer; Correlation; Causation; Complexity Science

（发表于《医学与哲学》，中文核心，IF：综述.25-27.）

动物抗寄生虫药物作用机理研究进展

张吉丽，李 冰，周绪正，张继瑜*

（中国农业科学院兰州畜牧与兽药研究所/农业部兽药创制重点实验室，兰州 730050）

摘 要：动物抗寄生虫药物主要用于防制动物寄生虫病，是保障畜牧业健康发展和公共卫生安全的有效手段。抗寄生虫药物的药效建立在药物分子与寄生虫或动物机体不同靶组织、靶细胞间相互作用的基础上。由于抗寄生虫药物种类多样，抗寄生虫机理复杂。随着新抗寄生虫药物的开发和科学研究的深入，尤其是化学合成和生物制药技术的发展，抗寄生虫药物的品种和数量都在不断的增加，新的药物结构和作用靶标不断发现，作用机理研究也不断深入。作者主要对抗寄生虫药物的作用机理研究进展进行综述。

关键词：动物；抗寄生虫药物；机理

Research Progress on Action Mechanism of Animal Anti-parasitical Drugs

ZHANG Jili, LI Bing, ZHOU Xuzheng, ZHANG Jiyu*

(Key Laboratory of Veterinary Pharmaceutical Development, Ministry of Agriculture/Lanzhou Institute of Husbandry and Pharmaceutical Sciences, CAAS, Lanzhou 730050, China)

Abstract: Animal anti-parasitical drugs are mainly used to prevent and control animal parasitosis, and also an effective way to protect the healthy development of animal husbandry and public health security. The efficacy of anti-parasitical drugs depends on the interaction between drug molecular and different target tissues and target cells of parasite or animal. Because of the variety of anti-parasitical drugs, the anti-parasitical mechanisms are complex. With the development of new anti-parasitical drugs and the depth of scientific research, especially the development of chemical synthesis and biological pharmaceutical technology, the variety and quantity of anti-parasitical drugs constantly increase. With the discovery of new drug structure and its action target, the study on the mechanism is constantly deep. The author mainly summarized the research progress on action mechanism of anti-parasitical drugs.

Key words: Animal; Anti-parasitical drugs; Mechanism

（发表于《中国畜牧兽医》，中文核心，IF：综述.242-247.）

细菌耐药性研究进展

朱阵，曹明泽，张吉丽，周绪正，李冰，张继瑜*

（中国农业科学院兰州畜牧与兽药研究所/农业部兽用药物创制重点实验室，兰州 730050）

摘 要：细菌抗生素类药物耐药性的产生是临床治疗感染性疾病的一大难题，已受到人们的广泛关注。细菌主要通过产生灭活酶或钝化酶获得耐药性，除此之外还有细胞壁的渗透障碍、外排泵的泵出作用、靶位改变等多种机制，这些机制相互作用共同决定细菌的耐药水平。随着新型抗生素的临床应用，新的耐药机制随之出现，耐药菌也越来越广泛。细菌耐药机制的研究对耐药菌的控制和新药开发具有指导性意义。文章从耐药性的起源、产生机理、耐药特性及耐药性的检测方法4个方面进行了阐述。

关键词：细菌；抗生素；耐药性；耐药机制

Research Progress on Bacterial Resistance

ZHU Zhen, CAO Mingze, ZHANG Jili,
ZHOU Xuzheng, LI Bing, ZHANG Jiyu*

（Key Laboratory for Veterinary Drug Innovation, Ministry of Agriculture/Lanzhou Institute of Husbandry and Pharmaceutical Science, Chinese Academy of Agricultural Sciences, Lanzhou 730050, China）

Abstract: Antibiotic resistance of bacteria is a major problem in the clinical treatment of infectious diseases, and it has received the extensive attention. Bacteria producing inactivated enzyme or passivated enzyme is a mainly way to acquire resistance. Besides, the mechanisms of cell wall permeability barrier, efflux pump and target changes are existed, these mechanisms interact to the level of the bacterial antibiotic resistance. With the clinical application of new antibiotic, the new resistant mechanisms emerge and the antibiotic-resistant bacteria are widespread. The study of bacterial resistance mechanisms has instructional meaning for antibiotic-resistant bacteria control and new drug development. We elaborated the bacterial resistance from the origin of the resistance, mechanisms, characteristics and test method in this article.

Key words: Bacteria; Antibiotics; Resistance; Resistance mechanism

（发表于《中国畜牧兽医》，中文核心，IF：综述.3 371-3 376.）

奶牛隐性子宫内膜炎诊断技术研究进展

闫宝琪[1]，董书伟[1]，王东升[1]，那立冬[1,2]，杨洪早[1,2]，张世栋[1*]，严作廷[1*]

（1. 中国农业科学院兰州畜牧与兽药研究所，兰州 730050；
2. 甘肃农业大学动物医学院，兰州 730070）

摘 要：奶牛隐性子宫内膜炎在以子宫内膜被嗜中性粒细胞（PMN）持续性的广泛浸润为主要特征，没有直观的临床症状，经常导致奶牛屡配不孕。在中国，由子宫内膜炎引起的奶牛不孕的比例达50.62%，且由隐性子宫内膜炎引起的奶牛不孕占其中的90%。由于隐性子宫内膜炎在发病初期不易被观察到，容易错过最佳治疗时间，造成巨大的经济损失，故如何对其做出正确、及时的诊断成了近年来的研究热点。目前，国内外奶牛隐性子宫内膜炎的诊断方法主要有阴道内窥镜法、细胞刷法、活体组织诊断法、超声诊断法及白细胞酯酶法等，但目前这些诊断方法没有统一的判断标准。本文对几种不同的奶牛隐性子宫内膜炎的诊断方法进行了综述，为更早、更准确地诊断隐性子宫内膜炎提供依据。

关键词：奶牛；隐性子宫内膜炎；诊断技术

Research Progress on Diagnosis Methods of Subclinical Endometritis in Dairy Cows

YAN Baoqi[1], DONG Shuwei[1], WANG Dongsheng[1], NA Lidong[1,2],
YANG Hongzao[1,2], ZHANG Shidong[1*], YAN Zuoting[1*]

（1. Lanzhou Institute of Animal Husbandry and Pharmaceutical Sciences, Chinese Academy of Agricultural Sciences, Lanzhou 730050, China; 2. College of Veterinary Medicine, Gansu Agricultural University, Lanzhou 730070, China）

Abstract: In the absence of clinical symptoms, dairy cows with subclinical endometritis are defined as sustainably extensive infiltration with the polymorphonuclear cells (PMN) in endometrial samples, and it often lead to dairy cow with repeated infertility. In China, the infertility proportion of dairy cows is 50.62% caused by endometritis, in which 90% is caused by subclinical endometritis. It is difficult to be observed during the early stage of subclinical endometritis, and the best treatment time is delayed in the most cases causing huge economic losses. So it has become a research hotspot to make correctly and timely diagnosis of this disease. At present, the main diagnosis methods of dairy cows subclinical endometritis are vagina endoscope method, cell brush, biopsy technology, ultrasonic diagnosis technology, leukocyte esterase method and so on. However there is no uniform criterion. Different diagnosis methods of subclinical endometritis were summarized, analyzed and discussed in this paper to provide the references for the subclinical endometritis diagnosis more accurately and earlier, and reduce the losses of dairy farms.

Key words: Dairy cow; Subclinical endometritis; Diagnosis technology

（发表于《中国畜牧兽医》，中文核心，IF：综述.683-688.）

羔羊腹泻细菌和病毒病原的研究进展

妥 鑫，刘永明*，黄美州，崔东安，王 慧，王胜义，齐志明

（中国农业科学院兰州畜牧与兽药研究所/农业部重点兽用药物创制重点实验室/甘肃省新兽药工程重点实验室/甘肃省中兽药工程技术研究中心，兰州 730050）

摘 要：导致羔羊腹泻的因素多样，细菌和病毒感染是最关键的因素，其危害也最严重。已有很多专家、学者应用不同的方法对引起羔羊腹泻的病原开展了大量研究，旨在探明引起羔羊腹泻的主要病原。近年来，对病原的研究主要集中于病原主要编码基因和功能蛋白在致病过程中的作用及其分子流行病学。作者从分子水平将近年来导致羔羊腹泻的主要细菌和病毒病原进行综述，并据此为羔羊腹泻病原学发展提出看法。

关键词：羔羊腹泻；病原学；分子流行病学

Research Progress on the Bacterial and Viral Etiology of Lamb Diarrhea

TUO Xin，LIU Yongming*，HUANG Meizhou，CUI Dongan，
WANG Hui，WANG Shengyi，QI Zhiming

(Engineering & Technology Research Center of Traditional Chinese Veterinary Medicine of Gansu Province/Key Laboratory of New Animal Drug Project of Gansu Province/Key Laboratory of Veterinary Pharmaceutical Development of Ministry of Agriculture/Lanzhou Institute of Husbandry and Pharmaceutics Sciences of CAAS，Lanzhou 730050，China)

Abstract: The etiology of lamb diarrhea disease is complex, however, bacterial and viral infection are the most critical factors and cause most serious harm. In order to explore the main pathogens of lambdiarrhea, many scholars and experts have undertaken extensive research on different aspects of viral and bacterial etiology of lamb diarrhea with various methods. Recently, the research about the etiology concentrates on the main code genes and functional proteins in the progress of pathogen invading host and molecular epidemiology. This paper reviewed the recent and main pathogens of lamb diarrhea caused by bacteria and viruses from the molecular level, and proposed some ideas about the development of etiology of lamb diarrhea in the future.

Key words: Lamb diarrhea； Etiology； Molecular epidemiology

（发表于《中国畜牧兽医》，中文核心，IF：综述.831-836.）

PRDM家族蛋白结构与功能研究进展

赵生军[1,2]，张勇[2]，阎萍[1]，郭宪[1*]

（1. 中国农业科学院兰州畜牧与兽药研究所/甘肃省牦牛繁育工程重点实验室，兰州 730050；2. 甘肃农业大学动物医学院，兰州 730070）

摘 要：PRDM家族蛋白是具有一个PR结构域和数量不等锌指结构（除了PRDM11），广泛存在于灵长类、啮齿动物、鸟类和两栖动物中能够调控表观遗传的蛋白。一部分PRDM蛋白家族成员的PR结构域具有组蛋白赖氨酸转移酶活性（HMT），影响表观遗传，PRDM蛋白锌指结构具有DNA、RNA、蛋白识别结合能力；另一部分PRDM家族成员虽然具有PR结构域，但不具有HMT活性，在癌症发生、原始生殖细胞特化、神经系统发育、造血、血管发展、胚胎发育中发挥重要作用。PRDM家族蛋白通过甲基转移或募集染色体结构重塑复合物，调控基因表达和修改染色体结构，在细胞特化和疾病调控中起到重要作用。论文综述了PRDM家族蛋白结构与功能最新研究进展。

关键词：PRDM家族蛋白；表观遗传；PR结构域；锌指结构；组蛋白赖氨酸转移酶活性

Advances on Structure and Function of PRDM Protein Family

ZHAO Shengjun[1,2], ZHANG Yong[2], YAN Ping[1], GUO Xian[1*]

(1. Key Laboratory of Yak Breeding Engineering of Gansu Province/Lanzhou Institute of Husbandry and Pharmaceutical Sciences, Chinese Academy of Agricultural Sciences, Lanzhou 730050, China; 2. College of Veterinary Medicine, Gansu Agricultural University, Lanzhou 730070, China)

Abstract: PRDM protein family which has a PR domain and varying amounts of zinc finger domains (except PRDM11), widely present in primates, rodents, birds and amphibians, involve in regulation of epigenetic proteins. Some PR domains possess transferase activity of histone lysine (HMT), affect epigenetics features and the Zinc finger domains may recognize and bind DNA, RNA, protein; The other PR domains of PRDM without HMT activity, play a key role in carcinogenesis, specification of primordial germ cell, neurogenesis, hematopoiesis, vascular development, and embryonic development. PRDM protein family participate in cell specialization, and disease regulation, regulating gene expression and modifying chromatin structure by methyl transfer, or recruited chromosome structure reshape compounds. This review summarized the latest research progress on structure and function of PRDM protein family.

Key words: PRDM protein family; Epigenetics; PR domain; Zinc finger; Histone lysine transferase activity

（发表于《中国畜牧兽医》，中文核心，IF：综述.1 188-1 193.）

牦牛瘤胃微生物降解纤维素及其资源利用的研究进展

苏贵龙，张景艳，王 磊，李建喜*

（中国农业科学院兰州畜牧与兽药研究所/甘肃省中兽药工程技术研究中心，兰州 730050）

摘 要：牦牛主要生活在中国青藏高原地区，以采食牧草为主。研究表明，牦牛消化纤维素的能力远高于黄牛和奶牛，其主要原因在于三者瘤胃内微生物种类存在着明显的差异。牦牛瘤胃内存在着庞大的可降解纤维素的微生物群体，通过分泌纤维素酶，从而对采食的牧草具有极强的降解能力。本文主要对牦牛瘤胃微生物的组成特征、降解纤维素机制、影响因素及对牦牛瘤胃微生物资源利用进行了综述。旨在为加强对纤维素菌的利用，将其更加高效地运用于中药发酵等现代化领域中提供理论依据。

关键词：牦牛；瘤胃微生物；纤维素代谢；应用；中药发酵

Research Progress on Cellulose Degradation by Yak Rumen Microorganism

SU Guilong, ZHANG Jingyan, WANG Lei, LI Jianxi*

（Lanzhou Institute of Husbandry and Pharmaceutical Sciences of CAAS/Engineering and Technology Research Center of Traditional Chinese Veterinary Medicine of Gansu Province, Lanzhou 730050, China）

Abstract: Yak mainly lived in the Qinghai-Tibetan plateau in Qinghai province of China and lived on grasses. Many studies had shown that yak's ability to digest cellulose was far higher than Yellow cattle and dairy cows. Because there were significant differences in the rumen microorganisms among these three kinds of animals. There was a large microbial population in the yak rumen that could degrade cellulose. In this paper, the characteristics of microorganism, influence factor and the mechanism of cellulose degradation in yak were summarized. In addition, the brief introduction of the yak rumen microbial's possible applications had carried on in order to provide theoretical basis and strengthen the use of cellulose degradation bacteria in the areas such as modernization of traditional Chinese medicine fermentation and so on.

Key words: Yak; Rumen microorganism; Metabolism of cellulose; Application; Traditional Chinese medicine fermentation

（发表于《中国畜牧兽医》，中文核心，IF：综述.659-699.）

繁殖毒理学研究进展

赵晓乐，孔晓军，李世宏，秦　哲，刘希望，杨亚军，李剑勇*

（中国农业科学院兰州畜牧与兽药研究所兽药重点实验室/
农业部兽用药物创制重点实验室/甘肃省新兽药工程重点实验室，兰州　730050）

摘　要：随着现代生物科学研究的不断深入，产科学、药理学和毒理学相互交叉渗透，使得繁殖毒理学逐渐发展起来。繁殖毒理学包括生殖毒理学和发育毒理学，是探究环境、药物等因素对亲代的生殖功能及子代发育过程是否有影响的一门科学。本文就繁殖毒理学所探讨的受试物对试验动物的生殖细胞发生、卵细胞受精、胚胎形成、妊娠、分娩、哺乳过程及初生幼仔发育的损害作用等进行综述。

关键词：生殖毒性；发育毒性；新技术

Research Progress on Reproductive Toxicology

ZHAO Xiaole, KONG Xiaojun, LI Shihong, QIN Zhe,
LIU Xiwang, YANG Yajun, LI Jianyong*

（Key Laboratory of New Animal Drug Project of Gansu Province/Key Laboratory of Veterinary Pharmaceutical Development, Ministry of Agriculture/Lanzhou Institute of Husbandry and Pharmaceutical Science, CAAS, Lanzhou 730050, China）

Abstract: With the deepening of modern biological science, obstetrics, pharmacology and toxicology intersect penetration with each other, thus gradually make the development of reproductive toxicology. Reproductive toxicology include two parts: Reproductive toxicology and developmental toxicology, it is a method focused on exploring whether there are influences of environmental and drugs factors on parental reproductive functions and the development of offspring. In this paper, the injury effects of test drugs on the reproductive cytogenesis, fertilization of egg cell, embryogenesis, pregnancy, childbirth, breast-feeding and development of newborn pups were summarized.

Key words: Reproductive toxicity; Developmental toxicity; New technology

（发表于《中国畜牧兽医》，中文核心，IF：综述.714-719.）

植物精油在畜禽生产中的应用效果研究进展

朱永刚,王 磊,崔东安,张景艳,林 杰,王旭荣,李建喜,杨志强[*]

(中国农业科学院兰州畜牧与兽药研究所/甘肃省中兽药工程技术研究中心,兰州 730050)

摘 要:动物养殖业目前面临着疾病、应激等各种不良因素的影响,添加抗生素等方法会导致肠道菌群失衡、药物残留和耐药性等多方面负效应,存在着威胁食品安全等诸多问题和潜在风险。因此,开发新型绿色饲料添加剂是畜牧业可持续发展的必然趋势。植物精油是从植物不同组织如果实、叶片、花和根中提取的具有强烈气味的挥发性液体,是一种"高效、安全、稳定、可控"的植物源性药物,具有抗菌、抗病毒、抗寄生虫、抗真菌、增强动物机体免疫机能、提高动物抗应激能力及改善动物生产性能等优点,是目前新型饲料添加剂研发的重点和热点之一。其作为抗生素的替代品运用到畜禽生产上具有很大的潜力。作者综述了植物精油的研究现状,包括其在抗菌、促生长、抗氧化、增强免疫力等方面的作用,旨在为植物精油生理功能的研究与应用提供参考。

关键词:植物精油;抑菌;免疫力;生产性能;抗氧化

Research Progress on the Application of Essential Oil in Animal Production

ZHU Yonggang, WANG Lei, CUI Dongan, ZHANG Jingyan,
LIN Jie, WANG Xurong, LI Jianxi, YANG Zhiqiang[*]

(Engineering & Technology Research Center of Traditional Chinese Veterinary Medicine of Gansu Province, Lanzhou Institute of Husbandry and Pharmaceutical Science of CAAS, Lanzhou 730050, China)

Abstract: The animal breeding industry is faced with a variety of adverse factors such as disease, stress and the negative effects of using antibiotics, like intestinal flora imbalance, drug residues and resistance. These influences have become a threat to food safety, and cause many other issues and potential risks. Therefore, the development of new green additive is inevitable for the sustainable development of animal husbandry. Essential oil is a kind of volatile liquid which is extracted from different organizations of plants such as fruit, root, leaf or flower, with strong fragrance and sweet smell. Essential oil is regarded as an "efficient, safe, stable and controllable" plant-derived drug, it has the functions of antibacterial, antiviral, anti-parasite, antifungal, enhancing the animal immune function, anti-stress ability and improving animals production performance, etc. Essential oil is one of the focus and hot spot of current research and development, it can be used as a substitute for antibiotics in the production of livestock and poultry. This review summarized the research status of essential oils, focused on their functions about antibacterial, growth promoting, anti oxidation and immunity enhancement. It is wished to prove reference for the research and application of essential oil.

Key words: Essential oil; Antibacterial; Immunity; Production performance; Antioxidant

（发表于《中国畜牧兽医》，中文核心，IF：综述.1 812-1 817.）

基于复杂性科学探讨中药安全性评价

罗超应[1]，罗磐真[2]，李锦宇[1]，王贵波[1]，谢家声[1]

（1. 中国农业科学院兰州畜牧与兽药研究所/甘肃省中兽药工程技术研究中心，兰州 730050；2. 户县中医医院，陕西户县 710300）

摘 要：针对近年来中药安全评价方法及其研究存在着与临床尤其是中医辨证施治临床实践相脱节之不足，从辨证施治与复杂性科学的角度探讨中药安全性评价，提出：①转变科学观念，改变中药安全性评价唯成分论的片面做法；②积极主动地抑毒减毒，而不是简单地弃毒；③严格控制适应症，尽量减少或消除因为误用而所造成的损失与对中药的负面影响；④有针对性的选择，而不是简单的禁用；⑤加强临床监测与研究，促进中药安全性评价的不断深入与完善。

关键词：中药；安全性评价；复杂性科学；辨证施治；临床

Safety Evaluation of Traditional Chinese Medicine Based on Complexity Science

LUO Chaoying[1], LUO Panzhen[2], LI Jinyu[1], WANG Guibo[1], XIE Jiasheng[1]

（1. Engineering & Technology Research Center of Traditional Chinese Veterinary Medicine of Gansu Province/Lanzhou Institute of Husbandry & Pharmaceutics Science, Lanzhou 730050, China; 2. Shanxi Huxian County Hospital of TCM, Huxian 710300, China）

Abstract: Aiming at the deficiency of the disjoint phenomenon between the safety evaluation and diagnosis and treatment based on syndrome differentiation of TCM, and based on the discussion on safety evaluation of TCM from diagnosis and treatment based on syndrome differentiation and complexity science. It puts forward viewpoints as follows: ①to shift the scientific ideas, and change the theory of the unique importance of class origin in safety evaluation of traditional Chinese medicine（TCM）; ②to reduce poisonousness actively rather than abandon poison simply; ③to control the indications strictly, and reduce or eliminate negative effects caused by misapplying TCM as far as possible; ④to select targeted rather than forbidden simply; ⑤to strengthen clinical monitoring and research, and promote the deepening and improvement of the safety evaluation of TCM.

Key words: Traditional Chinese medicine; Safety evaluation; Complexity science; Diagnosis and treatment based on syndrome differentiation; Clinic

（发表于《中华中医药杂志》，中文核心，IF：综述.2 929-2 932.）

不同种植年限紫花苜蓿种植地土壤容重及含水量特征

周 恒[1]，时永杰[1]，路 远[1]，胡 宇[1]，田福平[1]，陈 璐[2]

（1.中国农业科学院兰州畜牧与兽药研究所/农业部兰州黄土高原生态环境重点野外科学观测试验站，兰州 730050；2.甘肃农业大学草业学院，兰州 730070）

摘　要：对不同种植年限（1、3、10、15年）紫花苜蓿种植地土壤容重及土壤含水量进行了研究，结果表明：随生长年限的增加，相同土层不同年限之间和各年限不同土层之间的土壤容重均逐渐减小，15年最小，1年最高；而各年限土壤含水量在垂直方向也随土壤深度的增加而呈减小趋势；相同土层不同年限之间，土壤含水量随种植年限的增加而减小。相关分析表明，土壤容重与含水量在0.01水平上显著相关。

关键词：紫花苜蓿；土壤容重；土壤含水量；利用年限；分布特征

（发表于《江苏农业科学》，490-494.）

2001—2010年张掖市甘州区土地利用景观格局演变研究

李润林[1,2]，董鹏程[1,2]

（1.农业部兰州黄土高原生态环境重点野外科学观测实验站，兰州 730050；
2.中国农业科学院兰州畜牧与兽药研究所，兰州 730050）

摘 要：为了解张掖市甘州区土地利用及景观格局时空变化特征，以3期Landsat TM/ETM+遥感影像为数据源，利用RS、GIS技术和景观生态学方法对研究区2001—2010年土地利用和景观格局时空变化特征进行了分析研究，结果表明：在2001—2007年草地和建设用地面积增加，裸地和水体面积减少；在2007—2010年建设用地急剧增加，增加7.81 hm^2，裸地和耕地面积减少。从整个研究期间来看，草地和建设用地面积增加，耕地和裸地面积减少，但耕地和裸地面积减少比重较小。从景观水平上看，景观破碎化程度加剧，景观异质性和景观多样性增加，景观空间连接性逐渐减弱，人类活动对景观生态格局变化造成的影响愈加显著。从类型水平上看，建设用地面积增加，破碎化加剧；水体的自然景观特征增强，水体空间分布结构趋于均衡，裸地的破碎化进一步加剧。

关键词：土地利用；景观格局；绿洲城市

Study on Land Use and Landscape Pattern Changes in Ganzhou District of Zhangye City from 2001 to 2010

LI Runlin[1,2], DONG Pengcheng[1,2]

（1. Lanzhou Municipal Key Field Scientific Observation and Experimental Station for Ecological Environment in Loess Plateau Area, Ministry of Agriculture, Lanzhou 730050, China; 2. Lanzhou Institute of Animal Husbandry and Veterinary, Chinese Academy of Agricultural Sciences, Lanzhou 730050, China）

Abstract: In order to understand the temporal and spatial patterns characteristics of land use and landscape change in Ganzhou district of Zhangye city, using three Landsat TM/ETM images as the data source, analyzed the temporal and spatial variation features of land use and landscape by RS, GIS technology and landscape ecology methods from 2001 to 2010, the results showed that the grassland area and construction land significantly increased, the bare land and water area decreased during 2001—2007. construction land increased dramatically during 2007—2010, increasing the area of 7.81hm², the bare land and the arable land area decreased. In the entire study period, the grass and construction land area increased, the arable land and the bare land area decreased, but the reducing proportion of arable land and bare land area was small. At the level of landscape, landscape fragmentation degree, landscape heterogeneity and landscape diversity increased, the impact of human activities on the ecological landscape pattern change significantly increased. At the class level of landscape, construction land area increased and fragmentation intensified. Water enhanced natural landscape feature and the spatial distribution structure of water more balanced, fragmentation of bare land further exacerbated.

Key words: Land use; Landscape pattern; Oasis town

（发表于《江西农业学报》，109-113.）

酿酒酵母菌生长特性的研究

刘龙海[1]，李新圃[1]，杨　峰[1]，罗金印[1]，张　哲[1,2]，李宏胜[1]

（1. 中国农业科学院兰州畜牧与兽药研究所/农业部兽用药物创制重点实验室/甘肃省新兽药工程重点实验室，兰州　730050；2. 甘肃农业大学动物医学院，兰州　730070）

摘　要：为探究温度和pH值对酿酒酵母菌生长特性的影响，采用单因素实验设计，检测了不同温度、不同初始pH值对酿酒酵母菌生长的影响。结果表明，培养基pH值和温度对酿酒酵母菌的生长具有明显的影响，当pH值为3.0或者8.0，温度为20℃或高于40℃时，酿酒酵母菌的生长明显受到抑制；酿酒酵母菌在培养基初始pH值为4.0~6.0，温度在25~30℃时生长良好。

关键词：酿酒酵母；温度；pH值；生长特性

Effects of Temperature and pH on the Growth of *Saccharomyces cerevisiae*

LIU Longhai, LI Xinpu, LI Hongsheng, et al

(Key Laboratory of New Animal Drug Project of Gansu Province/Key Laboratory of Veterinary Pharmaceutical Discovery, Ministry of Agriculture/Lanzhou Institute of Animal Husbandry and Pharmaceutical Sciences, Chinese Academy of Agricultural Sciences, Lanzhou 730050, China)

Abstract: To investigate the effects of temperature and pH on the growth of *Saccharomyces cerevisiae*, the influence of different temperature and pH were determined on *Saccharomyces cerevisiae* growth by single factor design. The results indicated that the temperature and pH of mediu would significantly affected the growth characteristics of *Saccharomyces cerevisiae*. The *Saccharomyces cerevisiae* growth would be inhibited when exceed the temperature (20~40℃) or the pH (3.0~8.0). In this study, the optimum growth temperature of Saccharomyces cerevisiae is 25~30℃, the optimum growth pH is 4.0-6.0. This study laid a foundations for the application of fermentation process.

Key words: *Saccharomyces cerevisiae*; Temperature; pH; Growth characteristics

（发表于《中国草食动物科学》，38-41.）

记中国农业科学院兰州畜牧与兽药研究所"中兽医药陈列馆"

罗超应，王贵波，李锦宇，谢家声，王旭荣，王磊

（中国农业科学院兰州畜牧与兽药研究所，
甘肃省中兽药工程技术研究中心/兰州 730050）

摘 要：中兽医药学从远古走来，秉黄帝精髓，承伏羲心智，传师皇技艺，精深而博大，正走向世界。她曾经历过辉煌，也曾历经磨难；今逢盛世受重视，蓬勃发展振翅飞，医药探索硕果丰，机遇挑战勇面对。为弘扬中华传统文化，促进中兽医药学的继承与发展，在国家科技部基础工作专项资助下，在中国农业科学院中兽医研究所和中国农业科学院兰州畜牧与兽药研究所中兽医药陈列室的基础上，于2013年开始筹建中兽医药陈列馆。

关键词：中国农业科学院；中兽医药；研究所；陈列馆；兽药；畜牧；兰州；中华传统文化

（发表于《中国草食动物科学》，77-78.）

两种植物在不同生长期控制Na$^+$流入的差异

王春梅，张　茜，朱新强，贺洞杰，段慧荣，王晓力

（中国农业科学院兰州畜牧与兽药研究所，兰州　730050）

摘　要：控制Na$^+$进出植物根系的能力是研究植物耐盐机制的关键，但由于测量技术和条件的限制，目前关于植物根系Na$^+$内流速率受生长阶段的影响还无相关研究。该研究采用非损伤微测技术，选取小麦与野大麦为代表材料，研究了单子叶甜土植物与单子叶盐生植物在不同生长阶段下的根系Na$^+$内流速率差异。结果表明：2个盐浓度处理下野大麦和小麦根系的Na$^+$内流速率都随生长而逐渐增大：2~3叶期，2个材料之间Na$^+$内流速率差异不显著（$P>0.05$）；4叶期时，浓度为25mmol/L NaCl处理下野大麦比小麦低10%，但差异也不显著（$P>0.05$）；浓度为100mmol/L NaCl时野大麦比小麦高19%，差异显著（$P<0.05$）。可见，生长阶段能显著影响植物控制Na$^+$流入的能力，4叶期以后的野大麦比小麦吸收更多的Na$^+$。因此，在比较植物耐盐能力以及相关研究时应考虑生长阶段的影响。

关键词：非损伤微测技术；Na$^+$内流；甜土植物；盐生植物

Influence of Growth Stage on Na$^+$ Influx in Two Plant Species

WANG Chunmei，ZHANG Qian，WANG Xiaoli，et al

（Lanzhou Institute of Husbandry and Pharmaceutics sciences of CAAS，Lanzhou 730050，China）

Abstract: The ability of controlling Na$^+$ influx and efflux of plant roots is the key to study the mechanism of salt to lerance in plants. However, due to the limitation of measurement techniques and conditions, the influence of growth stage on the ability of Na$^+$ influx in plants remains elusive. Gram ineous halophyte wild barley and glycophyte wheat were chosen there, for testing the unidirectional Na$^+$ influxes of monocotyledonous at 2 leaves, 3 leaves and 4 leaves stages, respectively. The micro-electrode ion flux estimation technique experiments showed that Na$^+$ influx increased obviously with the plant growth stage in both species. There was no significant difference between wild barley and wheat under2-3 leave stage（$P>0.05$）. At 4 leaves stage, wild barley was lower by 10% than that of wheat under 25 m mol/L NaCl（$P>0.05$），but under 100 m mol/L NaCl，Na$^+$ influx of wild barley was significantly higher by 19% than that of wheat（$P<0.05$）.

These indicated that growth stage had obvious influence on the Na$^+$ fluxes in both species; and wild barley had stronger ability of Na$^+$ influx than wheat under 4 leaves stage. Therefore, plant growth stage should be considered in salt to lerance and relative researches.

Key words: Micro-electrode ion flux estimation technique; Na$^+$ influx; Glycophyte; Halophyte

（发表于《中国草食动物科学》，39-41.）

欧拉型藏羊生长发育曲线模型预测及趋势分析

王宏博[1,2]，梁春年[1,2]，包鹏甲[1,2]，
朱新书[1,2]，阎　萍[1,2]

（1. 中国农业科学院兰州畜牧与兽药研究所，兰州　730050；
2. 甘肃省牦牛繁育工程重点实验室，兰州　730050）

摘　要：试验旨在研究欧拉型藏羊的生长发育状况，初步探讨欧拉型藏羊的生长发育曲线理论模型并找出欧拉型藏羊养殖的最佳生长点，以期为欧拉型藏羊的选育和饲养管理的改善提供理论依据。选用甘肃省甘南玛曲县欧拉乡牧户的欧拉型藏羊羔羊60只，测定了0（初生）、2、7、14、21、28、42、56、70、84、98、112日龄的体重、体高和体长。采用Logistic、Richards、Exponential Association和Gompertz Relation 4种生长曲线模型对欧拉型藏羊羔羊的生长曲线进行拟合。结果发现：Richards曲线模型最适宜描述欧拉型藏羊体重增长模式，Exponential Association是拟合欧拉型藏羊羔羊体高和体长早期生长过程的最优模型。

关键词：欧拉型藏羊；生长曲线；非线性模型

Growth Curve Model Prediction and Trend Analysis of Oula Tibetan Sheep

WANG Hongbo[1,2], LIANG Chunnian[1,2],
YAN Ping[1,2], et al

(1. Lanzhou Institute of Husbandry and Pharmaceutical Sciences of CAAS,
Lanzhou 730050, China; 2. Gansu Province Key Laboratory of Yak Breeding
Engineering, Lanzhou 730050, China)

Abstract: This paper aimed to study the growth and development of Oula Tibetan sheep, and make a preliminary discussion on growth curve model and a trend analysis for improving the Oula Tibetan sheep breeding and the feeding management. 60 grazing Oula Tibetan sheep were chosen and the weight, height and body length of new born, 2-day-old, 7-day-old, 14-day-old, 21-day-old, 28-day-old, 42-day-old, 56-day-old, 70-day-old, 84-day-old, 98-day-old, 112-day-old were measured. The growth curves of Oula Tibetan sheep were analyzed with four kinds of non linear models (Logistics, Richards, Exponential Association and Gompertz

Relation). The results showed that the growth curves were appropriately fitted with Richards and Exponential Association.

Key words: Oula Tibetan sheep; Growth curve; Non linear model

（发表于《中国草食动物科学》，14-18.）

酮病治疗对奶牛总抗氧化能力的影响

李亚娟[1]，王　浩[1]，李　佳[1]，张培军[1]，胡俊菁[1]，
胡国平[1]，杜玉兰[1]，董书伟[2]，何宝祥[1]

（1. 广西大学动物科学技术学院，南宁　530005；
2. 中国农业科学院兰州畜牧与兽药研究所，兰州　730050）

摘　要：为探究奶牛酮病治疗对机体总抗氧化能力的影响，本研究选取南宁市某规模化养殖场的65头围产期奶牛作为研究对象，经血浆酮体定量检测，从中选取5头健康牛作为健康对照组，将检出的10头酮病患牛分为治疗组（5头）和阳性对照组（5头）。试验期间，对治疗组进行药物治疗，阳性对照组和健康对照组不做任何治疗。对各组奶牛分别于治疗前和停药后清晨空腹颈静脉采血，测定血浆酮体（改良水杨醛比色法）、血糖（氧化酶—过氧化物酶法）和TAC（铁还原比色法）水平。结果显示，治疗后，治疗组奶牛的血酮含量较治疗前极显著下降，血糖和总抗氧化能力含量极显著升高。阳性对照组和健康对照组的血酮和血糖含量、总抗氧化能力治疗前后变化不大，差异均不显著。结果表明，酮病治疗可提高奶牛血浆TAC水平，提示TAC的提高与酮病痊愈有关。

关键词：酮病；治疗；总抗氧化能力；奶牛

DOI：10.19305/j.cnki.11-3009/s.2016.08.009

（发表于《中国奶牛》，32-34.）

丹参酮乳房注入剂中有效成分含量测定

黄 鑫，梁剑平，郝宝成，高旭东，郭文柱*

（中国农业科学院兰州畜牧与兽药研究所/农业部兽用药物创制重点实验室/
甘肃省新兽药工程重点实验室，兰州 730050）

摘 要：采用高效液相色谱测定丹参酮乳房注入剂有效成分隐丹参酮和丹参酮ⅡA的含量。将供试品用正己烷萃取后挥干，甲醇溶解后检测；采用反相高效液相色谱，ZORBAX SB-C18键合硅胶柱（4.6mm×250mm，5μm）；柱温30℃；流动相为乙腈（A）-0.026%磷酸水溶液（B），梯度洗脱（0~20min：V（A）60→90，V（B）40→10；20~30min：V（A）：V（B）=90：10）；流速1.2mL/min；检测波长270nm。丹参酮乳房注入剂中隐丹参酮含量在500~1 750μg/g（r=0.999 9）范围内线性关系良好，平均加样回收率为101.15%，RSD为1.55%；丹参酮ⅡA的含量在1 125~3 937.5μg/g（r=0.999 9）范围内线性关系良好，平均加样回收率为100.89%，RSD为1.64%。结果显示，该方法简单、准确、灵敏，精密度、重复性好，可作为丹参酮乳房注入剂中有效成分含量测定的参考方法。

关键词：奶牛乳房炎；含量测定；丹参酮乳房注入剂；高效液相色谱

Determination of Active Ingredient Content in Tanshinone Injection Agent for the Treatment of Bovine Mastitis

HUANG Xin, LIANG Jianping, HAO Baocheng,
GAO Xudong, GUO Wenzhu*

(Lanzhou Institute of Husbandry and Pharmaceutical Sciences of CAAS/Key Laboratory of New Animal Drug Project, Gansu Province/Key Laboratory of Veterinary Pharmaceutical Development, Ministry of Agriculture, Lanzhou 730050, China)

Abstract: This study is aimed to establish a high performance liquid chromatography (HPLC) method to determine the active ingredient in tanshinone injection agent for the treatment of bovine mastitis. Samples were dried with hexane extraction and analyzed by reverse phase high performance liquid chromatography after dissolved with methanol. The target cryptotanshinone and tanshinone ⅡA were separated on a ZORBAX SB-C18 column (4.6 mm × 250 mm, 5 μm) using acetonitrile (A) and 0.026% phosphoric acid solution

(B)（0-20 min，V（A）60→90，V（B）40→10；20-30 min：V（A）：V（B）= 90：10）as the mobile phase by gradient elution，the flow rate was 1.2 mL/min with colomn temperature of 30℃ and detection wavelength of 270 nm. The content of cryptotanshinone in tanshinone injection agent for the bovine mastitis showed good linear relationship at the range of 500-1 750μg/g（r=0.999 9），the average recovery was 101.15% with the relative standard deviation of 1.55%. The other content of tanshinone ⅡA in it also displayed a good linear relationship at the range of 1 125-3 937.5μg/g（r=0.999 9），The average recovery was 100.89% with the relative standard deviation of 1.64%. The results showed that the method demonstrated advantage of simpleness，accuracy and sensitivity，and successfully applied to the analysis of cryptotanshinone and tanshinone ⅡA in tanshinone injection agent for the bovine mastitis.

Key words: Bovine mastitis; Determination of content; Tanshinone injection agent for the bovine mastitis; HPLC

（发表于《中国兽药杂志》，53-57.）

高效液相色谱法测定二氧化氯固体消毒剂中 DL-酒石酸的含量

张景艳[1]，张 宏[2]，陈化琦[1*]，徐继英[3]

（1. 中国农业科学院兰州畜牧与兽药研究所，兰州 730050；
2. 甘肃省兽药饲料监察所，兰州 730030；3. 西北民族大学，兰州 730050）

摘 要：为了测定二氧化氯固体消毒剂中DL-酒石酸的含量，采用C18柱（4.6mm×250mm，5μm），在室温下，以0.4mol/L磷酸二氢钾（用磷酸调pH至2.4）为流动相，214nm为检测波长，流速为1.0mL/min，建立了反相高效液相色谱法（RP-HPLC）测定方法。结果显示，DL-酒石酸在0.2~2.0mg/mL范围内线性关系良好，R2=0.999 9，回收率范围为97.0%~102.3%。本方法简便、准确、快速，可用于二氧化氯固体消毒剂中DL-酒石酸的含量测定。

关键词：DL-酒石酸；含量测定；反相高效液相色谱法

RP-HPLC Determination of DL-tartaric Acid in Solid Chlorine Dioxide Disinfectant

ZHANG Jingyan[1], ZHANG Hong[2], CHEN Huaqi[1*], XU Jiying[3]

（1. Lanzhou Institute of Husbandry and Pharmaceutical Science of CAAS, Lanzhou 730050, China; 2. Gansu Institute of Veterinary Drug and Feed Inspection, Lanzhou 730030, China; 3. Northwest University for Nationalities, Lanzhou 730050, China）

Abstract: A method for the determination of DL-tartaric acid in the Solid chlorine dioxide disinfectant Was developed by the reverse high performance liquid chromatography（RP-HPLC）. The C18 colum（4.6 mm × 250 mm, 5μm）was used with the mobile phase consisted of 0.4 mol/L KH_2PO_4（pH was adjusted to 2.4 by H_3PO_4）and the detection wavelength of 214 nm. The results showed that DL-tartaric acid had the good linear relationship at the range of 0.2-2.0 mg/mL

(R^2=0.999 9) and the mean recovery rate was 97.0%–102.3%. The method was simple, accurate and fast in the determination of DL-tartaric acid in solid chlorine dioxide disinfectant.

Key words: DL-tartaric acid; Determination; RP-HPLC

（发表于《中国兽药杂志》，40-43.）

两种定值方法测定阿司匹林丁香酚酯对照品的含量△

杨青青[1*]，刘希望[2]，杨亚军[2]，李剑勇[2#1]，查 飞[1#2]

（1. 西北师范大学化学化工学院，兰州 730070；2. 中国农业科学院兰州畜牧与兽药研究所/农业部兽用药物创制重点实验室/甘肃省新兽药工程重点实验室，兰州 730050）

摘 要：目的：采用质量平衡法与核磁共振法，测定阿司匹林丁香酚酯（aspirin eugenol ester，AEE）对照品含量。方法：采用高相液相色谱测定纯度，采用干燥失质量法测定水分和挥发性物质，炽灼残渣测定无机盐杂质，根据质量平衡法计算AEE对照品含量。建立核磁共振定量测定AEE含量的方法，仪器温度25℃，扫描次数（NS）16次，弛豫延迟时间（D1）15s。结果：质量平衡法测定的AEE对照品含量为99.82%；核磁共振定值结果为99.55%，AEE和内标物（1，4-二硝基苯）峰面积的比值与质量比在3~21mg范围内呈良好线性关系（R2=0.999），日内精密度的RSD位为0.04%（n=5），重复性的RSD值为0.19%（n=5），稳定性测定的RSD值为0.57%。结论：2种方法相互验证，测定结果基本一致。

关键词：质量平衡法；核磁共振法；对照品；阿司匹林丁香酚酯；方法验证

DOI：10.14009/j.issn.1672-2124.2016.11.007

Content Determination of Aspirin Eugenol Ester Reference Substance by Two Measuring Methods△

YANG Qingqing[1], LIU Xiwang[2], YANG Yajun[2], LI Jianyong[2], ZHA Fei[1]

（1. College of Chemistry and Chemical Engineering, Northwest Normal University, Gansu Lanzhou 730070, China; 2. Key Laboratory of New Animal Drug Project, Gansu Province/Key Laboratory of Veterinary Pharmaceutical Development, Ministry of Agriculture/Lanzhou Institute of Husbandry and Pharmaceutical Sciences of CAAS, Gansu Lanzhou 730050, China）

Abstract: [OBJECTIVE]: To determine the content of aspirin eugenol ester (AEE) reference substance by by the mass balance method and the nuclear magnetic resonance (NMR) method. [METHODS]: Chromatogram purity was determined by HPLC, the moisture and volatile substance were determined by weightlessness method and inorganic impurities was determined by the residue on ignition method. The content of AEE reference substance was calculated by the mass balance method. NMR quantification method for AEE was established with determine temperature of 25℃, scan times of 16 and the relaxation delay time of 15 s respectively. [RESULTS]: The results of content determination were 99.82% by the mass balance method and 99.55% by the NMR method. Fitting line between peak area ratio and weight ratio of sample AEE to inner standard (1,4-dinitrobenzene) ranged from 3 mg to 21 mg with correlative coefficient of 0.999. Method validation of the NMR about the precision RSD was 0.04% (n=5) and the repeatability RSD was 0.19% (n=5) and the stability of RSD was 0.57%. [CONCLUSIONS]: Two methods can verify each other and the determination results are basically identical.

Key words: Mass balance method; NMR method; Reference substance; Aspirin eugenol ester; Aspirin

（发表于《中国医院用药评价与分析》，1 466–1 468.）

中西兽药结合治疗安卡红种公鸡冠癣420例

张仁福[1]，王贵波[2]，李锦宇[2]，罗超应[2]，谢家声[2*]

（1. 甘肃省甘谷县畜牧兽医局，甘谷　741200；
2. 中国农业科学院兰州畜牧与兽药研究所，兰州　730050）

（发表于《中兽医学杂志》，7.）

宫康中水苏碱含量测定

苗小楼[1]，尚小飞[1]，王东升[1]，潘 虎[1]，王 瑜[1]，李升桦[2]

（1.中国农业科学院兰州畜牧与兽药研究所/农业部兽用药物创制重点实验室/甘肃省中兽药工程技术研究中心/甘肃省新兽药工程重点实验室，兰州 730050；
2.甘肃省通渭县农牧局畜牧中心，通渭 743300）

摘 要：目的：筛选宫康中水苏碱含量测定供试品溶液的制备方法，建立宫康中水苏碱HPLC-ELSD含量检测方法。方法：采用阳离子交换树脂、五氯酚、硫氰酸铬铵3种方法制备供试品溶液，采用HPLC-ELSD测定水苏碱含量。结果：采用硫氰酸铬铵制备的供试品溶液较好，水苏碱含量在2.20～11.00μg范围内线性关系良好（$R^2=0.999$），平均回收率为95.96%。结论：该法简便、样品纯净、灵敏、阴性无干扰，可以有效控制该制剂质量。

关键词：子宫内膜炎；宫康；HPLC-ELSD；水苏碱

（发表于《中兽医医药杂志》，69-71.）

中国藏兽医药数据库系统建设与研究

尚小飞,苗小楼,王 瑜,汪晓斌,张继瑜,潘 虎

(中国农业科学院兰州畜牧与兽药研究所/农业部兽用药物创制重点实验室/
甘肃省新兽药工程重点实验室,兰州 730050)

摘 要:作为我国传统兽医药学的重要组成部分,藏兽医药学为高原畜牧业的健康可持续发展做出了卓越贡献。"中国藏兽医药数据库系统"(http://zsyy.lzmy.org.cn/)的编制为研究者和应用者提供了一个传统藏兽医药较为集中的研究和信息共享平台,对促进传统藏兽医药的抢救、整理、保护、传承和发展具有现实意义。本文介绍了该系统的建设背景、意义、用途、特点及内容,并对该系统的未来发展进行展望,以期读者更加了解其功能和特点,促进中国藏兽医药数据库的广泛使用和进一步完善。

关键词:中国藏兽医药;数据库建设;系统发展

(发表于《中兽医医药杂志》,85.)

两种多糖对家兔眼刺激性试验

张　哲[1]，李新圃[1]，杨　峰[1]，罗金印[1]，李晓强[2]，刘龙海[1]，李宏胜[1]

（1.中国农业科学院兰州畜牧与兽药研究所/农业部兽用药物创制重点实验室/甘肃省新兽药工程重点实验室，兰州　730050；2.西藏奇正藏药股份有限公司，兰州　730000）

摘　要：为了检测低聚糖硫酸酯与右旋糖酐硫酸酯2种多糖对家兔的眼刺激性，评价其安全性，试验采用同体左右侧眼珠自身对比法，以生理盐水为对照，将不同浓度的两种多糖溶液滴入家兔眼睛，每天用药3次，连续使用7d。参照国家食品药监局颁布的眼刺激反应评分标准计算其总分值，评价2种多糖对家兔的眼刺激性。结果表明，2种多糖不同浓度给药组和对照组眼刺激反应平均分值均为0，表明2种多糖对家兔眼睛均无刺激性，临床使用安全。

关键词：低聚糖硫酸酯；右旋糖酐硫酸酯；奶牛乳房炎；刺激性试验

（发表于《中兽医医药杂志》，35-37.）

犊牛腹泻血液生化指标与主成分分析研究

妥 鑫，王胜义，王 慧，黄美州，崔东安，齐志明，刘永明

（中国农业科学院兰州畜牧与兽药研究所
农业部兽用药物创制重点实验室，甘肃兰州 730050）

摘 要：为考察由不同病原引起的腹泻对犊牛血液生化指标的影响，采集39例犊牛的粪便和血清，利用细菌鉴别培养、RT-PCR病毒检测试剂盒、胶体金多重病原检测试剂盒等对病原进行检测，采集的血清用全自动血液生化分析仪进行测定。检测结果表明，选择的39例12-14日龄的腹泻犊牛粪便中检测到单纯由细菌、或病毒、或寄生虫感染分别为17份、10份、7份，另外3份为混合感染，还有2份样品中未检测到病原；采用主成分分析法将12项血清生化指标转化为综合性更强的4种主成分，结果显示4种主成分的聚类结果与病原检测结果一致；对不同病原所致腹泻犊牛各项血清生化指标进行方差分析，发现病毒性腹泻的总胆红素CT-BIL、总胆固醇（CT）水平与其他病原所致的腹泻有显著性差异（$P<0.01$）。

关键词：犊牛腹泻；病原；血液生化指标；主成分分析

（发表于《中兽医医药杂志》，14-17.）

犊牛腹泻病原调查及与临床症状相关性分析

王胜义,黄美州,王 慧,崔东安,齐志明,刘永明

(中国农业科学院兰州畜牧与兽药研究所/农业部兽用药物创制重点实验室,兰州 730050)

摘 要:为了解我国主要奶业产区导致犊牛腹泻的主要病原,于2015年6月至2016年1月,采集华北、西北、东北主要奶业产区内的176份犊牛腹泻粪样,并对样品中的牛冠状病毒(BCOV)、牛轮状病毒(BRV)、牛病毒性腹泻病毒(BVDV)、大肠杆菌K99、隐孢子球虫、肠兰伯氏鞭毛虫和其他致病菌进行检测,同时结合临床收集的资料,分析病原与临床症状的相关性。结果表明,不同奶业产区内导致犊牛腹泻的病原存在一定差异,西北地区导致犊牛腹泻的病原主要是痢疾杆菌,而华北地区则是隐孢子球虫和肠兰伯氏鞭毛虫,东北地区主要是牛轮状病毒和肠兰伯氏鞭毛虫,且病原感染与犊牛腹泻的临床症状存在明显相关性。

关键词:犊牛腹泻;不同区域;病原;回归分析

(发表于《中兽医医药杂志》,7-10.)

博物学素养与现代医药学研究

杨亚军，刘希望，李世宏，孔晓军，秦　哲，杜文斌，李剑勇

（中国农业科学院兰州畜牧与兽药研究所/农业部兽用药物创制重点实验室/
甘肃省新兽药工程重点实验室，兰州　730050）

摘　要：博物学是各门学科综合的实用知识。现代医药学研究是以药物化学、药剂学、药理学与药物分析等学科为基础，同时融合了生物技术、计算机科学、生态学、信息科学等研究方法的系统性知识体系。研究人员应本着"博物洽闻"的博物学理念，积极学习其他学科的相关知识，拓宽自己的知识面，以期更好地从事医药学研究。

关键词：博物学；医药学研究

（发表于《中兽医医药杂志》，74-76.）

常山碱对小鼠巨噬细胞功能的影响

郭志廷，王 玲，衣云鹏，梁剑平

（中国农业科学院兰州畜牧与兽药研究所/农业部兽用药物创制重点实验室/
甘肃省新兽药工程重点实验室，兰州 730050）

摘 要：探讨常山碱对小鼠腹腔巨噬细胞免疫活性的影响。采用中性红试验和MTT法分别测定常山碱对巨噬细胞吞噬中性红作用及其代谢功能的影响。结果显示，常山碱浓度在0.01~5.00μg/mL对小鼠巨噬细胞吞噬中性红有较好的促进作用（$P>0.05$），对巨噬细胞代谢有显著的促进作用（$P>0.05$）。结果提示，常山碱可以提高小鼠的免疫功能。

关键词：常山碱；巨噬细胞；中性红；代谢

（发表于《中兽医医药杂志》，34-35.）

藿芪灌注液治疗奶牛卵巢静止和持久黄体临床试验

严作廷[1],王东升[1],张世栋[1],董书伟[1],苗小楼[1],那立冬[1],
闫宝琪[1],朱新荣[2],韩积清[3],王雪郦[4]

(1.中国农业科学院兰州畜牧与兽药研究所,兰州 730050;2.甘肃荷斯坦奶牛繁育示范中心,兰州 730080;3.甘肃省秦王川奶牛试验场,永登 730300;4.兰州市城关区奶牛场,兰州 730030)

摘 要:为了评价藿芪灌注液治疗奶牛卵巢静止和持久黄体的临床效果,将确诊为患持久黄体和卵巢静止的病牛各40头随机分为4组,分别为藿芪灌注液高、中(推荐剂量)、低剂量治疗组和对照药物治疗组,每组10头。试验组每头牛每次分别按150mL、100mL和50mL的剂量子宫灌注,隔日1次,4次为1个疗程;对照药物治疗组为促孕灌注液,每次100mL,隔日1次,4次为1个疗程。试验结果表明,用藿芪灌注液治疗奶牛持久黄体的高、中、低剂量组的治愈率分别为80%、70%和50%,总有效率分别为90%、90%和60%;对照组治愈率80%,总有效率90%,藿芪灌注液高剂量组、中剂量组与对照组的治愈率和总有效率均显著高于低剂量组($P<0.05$),高剂量组与中剂量组之间无显著性差异($P>0.05$);高、中剂量组与对照组之间无显著性差异($P>0.05$)。用藿芪灌注液治疗奶牛卵巢静止高、中、低剂量组的治愈率分别为90%、80%和60%,总有效率分别为100%、100%和70%;对照组治愈率80%,总有效率90%,高剂量组、中剂量组和对照组的治愈率和总有效率均显著高于低剂量组($P<0.05$),高剂量组与中剂量组之间无显著性差异($P>0.05$),高、中剂量组与对照组之间无显著性差异($P>0.05$)。

关键词:奶牛;持久黄体;卵巢静止;疗效

(发表于《中兽医医药杂志》,51-54.)

肉制品中药物残留风险因子概述

熊　琳[1,2*]，李维红[1,2]，杨晓玲[1,2]，高雅琴[1,2]

（1. 中国农业科学院兰州畜牧与兽药研究所，兰州　730050；
2. 农业部畜产品质量安全风险评估实验室（兰州），兰州　730050）

摘　要：肉品中药物残留问题严重影响着人体健康，近几年来肉品中兽药残留引起的食品安全问题层出不穷。随着对肉品中各种兽药残留监控的力度越来越严格和规范，研究人员发现了更多的药物残留风险因子，并开始对相关风险因子进行饮食安全性评估。本文综述了目前肉品中可能存在的抗生素、抗寄生虫、激素和抗菌化学药等4大类风险因子的具体物质类别、对人体的危害性以及国内外限量标准的比较。本文明确了肉品安全质量监控的着手点和监控的具体项目指标，并按照风险因子的危害程度和必要性来确定需要监控的风险因子。通过有效地监控和风险评估，为肉品质量安全检测和研究工作提供参考。

关键词：食品安全；肉制品；安全风险因子

Review of Risk Factors of Drug Residues in Meat Products

XIONG Lin[1,2], LI WeiHong[1,2], YANG XiaoLin[1,2], GAO YaQin[1,2]

（1. Lanzhou Institute of Husbandry and Pharmaceutical Sciences, Chinese Academy of Agricultural Sciences, Lanzhou 730050, China; 2. Laboratory of Quality & Safety Risk Assessment for Livestock Product (Lanzhou), Ministry of Agriculture, Lanzhou 730050, China）

Abstract: The problem of drug residues in meat seriously affects human health. The food safety problems brought by the veterinary drug residue in meat emerge endlessly in recent years. Along with the monitoring for various kinds of veterinary drug residues in meat more and more strict and standard, and more and more drug residues risk factors were found and evaluated. In this paper, we summarized the specific material categories of 4 kinds of risk factors in meat including antibiotics, parasitic resistance, hormone and antibiotic chemicals, and the hazard to health, as well as the comparison of limit standard at home and abroad. This article clearly reviewed the meat safety and quality monitoring of specific projects as starting point and monitoring indicators, and determined the monitoring risk factors according to the harm of risk factors and the necessity. By effective monitoring and risk assessment, it can provide a reference for safety inspection and research of meat quality.

Key words: Food safety; Meat products; Safety risk factor

（发表于《食品安全质量检测学报》，1 572–1 577.）

代谢组学在奶牛蹄叶炎研究中的应用前景

李亚娟[1,2]，董书伟[2]，王东升[2]，张世栋[2]，严作廷[2]，
杨志强[2]，杜玉兰[1]，何宝祥[1]

（1.广西大学动物科技学院，南宁 530005；2.中国农业科学院兰州畜牧与兽药研究所/农业部兽用药物创制重点实验室/甘肃省新兽药工程重点实验室，兰州 730050）

摘 要：代谢组学是研究生物体在内外源因素影响下，其内源性代谢物种类、数量变化规律的一门新兴学科，在兽医科研中应用也日益广泛。蹄叶炎是影响奶牛养殖业发展的重要疾病之一，但其发病机理目前尚无定论，文章对代谢组学和奶牛蹄叶炎的研究现状及其在奶牛蹄叶炎研究中的应用前景进行了探讨。

关键词：代谢组学；奶牛；蹄叶炎

Metabonomics and Its Application Prospect in Laminitis of Dairy Cow

LI Yajuan[1,2], DONG Shuwei[2], HE Baoxiang[1], et al

(1. College of Animal Science and Technology, Guangxi University, Nanning 530005, China; 2. Lanzhou Institute of Husbandry and Pharmaceutical Sciences of CAAS/Key Laboratory of Veterinary Pharmaceutical Development, Ministry of Agriculture/Key Laboratory of New Animal Drug Project, Gansu Province, Lanzhou 730050, China)

Abstract: Metabonomics is a new science and technology in recent years which refers to research the species and amounts change of endogenous metabolites when the organism was affected by the alteration of genes or environment. And metabonomics was applied widely in veterinary. Laminitis is one of the serious diseases which can restrict the development of dairy industry, however, the etiology of laminitis remains unclear. This paper reviewed the main progress in metabonomics and laminitis in dairy cow, and the prospect of metabonomics application in research on laminitis in dairy cow.

Key words: Laminitis; Dairy cow; Metabonomics

（发表于《中国草食动物科学》，49-53.）

奶牛乳房炎病原菌抗生素耐药性研究进展

张亚茹，李新圃，杨 峰，罗金印，王旭荣，张 哲，刘龙海，王 丹，李宏胜

（中国农业科学院兰州畜牧与兽药研究所/农业部兽用药物创制重点实验室/
甘肃省新兽药工程重点实验室，兰州 730050）

摘 要：奶牛乳房炎是影响世界奶牛养殖业发展的最重要疾病之一，病原菌感染，尤其是葡萄球菌、链球菌和肠道菌感染是其最主要的病因。目前，抗生素治疗奶牛乳房炎仍然是最常用及最有效的方法，然而，由于抗生素不合理及大量滥用，导致奶牛乳房炎病原菌产生耐药性，且耐药菌株呈不断增多趋势，甚至出现超级耐药菌株，为奶牛乳房炎的临床治疗带来了很大的挑战。因此，文章从奶牛乳房炎主要致病菌耐药现状、耐药变化规律、产生耐药性的原因及如何减少耐药性等方面进行了综述，以期为规范临床使用抗生素和进一步提高奶牛乳房炎治疗效果提供帮助。

关键词：奶牛；乳房炎；病原菌；抗生素耐药性

Progress in the Antibiotic Resistance to the Main Pathogenic Bacteria in Dairy Cow Mastitis

ZHANG Yaru, LI Xinpu, LI Hongsheng, et al

(Key Laboratory of New Animal Drug Project of Gansu Province; Key Laboratory of Veterinary Pharmaceutical Discovery, Ministry of Agriculture; Lanzhou Institute of Animal Husbandry and Pharmaceutical Science, Chinese Academy of Agricultural Sciences, Lanzhou 730050, China)

Abstract: The dairy cow mastitis is one of the most costly disease affecting the development of the world dairy agriculture, the main cause is pathogen infection, particularly staphylococcus, streptococcus and intestinal bacteria infection. At present, the antibiotic treatment is still the most commonly used and the most effective method to the dairy cow mastitis. However, due to the unreasonable and a lot of abuse antibiotic, causing cow mastitis pathogen resistance, and resistant strains showed a trend of increasing, even appear super resistant strain. It has brought great challenges for the clinical treatment of dairy cow mastitis. This article summarized from the status of main pathogenic bacteria drug resistance of dairy cow mastitis and the change rule of the drug resistance, the causes of drug resistance and how to reduce resistance for standardizing the clinical use of antibiotics and further improveing the effect of dairy cow mastitis treatment.

Key words: Dairy cow; Mastitis; Pathogen; Antibiotic resistance

(发表于《中国草食动物科学》, 40-44.)

苹果渣发酵生产蛋白饲料的研究进展

赵 萍，王晓力，王春梅，朱新强，张 茜

（中国农业科学院兰州畜牧与兽药研究所，兰州 730050）

摘 要：苹果渣是果汁生产中的废弃物，可以通过微生物转化成饲料，或提取乙醇、果胶、膳食纤维等多种产品。经微生物发酵处理后，苹果渣中的蛋白质、氨基酸、矿物质等营养物质含量显著提高，纤维素、果胶、单宁等不易被利用的物质明显减少。文章从苹果渣的营养价值、发酵菌种、发酵工艺及发酵产物的应用几个方面阐述了苹果渣用于生产蛋白饲料的研究现状，以期为大规模以苹果渣为主要原料生产生物饲料提供思路和参考。

关键词：苹果渣；蛋白饲料；发酵

Advance in the Protein Feed Production by Apple Pomace Fermentation

ZHAO Ping, WANG Xiaoli, ZHANG Qian, et al

（Lanzhou Institute of Animal Husbandry and Pharmaceutical Sciences, Chinese Academy of Agricultural Sciences, Lanzhou 730050, China）

Abstract: Apple pomace is the waste of fruit juice production, which has high nutrition values. Apple pomace can be turned into feed by microorganism or be used to extract a variety of different products, such as ethanol, pectin, and dietary fiber. Microbial fermentation could improve apple pomace's nutritive substance significantly, such as protein, amino acid, mineral and probiotics and decrease the materials which could not to be used easily, such as cellulose, pectin and tannin. This essay discussed the research status of apple pomace in producing protein feed in respect of nutritional value, fermentation, fermentation process and the product, in order to provide ideas and references for the large-scale production of biological feed by using apple pomace as main raw material.

Key words: Apple pomace; Protein feed; Fermentation

（发表于《中国草食动物科学》，54-59.）

奶牛健康养殖重要疾病防控关键技术研究*

课题承担单位：中国农业科学院兰州畜牧与兽药研究所
课题参加单位：吉林大学
　　　　　　　东北农业大学
　　　　　　　华中农业大学
　　　　　　　中国农业大学
　　　　　　　黑龙江八一农垦大学
　　　　　　　中国农业科学院哈尔滨兽医研究所
　　　　　　　四川省畜牧科学研究院
　　　　　　　江苏农林职业技术学院
课题负责人：严作廷

（发表于《中国科技成果》，30-33.）

中兽医学资源抢救和整理的重要性浅述

王贵波,罗超应,李建喜,李锦宇,谢家声,郑继方

(中国农业科学院兰州畜牧与兽药研究所,兰州 730050)

摘　要:中兽医药学是中华民族科学技术遗产中的一颗璀璨明珠,是人们几千年来与畜禽疾病斗争的实践经验和智慧结晶,拥有独特精湛的诊疗方法与技艺。然而,随着我国畜禽养殖模式的转变以及西兽医学技术和现代生物技术迅猛发展,给中兽医药学带来了严峻挑战,逐渐形成了对中兽医学认识与人才培养的严重异化、经典著作亡佚殆尽,有形资源标本亟待创新、补缺,无形技术遗产亟待抢救、整理。因此,开展中兽医药的资源抢救和整理,对发扬中兽医药在畜禽疾病群体防治高效技术研发、自身理论体系发展等方面具有科技前瞻性意义。

关键词:中兽医学资源;抢救;整理

(发表于《中兽医医药杂志》,77-80.)

丁香酚解热作用机理研究进展

杨亚军，刘希望，赵晓乐，孔晓军，李世宏，秦　哲，杜文斌，李剑勇

（中国农业科学院兰州畜牧与兽药研究所/农业部兽用药物创制重点实验室/
甘肃省新兽药工程重点实验室，兰州　730050）

摘　要：天然产物丁香酚在各种香精的调配、精细化工、材料科学等方面的研究与应用十分广泛。现代医药学对丁香酚的生物活性开展了深入而细致的研究。研究表明，丁香酚对正常体温和发热均有降温作用，并可能通过影响视前区—下丘脑前部和弓状核温度敏感神经元的放电活动，引起弓状核区脑组织中PGE_2和cAMP含量的降低，增加体内精氨酸加压素的含量以及调节肝微粒体细胞色素P450和细胞色素b5的含量等途径，而发挥降温解热作用。

关键词：丁香酚；解热作用；机理

（发表于《中兽医医药杂志》，24-27.）

疯草解毒复方中药制剂的药代动力学研究

郝宝成[1§],权晓弟[1§],叶永丽[2],高旭东[2],黄鑫[1],
王学红[1],刘建枝[3],王保海[3],梁剑平[1*]

(1. 中国农业科学院兰州畜牧与兽药研究所/农业部兽用药物创制重点实验室/甘肃省新兽药工程重点实验室,兰州 730050;2. 西北民族大学生命科学与工程学院,兰州 730030;3. 西藏自治区农牧科学院畜牧兽医研究所,拉萨 850000)

摘 要: 本试验旨在探讨疯草解毒用复方中药制剂在大鼠体内的药代动力学特性。在确定制剂中有效成分的基础上,采用紫外分光光度法检测大鼠经口服药物制剂后体内有效成分的吸收情况,并计算相应的药代动力学参数。本试验建立了适用于检测8种成分的紫外光谱分析法;制剂经口服给药后能在胃肠道被快速吸收,1h左右大鼠体内各组分的血药浓度达到最大,药效发挥迅速;党参皂甙、多糖和阿魏酸代谢速度快,甘草酸、柴胡皂苷、黄芪甲苷、升麻素苷的代谢产物在体内作用时间较长。复方中药制剂在试验大鼠体内的药代动力学过程基本符合二室模型(陈皮苷除外)。所选用的测定方法简便、准确度和稳定性较高,适用于该制剂的药代动力学研究。

关键词: 复方中药制剂;药代动力学;紫外分光光度法;疯草

Study on Pharmacokinetics of Compound Traditional Chinese Medicine Preparation for Locoweed Detoxication

HAO Baocheng[1§], QUAN Xiaodi[1§], YE Yongli[2], GAO Xudong[2],
HUANG Xin[1], WANG Xuehong[1], LIU Jianzhi[3],
WANG Baohai[3], LIANG Jianping[1*]

(1. Key Laboratory of New Animal Drug Project, Gansu Province/Key Laboratory of Veterinary Pharmaceutics Discovery, Ministry of Agricultural/Lanzhou Institute of Husbandry and Pharmaceutical Sciences of CAAS, Lanzhou 730050, China; 2. College of Life Science and Engineering, Northwest University for Nationalities, Lanzhou 730030, China; 3. Tibet Academy of Agriculture and Animal Husbandry, Lasa 850000, China)

Abstract: The test was conducted to study the pharmacokinetics of compound traditional Chinese medicine preparation for locoweed detoxication in rats. After oral drug preparations, the absorption circumstances of active ingredients in vivo were detected by ultraviolet spectrophotometry and then the relative pharmacokinetic parameters were calculated based on identifying the effective components. The results showed that a suitable ultraviolet spectrum analytical method was established for 8 compounds at first. It had shown that the effective components were absorbed into the blood quickly after oral administration, and the maximum plasma concentration of each composition was got after about 1 h, indicating that the efficacy developed rapidly. The metabolic rates of Codonopsis pilosula saponin, polysaccharide and ferulic acid were fast, while the action time of the metabolites of glycyrrhizic acid, saikosaponin, astraloside and prim-O-glucosylc-imifugin were longer. The results suggested that pharmacolinetic process of this medicine preparation in rats basically conformed with two compartment model (expect the hesperidin). The determination method was simple, and had high accuracy and stability, as well as suitable for the pharmacokinetic study.

Key words: Compound traditional Chinese medicine preparation; Pharmacokinetics; UV spectrophotometry; Locoweed

（发表于《中国畜牧兽医》，1 550-1 556.）

作者简介

杨志强（1957—），中共党员，学士，二级研究员，博士生导师。甘肃省优秀专家，甘肃省第一层次领军人才，中国农业科学院跨世纪学科带头人，甘肃省"555"创新人才，《中兽医医药杂志》主编。兼任中国毒理学兽医毒理学分会会长，中国畜牧兽医学会常务理事，中国兽医协会常务理事，中国畜牧兽医学会动物药品学分会副会长，中国畜牧兽医学会毒物学分会副会长，中国畜牧兽医学会中兽医学分会副会长，中国畜牧兽医学会西北地区中兽医学会理事长，农业部兽药评审委员会委员。长期从事中兽医药学、兽医药理毒理、动物营养代谢与中毒病等研究工作，是该领域内的知名专家，先后主持和参加国家、省、部级科研课题33项，其中主持20项，获奖9项，自主和参与研发新产品8个，获授权专利3项。先后培养硕士研究生20名，培养博士研究生10名。在国内和国际学术刊物上共发表学术论文100余篇，其中主笔发表论文80篇。主编和参与编写《微量元素与动物疾病》等学术专著13部。

刘永明（1957—），中共党员，大学文化程度，三级研究员，硕士研究生导师。先后担任中国农业科学院中兽医研究所党委办公室副主任、主任，中国农业科学院兰州畜牧与兽药研究所人事处处长、副所长、党委副书记和纪委书记等职务，2001年7月至今任中国农业科学院兰州畜牧与兽药研究所党委书记、副所长、工会主席，兼任《中兽医医药杂志》和《中国草食动物科学》杂志编委会主任、中国农业科学院思想政治工作研究会理事、中国兽医协会会员和兰州市科学技术奖励委员会委员等职务。主要从事动物营养与代谢病研究工作。先后主持国家科技支撑计划、公益性行业专项、科技成果转化基金项目、948项目以及省级科研课题或子专题12项，主持基本建设项目4项，获授权专利8项，取得新兽药证书1个、添加剂预混料生产文号5个；主编（主审）、副主编著作6部，参与编写著作6部。

张继瑜（1967—），中共党员，博士研究生，三级研究员，博（硕）士生导师，中国农业科学院三级岗位杰出人才，中国农业科学院兽用药物研究创新团队首席专家，国家现代农业产业技术体系岗位科学家。兼任中国兽医协会中兽医分会副会长，中国畜牧兽医学会兽医药理毒理学分会副秘书长，农业部兽药评审委员会委员，农业部兽用药物创制重点实验室常务副主任，甘肃省新兽药工程重点实验室常务副主任，中国农业科学院学术委员会委员，黑龙江八一农垦大学和甘肃农业大学兼职博导。主要从事兽用药物及相关基础研究工作，重点方向包括兽用化学药物的研制、药物作用机理与新药设计、细菌耐药性研究。带领的研究团队在动物寄生虫病、动物呼吸道综合症防治药物研究上取得了显著进展。在肠杆菌耐药机理、血液原虫药物作用靶标筛选的研究处于领先地位。先后主持完成国家、省部重点科研项目20多项，获得科技奖励6项，研制成功4个兽药新产品，其中国家一类新药一个，以第一完成人申报10项国家发明专利，取得专利授权5项。培养研究生21名，发表论文170余篇，主编出版《动物专用新化学药物》和《畜牧业科研优先序》等著作2部。

阎萍（1963—）中共党员，博士，三级研究员，博士生导师。2012年享受国务院特殊津贴，是中国农业科学院三级岗位杰出人才，甘肃省优秀专家，甘肃省"555"创新人才，甘肃省领军人才。曾任畜牧研究室副主任、主任等职务，2013年3月任研究所副所长职务。兼任国家畜禽资源管理委员会牛马驼品种审定委员会委员，中国畜牧兽医学会牛业分会副理事长，全国牦牛育种协作组常务副理事长兼秘书长，中国畜牧兽医学会动物繁殖学分会常务理事和养牛学分会常务理事等。阎萍研究员主要从事动物遗传育种与繁殖研究，特别是在牦牛领域的研究成绩卓越，先后主持和参加完成了科技部支撑计划、科技部基础性研究项目、科技部"863"计划、"948"计划、农业部行业科技项目、国家肉牛产业技术体系岗位专家、人事部回国留学基金项目、科技部成果转化项目、甘肃省科技重大专项计划、甘肃省农业生物技术项目等20余项课题。现为国家肉牛牦牛产业技术体系牦牛选育岗位专家，甘肃省牦牛繁育工程重点实验室主任。作为高级访问学者多次到国外科研机构进行学术交流。培育国家牦牛新品种1个，填补了世界上牦牛没有培育品种的空白。获国家科技进步奖1项，省部级科技进步奖5项及其他科技奖励3项。培养研究生15名，发表论文180余篇，出版《反刍动物营养与饲料利用》《现代动物繁殖技术》《牦牛养殖实用技术问答》《Recend Advances in Yak Reproduction》《中国畜禽遗传资源志—牛志》等著作。

董书伟（1980—），男，汉族，在读博士，助理研究员，毕业于西北农林科技大学。主要从事奶牛营养代谢病与中毒病的蛋白质组学研究。先后主持参加国家科技支撑计划、中央级公益性科研院所基本科研业务费专项资金项目、中国科学院西部之光项目、国家自然科学青年基金，发表学术论文20余篇，申请专利18项。

郭宪（1978—），中国农业科学院硕士生导师，博士，副研究员。中国畜牧兽医学会养牛学分会理事，中国畜牧业协会牛业分会理事，全国牦牛育种协作组理事。主要从事牛羊繁育研究工作。先后主持国家自然科学基金、国家支撑计划子课题、中央级公益性科研院所基本科研业务费专项资金项目等5项。科研成果获奖3项，参与制定农业行业标准3项，授权专利3项。主编著作2部，副主编著作5部，参编著作2部。主笔发表论文30余篇，其中SCI收录5篇。

郭志廷（1979—），男，内蒙古人，助理研究员，执业兽医师，九三学社兰州市青年委员会委员，中国畜牧兽医学会中兽医分会理事。2007年毕业于吉林大学，获中兽医硕士学位。近年主要从事中药抗球虫、免疫学和药理学研究。先后主持或参加国家、省部级科研项目5项，包括中央级公益性科研院所专项基金。作为参加人获得兰州市科技进步一等奖1项，兰州市技术发明一等奖1项，完成甘肃省科技成果鉴定4项，授权国家发明专利5项（1项为第一完成人），参编国家级著作2部。在国内核心期刊上发表学术论文80余篇（第一作者30篇）。

郝宝成（1983—），甘肃古浪人，硕士研究生，助理研究员，研究方向为新型天然兽用药物研究与创制。先后主持和参与了中央级公益性科研院所基本科研业务费专项资金、国家支撑计划、863项目子课题等项目6项，以第一作者发表论文23篇（其中SCI论文2篇，一级学报2篇），参与编写著作3部《中兽药学》《兽医中药学及试验技术》《天然药用植物有效成分提取分离与纯化技术》，以第一发明人获得国家发明专利3项，荣获2012年度兰州市九三学社参政议政先进工作者。

李宏胜（1964—），九三学社社员，博士，研究员，硕士生导师，甘肃省"555"创新人才。中国畜牧兽医学会家畜内科学分会常务理事。多年来主要从事兽医微生物及免疫学工作，尤其在奶牛乳房炎免疫及预防方面有比较深入的研究。先后主持和参加完成了国家自然基金、国家科技支撑计划、国际合作、甘肃省、兰州市及企业横向合作等20多个项目。先后获得农业部科技进步三等奖1项；甘肃省科技进步二等奖2项、三等奖2项；中国农业科学院技术成果二等奖3项；兰州市科技进步一等奖2项、二等奖2项。获得发明专利4项，实用新型专利14项，培养硕士研究生5名，在国内外核心期刊上发表论文160余篇，其中主笔论文60余篇。

李剑勇（1971—），研究员，博士学位，硕士和博士研究生导师，国家百千万人才工程国家级人选，国家有突出贡献中青年专家。现任中国农业科学院科技创新工程兽用化学药物创新团队首席专家，农业部兽用药物创制重点实验室副主任，甘肃省新兽药工程重点实验室副主任，甘肃省新兽药工程研究中心副主任，农业部兽药评审专家，甘肃省化学会色谱专业委员会副主任委员，中国畜牧兽医学会兽医药理毒理学分会理事，国家自然基金项目同行评议专家，《黑龙江畜牧兽医杂志》常务编委，《PLOS ONE》《Medicinal Chemistry Research》等SCI杂志审稿专家。多年来一直从事兽用药物创制及与之相关的基础和应用基础研究工作。曾先后完成药物研究项目40多项，主持16项。获省部级以上奖励10项，2011年度获第十二届中国青年科技奖；2011年度获第八届甘肃青年科技奖；获2009年度兰州市职工技术创新带头人称号。获国家一类新兽药证书，均为第2完成人。申请国家发明专利22项，获授权9项。发表科技论文200余篇，其中SCI收录22篇，第一作者和通讯作者15篇。出版著作4部，培养研究生15名。

王学智（1969—），研究员、博士，硕士研究生导师，科技管理处处长。主要从事科技管理工作和兽医临床科研工作。主要在中兽药、动物营养代谢病等兽医临床学方面开展基础应用研究，先后主持"国家公益性行业专项中兽药生产关键技术研究与应用课题防治螨病和痢疾藏中兽药制剂制备""科技基础性工作专项—传统中兽医药资源抢救和整理"和"甘肃省科技重大专项——防治奶牛繁殖病中药研究与应用"、甘肃省科技支撑计划项目等科研课题10项。参与完成"948项目""国家自然基金项目""国家科技支撑计划"等各类科研项目16项。先后获得省部院级科技成果奖励7项："奶牛乳房炎联合诊断和防控新技术研究及示范"获甘肃省农牧渔业丰收一等奖，"重金属镉/铅与喹乙醇抗原合成、单克隆抗体制备及ELLSA检测技术研究"和"新型中兽药饲料添加剂"参芪散"的研制与应用"获中国农业科学院科学技术成果二等奖,"归蒲方中草药饲料添加剂的研制与应用""禽用复合营养素的研制及其应用"和"非解乳糖链球菌发酵黄芪转化多糖的研究与应用"获甘肃省科技进步三等奖。在国家级学术刊物上先后发表学术论文15篇,其中SCI收录7篇,参编著作17部,主编5部,获得专利22个。

李维红(1978—),博士,副研究员。主要从事畜产品质量评价技术体系、畜产品检测新方法及其产品开发利用研究等。先后主持和参加了中央级公益性科研院所基本科研业务费专项资金项目、甘肃省自然基金项目等。主笔发表论文20余篇。主编《动物纤维超微结构图谱》,副主编《绒山羊》,参加编写《动物纤维组织学彩色谱》和《甘肃高山细毛羊的育成和发展》著作2部。获得专利4项,作为参加人获甘肃省科技进步一等奖一项、三等奖一项。

李新圃(1962—),博士,副研究员。主要从事兽医药理学研究工作。已主持完成省、部、市级科研项目7项。参加完成国际合作、省、部、市级科研项目三十余。参加项目"奶牛重大疾病防控新技术的研究与应用"获2010年甘肃省科技进步二等奖;"奶牛乳房炎主要病原菌免疫生物学特性的研究"在2008年获兰州市科技进步一等奖、中国农科院科学技术成果二等奖和甘肃科技进步三等奖;"绿色高效饲料添加剂多糖和寡糖的应用研究"获2005年兰州市科技进步一等奖;"奶牛乳房炎综合防治配套技术的研究及应用"获2004年甘肃省科技进步二等奖。已发表研究论文40余篇,其中5篇被SCI收录。获实用新型专利授权3个。

梁春年(1973—),硕士生导师,博士,副研究员,副主任,兼任全国牦牛育种协作组常务理事,副秘书长,中国畜牧兽医学会牛学会理事,中国畜牧业协会牛业分会理事,中国畜牧业协会养羊学分会理事等职。主要从事动物遗传育种与繁殖方面的研究工作。现主持国家科技支撑计划子课题"甘肃甘南草原牧区牦牛选育改良及健康养殖集成与示范"和国家星火计划项目子课题"牦牛高效育肥技术集成示范"等课题4项;参加完成国家及省部级科研项目30余项,获得省部级科技奖励7项。参与制定农业行业标准5项。参加国内各类学术会议30余次,国际学术会议4次。主编著作2本,副主编著作5本,参编著作6本,发表论文90余篇,其中SCI文章5篇。

孙晓萍（1962—），硕士生导师，学士，副研究员。主要从事绵羊遗传育种工作，先后主持的项目有：甘肃省自然科学基金项目——绵羊毛生长机理研究；甘肃省农委推广项目——绵羊双高素推广应用研究；甘肃省支撑计划项目——奶牛产奶量的季节性变化规律研究；甘肃省星火项目——肉羊高效繁殖技术研究；甘肃省支撑计划项目——肉用绵羊高效饲养技术研究。先后发表学术论文40余篇，主编参编著作11部，实用新型专利3个。获甘肃省畜牧厅科技进步二等奖1项，甘肃省科技进步二等奖1项，中国农业科学院科技进步一等奖1项，中华农业科技二等奖1项。

王东升（1979—），农学硕士，助理研究员。主要从事奶牛繁殖疾病的研究。主持"奶牛子宫内膜中天然抗菌肽的分离、鉴定及其生物学活性研究"和"狗经穴靶标通道及其生物效应的研究"2个课题，参加"十二五"国家科技支撑计划项目"奶牛健康养殖重要疾病防控关键技术研究"和"十一五"国家科技支撑计划项目"奶牛主要繁殖障碍疾病防治药物研制"等10多个项目。参与申请并获得发明专利4项，实用性新专利5项，取得三类新兽药证书1个，参加的成果获甘肃省科技进步二等奖2项和兰州市科技进步二等奖2项。参编著作《兽医中药配伍技巧》《兽医中药学》《奶牛围产期饲养与管理》和《奶牛常见病综合防治技术》等5部，主笔发表论文20余篇。

王宏博（1977—），博士，助理研究员，博士。主要从事动物营养与饲料科学。先后主持甘肃省科技厅项目2项，中央级公益性科研院所科研业务费专项资金项目3项，先后参加农业部公益性行业（农业）专项3项，"948"项目1项，获得国家发明专利2项，实用新型专利1项，参与完成国家发明专利和实用新型专利总计10余项。参与制定国家和农业部标准10余项。主编图书1部，副主编图书1部，参编图书3部。发表学术论文30余篇。

王磊（1985—），硕士，研究实习员。主要从事中兽医药理学、奶牛疾病防治和药物残留研究。先后参加各类研究课题8项，主持课题1项。2013年参加的"非解乳糖链球菌发酵黄芪转化多糖的研究与应用"获甘肃省科技进步三等奖、2014年参加的"重金属镉/铅与喹乙醇抗原合成、单克隆抗体制备及ELISA检测技术研究"获中国农业科学院科学技术成果二等奖；主笔发表科技论文4篇、其中SCI收录1篇，获得授权专利2项。

王晓力（1965—），副研究员。主要从事牧草资源开发利用方面的研究工作。近年来参加省部级课题多项，主持农业行业科研专项子课题2项，获全国农牧渔业丰收二等奖1项、内蒙古自治区农牧业丰收一等奖1项、2011年贵州省科学技术成果转化二等奖1项。发表论文30余篇，出版著作4部，参编4部，授权专利5项。

王旭荣（1980—），博士，助理研究员。主要从事分子病毒学和分子细菌学方面的研究。主持完成"犬瘟热病毒"的基本科研业务费项目1项；现主持"奶牛乳房炎病原菌"方面的农业行业标准项目1项和甘肃省农业生物技术项目1项；参与国家奶牛产业技术体系项目、948项目、国家科技支撑项目、国家自然基金、科技基础性工作专项等10余项。参与的研究项目在2011—2014年期间获得奖项3个，其中甘肃省科技进步三等奖1项，甘肃省农牧渔丰收奖一等奖1项、中国农业科学院科学技术成果二等奖1项。发表科技论文40余篇，主笔15篇；参编著作1部；实用新型授权20余项，第一发明人授权5项；申请发明专利（第一发明人）3项。

王学红（1975—），硕士，高级兽医师。主要从事天然药物提取研究。先后主持及参加国家级省部级药物研究项目20余项，包括"十二五"农村领域国家科技计划项目、国家支撑计划项目、甘肃省科技支撑计划项目及中央级公益性科研院所基本科研业务费专项资金项目等。获省部级科技进步奖4项，院厅级奖2项，国家发明专利10余个。在国内外学术刊物上发表论文20余篇，出版著作4部。

吴晓云（1986—），博士，研究实习员，主要从事牦牛低氧适应机制和牦牛肉品质形成遗传机制的研究。目前以第一作者发表论文11篇，其中7篇被SCI收录。

严作廷（1962—），硕导，博士，研究员。现为第五届农业部兽药评审专家、中国畜牧兽医学会家畜内科学分会常务理事、中国畜牧兽医学会中兽医学分会理事。主要从事工作奶牛疾病和中兽医学研究（先后主持国家自然科学基金、国家科技支撑计划等项目10多项，现主持"十二五"国家科技支撑计划课题"奶牛健康养殖重要疾病防控关键技术研究"课题）。主编《家畜脉诊》《奶牛围产期饲养与管理》等著作3部，参编《中兽医学》《奶牛高效养殖技术及疾病防治》和《犬猫病诊疗技术及典型医案》等著作6部。发表论文70余篇，获省部级科技进步奖6项，院厅级奖7项。获得国家新兽药证书2个，国家发明专利10个，实用新型专利5个。

张世栋（1983—），硕士，助理研究员。主要从事奶牛疾病与中兽医药研究工作。近年来，主持中央级基本科研业务费及其增量项目共3项，国家自然基金1项；参与国家"十一五""十二五"课题3项，甘肃省、兰州市科技项目多项。参与获得兰州市科技进步二等奖1项。发表论文10篇，获得专利6项，参与申请专利10多项。

李建喜（1971—），研究员，博士学位，硕士研究生导师。现任中国农业科学院兰州畜牧与兽药研究所中兽医（兽医）研究室主任，中国农业科学院科技创新工程中兽医与临床创新团队首席专家，甘肃省中兽药工程技术研究中心副主任，农业部新兽药中药组评审专家，国家自然基金项目同行评议专家，国家现代农业（奶牛）产业技术体系后备人选等。从事兽医病理学、动物营养代谢病与中毒病、兽医药理与毒理、奶牛疾病防治、中兽医药现代化等研究工作。完成国家和省部级科研项目40余项，其中主持24项，包括国家自然基金面上项目、国家科技支撑计划课题、国家公益性行业（农业）科研专项课题、国家科技基础性工作专项课题、农业部948计划项目、甘肃省农业科技重大专项、甘肃省生物技术专项、中国西班牙科技合作项目、中泰科技合作项目等。先后获科技奖励6项，其中获省部级奖励2项，院厅级奖励4项。研发新产品6个，获授权发明专利9项，以第一导师培养硕士研究生6名，共同培养硕士研究生17名，参与培养博士研究生8名。在国内和国际学术刊物上共发表学术论文99篇，其中以第一作者和通讯作者发表论文54篇，SCI收录6篇，编写著作7部。

梁剑平（1962—），博士，三级研究员，博士生导师。2005年获"农业部有突出贡献的中青年专家"称号。2005年分别获中国科协"西部开发突出贡献奖"、中央统战部"为全面建设小康社会做出贡献的先进个人"、甘肃省"陇上骄子"、九三学社甘肃省委、"十佳青年"称号。现任兰州畜牧与兽药研究所兽药研究室副主任、农业部兽用药物创制重点实验室副主任、中国农业科学院新兽药工程重点开放实验室副主任。是中国农业科学院二级岗位杰出人才，甘肃"555"创新人才。兼任中国毒理学会兽医毒理学分会及中国兽医药理学分会理事，农业部新兽药评审委员会委员,农业部兽药残留委员会委员,中国兽药典委员会委员,中国农业科学院学术委员会委员,中国农科院研究生院教学委员会委员、政协兰州市委常委、九三学社兰州市七里河区主任委员,九三学社兰州市副主委。梁剑平研究员主要从事兽药化学合成和中草药的提取及药理研究,先后主持和参加国家和省部级重大科研项目20余项,获奖8项,其中获国家科技进步二、三等奖各1项,省(部)级二等奖2项、三等奖2项,发明专利10项。

蒲万霞（1964—）博士后,四级研究员,硕士生导师,甘肃省微生物学会理事,政协兰州市七里河区第六届委员。从事兽医微生物与微生物制药研究,重点方向为兽用微生态制剂的研制及细菌耐药性研究,在金黄色葡萄球菌耐药性研究方面取得一定进展。先后主持农业部公益性行业科研专项"新型动物专用药物的研制与应用"子专题,国家科技支撑计划"奶牛主要疾病综合防控技术研究及开发"子专题,中央级公益性科研院所基本科研业务费专项资金项目、甘肃省科技成果重点推广计划项目、甘肃省农业科技创新项目、兰州市院地校企合作项目、兰州市科技局项目、农业部畜禽病毒学开放实验室基金项目等各级各类项目15项。获得各级政府奖7项。取得国家发明专利2项。主编《食品安全与质量控制技术》等著作5部,副主编著作1部,发表论文70多篇,培养硕士生11名。

朱新书（1957—）,学士,副研究员,主要从事牦牛藏羊品种资源开发和利用工作,主要主持和参加的项目有:公益性行业（农业）科研专项《放牧牛羊营养均衡需要研究与示范》;农业部重点攻关项目《大通牦牛选育与杂交利用》;甘肃省重大科技项目《甘南牦牛改良与选育技术研究示范》;国家肉牛牦牛产业技术体系《牦牛繁育技术研究与示范岗位科学家团队》等。发表学术论文20余篇,主编参编著作3部,荣获农业部科技进步三等奖1项。

岳耀敬（1980—）,在读博士。主要从事绵羊繁殖、羊毛和高原适应性等重要经济、抗逆性状的分子调控机制研究。国家绒毛用羊产业技术体系分子育种岗位团队成员,兼任中国畜牧兽医学会养羊学分会副秘书长、世界美利奴育种者联盟成员,曾先后到法国、澳大利亚、新西兰等国学习考察细毛羊育种工作。现主持国家自然青年基金、甘肃省青年基金等项目3项;申报发明专利5项,授权发明专利2项、实用新型2项;参编著作5部,发表学术论文38篇,其中SCI收录5篇。

谢家声（1956—），大学专科学历，高级实验师。主要从事猪、禽、奶牛等动物疾病的中兽医临床防治研究，主持完成了甘肃省自然基金—"复方中草药防治畜禽病毒性传染病新制剂的研究"、甘肃省科技攻关项目—"治疗奶牛胎衣不下天然药物的研制"等项目。完成国家"七五""八五"攻关项目、国家"十五"及"十一五"奶业专项、国家"十一五""十二五"支撑计划以及省市各类研究项目30余项。发表论文60余篇，合编专著8部，获得国家发明专利3项。1990年以来先后获甘肃省科技进步三等奖1项；甘肃省科技进步二等奖3项；中国农业科学院科技进步二等奖4项；兰州市科技进步二等奖3项；甘肃省农牧厅二等奖1项；甘肃省天水市科技进步二等奖2项。

尚小飞（1986—），硕士，助理研究员。主要从事藏兽医药的现代化研究。目前主持中央级公益性院所基本科研业务费项目一项，参与多项国家及省部级课题。在 *Journal of Ethnopharmacology*，*Veterinary Parasitology* 等SCI杂志发表文章8篇，参与编写著作1部。

杨红善（1981—），硕士，助理研究员。主要从事牧草种质资源搜集与新品种选育研究工作，主持在研项目3项，其中甘肃省青年基金项目1项、甘肃省农业生物技术研究与应用开发项目1项、中央级公益性科研院所基本科研业务费专项资金项目1项，参加国家支撑计划子课题等各类项目共计3项。工作期间以第一完成人或参加人审定登记甘肃省牧草新品种4个。获中国农业科学院科技进步二等奖1项（第三完成人）。参加编写《高速公路绿化》著作1本（副主编）。在各类期刊发表论文20篇，其中主笔13篇。

刘宇（1981—），硕士，助理研究员。2007年7月来中国农业科学院兰州畜牧与兽药研究所工作至今，主要从事有机及天然药物化学研究。先后主持及参与国家级省部级药物研究项目10余项，包括"十二五"农村领域国家科技计划项目、国家支撑计划项目、甘肃省科技支撑计划项目、甘肃省自然科学基金项目及中央级公益性科研院所基本科研业务费专项资金项目等。获省部级科技进步奖1项，院厅级奖2项，国家发明专利4个。在国内外学术刊物上发表论文20余篇，出版著作2部。

程富胜（1971—）博士，副研究员，硕士生导师。主要从事天然药物活性成分免疫药理学研究及其制剂开发研制。先后主持和参加国家、省、部级科研课题20项。参加完成的国家支撑项目"中草药饲料添加剂'敌球灵'的研制"获农业部科技进步二等奖；国家自然基金项目"免疫增强剂-8301多糖的研究与应用"成果获中国农业科学院科技进步二等奖；"蕨麻多糖免疫调节作用及其机理研究与临床应用"成果已分别获中国农科院、兰州市科技进步二等奖。主持完成的"富含活性态微量元素酵母制剂的研究"获兰州市科技进步二等奖。在国内和国际学术刊物上共发表学术论文50余篇，其中主笔发表论文30余篇。

李冰（1981—），女，硕士研究生、助理研究员，研究方向为兽药新制剂研制与安全评价。主要从事新兽药制剂质量标准制定、体内药代动力学研究与残留研究。中国畜牧兽医学会兽医药理毒理学分会理事。主持在研项目2项，其中公益性行业专项子课题1项、中央级公益性科研院所基本科研业务费专项资金项目1项；参加国家支撑计划项目、国家高技术研究发展计划（863计划）、甘肃省科技重大专项、国家自然基金项目、国家基础专项等各类项目十余项。以主要完成人申报国家新兽药1个。前三以前作者发表论文30余篇，主笔核心期刊3篇，主笔SCI收录2篇，参编著作2部。获2011年兰州市科学技术进步奖一等奖，第6完成人；2012年中国农业科学院科技进步二等奖，第3完成人；2013年甘肃省科学技术发明奖一等奖，第3完成人；2013年兰州市科学技术发明奖一等奖，第3完成人。

杨博辉（1964—）博士，四级研究员，博士生导师。现为"国家绒毛用羊产业技术体系"分子育种岗位科学家，中国农科院兰州畜牧与兽药研究所"绵羊新品种（系）培育"科技创新团队首席专家。兼任中国畜牧兽医学会养羊学分会副理事长兼秘书长，中国农科院学位委员会委员，中国博士后基金评审专家，中国畜牧业协会羊业分会特聘专家，中国农业科学院三级岗位杰出人才。主要从事绵羊新品种（系）培育、分子育种及产业化研发。先后主持完成国家863计划、国家支撑计划、国家基础专项及国家（部）级标准等项目20余项。制定国颁标准6项，部颁标准5项。获国家发明专利2项，实用新型专利8项。已培育出"大通牦牛"新品种、西北肉羊新品群，基本培育出高山美利奴羊新品种。发表论文120篇，其中7篇SCI，获国际论文一等奖1篇。主编《甘肃高山细毛羊的育成和发展》与《中国野生偶奇蹄目动物遗传资源》著作2部，副主编著作4部，参编著作4部。培养博、硕士研究生21名，其中国际留学生1名。

王慧（1985—），硕士，研究实习员。主要从事动物营养代谢病研究。现主持"中央级科研院所基本科研业务费专项资金项目（NO.1610322013003）。2013年获农业部全国农牧渔业丰收二等奖（第7完成人）；兰州市科学技术进步二等奖（第9完成人）。目前第一作者发表论文16篇，其中SCI收录8篇。申请专利2项。

尚若锋（1974—），博士，副研究员。主要从事兽用药物研发工作。主持或参与国家支撑计划、"863"计划以及其他国家级和省部级的科研项目20余项。获得省部级科研奖励4项，国家发明专利12项，以第一作者或通讯作者发表文章30余篇，其中SCI文章12篇。

刘建斌（1977—），甘肃农业大学动物生产系统与工程专业博士，副研究员，主要从事绵羊现代遗传学理论与育种，动物种质资源优良基因发掘和重要经济性状分子遗传机理研究。先后主持省部级科研项目5项。获甘肃省科技进步二等奖1项、中国农业科学院科技进步二等奖1项、中华神农科技三等奖1项。主编出版《现代肉羊生产实用技术》和《绒山羊》著作2部，副主编出版《甘肃高山细毛羊的育成和发展》《中国野生偶奇蹄目动物遗传资源》《羊繁殖与双羔免疫技术》《适度规模肉羊场高效生产技术》和《优质羊毛生产技术》著作5部，参编出版著作5部；主笔发表专业论文30余篇，其中SCI论文7篇；申报国家发明专利6项，授权发明专利3项，合作授权国家实用新型专利30项。

辛蕊华（1981—），硕士，助理研究员。主要从事中兽医药物学工作，先后主持和参与过中央级公益性科研院所基本科研业务费专项资金项目、公益性行业（农业）科研专项、国家科技支撑项目等；发表论文12篇，参与著作两部。参加的"富含活性态微量元素酵母制剂的研究"科研成果获得兰州市科学技术二等奖。

杨峰（1985—），助理研究员，长期从事奶牛乳房炎的诊断、预防和治疗工作。主持2项中央级公益性科研院所基本科研业务费专项资金和1项甘肃省科技计划项目，先后参与国家自然科学基金、国家支撑计划、省部级等项目10多项。以第一完成人获得授权实用新型专利8项。

罗超应（1960—），学士，研究员，中西兽医药学结合研究（主持科技部科研院所开发研究专项"新型中兽药射干地龙颗粒的研究与开发"、科技部基础工作专项"华东区传统中兽医学资源抢救与整理"与国家"十一五"科技支撑子项目"中兽药中试及其生产工艺研究"等）。主编、副主编与参编出版《牛病中西医结合治疗》等著作16部，共计679余万字；主笔发表"Variability of the Dosage, Effects and Toxicity of Fu Zi（Aconite）From a Complexity Science Perspective" "以复杂科学理念指导中西医药学结合" "奶牛乳房炎的复杂性及其对传统科学观念的挑战"等学术论文近100篇，其中英文期刊文章5篇；专利12项，成果奖励5项。

冯瑞林（1959—），本科毕业，助理研究员，研究方向为羊繁殖育种。主要从事羊双羔免疫技术研究工作，参与甘肃省重大专项"甘肃超细毛羊新品种培育示范与推广"、中国农业科学院"羊绿色增产增效技术集成与示范"等十余项。工作期间参与申报细毛羊新品种1个。获国家科技进步三等奖1项（第九完成人），甘肃省科技进步一等奖1项（第九完成人）。主编出版著作《羊繁殖及双羔免疫技术》1部，参加编写《甘肃高山细毛羊的育成与发展》（副主编）、《绒山羊》（副主编）等著作5部。在各类期刊发表论文50余篇，其中以第一作者发表论文12篇。

郑继方（1958—），研究员，硕士生导师。现任甘肃省中兽药工程技术研究中心主任，中国农业科学院兰州畜牧与兽药研究所中兽医药创新团队首席科学家，中国农业科学院兰州畜牧与兽药研究所学术委员会委员，《中兽医医药杂志》编委，亚洲传统兽医学会常务理事，中国畜牧兽医学会中兽医分会理事，中国生理学会甘肃分会理事，西北地区中兽医学术研究会常务理事，中国畜牧兽医学会高级会员，农业部项目评审专家，农业部新兽药评审委员会委员，

科技部国际合作计划评价专家,西南大学客座教授。从事中兽医药学的研究工作。先后主持国家自然科学基金、国家科技攻关、国家支撑计划、省部级等项目20多项。获省部级科技进步奖3项,院厅级奖4项,国家级新兽药证书两个,国家发明专利十余项。主编出版了《兽医中药学》《中兽医诊疗手册》《兽医药物临床配伍与禁忌》《常用兽药临床新用》《中草药饲料添加剂配制与应用》《甲型H1N1流感防控100问》和《兽医中药临床配伍技巧》等专著10部,先后在国内外各学术刊物发表论文80余篇。

刘希望(1986—),硕士,助理研究员。2007年毕业西北农林科技大学环境科学专业,2010年获西北农林科技大学化学生物学专业硕士学位,同年参加工作至今。曾主持农业部兽用药物创制重点实验室开放基金项目1项,现主持中央公益性科研院所基本科研业务费1项。以第一作者发表SCI论文4篇,主要从事新兽药研发,药物合成方面的研究工作。

王华东(1979—),硕士,助理研究员。主要从事科技期刊编辑工作,先后出版著作3部(副主编1部,参编2部),主笔发表学术论文5篇。编辑、出版的《中兽医医药杂志》发行范围涵盖中国、北美、澳洲、西欧、韩国、日本、新加坡等海内外15个国家和地区。随着中兽医药在畜禽疫病防治及公共卫生安全等方面的独特作用的日益凸显,作为中兽医领域唯一的国家级学术刊物,《中兽医医药杂志》全面、详实地报道了近年来该领域的最新研究成果,并力促其推广应用与转化,为继承和传播祖国兽医学诊疗技术、促进中西兽医结合、繁荣学术、服务"三农"等做出了重大贡献,取得了显著的经济效益、社会效益和生态效益。

熊琳(1984—),硕士,助理研究。主要从事农产品质量安全的研究。发表相关科研论文4篇,其中SCI收录1篇,授权专利8项,参与制定国家标准1项。

魏小娟（1976—），女，硕士研究生、助理研究员，研究方向为分子药理学。曾先后主持国家自然科学基金1项、甘肃省青年基金项目1项、中央级公益性科研院所基本科研业务费专项资金项目1项；先后参加国家自然科学基金项目，863项目、国家支撑计划项目、公益性行业专项、甘肃省重大专项等项目10余项。工作期间获甘肃省科技进步一等奖1项，中国农业科学院科技进步二等奖2项，兰州市科技进步二等奖2项。参加编写著作5部。在各类期刊发表论文30篇，其中SCI 5篇。

崔东安（1981—），博士，主要从事奶牛胎衣不下方证代谢组学研究（奶牛胎衣不下血瘀证的代谢组学研究No. 1610322015006），发表文章4篇，其中3篇SCI文章。

路远（1980—），硕士，副研究员。主要从事牧草新品种选育及植物组培研究。自毕业以来，主持院所长基金项目"美国杂交早熟禾引进驯化及种子繁育技术研究"、和"黄花矶松驯化栽培及园林绿化开发应用研究"，曾参与并完成了973合作子课题项目"气候变化对西北春小麦单季玉米区粮食生产资源要素的影响机理研究"，全球环境基金（GEF）项目"野生牧草种质资源应用研究""放牧利用与草原退化关系研究"和国家科技基础条件平台工作项目子课题"牧草种质资源的实物共享及标志性数据采集""牧草种质资源的标准化整理和整合"、国家科技支撑计划项目子课题"西北优势和特色牧草生产加工关键技术研究与示范"等项目20余项。发表论文20余篇，主编著作1部，或甘肃省科技进步二等奖1项，选育牧草新品种1个，获专利5项。

高雅琴（1964—），研究员，硕士研究生导师，主要从事畜产品加工及动物毛皮质量评价研究工作。现主持农业部畜产品质量安全风险评估项目中牛羊肉质量安全风险评估子项目。曾主持的国家公益类科研项目"动物纤维毛皮产品质量评价技术的研究"，2009年获甘肃省科技进步三等奖；主持国家标准制定项目"GB/T 25885—2010羊毛纤维直径试验方法——激光扫描仪法""GB/T 25880—2010毛皮掉毛测试方法——掉毛测试仪法""GB/T 26616—2011

裘皮獭兔皮"均已颁布并实施；主持的农业行业标准制定项目"NY 1164—2006裘皮蓝狐皮""NY 2222—2012动物纤维直径及成分分析方法——显微图像分析仪法"已颁布实施，"动物毛皮各类鉴别方法——显微镜法"上报农业部审定。参与多项国家标准和农业行业标准"GB/T 25243—2010甘肃高山细毛羊""河西绒山羊"和"大通牦牛""天祝白牦牛""NY 1173—2006动物毛皮检验技术规范"等。主编的《动物毛纤维组织学彩色图谱》《动物毛皮质量鉴定技术》；参与出版著作5部。发表科技论文150余篇，其中主笔发表40余篇。